DAMAGE TOLERANCE AND DURABILITY OF MATERIAL SYSTEMS

DAMAGE TOLERANCE AND DURABILITY OF MATERIAL SYSTEMS

KENNETH L. REIFSNIDER

SCOTT W. CASE

WILEY-INTERSCIENCE

A JOHN WILEY & SONS, INC., PUBLICATION

This book is printed on acid-free paper. ♾

Published simultaneously in Canada.

For ordering and customer service, call 1-800-CALL-WILEY.

Library of Congress Cataloging-in-Publication Data is available.

ISBN 0-471-15299-4

CONTENTS

PREFACE

This book has two major elements. The first is an element of new philosophy, based on the idea that a useful definition of *damage* as related to the life of a functional component can be constructed on the idea of remaining strength as an equivalent-damage precept. The second element is an assemblage and interpretation of data and experience over a generation of research in the general field of durability, damage tolerance, and fatigue.

Damage is an abstraction. Anecdotal observations of physical details such as microcracks are evidence of damage, but defined only in the context of conveying the damaging effect of certain physical events on properties and performance. Strength is an observable, an independent variable that can be measured with clear and familiar engineering definitions. As a damage metric, it identifies the specific consequence of damage that is of interest and defines how to measure it. Moreover, we have advanced the claim that strength is a state variable that can represent the evolving state of the material in the presence of damage. This postulate is the basis for our development of a new mechanistic philosophy and related methodology that predicts remaining strength and remaining life. This philosophy associates the local physical changes in material states and stress states caused by time-variable applied mechanical, thermal, chemical, and electrical environments with global properties and performance. The result is a predictive methodology that relates how a composite material system is made to how long it will last under prescribed conditions.

But philosophy alone is sterile. The foundation of the message carried by this book is the physical behavior recounted therein and the interpretations and understandings of that behavior recorded in the following pages. This information comes at great cost, especially the cost of extensive experimental time and effort. In this book, we attempt to represent, relate, and distill a generation of such effort, beginning with the work of the authors and including as much of the significant literature as space allowed. These experimental findings provide the authority for our interpretations and models.

Finally, it should be noted that there is a guiding vision to this book. We are approaching an entirely new age of enlightenment, which holds the promise of constructing fundamental models of materials that span many scales of space and time. This age will bring many advances in science such as the prediction, from first principles, of physical properties and their dependence on independent variables such as temperature (a philosophy now only sparsely populated by phenomenological constitutive theory) and advances in engineering such as virtual design wherein design over the full range from materials to structural components is done with robust simulation. These advances are being driven not only by our inquisitive spirit but also by the remarkable cost savings they can bring to the creation and application of new materials and material systems. This book is a step in the direction of that vision.

Kenneth Reifsnider
Scott W. Case

INTRODUCTION: BASIC THESIS

A predictive, mechanistic model of damage tolerance and durability of fibrous composite systems can be constructed using familiar concepts of strength in conjunction with the mechanics and kinetics of stress state and material state evolution. The present approach incorporates the details of constituent properties, behavior, geometry, and interaction to achieve a description of remaining strength and life in terms of how the material system is designed and manufactured.

I.1 ELEMENTS OF THE APPROACH

Durability and damage tolerance are not mechanics problems per se. They are physical characteristics that require the combined disciplines of mechanics, materials science and engineering, chemistry, and physics to achieve a complete understanding and representation of the physical behavior that defines them. The principal objective of the present text is to bring those disciplines together with fundamental philosophy and mechanistic methodology to construct a predictive capability for remaining strength and life.

The central theme of this book is the effect of changes in material states over time under the influence of applied environments on the remaining strength and life of materials and material systems. As suggested by Figure I.1, these changes may include cycle-dependent events and processes such as defect initiation and growth, kinetic processes such as diffusion and migration, chemical processes such as oxidation, and thermodynamic activity such as changes in microstructural configurations. In this book, we will relate the action and interaction of such material

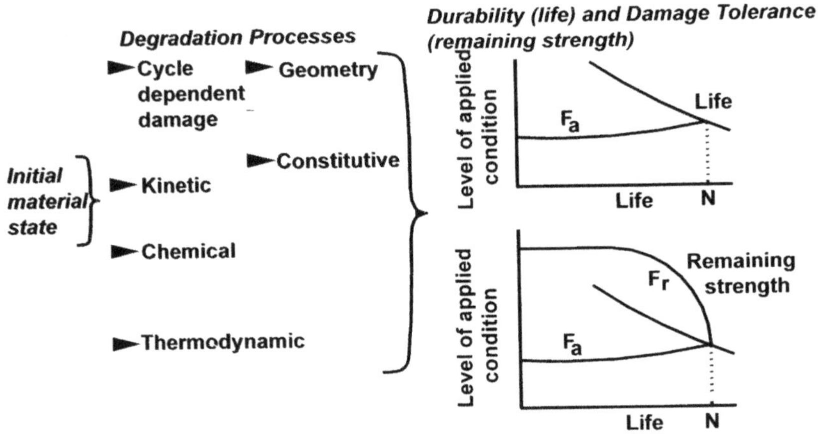

Figure I.1 Damage tolerance and durability in composite systems that degrade by multiple, interacting progressive degradation processes under mechanical, thermal, and chemical applied environments.

state changes to the remaining strength (damage tolerance) and remaining life (durability) of materials and material systems, including structures used in engineering applications.

The fundamental approach to be used is a combination of concepts, combined in a consistent, comprehensive philosophy that can be used to make engineering estimates of remaining strength and life of composite materials and structural systems. The most basic premise of that philosophy is the concept of remaining strength as a damage parameter. This concept is illustrated in Figure I.2.

The fundamental premise is that the number of cycles of load application (or generalized time of the application of any applied environment) at level 1 (in Fig.

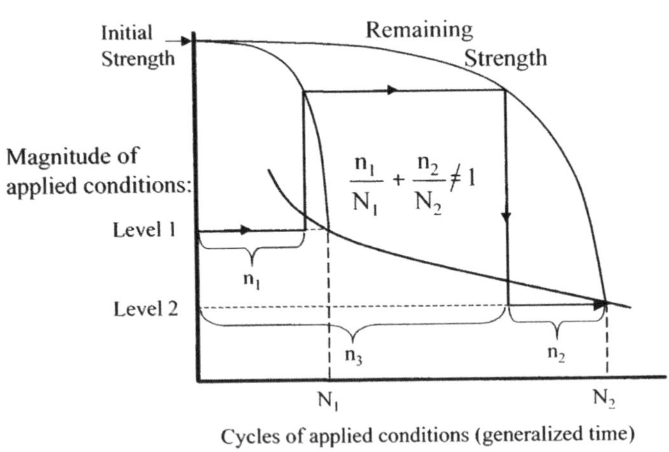

Figure I.2 Concept of remaining strength as a damage metric.

I.2) produces an amount of "damage" that is equivalent to the number of cycles of load application at level 2 if the remaining strength after application of the conditions for their respective levels and increments is the same; that is, equivalent damage is defined by equivalent remaining strength. Hence, in Figure I.2, n_1 cycles of loading at level 1 is equivalent to n_3 cycles of loading at level 2, according to this definition. Since remaining strength is a well-defined and easily measured engineering quantity, this definition is easily validated. If the remaining strength curve has been determined for constant-amplitude loading, as sketched in Figure I.2, then the number of cycles of life (or the remaining strength) can be predicted for any combination of applied load levels and cycles at those levels. An example for two load levels is illustrated in the figure. The experimental success of this premise has been the foundation of the present approach. Since strength is influenced by many processes, as discussed in Figure I.1, it is a natural integrator of the influence of time- and cycle-dependent material state change processes. As Figure I.2 also shows, the resulting "damage summation" for this metric is highly nonlinear and depends on the sequence of load or applied environment applications.

It is a basic premise of the present approach that the strength and life of composite material systems is controlled by

1. the statistical accumulation of damage and the subsequent interaction of the consequences of that damage,
2. large changes in material states and stress states, and
3. microstructure and microlevel events and behavior, especially anisotropy, inhomogeneity, and brittleness.

Fundamental features of the present approach include the following:

1. Boundary-value problems are set on the state of stress and state of material defined by incipient fracture for a given failure mode, that is, at the end of life.
2. Equivalent damage is defined by equivalent remaining strength.
3. Critical elements and subcritical elements as subdivisions of the representative volume chosen for analysis are defined for each failure mode.
4. Mechanistic micromechanics representations of principal material strengths and stiffnesses relate the microdetails of properties, performance, geometry, and arrangement of constituents and interphase regions to global performance.

The purpose of this introductory text is to lead the reader through an outline construction of the logic and methodology used to create the unified philosophy that is the subject of the book and the basis for the MRLife code that has been used in the applications community since 1981. Each of the chapters of the book elaborate on fundamental elements of that philosophy, the physical behavior associated with those elements, and the interpretations and characterizations that make the application of the philosophy successful.

I.2 BASIC CONCEPTS

Several of the basic elements and essential concepts of the present philosophy are extracted from familiar, well-established engineering practice. These concepts are expanded and interpreted in the context of the present problem. We begin with the concept of engineering strength.

Let us first consider the idea of strength in a unidirectional continuous-fiber composite under uniaxial loading. Defining strength in fibrous composite materials cannot generally be done by simply identifying a single *stress level* that causes failure. Such composites are generally anisotropic and inhomogeneous, so that the stress state in the interior of the material is nearly always complex, even when only one stress is applied at the global level. In addition, the material strength at the local level is generally anisotropic and spatially nonuniform. One must select, then, specific stress values and compare them to the correct corresponding strength values to construct a proper "strength" concept. The most common engineering method for doing this is to define a *failure function*. A failure function is a scalar, generally written as $\text{Fa} = \text{Fa}(\sigma_{ij}/X_{ij})$, where Fa lies between 0 and 1, with $\text{Fa} = 1$ defining failure of the material. A number of different expressions based on this form have been proposed over the years. A "natural" choice is to compare the stresses in the composite directions (1: fiber direction; 2: perpendicular or matrix direction) to their respective strengths (*X*: axial strength; *Y*: transverse strength; *S*: shear strength). For a lamina subjected to in-plane loading, a *maximum stress criterion* (in two dimensions) can be written as $\text{Fa} = [\sigma_1/X \text{ or } \sigma_2/Y \text{ or } \tau_{12}/S]$, depending on which ratio is maximum. In this failure function, no interaction between the failure strengths *X*, *Y*, and *S* is considered. In other words, axial, transverse, and shear failures are presumed to occur independently. But a stress applied in only one direction at the global level generally results in a complex stress state at the local level. For "brittle" polymer and ceramic composites, the failure strain of the matrix may be smaller than the failure strain of the fiber. For a uniaxial lamina loaded at an off-axis angle (angle between the loading direction and the fiber direction), matrix cracking in planes parallel to the fiber direction is likely because of high normal stresses perpendicular to the fibers. Typically, between 0° and 12° or so for a polymer matrix composite, τ_{12} at the fiber–matrix interface approaches the local shear strength, leading to interfacial sliding or local yielding. Between 12° and 90°, τ_{12} decreases, σ_2 increases, and the growth of interfacial cracks in an opening mode takes place. But for any angle, final failure is directly dependent on the combination of σ_1, σ_2, and τ_{12}. Failure functions like the Tsai–Hill function (discussed in Chapter 2), for example, are able to consider the net effect of a complex combination of stresses. For a unidirectional composite lamina with in-plane loading, it is written as

$$\text{Fa} = \frac{\sigma_1^2}{X^2} - \frac{\sigma_1 \cdot \sigma_2}{X^2} + \frac{\sigma_2^2}{Y^2} + \frac{\tau_{12}^2}{S^2}$$

Such expressions are widely used and typically show reasonable agreement with experiments.

Figure I.3 illustrates the fundamental concepts we wish to take from these engineering practices to construct a more general philosophy:

1. Strength will refer to fully multiaxial, normalized failure functions defined in a manner consistent with classical engineering concepts.
2. These functions will be used to combine the effects of degradation processes and events associated with multiple damage and failure modes that control the state of stress in the numerator of the failure functions and the material states that are represented in the denominator of those functions.

We will apply this interpretation of strength to all classes and types of composite materials and structures, not just to the single-ply laminae used to motivate the present discussion.

The next step in our construction is to broaden the concept of strength to composites in which the degradation processes that lead to failure involve significant changes in the stress state or material state on a pointwise or regional basis in the composite. To motivate this discussion, we recall that many continuous-fiber composites are made up of plies or laminae that are bonded together to form a laminate. The discussion of strength for each of those laminae is essentially the same as the previous section, with the addition of the effect on ply stresses of the accommodation each ply must make to deform with the same strain as its neighbors. (There are also edge effects, discussed later in the text.) For the laminate, each ply may fail in various ways and the global failure of the laminate can be said to be defined by *last-ply failure.*

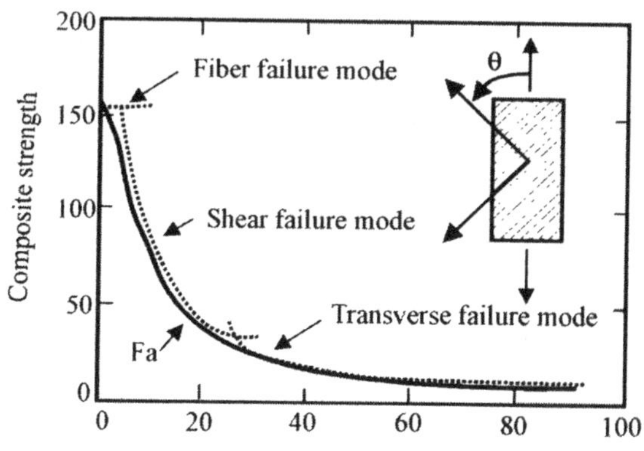

Figure I.3 Failure envelope (Fa) for a unidirectional lamina loaded uniaxially at various angles to the fiber direction, compared to physical strength observations for several modes of failure.

The most typical engineering approach to the definition of the strength of laminates is to address the failure of each ply (or other basic building element of the composite) with increasing load level or time. In that sense, one can define *first-ply failure* and subsequent ply failures as the process of internal stress state changes due to local stress redistribution. There is an accepted practice in the composite engineering field for calculating the progressive internal stress state by reducing (or "discounting") the stiffness in damage-affected directions in those plies that fail. This "ply discount" method assumes that stress in the damaged regions (often the matrix) of those plies is redistributed, with the effect of increasing various stress components in the other plies. If one calculates the strength of the last ply to fail as a function of this redistribution process, then it is clear that the failure function calculated in that ply will be increasing, even for a constant applied global loading. Estimation of laminate strength based on such a methodology is common practice and typically accurate to within a standard deviation or so of average observed values.

If global failure is defined by fracture, for example, the last ply (or element) to fail can be called a *critical element*, in the sense that global failure is defined by the local failure of that ply. For other failure processes, other critical elements could be identified and defined. The length scale of the critical element is determined by the region of influence of the local process that determines final fracture at the global level. It may be at the fiber–matrix level, the ply level, the sublaminate level, or any other appropriate scale. Then one might consider the failure function in that critical element as a canonical parameter for the definition of strength, or the remaining strength in the process of progressive failure of the plies in the laminate, and global failure can be defined in a more general sense.

This is, in fact, the fundamental foundation of the *critical-element theory* that we have developed (Reifsnider, 1992, 1996; Reifsnider and Stinchcomb, 1986). The concept of a local failure function in a critical element changing with loading history, or more precisely, with changes in internal stress state (the numerator of the failure function) and material state (the denominator of the failure function) is an important part of the foundation of the remaining strength philosophy that we have constructed. Hence, we add the following fundamental concepts to our prior list:

3. The failure function of interest is calculated in critical elements, which are the local material elements whose failure defines global failure. It should be noted that global failure implies a specific failure mode. Hence, a critical element is defined and the remaining strength is determined for each independent global failure mode.
4. Property changes (such as pointwise changes in stiffness or strength) are rigorously followed, and consequences (such as internal stress redistribution or local material state changes) are directly and continuously used to update the instantaneous value of the failure function in the critical element for a given failure mode, as a function of progressive degradation during constant or time-variable applied conditions.

Before explaining how these concepts are used for fatigue (and for combined fatigue, creep, and stress rupture), we introduce one more fundamental feature of the philosophy that is critical to the more general case of nonuniform stress states.

I.3 NONUNIFORM STRESS STATES: CHARACTERISTIC MATERIAL DIMENSIONS

So far we have examined the concept of strength in laminae, and strength in laminates, subjected to remote loading at the global level. However, for those discussions, it was assumed that the local stresses are uniform. Most applied conditions involve nonunifrom stress states and nonuniform strength as a point function. To extend our development of the critical-element concept to those situations, we will discuss an illustrative example, the familiar problem of a coupon laminate plate with a center hole. A uniaxial loading is applied to the coupon. We consider stresses at the ply level, exactly as we did for the uniform-stress case, and form a failure function at that level in the critical element. However, as Wu (1978), Waddoups et al. (1971), and subsequently Whitney and Nuismer (1974) have shown, the strength of notched laminates is controlled by a characteristic volume of material. It is necessary for us to write our failure function in this critical element or characteristic volume of material near the notch.

A convenient method of doing that has been suggested by Whitney and Nuismer (1974), although their methodology has been applied at the lamina level in the present approach rather than at the laminate level as they described, as we will explain in Chapter 7. Whitney and Nuismer suggested that failure occurs when the average stress $\bar{\sigma}_x(y)$ over some characteristic material distance A along the y direction (perpendicular to the loading axis, at the hole) is equal to X_0, the strength of the unnotched material. When applied at the lamina level, this characteristic distance is a material property independent of laminate geometry and stress distribution. The failure criterion is written by Whitney and Nuismer as $\sigma_{ap}/X_0 = 2(1 - \xi)/(2 - \xi^2 - \xi^4)$, where σ_{ap} is the loading applied on the laminate in the x direction, $\xi = r/(r + A)$, and r is the hole radius. This form also shows that the notched laminate strength depends on the hole radius r. The larger the hole, the larger the volume of material subjected to high stress. The probability of having a large flaw in this volume is greater, and as a consequence the strength is lower.

The characteristic material length (or volume) concept will be used in the present approach, universally, for nonuniform stress states. Moreover, the local material stress state will change as damage develops, as suggested by the data in Figure I.4.

I.4 STRENGTH EVOLUTION

The preceding concepts and methods can be extended to the most general case of the calculation of remaining strength and life of composite materials and structures

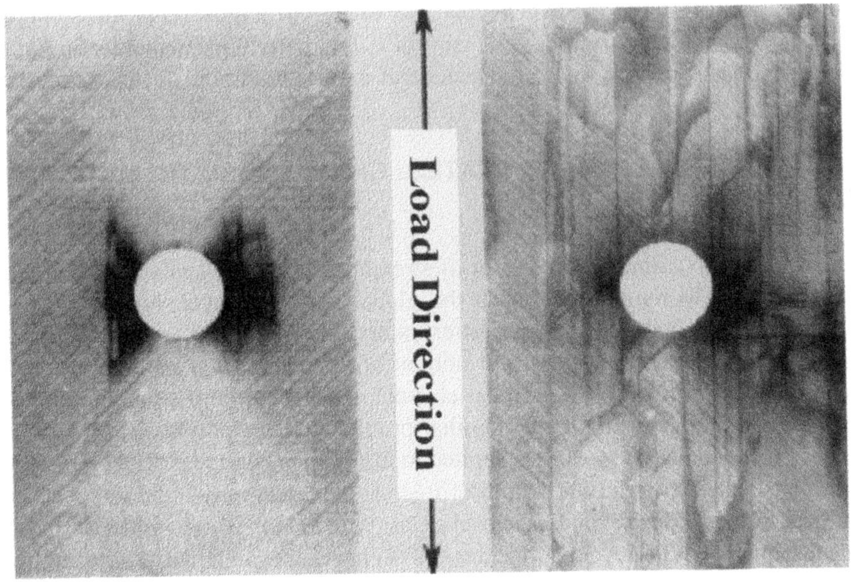

Figure I.4 Examples of damage development near a notch in a carbon–epoxy laminate.

under mechanical, thermal, and environmental applied conditions that produce combinations of fatigue, creep, and stress rupture (time-dependent failure). The first step in the philosophy is to carefully identify the failure mode that is induced by the applied conditions using experimental methods. Then, using the precepts described above, a failure function form is selected to describe the final failure event (e.g., fiber failure in a critical element). Then all of the processes that cause changes in the stress state or material state in that critical element are characterized by rates as a function of the applied conditions and (generalized) time.

Failure modes are primarily (but not exclusively) a function of the stress state in the critical element, under the thermal and chemical conditions expected in service. If the fibers in the 0° plies near a notch over some characteristic distance control the quasi-static failure in a notched coupon, for example, they (collectively) will probably be the critical element under long-term conditions. There are important exceptions to this generality that we will discuss below. A major reason for the validity of the generality is the fact that most engineering composites are designed to be fiber controlled; that is, the stiffness and strength of a "well-designed" composite are usually dominated by the fiber reinforcement. This, in turn, is driven by the fact that the fiber reinforcement is often comparatively insensitive to temperature, time, fatigue, and other long-term environments. However, laboratory results provide the surprising fact that the tensile strength in the fiber direction of a polymer composite coupon, for example, can change by 15–34% when the matrix properties or the fiber–matrix coupling changes due to temperature or local constituent variations, even though the fibers are unaffected by those changes (Mahieux et al., 2001). Hence, we must add that failure modes can change due to applied environments such as temperature, chemical agents, and time or cycles. For the present philosophy, the failure mode must be determined for the conditions to be modeled, preferably by experimental characterization. If that is not possible, experience available in the literature can often be used to anticipate that failure mode.

However, the damage development for a few cycles of loading (see Fig. I.4*a*) and for about 100,000 cycles of loading at the same stress level (Fig. I.4*b*) illustrate another essential feature of this philosophy. The initial stress state and material state of the material are greatly altered by the history of loading. *In fact, for the low stress levels applied for practical engineering applications, the amount of damage that occurs before failure is very great, much larger than the damage that occurs in laboratory tests for shorter times at larger stress levels.*

The central theme of the present approach, then, is to track the changes in the stress state and material state in the critical element as functions of the duration and history of the applied conditions as well as the constituent properties and the rates of the degradation processes that act to change properties. This must be done for the combined effects of all damage processes and events, not just for the microcracking illustrated in Figure I.4. Our problem, then, is to combine those elements in a self-consistent way such that the interactions and collective effects of, say, fatigue, creep, stress rupture, and other phenomena are retained, in the spirit of the critical-element philosophy.

Elements of this approach have been described in earlier publications (Reifsnider, 1992, 1995, 1996; Reifsnider and Case, 1998). A much-abbreviated outline will be presented here; we will expand upon it in Chapter 3. In general, we might claim that the state of the material can be represented by a free-energy function of some type, U, which has the form $U = U(\varepsilon,\ T,\ \Psi)$, where ε is a point function of strain, T is temperature, and the damage scalar Ψ is typically called the "continuity" of the material and has values that vary from unity when the material is "undamaged" to zero when the material has lost all capability to sustain loading. However, one can see that these are exactly the limits (in reverse) that were assigned to our failure function Fa defined earlier. Hence, we take, for a first approximation, the following form for the continuity scalar: $\Psi = 1 - \text{Fa}(\sigma_{ij}/X_{ij}$. It is understood that this simple form is a first approximation at best and may be regarded as the first terms in an expansion of the scalar function Ψ in terms of the instantaneous failure function Fa. With these foundations and a modest appeal to kinetic theory, it can be shown that the remaining strength at the global level can be written in terms of these variables as

$$\text{Fr}(t) = 1 - \int_0^t \left[1 - \text{Fa}\left(\frac{\sigma_{ij}(t)}{X_{ij}(t)}\right)\right] j \left(\frac{n}{N(t)}\right)^{j-1} d\left(\frac{n}{N(t)}\right) \tag{I.1}$$

The details of this derivation are found in Reifsnider (1995) and will be described in Chapter 3. In Equation (I.1), it is understood that the instantaneous failure function is evaluated in the critical element, that is, that part of a representative volume of the material that fails last in the sequence of damage processes and, therefore, defines global failure on the basis of a local microevent. The implications of this definition are rather substantial. For example, under tensile loading of a unidirectional material, the local critical element might be a small bundle of fibers that, if they rupture together in a given locality, will initiate global failure. However, under compressive loading of the same material, the local critical element might be a ply of a laminate for a region of the material that microbuckles, causing a local instability that propagates to cause global failure. These two failure modes are just two of many such modes that must be described individually and analyzed separately. It is important to note that we will assume that each failure mode will define a separate and specific remaining strength analysis that must be addressed with the proper associated models and characterization data. The effects of all engineering phenomena typically characterized (e.g., fatigue, creep, microdamage, stiffness change, stress rupture) are reduced to either a change in stress state (from pointwise stiffness variations) or a change in material state from point to point as a function of generalized time. These effects are combined in a standard engineering manner by the failure function, $\mathrm{Fa}[\sigma_{ij}(t)/X_{ij}(t)]$, calculated in the critical element. The effect of that function on the remaining strength is controlled by the nature of the integral in Equation (I.1). In addition to the failure function, which we have discussed, the integral involves a *characteristic time*, written as $N(t)$ in the present case. (We will explore other forms in Chapter 3.) The characteristic time has the analytical character of the current instantaneous value of life of the critical element under the current local conditions. In the simple illustration discussed above, it might be the fatigue life (at room temperature) of a unidirectional lamina (if that is the critical element) loaded in the fiber direction at room temperature. The functional form of $N(t)$ could have been written as

$$N(t) = N\left(\mathrm{Fa}\left[\frac{\sigma_{ij}(t)}{X_{ij}(t)}\right]\right) \tag{I.2}$$

since the characteristic life of the critical element is a function of the current value of the failure function at the local level. With this observation, we see that the strength evolution integral is completely controlled by the instantaneous failure function in the critical element and the history of changes of that function.

I.5 OUTLINE OF THE METHODOLOGY

The basic methodology of the present approach is illustrated in Figure I.5.

In principle, the present method is quite simple to apply. (The basic steps are outlined in Fig. I.5.) One begins by conducting a quasi-static test to failure using the same applied conditions (e.g., stress state, temperature) expected in service. The failure of the specimen is analyzed carefully, at the local level, to determine the exact manner in which the failure process occurred. This is the most difficult and important step in the analysis. If the failure is very local, the representative volume may be very small, and it may be very difficult to determine the nature of the failure at that

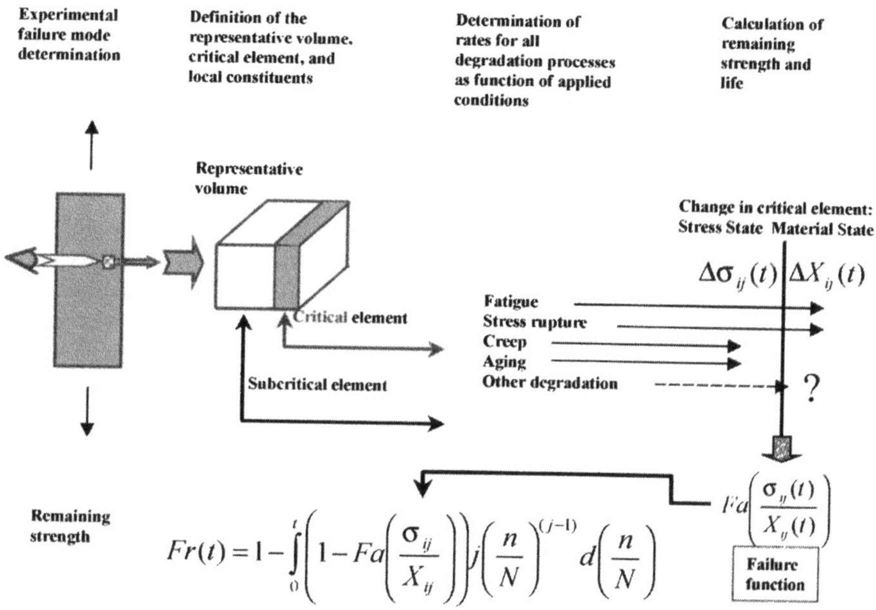

Figure I.5 Present approach to estimation of remaining strength and life.

scale. For uniform stress states in a laminate, for example, the representative volume may be something as simple as the cross section of the laminate, and the critical element may be the ply having the smallest angle with a principal loading direction. Indeed, as we will see in this book, the critical element in many continuous-fiber composite components is often a bundle of fibers in such a ply. But if matrix failure determines global failure, the critical element may be a volume of matrix material. Many other failure modes will be discussed.

The next step is to find the stress state in the representative volume and the critical element. Generally, this will be a local stress state in a global component, so some analysis method must be used to determine local conditions in terms of global applied variables. If the critical element is something as simple as the low-angle ply in a laminate, for example, classical laminate analysis may be adequate for this purpose. If the local stresses are nonuniform, however, an approach like finite-element analysis may be necessary. If there are geometric or material nonlinearities that affect the stress state, they must be considered carefully. It is essential to determine the local stress state in the critical element correctly. The accuracy of the remaining calculations hinges on the validity of this calculation.

A proper failure function must then be selected. For fiber-controlled failure, for example, the failure function may be as simple as the ratio of the fiber direction stress in the critical element divided by the fiber direction strength. In other cases, the failure process may be driven by multiaxial stresses. The selection of a proper failure function can be tested by comparing quasi-static strength predictions with observed values, as input variables (e.g., multiaxial stress components) are varied. The accuracy of this comparison and the modeling of the failure event will determine the accuracy of subsequent remaining strength and life estimates.

If quasi-static strength is properly represented for the failure mode of interest, then degradation processes must be identified, and the rate of those degradation processes must be measured. Typical processes include creep, fatigue, stress rupture, microcracking, delamination, corrosion, microstructural changes such as crystallization, aging, changes in molecular bonding, and a variety of other time- and cycle-dependent processes. These processes must be characterized as functions of the external variables that control their rate of influence, especially temperature and rate of mechanical loading. As Figure I.5 shows, the present approach reduces the effects of all such processes to changes in (local) stress state or material state. Creep, for example, is represented in terms of changes in the material stiffness or compliance tensor components, and those changes result in variations in the local stress state in the representative volume. A typical result of these changes is the redistribution of stress from subcritical elements to critical elements as the process continues. Stress rupture (time-dependent failure) is interpreted as a reduction in material strength, that is, changes in the tensor values of principal material strengths. All other processes are interpreted in a similar way. Consequently, all degradation processes are ultimately represented and combined by changes in the local failure function in the critical element as a function of generalized time.

As part of the rate characterization procedure, a characteristic time is also identified for the controlling process. In the equation shown in Figure I.5, the characteristic time is the number of cycles to failure in fatigue, $N(t)$, for the critical element under the local conditions that exist at time t. If the dominant or controlling process were creep or stress rupture, the characteristic time could be the time to failure of the critical element under instantaneous local conditions. As we will show in Chapter 3, the choice of characteristic time is generally not unique; the success of the predictions are assured if the selected characteristic time is a function of the failure function Fa(t). If it has that character, it will be driven to correct limits by Fa(t) as the conditions in the critical element become severe. If the characteristic time is specified in terms of fatigue, for example, one may find an equation such as

$$\mathrm{Fa}\left(\frac{\sigma_{ij}}{X_{ij}}\right) = A + B\log(N)^{p} \tag{I.3}$$

so that the characteristic time (life) N is a function of the instantaneous local applied conditions, as represented by Fa. The constants in Equation (I.3) and the other rate equations must be determined from independent experiments. For Equation (I.3), for example, room-temperature fatigue characterization must be conducted on the material that corresponds to the critical element to determine the material constants in that relationship.

These relationships are then entered into Equation (I.5) to calculate the remaining strength as a function of time or cycles. For each time increment, the local stress state in the critical element must be reevaluated, since phenomena such as creep are redistributing stress in the interior of the material. That calculation can be very demanding and is a key feature in the accurate modeling of the failure of the remaining strength and life of the material or component. Changes in the internal stress state may involve nonlinear material behavior, nonlinear geometric effects, and transients of various types. It is possible to have local material properties that depend on dependent or independent variables; thermal conductivity (in a thermal stress problem) may depend on temperature, for example, so that the thermal (and therefore the stress) distribution must be determined from a coupled mechanics and thermal problem, solved by an iterative method, for example. This aspect of the problem must be very carefully addressed. Chapter 8 will highlight some of these challenges.

Life is generally defined as the time (or cycles) at which the value of Fr just equals the current value of Fa. Remembering that all of these quantities are normalized, this statement implies that the remaining strength just equals the level of applied conditions, interpreted at the critical-element level. This follows the physical interpretation that failure occurs when the local strength just equals the local level of stress.

There are many ways to validate this representation. The prediction of initial quasi-static strength is a good method of checking the selection of the correct Fa function. Also, since remaining strength is predicted, tests can be stopped before failure and remaining strength measured to determine if the model is following the degradation correctly. Input parameters such as temperature and applied stress level can be varied to test the robustness of the model. Extensive testing and validation were done to construct the present modeling approach with the help of many industrial and university partners over a period of years.

I.6 VIRTUAL DESIGN

The level of representation is also variable in the model. If one has micromechanical models for stiffness and strength and constituent properties and performance data are available, it is possible to represent the entire model in terms of the constituent geometry, arrangement, and characteristics. The advantages of that approach include the capability to discuss processing and manufacturing in the analysis of durability and damage tolerance. One can estimate, for example, the effect of changing the fibers, matrix, interface, or other manufacturing variables on the long-term strength and life of the material and components made from the material without the cost of actually making and testing the material. The savings in cost from such "virtual designs" can be very great. The present model can be (and has been) used as an engine for virtual design.

REFERENCES

Mahieux, C. A., Reifsnider, K. L., and Case, S. W. (2001), "Property Modeling across Transition Temperatures in PMC's: Part I. Tensile Properties," *Appl. Composite Mater.*, Vol. 8, pp. 217–234.

Reifsnider, K. L., and Stinchcomb, W. W. (1986), "A Critical Element Model of the Residual Strength and Life of Fatigue-loaded Composite Coupons," in *Composite Materials: Fatigue and Fracture*, ASTM STP 907, H. T. Hahn, Ed., American Society for Testing and Materials, Philadelphia, pp. 298–313.

Reifsnider, K. L. (1995), "Evolution Concepts for Microstructure–Property Interactions in Composite Systems," in *Proceedings of the IUTAM Symposium on Microstructure–Property Interactions in Composite Materials, Aalborg, Denmark*, R. Pyrz, Ed., Kluwer, New York, pp. 327–348.

Reifsnider, K. L. (1992), "Use of Mechanistic Life Prediction Methods for the Design of Damage Tolerant Composite Material Systems," in *Composite Materials: Fatigue and Fracture*, ASTM STP 1157, M. R. Mitchell and O. Buck, Eds., American Society for Testing and Materials, Philadelphia, pp. 205–223.

Reifsnider, K. L. (1996), "A Micro-Kinetic Approach to Durability Analysis: The Critical Element Method," in *Progress in Durability of Composite Systems*, A. H. Cardon, K. L. Reifsnider, and H. Fukuda, Eds., Balkema, Rotterdam, pp. 3–11.

Reifsnider, K. L., and Case, S. W. (1998), "Strength-Based Predictive Methods for Large-Scale Combined Degradation Effects," in *Proc. Structures, Structural Dynamics and Materials Conf.*, Vol. 1, AIAA-98-268, American Institute for Aeronautics and Astronautics, Reston, VA, pp. 352–362.

Waddoups, M. E., Eisemmann, J. R., and Kaminski, B. E. (1971), "Macroscopic Fracture Mechanics of Advanced Composite Materials," *J. Composite Mater.*, Vol. 5, pp. 446–584.

Whitney, J. M., and Nuismer, R. F. (1974), "Stress Fracture Criteria for Laminated Composites Containing Stress Concentrations," *J. Composite Mater.*, Vol. 8, pp. 253–265.

Wu, E. M. (1978), "Failure Analysis of Composites with Stress Gradients," Report No. UCRL-80909, Lawrence Livermore Laboratory, Livermore, CA.

CHAPTER 1

PHYSICAL BEHAVIOR

Thesis: Damage accumulation is the physical process that reduces the strength and life of fibrous composite systems during service in the presence of applied mechanical, thermal, and chemical conditions.

1.1 CONTINUOUS-FIBER COMPOSITE MATERIALS

1.1.1 Basic Concepts and Conventions

Most of the data presented in this book, as well as the models and philosophy associated with the data, will be concerned with the properties, performance, and behavior of continuous-fiber composites. The general geometric features of the composites we will address are shown in Figure 1.1.

We will assume that there are *principal material directions* defined by the directions of alignment of continuous fibers, as shown in Figure 1.1*a*. Generally, these principal material directions will be indicated by orthogonal axes with numerical indicators; that is, the "1" direction will refer to the direction of the fiber alignment in a single ply of unidirectional material (Fig. 1.1*a*); the "2" fiber alignment direction will be perpendicular to that axis, in the plane of the lamina; and the "3" axis direction will be out of the plane of the lamina, forming an orthogonal set. Exceptions to this notation will be noted as they occur.

In many cases, laminae (or "plies") will be bonded together to form a laminate, as shown in Figure 1.1*b*. Except when noted, we will assume that the bonding is mechanically perfect and spatially uniform. We will also assume that the laminate has a global reference axis, generally indicated by (x, y, z), and the orientation of

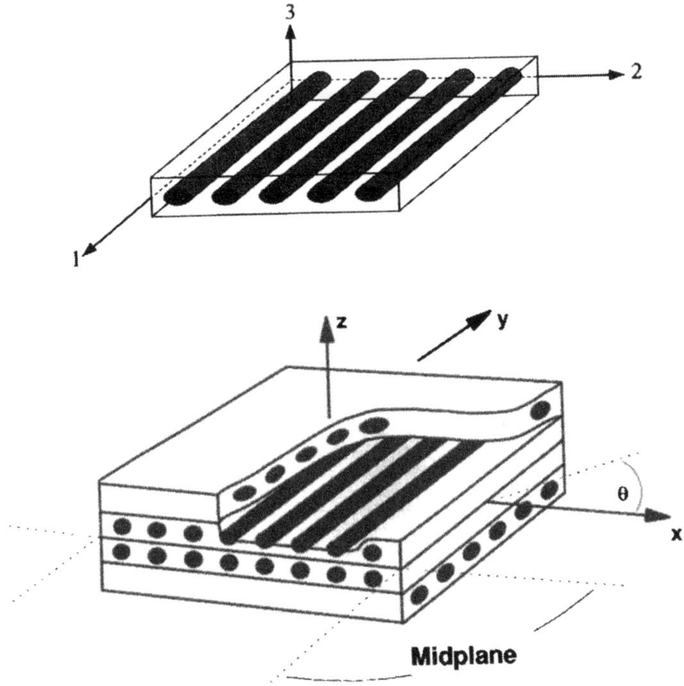

Figure 1.1 Basic coordinate system for (a) unidirectional composites and plies of a laminate and (b) laminates at the global level.

the plies in a laminate will be given relative to this (x, y, z) system. A 10° *ply*, for example, will have an orientation rotated 10° *counterclockwise* from the x axis (towards the y axis). The direction of positive rotation is very important and should be carefully noted when other results from the literature are compared to the present discussions.

The usual notation for laminates will be used throughout this book. A $[0, 90]_n$ laminate is an unsymmetric cross-ply laminate with $2n$ plies; $[0_n, 90_n]$ is the same laminate with all plies of each orientation grouped together. A $[0, \pm 45, 90]_s$ laminate is a quasi-isotropic symmetric laminate, and $[0, \pm 45, \overline{90}]_s$ is the same laminate with only one 90° ply that is shared across the centerline of the thickness of the laminate.

We will also be concerned with woven materials. The general directional definitions defined above will also be used for these materials, but the definition of plies and laminae may change. A lamina of woven material is usually taken to mean one thickness (or ply) of the woven material, and it may contain several fiber orientations. A plain weave of warp and fill fibers, for example, consists of 0° and 90° fibers, as shown in Figure 1.2*a*, and is sometimes written as a "woven [0, 90]"

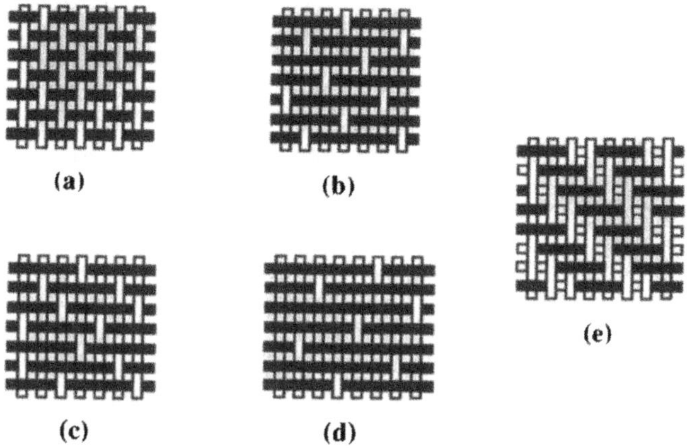

Figure 1.2 Typical weave patterns: (*a*) plain weave; (*b*) 5 harness satin; (*c*) crowfoot; (*d*) 8 harness satin; (*e*) 2/2 twill.

lamina or a "woven $[0, 90]_n$ laminate" with *n* plies. The other definitions mentioned generally apply.

Standard notation will also be used for the other fiber architectures commonly used to create woven materials. Figure 1.2 illustrates those definitions with some typical examples. Many of these weaves have historical origins. Satin weaves, for example, were created to achieve a special beauty resulting from a very smooth surface, but it is not surprising that the mechanical properties (especially strength and long-term behavior) depend on the subtleties of these weave patterns. The fatigue performance of a $[0, 90]_s$ laminate (made from unidirectional plies) is very different from that of a woven $[0, 90]_s$ laminate, for example.

The diagrams in Figures 1.1 and 1.2 are highly idealized. The single plies, or laminae, of unidirectional material usually consist of more than one fiber diameter through their thickness and are usually determined by manufacturing methods. Typical dimensions are 5–10 μm. Table 1.1 (see Appendix 1) shows some dimensions of fibers commonly used in composites. In many cases, there may be hundreds or thousands of fibers through the thickness of a given ply, although, in a few cases (especially for ceramic fibers), the diameter of the fibers is so large there is only one fiber through the thickness (much like Fig. 1.1*a*).

Figure 1.3 shows micrographs of some typical composites. We can see that the idealized diagrams sketched in Figure 1.1 (frequently used for analysis) do not represent reality very well in many instances because fiber spacing and fiber diameters may not be uniform. Fibers are often not straight and are usually handled and used in bundles or tows of hundreds or thousands of fibers. These and other such details will turn out to be very important for the subject at hand; indeed, in some cases, these deviations from textbook ideality control the strength and life of composites. Several other details should be noted. One important feature is illustrated in Figure 1.3*a*, which shows the cross-section of a Nicalon–SiC composite. The image,

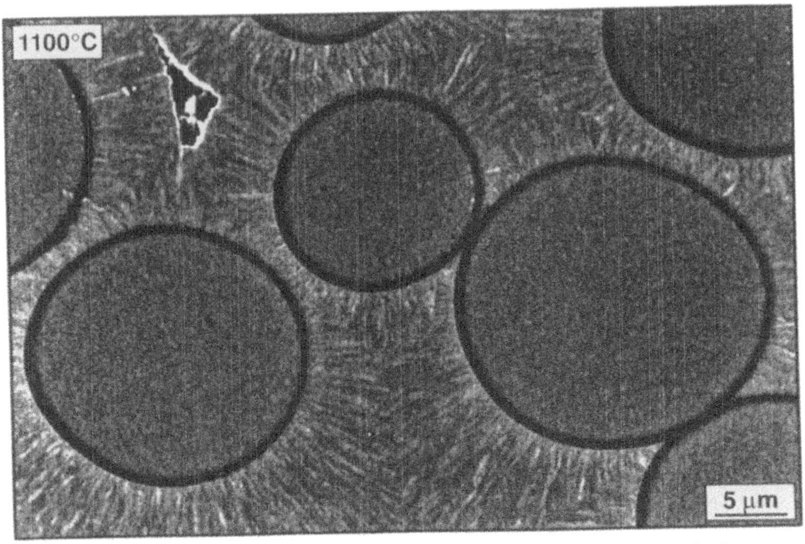

Low magnification SEM image of the general grain morphology in a Nicalon/SiC composite processed at 1100°C.

(a)

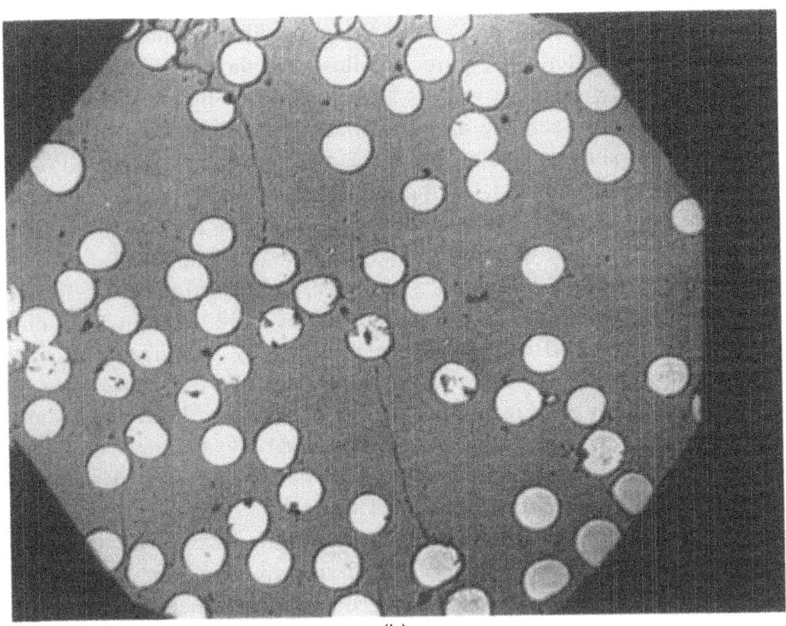

(b)

Figure 1.3 Scanning electron microscopy (SEM) micrographs of composite microstructure: (*a*) low-magnification cross-sectional image of general grain morphology in Nicalon–SiC composite processed at 1100°C (compliments of Rick Lowden, Oak Ridge National Laboratory); (*b*) Nicalon–calcium alumino silicate (CAS) composite cross section with matrix cracking (continued)

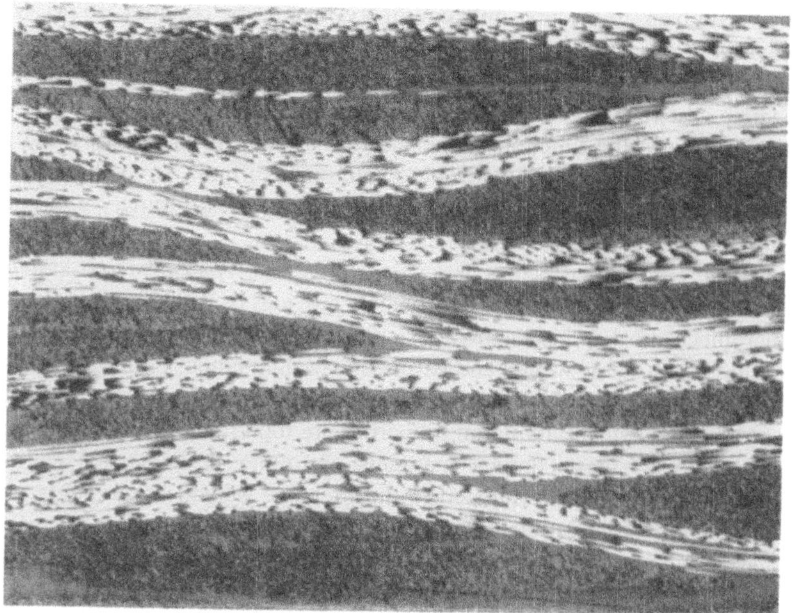

(c)

Figure 1.3 *Continued*. Scanning electron microscopy (SEM) micrographs of composite microstructure: (*c*) longitudinal section of plain-weave Nicalon–SiC system.

from an atomic force microscope, shows that the regions between the fibers (the matrix) are not uniform; instead, they consist of grains of material that are different as a function of position relative to the fibers. Moreover, there is a clearly defined region between the fibers and the matrix, typically called the interphase region, that, as we will see, often plays a major role in the strength and life of composite materials.

Finally, Figure 1.3*c* shows a micrograph of a woven composite in which one can see at least three types of primary regions in the material. The fibers and matrix are present in clearly defined tows, but there are also regions of pure matrix that do not have any reinforcement. The micrograph shows that some fibers may touch or rub each other during loading; this is another important feature we will address.

1.1.2 Composites as Material Systems

Tables 1.1–1.3 (see Appendix 1) show some typical properties of fibers, matrix, and composite materials and laminates. A careful comparison of these numbers will give us a physical perspective on the idea of material systems.

Engineered fiber–matrix systems have been in use for about 3000 years. Straw-reinforced bricks and mummy cases made from papyrus and resin were common in Middle Eastern Europe as early as 1000 B.C. Plant fibers were used by Inca and Maya native Indians to reinforce pottery around 500 B.C., and laminated steel tools and weapons date back to the first millennium B.C. in Europe (Knauth, 1974). Modern fiber–matrix material systems include fiber-reinforced concrete, linoleum, and fiber-reinforced rubber. Our primary concern, however, will be with high-performance engineering systems wherein the systems aspects of the fiber–matrix interaction play a dominant role in the function of the material system. In these cases, the micromechanics we will discuss can be used to determine stiffness, strength, and damage tolerance, the properties and performance that usually are of primary importance to the utility of a given material system.

We will refer to these systems as *advanced fiber–interphase–matrix systems*. An understanding of the nature of the constituents provides us with a basis for choices in our micromechanical representations. Many of the constituents are chosen on the basis of their strength-to-weight and stiffness-to-weight ratios. Table 1.1 (see Appendix 1) lists several types of reinforcing fibers and, for comparison, the properties of a common structural steel (aluminum properties are found in Table 1.2, Appendix 1). The properties of these fibrous materials are quite desirable for engineering purposes, especially when compared to their relative weight. Surprisingly, one can obtain aluminum oxide, boron, carbon, and silicon carbide fibers that are stiffer than steel and aluminum oxide, aramid, boron, carbon, glass, silicon carbide, and even organic fibers like Spectra™ that are stronger than steel, on an absolute basis. When those properties are normalized by the specific gravity of the materials, improvements of more than an order of magnitude over steel can be obtained.

But the limitations of these reinforcing materials are just as important, especially for the subject at hand. High strength and stiffness in natural materials are usually accompanied by low toughness and ductility and by difficulty in manufacturing. For example, certain glass fibers can be made with remarkable strength (e.g., far exceeding steel), and the strain to failure of such fibers is reasonable, compared to other fibers shown in Table 1.1 (Appendix 1). It is not feasible, however, to make high-performance engineering structures out of homogeneous glass because of its well-known brittleness; even the windows of vehicles are often laminar composite systems reinforced with an interlayer of plastic to help control that failure mode. In composite systems, glass and other fibers often break, even in the manufacturing process. (Our own research suggests that typically 10% of the fibers in any given cross section are broken in the manufacturing process.) In a composite, however, the matrix material allows a quick transfer of local stress back into the fibers with little loss of function. This "composite effect" overcomes many limitations of the materials listed in Table 1.1 (Appendix 1) and makes it possible to use them in material systems with great attendant gains in properties and performance. It also motivates the present discussion of the micromechanics that controls the composite effect.

Another important feature of fiber reinforcement is the length of the fibers. In some cases, fiber reinforcement is continuous; that is, the fibers extend over distances of many thousand fiber diameters. It is often easier, however, to process fibers that are shorter, 20–50 mm in length for glass, for example, to make short fiber composites such as the sheet-molding compound (SMC) that is common in the automotive industry. Finally, whiskers are also used (Piggott, 1980). These fibers are usually crystalline, of the order of 1–50 μm in diameter and 1–100 mm in length.

Because of their small volume and crystalline structure, whiskers can be grown (usually from a vapor phase) with near-theoretical strength (the greatest of all fibrous materials), but processing difficulties make them available only in limited quantities. Short fibers are, nevertheless, an important part of the technology. In fact, short-fiber micromechanics is also important in the context of fiber fracture, since long fibers become short fibers when they break. The influence of fiber fracture on stiffness, strength, and life is closely related to the subject of short-fiber composite behavior.

Matrix materials are just as critical to the fiber–matrix system. Many composites are made with fibers as the primary load-bearing constituent; the matrix material must distribute tensile loads and sustain shear loading, especially at the local level. Hence, tough matrices are common choices. Table 1.2 (Appendix 1) lists some of the properties of such materials. Synthetic polymeric resins have the advantage of possessing a remarkable range of properties, especially with respect to resisting environmental agents such as solvents, chemical elements, and even temperature. It is no surprise that polymer matrix composite systems have found so many commercial applications. However, ceramic and metal matrix systems are growing in use and importance. In the context of the present topic, we will be concerned mostly with polymer and ceramic matrix composites, primarily because of our base of experience and the need to limit the scope of our text.

We can see that reinforcing fibers have some very good engineering properties because they are very strong, stiff, and often low in density. We cannot use these fibers alone, however, to make engineering components, and the manufacturing methods used to make the fibers cannot be used, generally, to make monolithic engineering structures with the same properties (some exceptions apply). Moreover, the fibers often have some less desirable engineering properties, like low strain to failure (brittleness) and low energy of fracture (toughness). If we add a matrix to make a composite, however, many of the shortcomings of the fibers can be overcome. This result is achieved not just by adding the properties of the matrix, as one would do if we had a simple mixture of fibers and matrix. The composite properties and performance are greatly influenced by the *interaction* of the fibers and the matrix and the interaction of the fibers with each other via information transmitted through the matrix. This interaction makes the composite a material system, not just a material mixture. This basic concept is so important to our subject that we will develop this idea in greater detail before proceeding.

Material systems are an invention of nature and are commonly found in natural biological structures. Haversian bone, for example, is composed of cells (osteoblasts, osteoclasts, and osteocytes) in an extracellular matrix, which consists of an organic phase (osteoid consisting of collagen and glycosaminoglycans) and an inorganic phase (consisting primarily of calcium phosphate crystals) (Cowin, 1989). Figure 1.4 illustrates the arrangement of the constituents. The osteons are lamellar, and the lamellae (which are 5–7 μm thick) consist of collagen fibers that are spirally wound about the axis of the osteon. The collagen fibers, themselves, are inhomogeneous, and include regularly spaced hydroxyapatite and other mineral crystals about 20–40 nm long. The osteons are typically about 200 μm in diameter, about the same thickness as the laminae in laminar bone, and about 10–20 mm long. This system

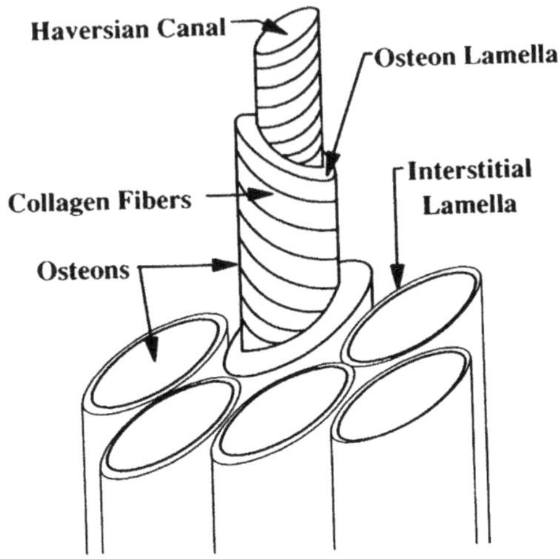

Figure 1.4 Microstructure of Haversian bone.

functions to achieve remarkable properties and behavior. The collagen fibers (about one-third by volume) provide tensile strength (over 100 MPa is typical in humans), and the mineral constituents (also about one-third of the volume) provide compression strength (nearly 200 MPa in humans in some cases). Various cavities and canals in the bone bring nutrients to the bone; no point in the tissue is more than 100 μm away from the blood supply. Bone is anisotropic and is generally characterized as orthotropic (with nine independent elastic constants) or transversely isotropic (with five independent constants). Even finer substructure than that shown in Figure 1.4 can be found, at a scale still above the molecular scale (Glimcher, 1984).

At a larger scale level, bone arranges itself in a cellular structure along the lines of principal stress in some cases to form the familiar cancellous bone structure typically found in the head of the femur, for example. This material system provides many lessons for our problem definition and shows that several levels of order and arrangement may occur and may require modeling. The material system also indicates that the geometry (size and shape) of the constituents, as well as their arrangement and the regions between them, may be important to the definition of a fiber–matrix system. This system also serves to suggest the scale of internal structure that may be appropriate for analysis based on the functions of the elements of the bone structure, at least to the extent that the basic structural elements are found to be of the order of 1–10 μm (e.g., not at the atomic or molecular level).

Further guidance in defining problems of interest, and the possible applications of the subject matter of our book, is derived from tendon tissue and from the structure of the intestines and the cornea of the eye. Figure 1.5 shows the hierarchical structure of tendon tissue (Baer et al., 1980). The substructuring is common in nature. We also note other important features. The geometric similarity of fibrils, subfibrils, microfibrils, the collagen fibers themselves, and the fascicle or tendon is apparent in this tensile member. The differences are, however, also distinct and important. The

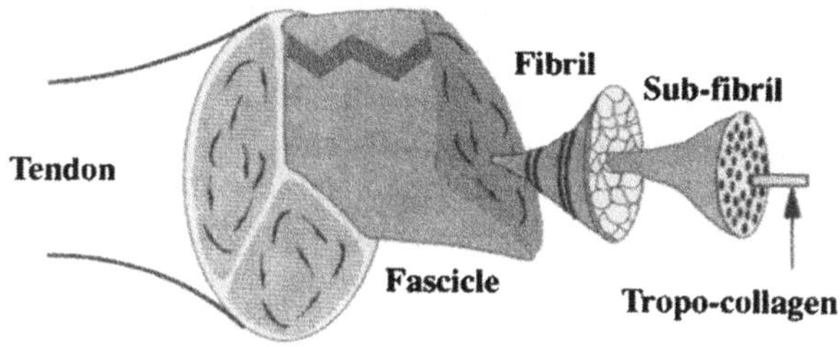

Figure 1.5 Substructuring in tendons.

fascicle is not circular, for example, and at this level (of the order of 100 μm), a wavy pattern is strongly evident. The wavy pattern contributes an important nonlinear feature to the tensile response of the tendon. The multilevel organization makes the tendon tough, in the sense that excessive stress or damage may cause individual subelements at different levels in the hierarchical structure to fail independently; this process absorbs energy and protects the tendon as a whole from catastrophic failure (McNicol and Strahlman, 1989). The fascicular membrane contributes to this isolation. Hence, we might expect to find it necessary to consider descriptions at several levels of resolution, and it might also be necessary to include, explicitly in our calculations, the properties and behavior of finite regions between those levels. Finally, in a given cross section of a fibrous material system, the properties may not be constant with respect to position in any given direction or at any level of resolution important to determining properties and performance. This last point will be of particular importance in the presentation of the material in this monograph. Continuous spatial variations of properties are rarely considered in the formulations of elasticity problems; the present monograph will consider this class of problems explicitly (a rather new feature in the literature).

A similar substructure is found in the structural hierarchy of an intestine. From the molecular through the fibular level, the structure is identical to that of a tendon. The two tissues differ only in how the collagen fibrils are arranged at the largest level, wherein fascicles are replaced with bundles of fibrils wound around the intestine in a helical fashion, forming a structure with fiber bundles at, say, $+30^\circ$ and -30° to the axis of the intestine (in the case of humans). In fact, these bundles form layers having those angles; that is, they form a laminate. Here, again, we learn that in nature, fibers, bundles of fibers, and now layers have important distinctive mechanical functions as well as definitive properties and geometric features. The structure of the intestine is very compliant; it sustains multiaxial forces and allows for the transport of fluids through numerous interstices of the structure. The wavy substructure of the fiber bundles seen in the tendon are also present in the intestine, but the crimp angle is larger to allow larger extensibility required for its function (Hiltner et al., 1983).

Engineered fibrous material systems are not yet as complex or sophisticated as those found in nature. Even the tiny sample above, from a very great population of such examples in nature, points to the future of the enterprise of making such systems and suggests the direction and scope of our current efforts. We will begin with engineered fibrous systems and be ever mindful of the future, which most certainly will bring examples of substructuring, multifunctionality, integration, and even "local intelligence." "Smart" and "intelligent" material systems are already, of course, being widely discussed, and embedded functional elements in materials are commonplace. Finally, our analysis must also address the details of material systems in nature itself; no other pursuit could be more relevant or important.

For composites, this system effect is also easily identified. For elastic response, a convenient reference point is the rule of mixtures (ROM). The physical arguments used to support the ROM are easily illustrated with the following discussion. Suppose we pull on the lamina shown in Figure 1.1*a* with a force P in the 1 direction. We assume that the lamina is in static equilibrium and write

$$P = A_f \sigma_f + A_m \sigma_m \tag{1.1}$$

where A_f, A_m are the cross-sectional areas of the fibers and matrix and σ_f, σ_m are the corresponding normal stress components on those areas. Then, if and only if we assume that the extension of the fibers and matrix is the same in the load direction, we can write

$$\varepsilon_f = \varepsilon_m = \frac{\sigma_f}{E_f} = \frac{\sigma_m}{E_m} \tag{1.2}$$

where we have also assumed that the stress state is uniaxial and ε_f, ε_m are the constituent strains in the fiber direction.

Combining Equations (1.1) and (1.2), we find that the elastic Young's modulus of the composite material can be given in terms of the constituent moduli by the expression

$$E_c = v_f E_f + v_m E_m \tag{1.3}$$

where v_f and v_m are volume fractions of fiber and matrix under the assumptions we have made in the previous paragraph. This expression implies that each of the constituent materials contributes to the global composite stiffness in the exact proportion to their volume fraction in the composite; the expression implies that the composite is a mixture. This ROM suggests that each constituent contributes its properties in proportion to its presence, a simple concept that represents a first approximation of the behavior of the material, at least under the assumptions we have made to derive the formula. These assumptions, however, are not rigorously correct. For example, we have assumed that no local stresses exist on the fibers in the direction transverse to pulling, which is not correct since the Poisson ratio of the constituent materials is different. We have also assumed that the geometric shape of the constituents and the surfaces of the component being tested do not significantly influence the internal stress state, which is certainly not true. The composite properties shown in Table 1.3 (Appendix 1) do not, in general, result from a simple mixture of the fiber and matrix properties shown in Tables 1.1 and 1.2 (Appendix 1), for example.

For the elastic response of composite materials, the coupling we have discussed between individual constituent behaviors can be determined from more complete solutions to the local stress analysis problem created by the fibers and the matrix acting together. Several versions of such treatments have appeared in the literature (Christensen, 1979; Russel, 1973). Figure 1.6 shows examples of calculations from a more complete solution of this problem (Carman and Reifsnider, 1992). First of all, the elastic Young's modulus in the direction of the fibers under unidirectional loading is essentially linear. However, Figure 1.7 shows evidence of nonlinearity for the shear modulus of a composite subjected to in-plane simple shear loading. The behavior depicted in Figure 1.7 results from Equation (1.4), which was derived by Hashin and Hill,

$$\frac{G_{12}}{G_m} = \frac{G_f(1+c) + G_m(1-c)}{G_f(1-c) + G_m(1+c)} \tag{1.4}$$

where G_{12} is the shear modulus in the fiber direction, G_f, G_m are the shear moduli of the fiber and matrix, respectively, and c is the volume fraction of reinforcement (Christensen, 1979). In fact, the subject of the effective moduli of composite materi-

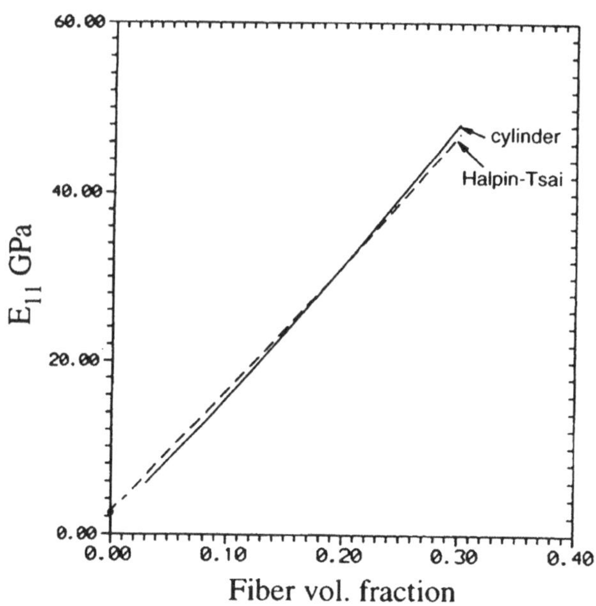

Figure 1.6 Micromechanical estimates of stiffness in direction of fibers in unidirectional composite as function of volume fraction of fibers using modified concentric cylinders model (Carman and Reifsnider, 1992) and Halpin–Tsai equations.

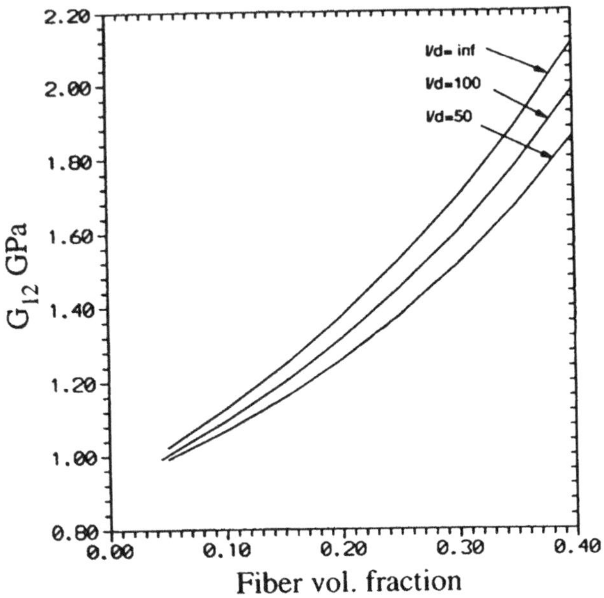

Figure 1.7 Micromechanical estimates of in-plane shear stiffness of unidirectional composite as function of volume fraction of fibers using modified concentric cylinders model (Carman and Reifsnider, 1992) and Halpin–Tsai equations.

als in general has received a great deal of attention in the literature and is well developed at this writing. The reader is referred to the literature for further discussions of this topic for elastic response (Christensen, 1979; Eshelby, 1957; Hashin, 1965, 1979; Hill, 1964).

Many of these treatments have been developed for "dilute solutions" in which the constituent interactions are explicitly ignored, usually in an effort to achieve feasibility of the solution to the problems involved. Notable recent additions to the literature relax that constraint to achieve better estimates of the limits on the elastic moduli when fiber interactions and other constituent coupling are included. One example of such a prediction is the analysis, generated by Mori and Tanaka (1973), that uses an approximate method to account for fiber–fiber interaction and produces a correction for the elastic moduli to account for that coupling.

In situations when the material is not ideal in the sense that fibers may not be uniform in geometry, or in their arrangement, or possibly in their length, additional coupling at the fiber–matrix level contributes to a greater number of departures from the ROM concept. This problem has been widely discussed in the literature as well, and two typical examples of the results of that work are shown in Figures 1.8 and 1.9. (Data for both figures were taken from Carman and Reifsnider, 1992. Validation of the estimates appears in that reference.)

Figure 1.8 shows estimates of the elastic Poisson ratio in the plane perpendicular to the fiber direction in a unidirectional sample for a carbon–epoxy system as a

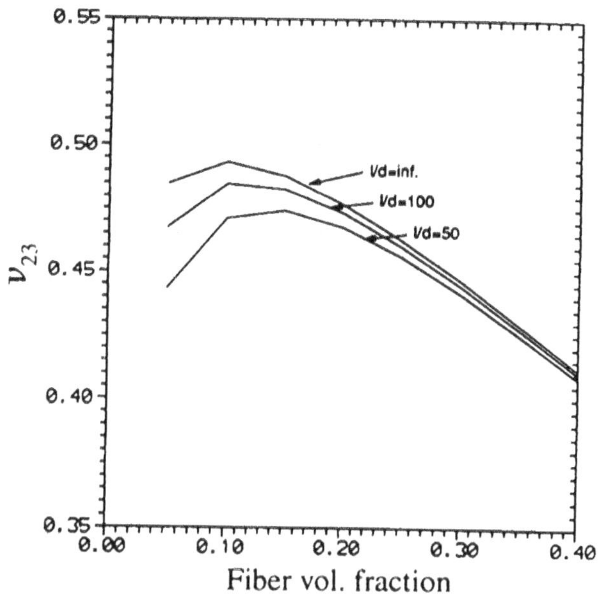

Figure 1.8 Poisson ratio estimates as function of fiber volume fraction for several fiber length-to-diameter ratios for carbon–epoxy system (Carman and Reifsnider, 1992).

function of volume fraction for fibers of finite length (for several length-to-diameter ratios). These results are highly nonlinear, showing an initial rise as the fibers influence local deformations, followed by a decrease toward the value of the fibers themselves as fibers become a more dominant part of the volume. Figure 1.9 shows the elastic Young's modulus of a composite consisting of fibers with variable lengths [results quite similar to an approximate analysis developed many years ago for application to the paper industry by Cox (1952)]. One can clearly see that the length of the fibers and the volume fraction of fibers interact to greatly influence the stiffness in the fiber direction. The nonlinearity of this plot suggests that the internal interactions and constituent coupling that account for this behavior are complex and play a major role in the determination of this elastic response. The ratio of the stiffness of the fibers to the stiffness of the matrix also plays an essential role in the effect of the fiber reinforcement on the composite stiffness (Russel, 1973). Composite stiffness and strength can be a highly nonlinear function of the geometry, arrangement, volume fraction, and coupling of the composite constituents; the dependence on those local characteristics does not necessarily comply with simple rules of mixture concepts.

The interaction of constituents, and the resulting dependence of properties and performance on that interaction, is typically referred to as the *composite effect*. We have illustrated above that this composite effect is apparent in the elastic response of the composites. There is a substantial body of literature that can be used to further illustrate this point, but that is not the subject of this book. Even more remarkable is

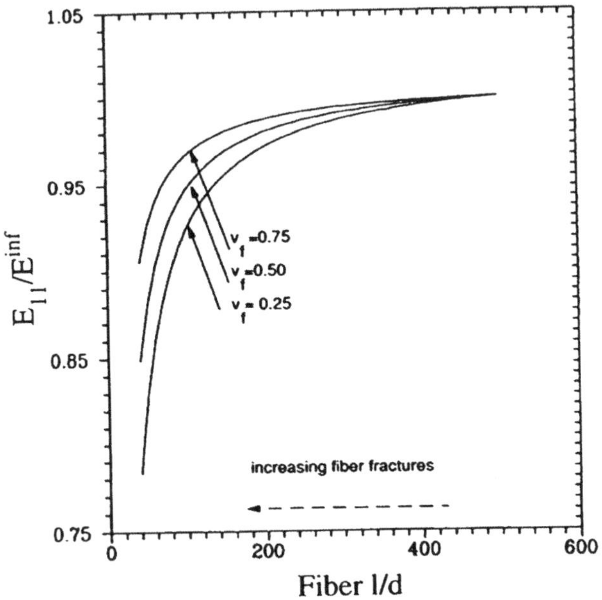

Figure 1.9 Young's modulus estimates as function of fiber length-to-diameter ratio for several fiber volume fractions (Carman and Reifsnider, 1992).

the fact that this composite effect, or systems effect as we have been calling it, is often more influential on the strength and life of a composite material than it is on the elastic response. Indeed, we will find that the systems effect in composite materials often controls the durability and damage tolerance of those materials. The details of this effect will form a major part of the text that follows. For the moment, we will cite a few examples of these effects as a foundation for that discussion.

Figures 1.10–1.12 show the influence of fiber–matrix coupling as it is altered by changing the interphase region between the fiber and the matrix on several properties and performance characteristics of polymer matrix composites. Laboratory data indicate that changes in the fiber–matrix coupling brought about by coating the fibers with a very small amount of interfacial material can alter the transverse strength by nearly 100%, the transverse strain to failure by as much as 400%, and the life (for notched coupons) under fully reversed tension–compression fatigue by two orders of magnitude (Lesko, 1994). Changes in other characteristics are also possible, and these topics will be discussed in detail in later sections. The point to be made here is that the systems effect in composite materials plays a major role in the long-term performance of those materials, especially for continuous-fiber materials such as those under discussion. This systems effect is substantial and can be altered by changing a very small part of the composite material, namely a few percent of the volume of the composite that controls the interaction between the fiber and the matrix. This interaction region is often created by coating the fibers in some way and has been called the interphase region by Sharpe (1972).

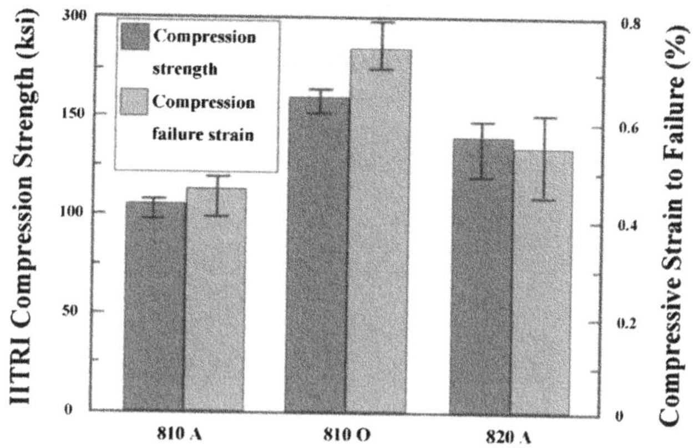

Figure 1.10 Effect of three different fiber coatings on compression strength and compression strain to failure in carbon–epoxy composite.

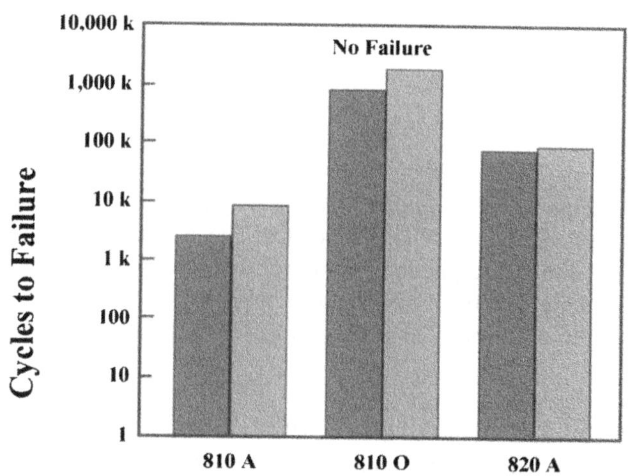

Figure 1.11 Effect of three different fiber coatings on notched fatigue life ($R = -1$) in carbon–epoxy composite.

1.2 DAMAGE TOLERANCE AND DURABILITY

1.2.1 Phenomenological Definitions

In the previous section, we have defined the fundamental nature of the composite systems we wish to consider and have identified the fundamental nature of the behavior of those systems that forms the foundation of the subjects of durability and damage tolerance. We know that the properties and characteristics of the consti-

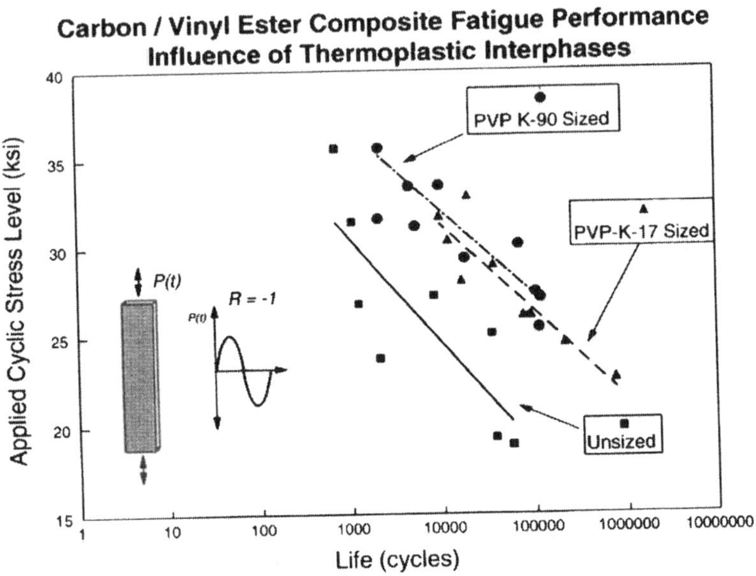

Figure 1.12 Effect of fiber sizing (with different molecular weight) on unnotched fatigue life ($R = -1$) in carbon–vinyl ester composite.

tuents are important, and their shape, their arrangement, and especially their interactions under the application of applied loading play a major role in the elastic and inelastic response of continuous-fiber composites. The subjects of this book, durability and damage tolerance, are phenomenological definitions based on the engineering response of materials, in most cases, under *long-term* conditions. In general, this subject addresses the philosophy associated with answers to questions such as the following:

1. How long will this engineering component last?
2. How safe is this engineering component after some period of service?
3. How reliable is this engineering component?
4. What is the remaining strength of this engineering component after some history of service?

There are several specific phenomena and physical behavior categories generally discussed under the topic of damage tolerance and durability, and we will present specific discussions of these topics in the sections that follow. It is useful, however, to define the terms *durability* and *damage tolerance* in a more specific way for the purpose of the rest of our discussion. Figure 1.13 serves this purpose. We will assume that durability has to do with the life of a component or material (or material system), and damage tolerance is a discussion of how much strength is left after some period of service and history of load application. These are typical definitions,

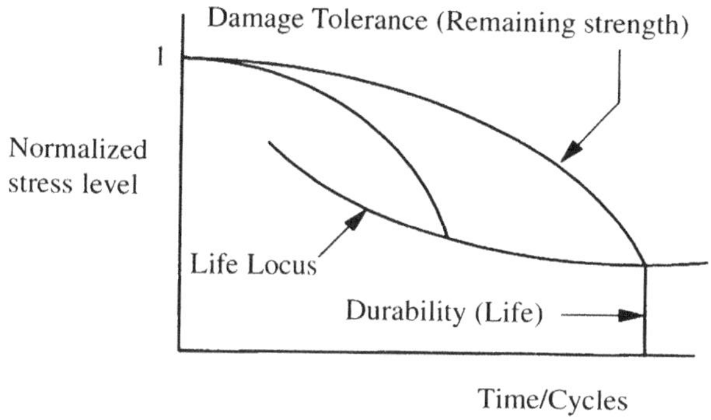

Figure 1.13 Concepts and definitions associated with durability and damage tolerance.

not entirely universal ones. Figure 1.14 is a further illustration of the problem we wish to address, and it demonstrates the most important fundamental assumption made in the discussion of technical material to be presented here.

It is our fundamental premise that durability and damage tolerance can be discussed entirely in the context of strength and remaining strength. Therefore, as illustrated in Figure 1.14, we will be concerned with the problem of how the initial strength of the composite material is altered by damage processes over a period of time, that is, during the service life and loading history of the material or component. As indicated in Figure 1.14, the initial strength may be altered by a variety of such processes, perhaps the most familiar being microdamage development events such as microcracking, delamination, debonding, and fiber fracture. Microdamage is widely discussed in the literature and accounts for a great deal of the information currently available on this subject (Horton and Whitehead, 1988; Reifsnider, 1982, 1991). In many instances, other degradation processes are equally important and may be dominant in the determination of remaining strength and life. Under certain conditions, especially high-temperature conditions, there may be mass transport of atomic or molecular species from point to point in the material; that is, there may be kinetic activity in that sense. There may also be chemical activity, such as corrosion or the

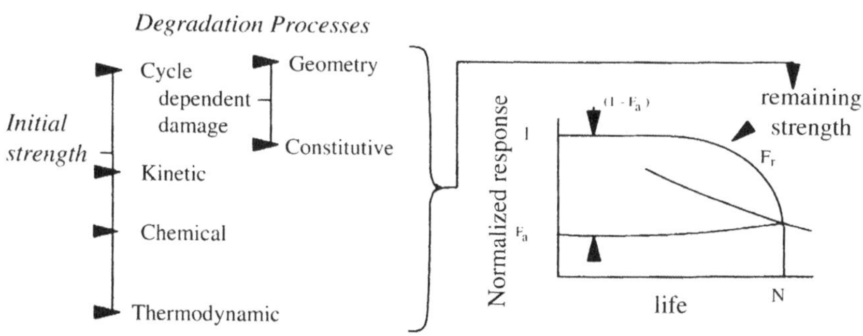

Figure 1.14 Definition of scope of problem of durability and damage tolerance.

chemical interaction of constituent species in the composite to form new compounds or to disassociate existing ones. Finally, there may be a variety of thermodynamic events that contribute to the degradation of the material or alter its remaining strength and life in various ways. Such events may include changes in microstructure that are driven by various activation energies such as temperature or deformation energy. In the following chapters, we will see that these physical events combine in various ways to create phenomenological behavior such as creep, stress rupture, and fatigue, which are widely discussed for all material systems.

In this context, durability is often discussed in terms of damage resistance or susceptibility to damage initiation. Both concepts imply that the subject component is being exposed to applied conditions such as mechanical loading and environments such as temperature and chemical agents over long periods of time considered typical of the projected service of the component. It is further assumed that these applied conditions induce degradation processes, which result in changes in the properties and performance of the composite materials in question. As mentioned, these changes are usually characterized by several categories of phenomenological response that are defined on the basis of behavior that is of special interest to engineering applications.

1.2.2 Creep Response

Perhaps the simplest type of engineering function is the requirement that a structure support a load for a long period of time. In some cases, such as in civil structures like buildings and bridges, the load sustained can be constant in time (or very nearly constant). This situation is also typical of many other weight-bearing structures; however, the response of a material to sustained loading may not be independent of time, even when the loading is constant. This fact is especially true when the material supports a load at elevated temperatures, in particular, when those temperatures are of the order of 60% of the melting or softening point. If the material does deform as a function of time when the applied load is constant, then the material sustains what is referred to as creep deformation.

An illustration of typical creep response for a material is shown in Figure 1.15. Quite often, there is an early period of transient creep, which is fairly rapid but

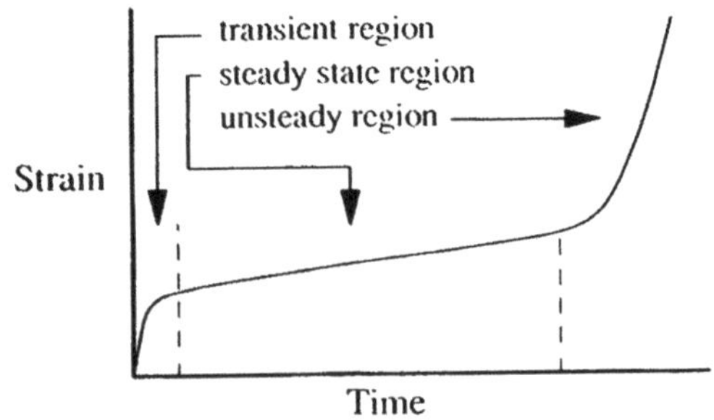

Figure 1.15 Typical creep response of an engineering material.

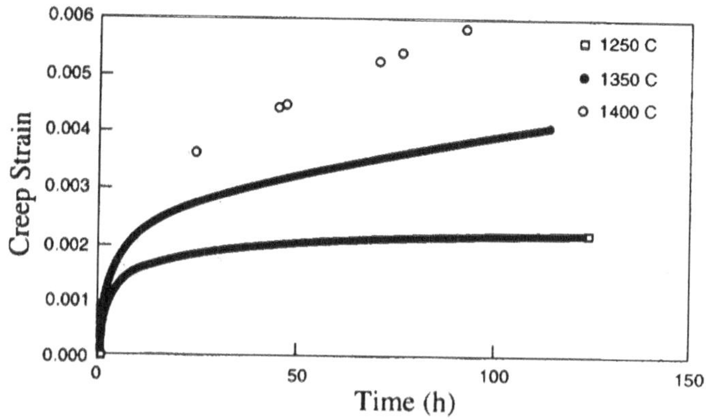

Figure 1.16 Creep behavior of carborundum alpha SiC fiber at an applied stress of 278 MPa in argon. (Reprinted with permission from Giannuzzi et al., 1992.)

generally small in amplitude, followed by a nearly linear region in which the creep rate is essentially constant over some period of time. Finally, the response is terminated by a so-called tertiary creep region in which the creep rate of the material is accelerating and failure ensues. We will learn in a later chapter that there are a variety of physical mechanisms that can produce such deformation behavior. Although creep deformation is frequently associated with viscoelastic behavior in polymers, and there is also a large body of literature that discusses the creep of metals at high temperatures, it is, perhaps, surprising to find that creep may also be an important deformation mode for ceramic materials at sufficiently high temperatures.

Figure 1.16 shows an example of such creep response for a ceramic fiber typically used in ceramic composites. Also, the influence of composite constituents on the creep response of the composite itself is a highly complex process. The constituents may contribute to composite creep by their own creep behavior (including microfracture), by the interaction of their creep behaviors, or by a change with time of the interphase regions, which control the coupling of the constituent behavior. Figure 1.17 shows the creep of a composite material using the fibers discussed in Figure 1.16. This creep response is significantly different than one would expect from the fiber response alone.

1.2.3 Stress Rupture

Another phenomenon closely related to creep deformation is called *stress rupture*, sometimes termed *creep rupture*. Like the response discussed in the previous paragraph, the loading for this kind of behavior is constant in time or nearly constant in time. After some period of time, however, the material or component fractures abruptly due to an accumulation of creep-induced damage. Hence, the data reported generally consist of an applied stress or strain level and an observed time to failure.

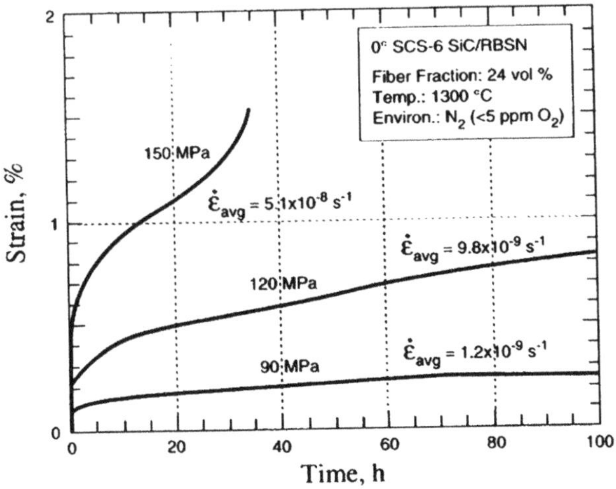

Figure 1.17 Creep response of SiC fiber–RBSN matrix ceramic composite at selected strain rates and temperatures. (Reprinted with permission from Hilmas et al., 1993.)

Figure 1.18 shows data of that type where the applied strain (as a fraction of the ultimate strain to failure) is plotted against the time to failure in a bending test of unidirectional polymer matrix–carbon fiber composite specimens.

As we will see in a later chapter, there are many mechanisms that contribute to this kind of behavior and many phenomenological characterizations and models that have been applied to the characterizations of those mechanisms generally referred to as rate processes in the fundamental sense. These processes are usually quite sensitive to temperature and other driving forces for the activation processes, such as stored energy or dissipated energy associated with deformation. Typically, very small changes in temperature, especially at elevated temperatures, may cause dramatic changes in the life of the component at a given stress level, as suggested by the

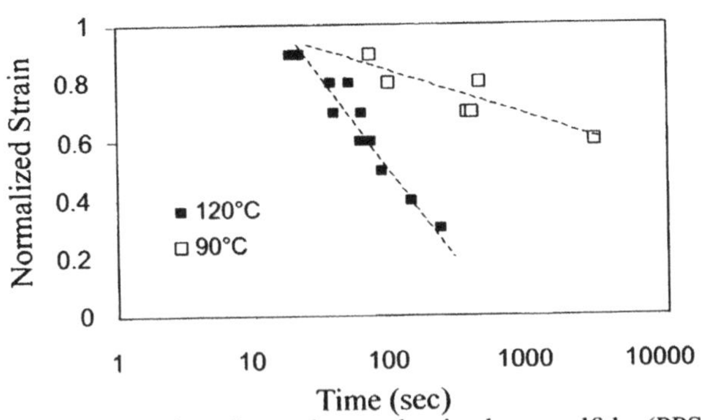

Figure 1.18 Stress rupture data for carbon–poly-phenlyene-sulfide (PPS) unidirectional composite under end-loaded bending compression at constant strain levels, for two temperatures.

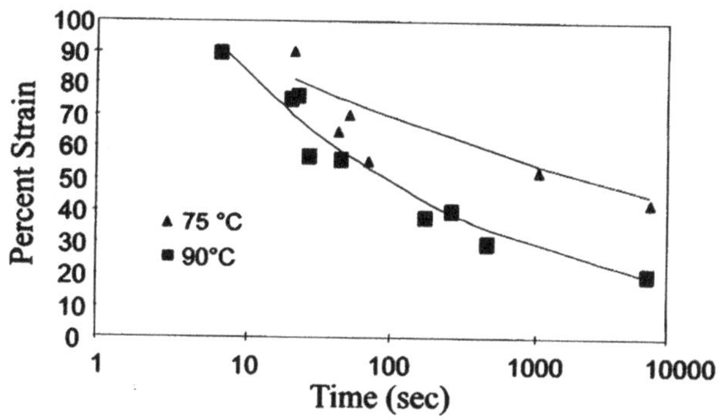

Figure 1.19 Tensile stress rupture behavior of a unidirectional carbon–PPS composite at two temperatures.

data in Figure 1.18. Figure 1.19 shows comparable data for tensile loading (to a fixed percent of tensile strain to failure) for the same material system that produced the data shown in Figure 1.18. The surprise here is that even though the carbon-reinforced unidirectional material shows essentially no time-dependent (creep) deformation upon loading in the fiber direction, time-dependent fracture (stress rupture) at elevated temperature occurs.

Like the creep behavior discussed earlier, stress rupture behavior is also very sensitive to the melting point or softening of the materials involved. Materials with high melting points frequently have less creep deformation and longer stress rupture lives at room temperature or elevated temperatures significantly below their melting point than do low-melting-point materials. Generally, one thinks of materials such as polymers (and polymer matrix composites) as being more susceptible to creep and creep rupture. However, most composite material systems are designed in such a way that the fibers dominate the mechanical response; they are said to be fiber dominated and are designed to be that way. Hence, polymer composites, which are often reinforced with carbon fibers, frequently do not demonstrate substantial creep behavior or stress rupture behavior when fiber-dominated laminates are constructed. In contrast, ceramic composite materials, which are inherently less likely to creep at room temperature, are often controlled by their creep and stress rupture behavior since they are more often used at temperatures close to their melting point, which induces creep and stress rupture behavior in many instances. Even in fiber-dominated polymer matrix materials, the matrix may play a much more important role in stress rupture than it does in creep. The data in Figures 1.18 and 1.19 are for unidirectional polymer matrix materials, which obviously show time-dependent failure behavior. Designing a composite to have fiber-dominated quasi-static stiffness and strength may not ensure that the composite does not demonstrate time-dependent deformation and fracture.

Time-dependent degradation phenomena are important to our study of the long-term strength and life of composite material systems. Indeed, in many weight-bearing situations, time-dependent deformation is an important feature of the design limitations and specifications for an engineering component. As it happens, this is especially true for devices such as jet engine parts and other machine and engine components that serve under high stress and high temperatures during their loading history. Since composite materials are natural choices for many of those applications, the subject of time-dependent deformation and degradation is an important one for many of these applications.

1.2.4 Fatigue Behavior

Many of the loads applied to engineering components during their service life are not constant with time; in many instances, loads are applied in a dynamic or repeated fashion. In the transportation industry, for example, it is easy to find cases of repeated loading resulting from such situations as an automobile driving down a bumpy road or an airplane flying through turbulent air. Other common examples are associated with rotating or reciprocating parts, including automobile drive shafts and piston connecting rods in combustion engines.

Wind loads on buildings, traffic over bridges, and the pressurization/depressurization of pressure vessels are other common examples of time-dependent degradation. It was noticed in the 1840s that many materials fail abruptly, and with little or no change in properties prior to that failure, when a mechanical load is applied cyclically over a long period of time. This phenomenon eventually became known as "fatigue." It is difficult to find a way to characterize such material behavior systematically, but the engineering approach usually adopted is the idealization of the dynamic or repeated loading as a cyclic loading with a constant amplitude. The susceptibility or resistance of a given material to this cyclic loading is typically discussed in terms of the number of cycles to failure, that is, the life N of the material under a given cyclic stress amplitude S, and is generally represented on an S–N curve, such as the one shown in Figure 1.20.

Although fatigue behavior occurs over time (especially over long periods of time and under the application of many cycles of loading), the phenomenon is not time dependent; fatigue is a *cycle-dependent* phenomenon. The cycles of load application

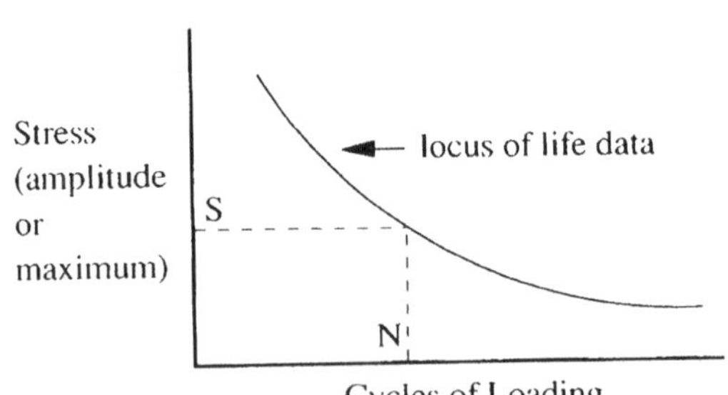

Figure 1.20 Classical S–N diagram for description of fatigue life as function of applied stress level.

may be applied quickly over a long period of time or discontinuously (with periods of no loading interspersed), with no change in the number of cycles to failure (the fatigue life). In engineering, the number of cycles to failure is the defining feature of this phenomenon. Of course, as we will discuss in later chapters, the material behavior may not be strictly independent of time in many practical situations. It is possible for fatigue and creep (as well as stress rupture) to occur at the same time, for example. In such a case, the physical damage mechanisms that cause creep and fatigue may or may not be independent; some mechanisms may contribute to both cycle- and time-dependent changes in properties and performance. As we will see, there may be significant interactions between time- and cycle-dependent mechanisms.

Composite materials are remarkably resistant to fatigue; in general, it is one of their best features. Figure 1.21 shows some fatigue curves for various composite materials and laminates normalized by the ultimate strength in tension of those systems. As seen in Tables 1.1–1.3 (Appendix 1), composites typically have an even greater advantage when their density is considered; they are typically very light, strong, and resistant to fatigue. In fact, the fatigue performance of most engineering materials (including metals and composites) frequently scales to the ultimate strength of those materials, so the advantages seen in Tables 1.1–1.3 (Appendix 1) for the quasi-static properties of composite materials can be expected to translate into improved fatigue performance in many cases.

In the context of our discussion of the physical mechanisms and the behavior that underlies the subject of durability and damage tolerance, the subject of fatigue introduces another important topic, namely, the statistical nature of the problem. Like most specific topics raised in this chapter, we will revisit this subject many times in subsequent chapters because it is especially important. In fact, in many cases, the statistical nature of the microevents that control remaining strength and life is a canon of the philosophy and modeling that describe and explain the resulting properties and performance. At this point, we will only mention a few salient features to support our conclusions about physical behavior. The statistical variation of the characteristics of composites is generally significantly larger than that observed for homogeneous materials. Figure 1.22 presents some typical features of this situation; the probability of failure as a function of load level for the static loading of fibers has a much larger statistical spread than the distribution of strength for a common metal. Moreover, the statistical distribution of life for a composite has a large statistical spread. Finally, the statistical spread of the strength for a composite is, typically, less than the spread of the fibers. The details of this subject are complex for composites, and the specific nature of these details is important. For example, there is a statistical distribution of strength from point to point in any homogeneous material (including the fibers, especially due to surface defects and their large surface-to-volume ratio), but that does not explain the statistical behavior of composite materials (which are inhomogeneous) in which many damage events occur.

Indeed, these facts bring us to two of the most important points of discussion in this chapter. First, the local fracture of a constituent in a composite rarely causes the

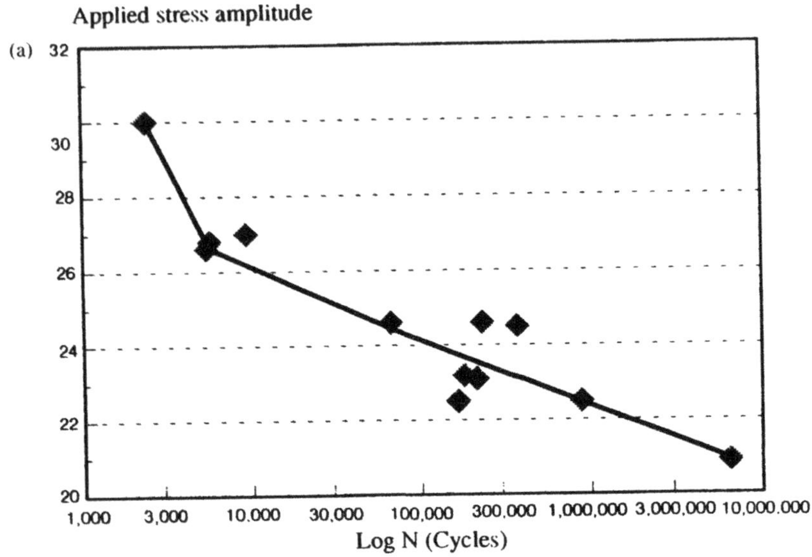

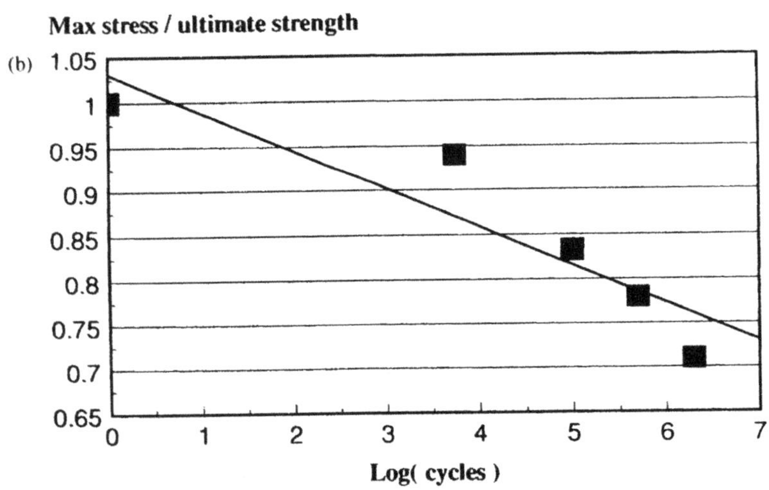

Figure 1.21 Typical tensile fatigue behavior of (*a*) unidirectional carbon-reinforced polyeithereitherketone (PEEK) composite (Picasso and Priolo, 1988) and (*b*) unidirectional carbon-reinforced epoxy composite.

global failure of the composite. Constituents typically fail many times, from point to point, and well-designed composites usually do not fail as a result of any single flaw or single flaw growth. Second, composites typically fail as a result of the *statistical accumulation* of microdefects rather than as a result of the *statistical distribution* of defects or flaws.

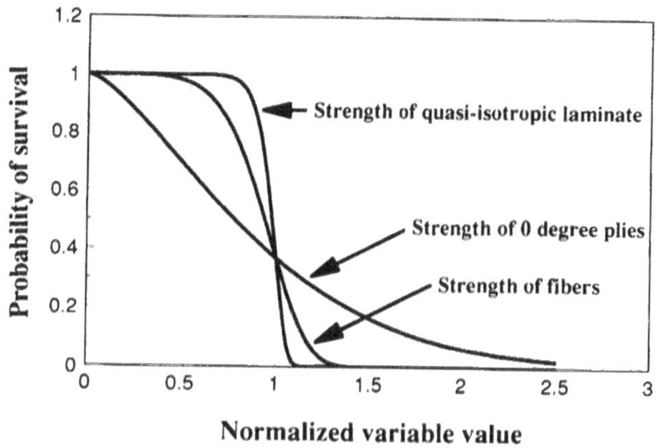

Figure 1.22 Characteristic strength distribution shapes for fibers, unidirectional composites, and quasi-isotropic composite laminates.

To demonstrate these points and to substantiate their consequences, we will explore several of the physical mechanisms that contribute to the defect distributions referred to previously.

1.3 DAMAGE MODES AND FAILURE MODES

1.3.1 Microcracking

The most pervasive damage mode in composite materials is microcracking, most often in the matrix material. Figure 1.23 shows a radiograph of a $[0, 90]_s$ laminate made from glass–epoxy that has been subjected to nearly a million cycles of tensile fatigue loading in tension with a maximum stress of about 65% of the tensile ultimate strength.

Several features of this radiograph are distinctive. First, the cracks extend through the thickness of the plies and follow the fiber directions. Second, they are very densely distributed; that is, they occur in great numbers throughout the specimen. Third, the cracks do not develop and grow independently; the stress fields associated with the cracks interact. Indeed, as we will show in a later chapter, the density of the cracks in a given layer is uniquely and precisely determined by the mechanical properties of the plies in the laminate, their geometry, and their stacking sequence as they influence the stress field interaction of matrix cracks in the laminate.

Another basic feature associated with microcrack development is the change in properties associated with these defects. In general, moderate amounts of microcracking have a small effect, or even no effect, on the strength of a lamina or laminate. This surprising observation is a result of the simple fact that the initial strength of the material is measured by extending (or compressing) the material to the ultimate load, and during that defining experiment, microcracks generally

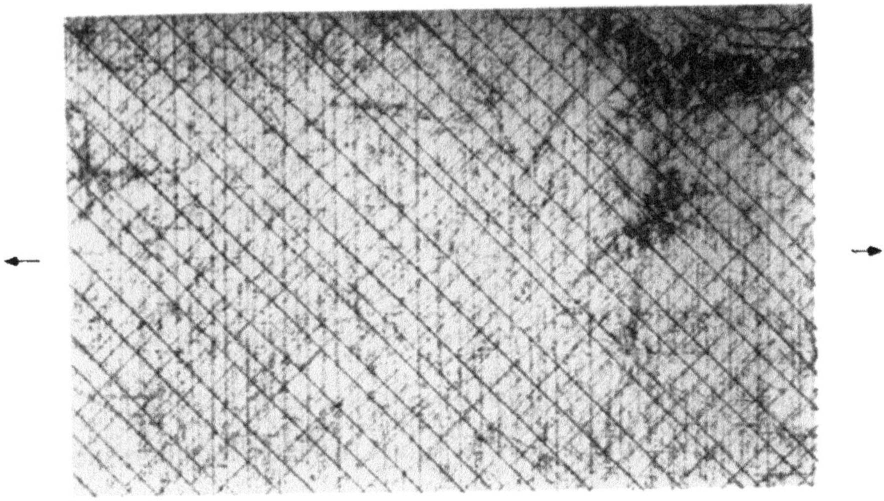

Figure 1.23 Radiograph of quasi-isotropic laminate after it has been subjected to fatigue loading at about 60% of ultimate tensile strength ($R = 0.1$), showing microcrack formation in all plies.

develop (the initial strength of the material is *defined* by experiments that involve the formation of microcracks). However, other properties and performance may be affected; for example, global stiffness may be altered significantly by the formation of microcracks.

Figure 1.24*a* demonstrates the basic nature of this effect. It shows the extension of a $[0, 90]_s$ laminate and illustrates the typical result of that extension, that is, the formation of microcracks in the 90° plies. As each of these cracks forms, there is a region of material near the cracks in which the stress relaxes, or unloads. This

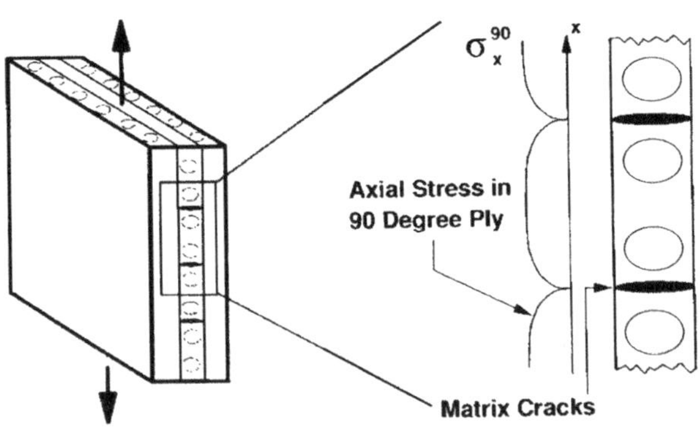

Figure 1.24 (*a*) Nature of transverse (to the plane of the ply) crack formation in off-axis plies and (*b*) local relaxation of load direction normal stress in region of such cracks.

internal redistribution of stress results in a change in the global stiffness of the laminate, as suggested by the figure. Several features of this global change are important to the current discussion, and there is a stress level at which the initiation of cracking in a given ply begins. We will identify many of these damage threshold stress (or strain) levels in our later discussions. For the situation shown in Figure 1.24*b*, the stiffness change does not involve any change in the laminate shape when the load is removed (i.e., there is no residual strain or permanent deformation associated with this damage initiation), an unusual result. The reason there is no change is that the 0° plies are not damaged, and since they are bonded to the 90° plies, when they return to their original length as the load is removed, they bring the 90° plies back to their initial length as well. This event is one of the first clear demonstrations of the concept of *fiber-controlled behavior* or *fiber-controlled composites*, as we will be calling them. Composites are typically designed to be fiber controlled, since it is an excellent way to control the initiation and growth of damage. This fact will be of primary importance to our development of philosophy and models and to the utility of composite materials in general.

The changes in stiffness caused by matrix cracks are not uniform as a function of the number of cycles of loading (for fatigue loading) or with time (for time-dependent behavior). Figure 1.25 shows an example of a typical distribution of the density of cracks as a function of the level of applied quasi-static tensile stress for a glass–epoxy laminate. As one might expect, the initial stiffness change is large, followed by a region of modest change, with a final region of rapidly accelerating stiffness change as fracture becomes imminent.

For cyclic loading, most of the change in stiffness caused by matrix cracking occurs in the early part of the specimen life. This is easily seen from Figure 1.26, which shows the variation of laminate stiffness and the number of cracks in the off-axis plies as a function of the number of cycles of fatigue loading (Stinchcomb and Bakis, 1991). By adjusting the scales on the two dependent variables, the variation in stiffness closely follows the variation in the crack density. In addition, both the crack density and the stiffness change reach an equilibrium state after some fraction of the

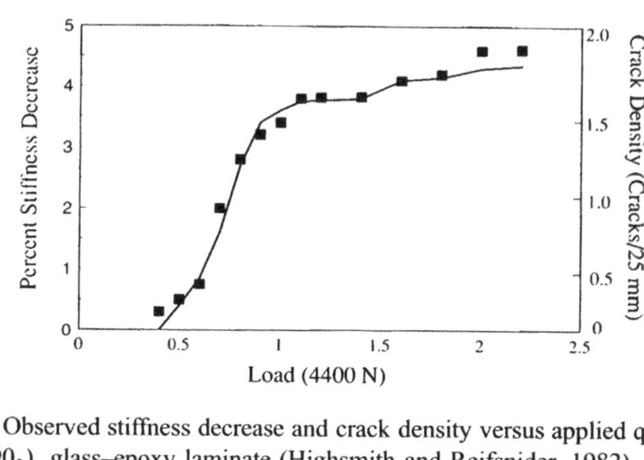

Figure 1.25 Observed stiffness decrease and crack density versus applied quasi-static stress level for $(0, 90_3)_s$ glass–epoxy laminate (Highsmith and Reifsnider, 1982).

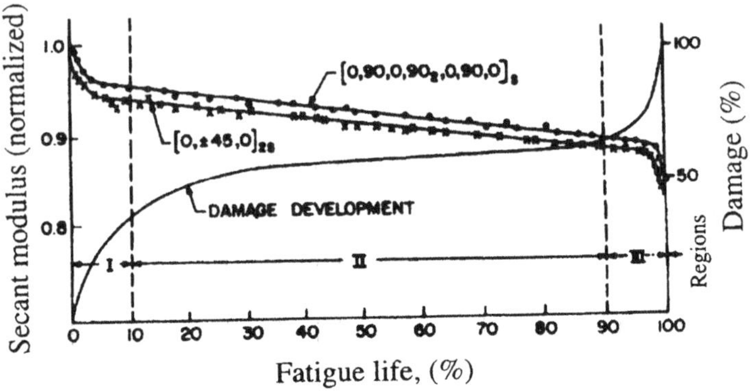

Figure 1.26 Typical (normalized) stiffness change as function of percent of life in two quasi-isotropic laminates subjected to fully reversed cyclic loading.

fatigue life. We will discuss the nature of this "characteristic damage state" in subsequent chapters. The comparison of Figures 1.26 and 1.25 shows that the initial drop in the quasi-static stiffness of the laminate corresponds to the matrix crack development shown in Figure 1.26. However, the present discussion does not illustrate what causes the second drop in stiffness. We will examine those details later, but for now, we will mention that the second (and final) drop in stiffness can be related to a second important damage mode, called delamination.

1.3.2 Delamination

Delamination is driven by the fact that local regions of the composite (such as different plies) would deform differently in response to the local loads if they were not bonded together in the composite. Hence, stored energy is released if they separate, and that energy drives the separation process. Figure 1.27 shows a schematic of the physical process when it occurs near the edge of a plate (or "coupon") specimen. For reasons we will discuss later, this is a typical situation; delaminations often start and spread from the edge of a laminate. As suggested in Figure 1.27, it is clear that 90° plies under uniaxial tensile loading would like to change their shape quite differently from the 0° plies under the same loading. (The Poisson ratio of the 90s is less than one-tenth that of the 0s, for example.) If they are bonded together in a laminate, however, they are forced to have the same compromise deformation. As we will show later, the in-plane stresses that develop as a result of this compromise give rise to interlaminar stresses near any external or internal boundary of the laminate and can cause the laminate to separate at those locations and form a delamination, as shown. Also, since the deformation of the plies of the laminate are different when they are free from when they are bonded together, the collective stiffness of the plies is different under those two conditions. Hence, there is a stiffness change associated with the delamination process that is important for at least two reasons. First, the stiffness change associated with delamination can be

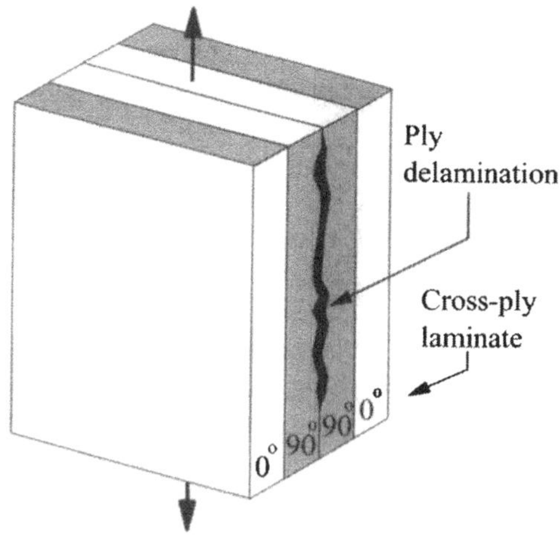

Figure 1.27 Delamination formation in cross-ply laminate.

large, especially when compared to the stiffness change caused by other damage modes such as matrix cracking and fiber fracture. Second, the stiffness change is directly proportional to the change in stored energy during a given deformation of the laminate; this energy change (release) drives the growth of the delamination. Delamination growth is one of the few examples of self-similar crack (or cracklike) growth typical of composite materials.

But delamination is very often *not* an independent damage mode; in many cases, delamination is nucleated by matrix cracking (O'Brien, 1982; Reifsnider, et al., 1979). Figure 1.28 shows an edge view of two laminates with typical matrix crack patterns and delaminations that have begun to nucleate from the ends of

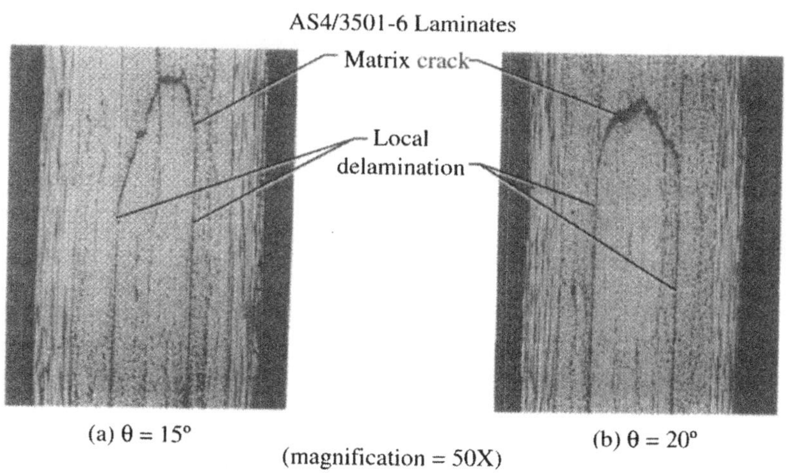

Figure 1.28 Replicas of edge crack delamination patterns for two $(0, +\theta, -\theta)_s$ laminates: (*a*) $\theta = 15°$; (*b*) $\theta = 20°$. (Magnification 50X.)

several of the microcracks. We will find that there are interlaminar normal stresses in the interfacial regions near the end of matrix cracks at the edge of the specimen that cause the delaminations to initiate, but this type of initiation is also possible in the interior of a laminate.

Several examples of this behavior can be found in Figure 1.29*a*. When the matrix cracks in one ply reach the interface between the cracked ply and the next ply (which has a different fiber orientation), the local stress state is quite similar to the edge situation shown in Figure 1.28 (Highsmith, 1984). Under these conditions, internal delaminations can nucleate and grow. Several of the dark regions, such as the one indicated by the arrow in Figure 1.29*a*, are examples of such a delamination. Internal delamination is common in laminated composites and can lead to the development of still other damage modes, as we will discuss below.

Figure 1.29*b* shows edge delaminations that have grown in a coupon specimen subjected to fully reversed cyclic loading. We mentioned that the stiffness changes

(a)

(b)

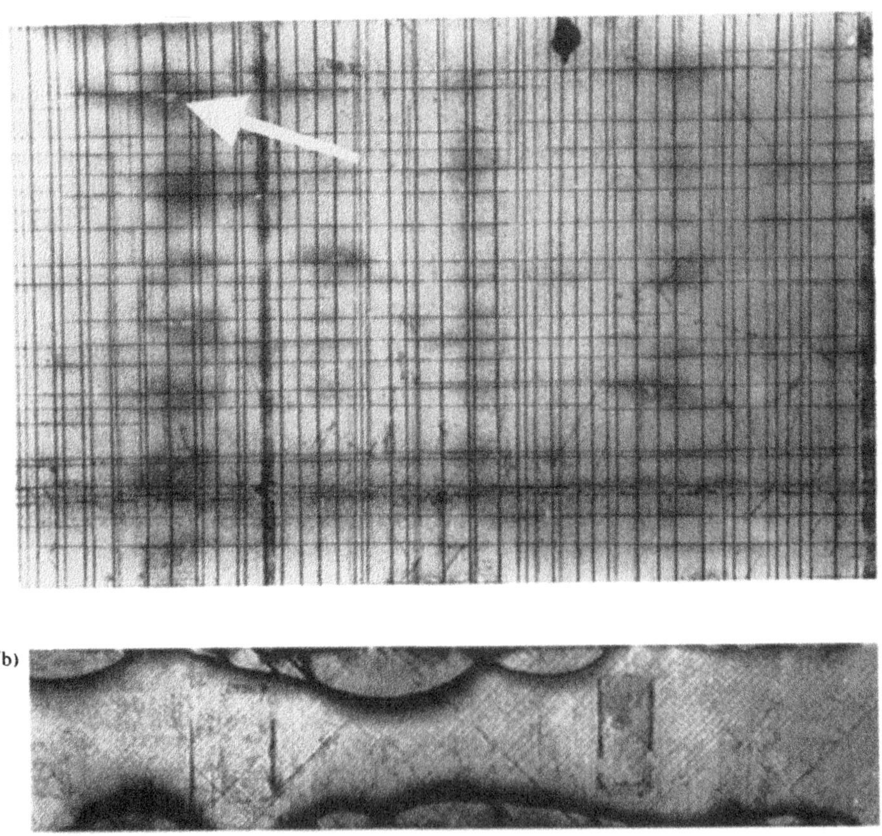

Figure 1.29 Radiographs of (*a*) internal delaminations between plies where matrix cracks cross in a cross-ply glass–epoxy specimen and (*b*) edge delamination growth in a quasi-isotropic carbon–epoxy laminate under cyclic loading.

caused by this growth may be large and, in some cases, significantly larger than those caused by matrix cracking. This growth causes an internal stress redistribution that may also be large. When that happens, some of the material in the interior that has not changed in stiffness is obliged to sustain more of the load when other material unloads. If there is a ply with fiber orientations close to the principal loading direction, for example, then the stress in those plies typically increases with the growth of the delamination, sometimes by substantial amounts. Hence, unlike some matrix cracking, delamination may change the strength of the laminate. This leads us to an important distinction that we must make in the study of the subject at hand. Delamination is rarely, if ever, a *failure mode*; instead, it is a typical *damage mode*. This distinction between damage modes and failure modes will be very important as we develop the subject of microdamage and the resulting damage tolerance and durability.

Delamination *is* an important damage mode, and the stress redistribution associated with delamination growth can be an important feature of the change in strength associated with composite systems in long-term service. Figure 1.30 shows an especially important example of this stress redistribution: a radiograph of the damage pattern that has developed in a graphite–epoxy laminate with a center hole under fully reversed cyclic loading (Stinchcomb and Bakis, 1991).

Many matrix cracks have developed in the region around the center hole; however, there is also a large region of delamination in that area. Note that a notch or hole (or "cutout," as it is often called) in a laminate is a stress concentrator in the usual sense, and it is also an *edge* of the laminate. Hence, one would expect to find delaminations growing in that region if the corresponding internal interlaminar stresses are present. For the specimen in Figure 1.30, these stresses are present, and the delamination has grown extensively. The corresponding stress redistribution is

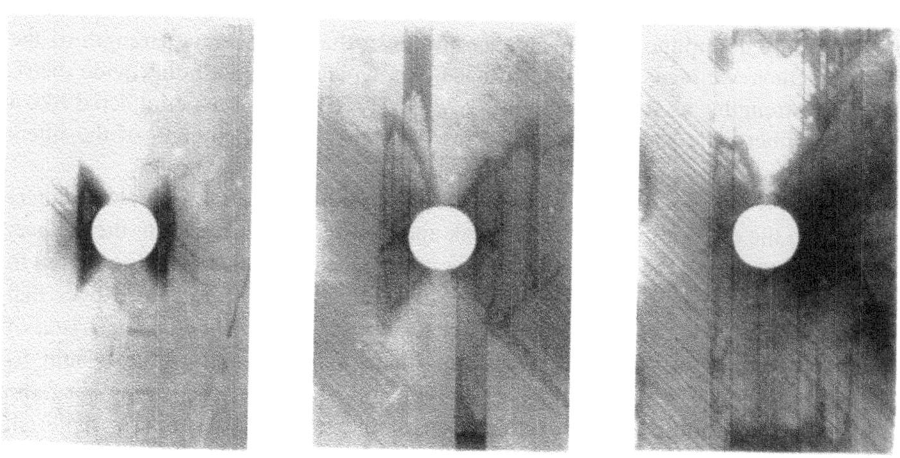

(a) (b) (c)

Figure 1.30 Radiographs of center-notched, AS4/1808, (0, 45, 90, −45)$_{s4}$ specimens during low-load fatigue; (*a*) early, (*b*) mid, and (*c*) late in life of specimen.

great, and we will see that the tensile strength of the notched laminate will typically increase (sometimes by more than 30%) as a function of loading cycles in the early part of the fatigue life of the laminate due to delaminations reducing the local stress concentration near the hole during that period. Obviously, any model that purports to predict the remaining strength of the coupon must account for this stress redistribution, but because of the profusion of interacting damage modes in the region of a notch, this problem turns out to be very complex, as we will discuss in later chapters.

1.3.3 Fiber Fracture

Another damage mode, and often a failure mode as well, is fiber fracture. We mentioned that many well-designed composite materials are fiber controlled, in the sense that they depend primarily on the fibers for their stiffness and strength. The degradation of fibers, therefore, is an important damage and failure mode because fiber fracture is difficult to detect and has been studied less completely than many other modes of damage. Nevertheless, a considerable body of data has been collected (Lorenzo and Hahn, 1986; Razvan and Reifsnider, 1991; Razvan et al., 1990). Several important features of fiber fracture control remaining strength and life in many cases. One of these features is the fact that an individual fiber can break many times inside a composite material system; this is a key composite effect for fibrous composites. If we have a bundle of fibers (with no matrix material), and we pull on that bundle, each fiber in the bundle can break only once. We will derive the strength of such a bundle in a later section and find that the overall strength is never as great as the average strength of the fibers in the bundle, since the load carried by each fiber that breaks under increasing applied load is redistributed to the other fibers (equally, in this case). Unless all fibers have identical strength, as the weakest fibers in the distribution break, they increase the load on the remaining fibers and cause bundle failure at applied load levels that are lower than the average strength of the fibers together. We have already seen that fibers typically have a rather wide distribution of strengths, as shown in Figure 1.22, so if a composite material acted like a bundle of fibers, the strength would be less than the average strength of the fibers themselves.

But that is not the case for a composite. The matrix material around the fibers holds on to the ends of the fibers that break and transfers stress back into these fibers by shear forces along the length of the fiber over a distance that is usually referred to as the ineffective length, since the fiber is less effective in bearing load over that distance near a broken end. This ineffective length is a fundamental characteristic of a fiber-reinforced material and of great importance to the strength of the system, as we will show. It is obvious that this length and, therefore, the strength of the composite are controlled by the interaction of the fiber and the matrix material, that is, by the composite effect, or the systems effect, as discussed.

Figure 1.31 shows fiber fracture in a calcium aluminosilicate (CAS) glass–ceramic matrix material reinforced with Nicalon fibers, which are principally silicon carbide (SiC) (see Table 1.1, Appendix 1). In this figure, a cross-ply ($[0, 90]_s$) laminate was subjected to fully reversed fatigue loading for many cycles, and the matrix was removed by etching to reveal the broken fibers. The figure shows that the ineffective length of the broken fibers is finite and that it is rather short for this composite system with a stiff matrix. Hence, the fibers can break many times in the composite without a significant effect on the strength of the system. In fact, the composite system may be stronger than the corresponding bundle of fibers because of this capability to have multiple breaks in the fibers. Over long periods of loading, especially cyclic (fatigue) loading, these fiber fractures may cause significant stiffness and strength reductions, but this process is very different in different composite systems. It is one of the more challenging subjects of our book.

Figure 1.32 shows another important aspect of fiber fracture, the influence of one fiber fracture on the tendency of fibers around it to fracture. The fiber fractures in this figure occurred in a graphite–epoxy composite. The matrix was removed after cyclic loading to reveal fiber fractures, in a manner similar to the micrograph in Figure 1.31. When the matrix transfers stress back into a broken fiber over the ineffective length, it creates a stress concentration in the neighborhood of the ends of the broken fiber that tends to cause the neighboring fibers to break, depending on their relative strengths in the distribution of fiber strengths. This "fiber–fiber" interaction is controlled by the bonding of the fibers to the matrix, the relative stiffness of the matrix and the fiber (and the interphase region between the matrix and fiber, if present), and the local geometry and arrangement of the constituents. It is further evidence of the system's effect in these composites.

Many other kinds of local interactions influence fiber fracture. Matrix cracking also contributes to the initiation of fiber fractures; Figure 1.33*a* shows some of the evidence of that fact. In this figure, the matrix has been removed from a graphite–epoxy $[0, 90]_s$ laminate after cyclic tensile loading at about 60% of the static ultimate tensile strength (UTS). Also, the 0° plies have been separated from the rest of the laminate, so that the fiber fractures on the plane of that ply adjacent to the 90° ply are viewed. By marking the cracks in the adjacent 90° ply with a penetrant material (that survives the burning off of the matrix), it is possible to determine that the fiber fractures lined up in Figure 1.33*a* are coincident with the position of the matrix crack in the neighboring 90° ply. Hence, in another case of defect and constituent interaction, we see that the stress concentration from the matrix cracks causes preferential fiber fractures in the primary load-bearing ply of the laminate. Figure 1.33*b* shows yet another example of damage mode interactions involving fibers in a region from a 0° ply prepared in the same way as the one shown in Figure 1.33*a*. But the adjacent ply was a 45° ply, and the location shown is close to a center hole in a coupon specimen, such as the one shown in Figure 1.30. In this case, we determined that part of the interface region shown in the figure was covered by a delamination between the ply shown and the neighboring one. In fact, the delaminated region shows many more fiber fractures than the rest of the observation area. Hence, delamination, and the attendant local stress redistribution, can be a driver for the initiation of fiber fracture. Fibers can also be broken as they attempt to bridge the opening surfaces of a delamination.

An important point to remember from the basic discussion of fiber fracture outlined above is that, in general, fiber fracture in a continuous fiber-reinforced

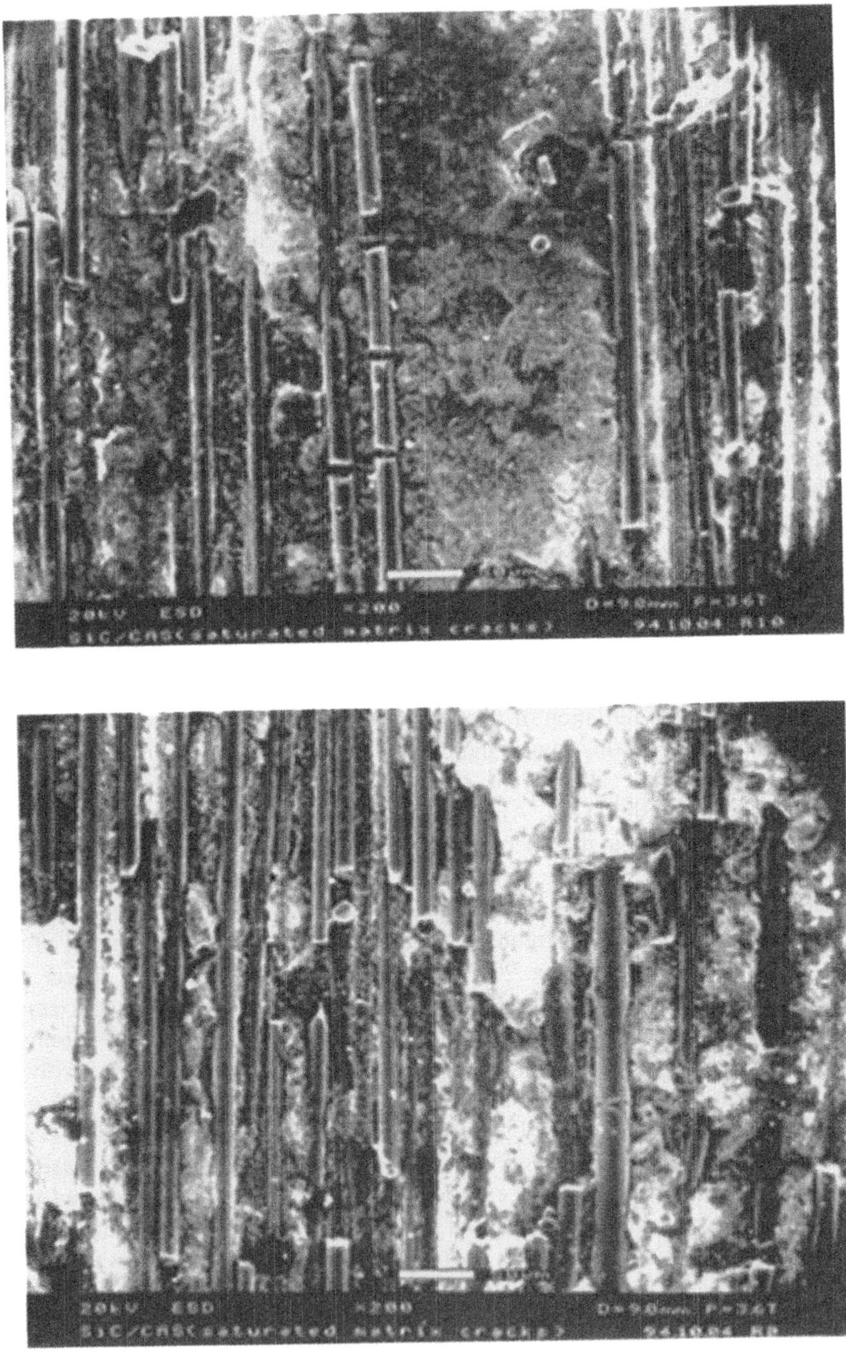

Figure 1.31 Micrographs of fiber fractures in SiC–CASs ceramic composite (in which matrix was etched away after quasi-static loading) showing multiple breaks in individual fibers and adjacent fiber fractures.

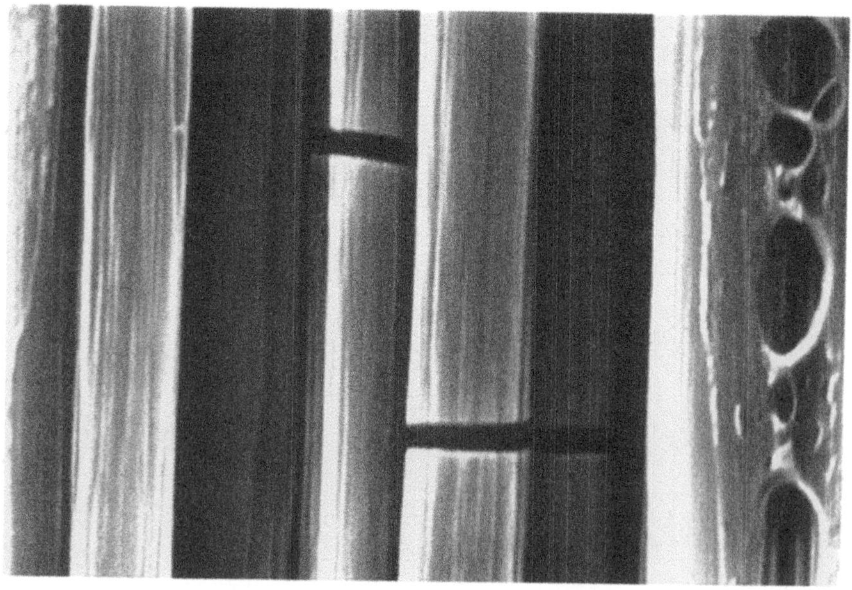

Figure 1.32 Adjacent fiber fractures in interior of carbon–epoxy composite.

composite is generally a highly coupled or interactive process. Fiber fracture is interactive with other fiber fractures, matrix cracking, and delamination, and that interaction is greatly influenced by the coupling between the fibers and the matrix materials.

Figure 1.34 shows an example of the development of fiber fractures in a unidirectional and cross-ply glass-reinforced polymer composite during quasi-static loading. It should be mentioned that some fibers are broken during manufacturing, before the application of any loading, often of the order of 10% over a given cross section (Jamison and Reifsnider, 1986, 1982). Figure 1.34*a* shows that fiber fractures in the unidirectional material develop slowly, with a rate that increases very sharply for applied stress levels near the UTS. This occurs in the last 15% of loading for the example shown (from the data of Jamison and Reifsnider, 1986, 1982). That figure also shows that the rate of fiber fractures increases earlier (over about the last 30% of loading) for the cross-ply laminate and that the number of fiber fractures is substantially higher for the cross-ply laminate than for the unidirectional material. The data suggest that, as we have seen in Figure 1.33, matrix cracks cause local stress concentrations near their termination points at ply boundaries, which induce fiber fractures in adjacent plies. Figure 1.34*b* provides further verification of this fact by showing a close correlation between the incidence of matrix cracking and fiber fracture for the cross-ply laminate.

Figure 1.35 shows several interesting features introduced by long-term cyclic loading (Razvan, 1992). In the case of tensile fatigue in which the maximum stress is 65% of the ultimate strength, most of the fiber fractures occur in the early part of life, and very few are observed after that point up to 90% of life (the

(a)

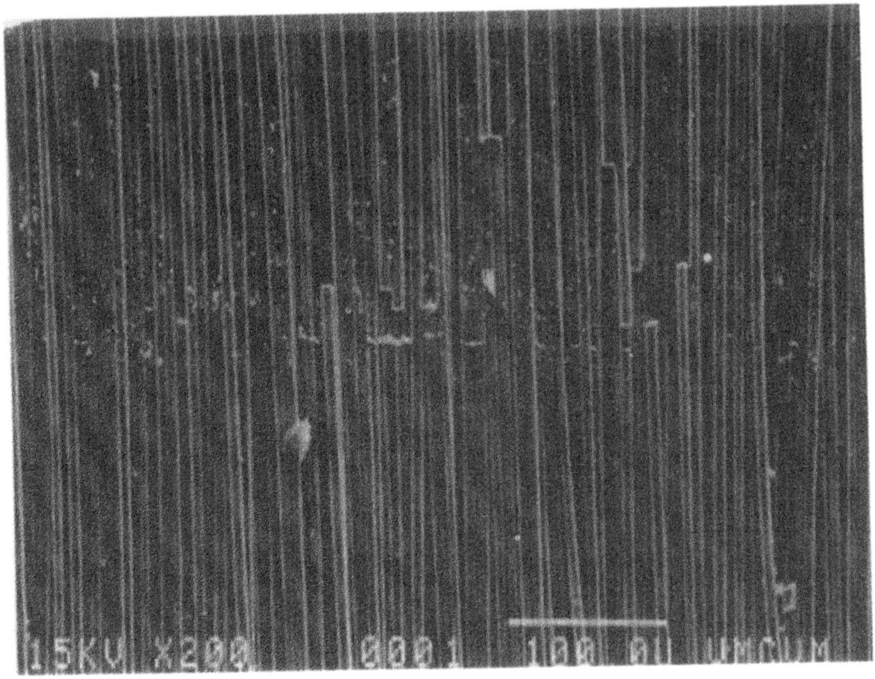

(b)

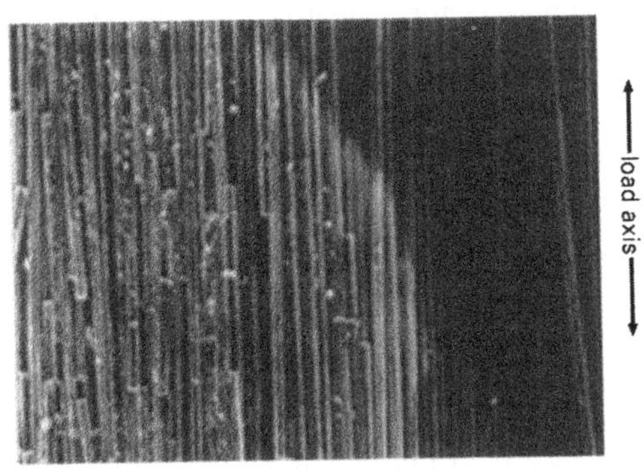

Middle ply ,$[O_2]$, of T300/5208. $[0/45/90/\ -45]_{s4}$, specimens under *Low* amplitude loading at impending failure

Figure 1.33 Scanning electron micrographs showing fiber fracture patterns caused by (*a*) adjacent matrix cracks after long-term cyclic loading of cross-ply carbon–epoxy laminate and (*b*) passage of local delamination between plies in quasi-isotropic laminate near notch.

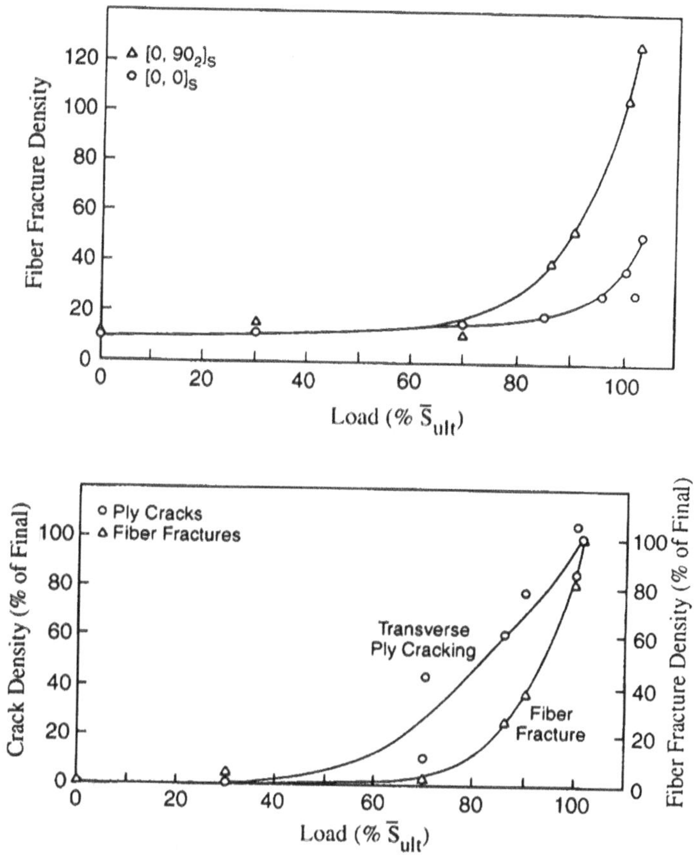

Figure 1.34 (*a*) Fiber fracture density in viewing area for unidirectional and cross-ply glass–epoxy laminates as function of quasi-static load level (as fraction of ultimate strength) and (*b*) correspondence between transverse ply crack development and fiber fracture in cross-ply laminate.

last data point in the figure, for both loading levels, is taken at 90% of life). At 70% of ultimate strength, however, substantial increases in the fiber fracture count are seen over the life of the specimen, prior to 90% of life. In addition to the fiber fracture caused by the first cycle of loading to a given level (of the type shown in Fig. 1.34), additional fiber fracture can be caused by cyclic loading, especially if that loading reaches a stress level known to cause substantial fiber fracture in quasi-static loading.

Figure 1.35 also shows that the formation of adjacent fiber fractures follows essentially the same rates as the formation of the singlets; in the case shown, the number of doublets is smaller but comparable to the number of singlets. The coupling process also forms multiple-fiber-fracture sites, or "i-plets," as defined by Batdorf (1982) ("i-plets" may be singlets, doublets, triplets, fourplets, etc., up

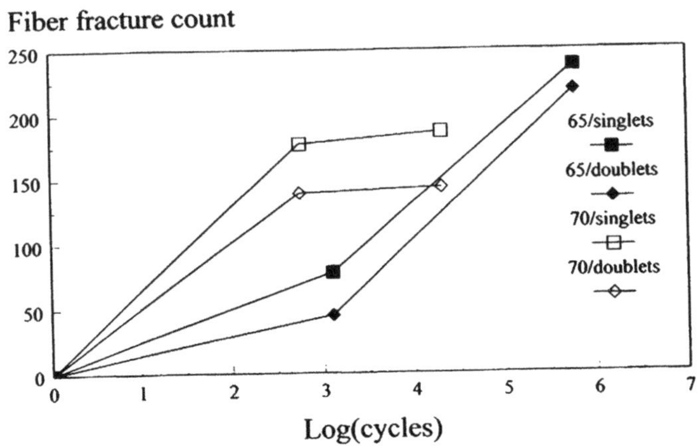

Figure 1.35 Single fiber fracture counts ("singlets") and two adjacent fiber fracture counts ("doublets"—in fixed area of observation) under tensile fatigue loading ($R = 0.1$) at 65 and 70% of tensile strength for unidirectional carbon fiber-reinforced epoxy system.

to n-plets for one, two, three, four, up to n fiber fractures in the neighborhood of each other, or at least within the distance of one ineffective length). As we can see, part of the definition of progressive damage in these materials is an increase in the coupling of damage events, a localization of fiber fractures in this case.

Further evidence of this progressive nature is shown in Figure 1.36 (Lee and Stinchcomb, 1996). In this figure, one can see a complex pattern of matrix cracks and fiber fractures in a glass–ceramic composite reinforced with SiC-based fibers. Matrix cracks that form in this material in the *"on-axis plies"* are perpendicular to the loading direction (which is in the direction of the fibers in this case). These matrix cracks typically form at load levels of the order of one-third to one-half of the failure load of the lamina. For loading above that level, the matrix transfers stress back into the fibers in regions surrounding fiber breaks that do not occur in the matrix-crack plane. When a sufficient number of fibers break, the *accumulation* of fiber fractures creates an unstable condition that leads to fracture at the global level, with broken fiber fragments pulling out of the matrix, as shown in Figure 1.36. In a later chapter, we will discuss various detailed models of this complex process.

One such model shows that the number of fiber fractures that controls the amount of remaining strength increases in a nonlinear fashion with the number of applied cycles. Figure 1.37 shows some predictions for an oxide–oxide ceramic composite (Liao, 1994). Here we see evidence of the importance of the statistical nature of the constituent strength (shown by the strong dependence on the Weibull shape factor of the fiber strength given by different values of m in Fig. 1.37). If the static fiber strength has a wide distribution as a function of applied stress, then fiber fractures start early in the quasi-static loading process, and a large fraction of fibers break before the specimen ruptures, while a narrow distribution of fiber strengths delays the onset of fiber fracture but causes a smaller number of fiber fractures before

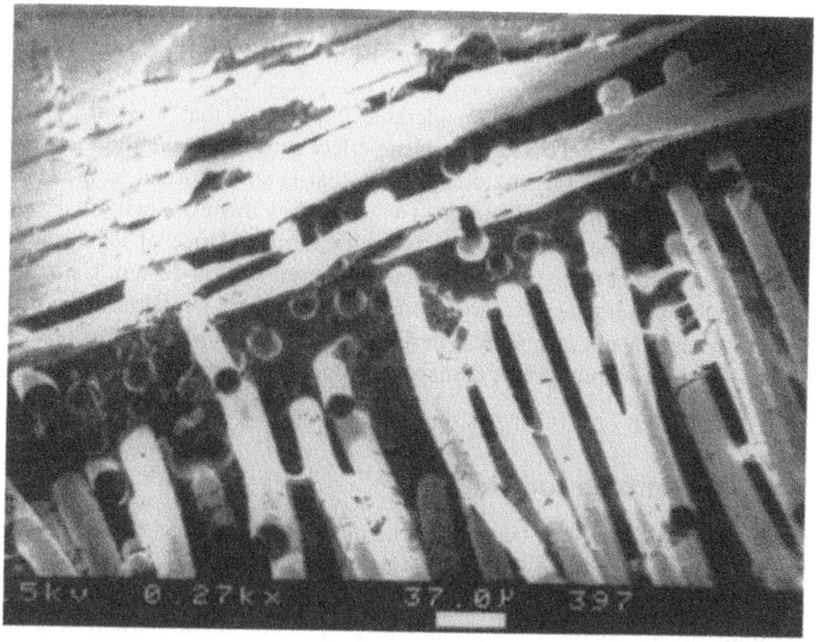

Figure 1.36 Fracture surface of $[0, 90]_{4s}$ Nicalon–SiC laminate loaded with tensile cyclic load ($R = 0.1$) to 85% of ultimate strength until failure.

specimen rupture and a lower strength. The last situation is based, in large part, on the fact that if essentially all of the fibers have the same strength, the increased stress in a fiber adjacent to a broken fiber, caused by the stress concentration around a broken fiber, is very likely to be higher than the strength of that adjacent fiber. In that case, when local stress concentrations occur, they tend to cause immediate propaga-

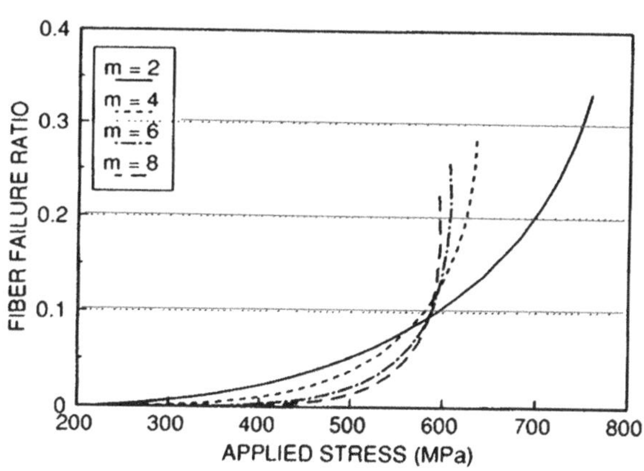

Figure 1.37 Predicted fiber fracture ratios (fraction of broken fibers) in oxide–oxide composite as function of applied tensile strength for several values of Weibull shape parameter m.

tion of the early fiber fractures across a given cross section to induce a brittle fracture with low strength. In this case, coupling is bad, since it can make the composite weaker.

This raises a general point of considerable importance that we will develop in greater detail in later chapters. The composite effect caused by coupling between the properties and performance of the constituents that is so important to the strength, stiffness, and life of composite materials (which makes them of such great value to engineering applications) usually depends on a careful balance of local characteristics that work together to create the optimum material system design. Using proper mechanics analysis methods, we can determine the best combination of constituent geometries, arrangements, properties, performance, and coupling to make the best composite. From the standpoint of damage tolerance and durability, that is an objective of this book.

1.3.4 Fiber–Matrix Decoupling

Another important damage mode occurs in materials that develop damage in the manner suggested by Figure 1.36. We have emphasized the importance of the interphase region between the fibers and the matrix, especially from the standpoint of providing the mechanical coupling that plays such an important role in the durability and damage tolerance of composites. We noted the influence of the interphase region, and that coupling is often of much greater importance for long-term behavior than for quasi-static properties (cf. Figs. 1.11 and 1.12). For ceramic composite systems, this influence is often altered as a function of time by a wear process, that is, by relative sliding of the fibers in the surrounding matrix with increasing and decreasing load during cyclic loading. We will show in a later section that this process is progressive and often depends on the accumulation of wear particles in the fiber–matrix interphase region.

Figure 1.38 shows a typical degradation of the shear strength of the fiber–matrix interface as a function of cycles for a SiC-based fiber-reinforced CAS glass ceramic composite from data reported by Holmes and Cho (1992). This damage mode directly affects the coupling in a composite system and may greatly influence remaining strength and life. It is also, in fact, a more general damage mode than the specific case we have introduced. Fiber–matrix debonding, in general, may occur for many reasons and may be driven by such things as fiber fracture, a matrix crack that encounters a fiber, or a wide variety of thermochemical events and processes. This debonding process is the subject of much study to develop a means to control it. We will find, in some cases, that fiber debonding can substantially *help* the properties and performance of the composite.

This idea of helpful damage should be broadened at this point. Damage has the connotation of "danger," or more generally of degradation. The physical events that are generally discussed as damage in composites, especially the damage modes discussed above, may or may not decrease the strength and life of the composite. In some important cases, one type of damage development in composites is used to *control* the rate and extent of growth of another type of damage, with the net effect

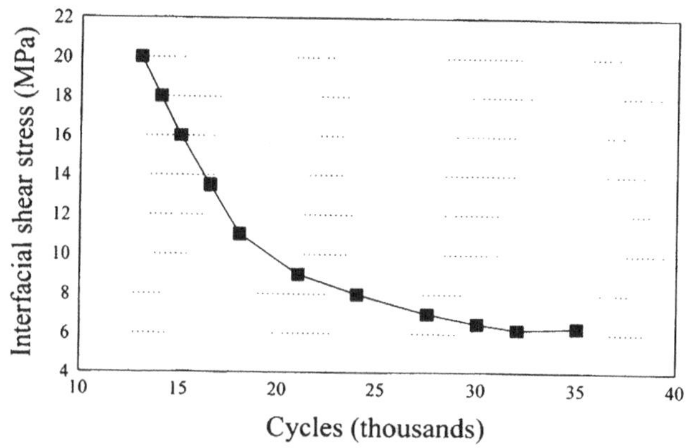

Figure 1.38 Decrease in interfacial sliding stress in ceramic composite caused by fatigue loading (Holmes and Cho, 1992).

of saving the composite from catastrophic failure due to the unchecked progress of a single damage mode. Although we will not attempt to alter the historical usage of the word *damage*, the reader is well advised to avoid the assumption that the word is synonymous with degradation; that will not always be the case.

1.3.5 Microbuckling

A final damage mode having to do with fiber fracture is called microbuckling. A micrograph of such behavior and a diagram of the idealized condition are shown in Figure 1.39. Microbuckling is an instability at the local level, in which a region or zone of fibers under compressive loading buckles together to form a kink in the fibers, as shown in the figure. This damage mode may be a failure mode, in principle, but it is very difficult to be sure of this since microbuckling is nearly always accompanied by other damage modes like fiber–matrix shear and fiber fracture in the kinked region. Microbuckling is distinct in many ways; it is literally an instability, so that it is highly dependent on the *elastic* properties of the constituents, their geometry, and their arrangement. One can imagine, therefore, that such things as the inevitable irregularities induced by manufacturing and processing may have a significant influence on the initiation of microbuckling; this influence is easily found in experimental data. The effect of this damage mode on composite properties and performance is greatly influenced by the manner in which the composite is made. Microbuckling is also distinct in the sense that it is usually defined by the coupling of the action of many fibers as they buckle together to form a "kink band," as shown in Figure 1.39. Since the microbuckling process is an instability, the kink band is generally the only observable evidence of this damage mode. Kink-band formation, however, involves (at least) a shear of the matrix or the fiber–matrix interface (or interphase region) and usually involves fiber fracture. We shall see in a later chapter that the damage process usually associated with this damage mode is, in actuality, a

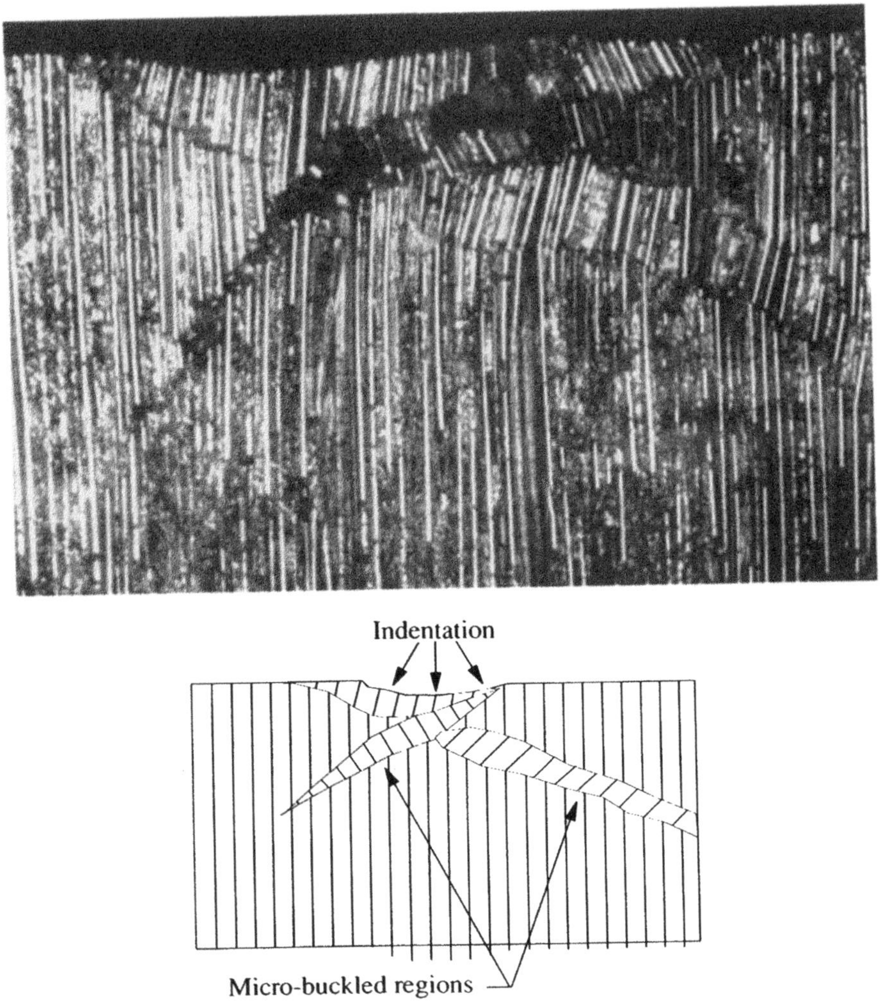

Figure 1.39 (*a*) Photomicrograph and (*b*) schematic of microbuckled region in carbon–epoxy system created by contact deformation around spherical indenter applied in direction of fibers in unidirectional specimen.

rather complex combination and sequence of events, several of which may correspond to the maximum load-carrying capability of the local region of material, so that the strength of a composite in compression, for example, may not be defined by the actual microbuckling event itself. Nevertheless, the microbuckling event is generally the trigger and, therefore, is the rightful object of attention when this damage mode is studied or considered during efforts to design a material system for maximum strength in compression.

1.4 SUMMARY OF CONCEPTS

1.4.1 Damage Accumulation

With the primary elements of the physical subject we wish to address before us, what are the basic canons on which the philosophy of our subject must rest? For the behavior of present interest, the strength (and remaining strength) and life of continuous-fiber composite systems (as we have defined them) are controlled by the following factors:

1. the statistical *accumulation* and subsequent *interaction* of microdamage;
2. large changes in material states and stress states, especially at the local level; and
3. microstructure and microbehavior, especially anisotropy, inhomogeneity, and brittleness.

These are the principal subjects of this book and the basis for discussions in subsequent chapters. The subjects derive from the physical details presented in this chapter. We will summarize those physical features below.

Damage accumulation is an inherently different process than damage initiation and damage growth. Perhaps the most prominent difference is the fact that accumulation as a failure mode is often defined by the interactions of microdefects and microevents. If we make the claim that the initiation or growth of a single defect does not cause failure but that when a sufficient number of those defects accumulate fracture ensues, then the fracture must have been caused by the net effect of the individual defects (such as the reduction in load-carrying area of a cross section by the introduction of microcracks) or by the collective condition caused by the interaction of the defects or microevents when they have accumulated in sufficient numbers to interact with one another. The data we have examined support the later case as the norm, in most instances, for the material systems studied here. Matrix cracking achieves a *saturation density*, most often called a *characteristic damage state* (CDS), for matrix cracking that results directly from the nature of the interaction of the cracks with each other to create equilibrium spacing. As a result of local stress concentrations caused by these matrix cracks, additional fiber fracture is induced, especially under long-term conditions. Matrix cracks also "nucleate" delamination, especially under cyclic loading. Delamination growth can cause fiber fracture in the plane of the delamination and also generally causes significant local stress redistribution, which can lead to local fiber overload and fracture. Fiber fracture can cause the fracture of neighboring fibers, depending on the fiber–matrix and fiber–fiber interactions in a composite. These interactive damage modes are typically time or rate dependent, and they have rates of occurrence that may depend on stress level, temperature, loading rate, and many material parameters.

The fact that interactive damage accumulation controls the long-term response of continuous-fiber composite systems in many cases has strong consequences in our choice of analysis tools to represent these events. For example, the popular dilute solution approximations widely used in a great body of distinguished literature for the estimation of quasi-static properties such as composite stiffness (with good success) are generally inappropriate and inadequate to represent situations where interactions of constituents and microdefects define the long-term performance. Also, we must introduce the concept of time or rate into our analysis and physical characterization. Finally, the homogenization methods that are so convenient for the estimation of composite stiffness are often misleading and sometimes useless for the present discussion since, as we have seen, the details of the local stress states often control the performance we are concerned with, and the effect of those details on properties and performance is usually not correctly represented by averaged values of properties, geometry, or local stresses. Hence, if microdamage accumulation and interaction are the controlling features of our problem, we must choose mechanics formulations that can represent these details well and that can support the need to discuss progressive damage development and accumulation with rates that are related to applied conditions and material (system) properties.

1.4.2 Evolution of Stress States and Material States

Our second canon brings to light the importance of large changes in material states and stress states, especially at the local level. These changes are a distinctive feature of composite systems requiring some new concepts and methods of analysis. First of all, some change in the properties of engineering materials under long-term use is generally expected; however, some composites may change their stiffness by as much as 30 or even 40% without any appreciable loss in strength. Such changes are quite uncommon in other materials. Although, as we will see, it is not always possible or advisable to design (with composites) for the *intentional occurrence* of such changes, it is possible to design for the *possible presence* of such changes with sufficient confidence to ensure the safety and reliability of the parent engineering component. Moreover, some of the most valuable benefits of the use of composite systems (especially for high-temperature applications) can best be realized by designing in the presence of damage development. This is a major thrust of this book.

The designers of high-performance composite components often tell us that the state of the material at the end of life (not at the beginning of life) controls durability and damage tolerance. Hence, in general, we will set our analytical representations on the damaged material state, rather than on the initial pristine condition. Even with current computational capabilities, it is not yet possible to model the full local complexity of the final damage state, such as the remarkably complex situation shown in Figure 1.30 for a notched coupon under cyclic loading. But it is essential to represent as much detail as needed to capture the controlling mechanics and physics of the processes that determine remaining strength and life.

1.4.3 Influence of Microprocesses

Finally, we consider the third canon, which claims that the remaining strength and life of continuous-fiber composite systems are controlled by microstructure and microbehavior and especially by local characteristics associated with anisotropy, inhomogeneity, and brittleness. This behavior was very evident in our earlier discussion of damage and failure modes where we saw that global stiffness changes closely followed matrix cracking (Fig. 1.26) and delamination (Fig. 1.25). We also saw that the fracture of a small number of fibers in a small region (Fig. 1.32) or the microbuckling of a small number of local fibers can trigger global failure.

In addition, we understood that changing the coating on fibers, typically a change of less than 1% of the composite by volume, could alter strength and strain to failure by hundreds of percent and the life of the material under cyclic loading by orders of magnitude. Figure 1.40 shows that such an alteration (brought about, in this case, by using a different binder during the powder processing of a polymer composite) can greatly alter the *failure mode* of the composite, from a (brittle) transverse mode to a (stronger) longitudinal mode. Figure 1.41 shows the same result for a ceramic composite system, in which the interphase region was adjusted, a familiar enterprise

Figure 1.40 Fracture mode transitions caused by fiber coatings in carbon–PEEK system with (*a*) weak fiber–matrix interface and (*b*) strong fiber–matrix interface.

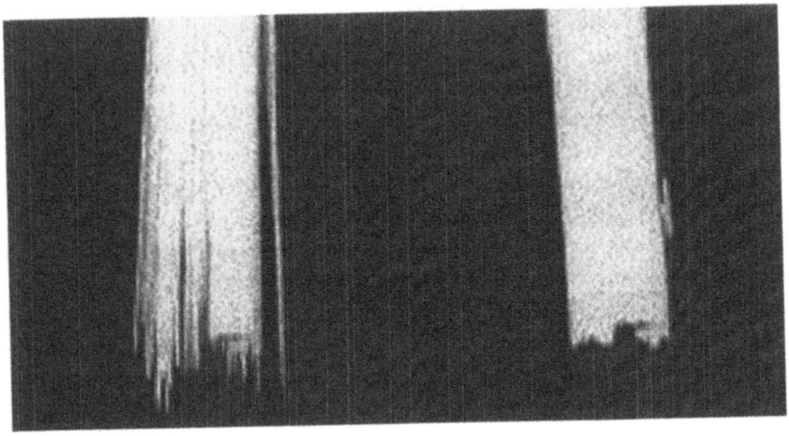

(a) (b)

Figure 1.41 Fracture mode transitions caused by fiber coatings in oxide–oxide ceramic matrix composite system with (*a*) weak fiber–matrix interface and (*b*) strong fiber–matrix interface.

for those systems. Looking at Figures 1.40 and 1.41, we are drawn to the conclusion that the ductile failure modes (with the more brushy appearance in both cases) are better for engineering applications than the more brittle failure mode (for which the fracture plane is more nearly straight across the section). For the ceramic matrix composite, that conclusion is generally correct. In the case shown in Figure 1.41, the strain to failure increased from about 0.001 to 0.0027 and the strength nearly doubled by making the composite more ductile (by making the interface weak). But for the polymer system shown in Figure 1.40, the tensile strength in the direction of the fibers in the unidirectional specimens was increased by more than 30% by making the interface *stronger* and the composite (slightly) more brittle. In a later chapter, we will see that these changes can be correctly modeled and predicted based on the micromechanics of the fracture process (Subramanian et al., 1995a,b).

Generally, the internal details of a composite material system, that is, the properties, geometry, and arrangements of the constituents and the interphase regions between them, along with the changes of these details with loading history are critical to the determination of the durability and damage tolerance of the system. In this context, one would expect to find a dependence on anisotropy and inhomogeneity, but the nature of the dependence on brittleness is unusual (Clyne, 1996). It is, of course, not necessary for a continuous-fiber-reinforced composite to be brittle. (Under certain conditions, metal matrix composites, for example, have significant ductility.) Many of the composite material systems that have found successful use in engineering applications at the present time, however, are brittle or quasi-brittle at the global level; that is, they have small strains to failure (a few percent) compared to common metals (tens of percent). In any case, this book will consider systems for which brittle behavior is an important consideration.

In the chapters that follow, we will develop the concepts and analyses needed to represent the physical situations we have discussed in this chapter in such a way that we can assess the collective effect of the long-term application of applied conditions—mechanical, thermal, and environmental—in terms of remaining strength (damage tolerance) and life (durability) using the problem definition shown in Figure 1.14.

REFERENCES

Baer, E., Kastelic, J., and Palley, I. (1980), *Biomechanics*, Vol. 13, p. 887.

Batdorf, S. B. (1982), "Tensile Strength of Unidirectionally Reinforced Composites," *J. Reinforced Plastics Composites*, Vol. 1, p. 153.

Carman, G., and Reifsnider, K. (1992), "Micromechanics of Short-Fiber Composites," *Composites Science and Technology*, Vol. 43, pp. 137–146.

Christensen, R. M. (1979), *Mechanics of Composite Materials*, Wiley, New York.

Clyne, T. W., Ed. (1996), *Interfacial Effects in Particulate, Fibrous, and Layered Composite Materials*, Trans Tech Publications, Zuerich-Uetikon, Switzerland.

Cowin, S. C. (1989), *Bone Mechanics*, CRC Press, Boca Raton, FL.

Cox, H. L. (1952), "The Elasticity and Strength of Paper and Other Fibrous Materials," *Br. J. Appl. Phys.*, Vol. 3, pp. 72–79.

Eshelby, J. D. (1957), "The Determination of the Elastic Fields of an Ellipsoidal Inclusion, and Related Problems," *Proc. Roy. Soc. Lond. Ser. A241*, Vol. 241, no. 1226, August 20, p. 376.

Giannuzzi, L. A., Rugg, K. L., and Tressler, R. E. (1992), "High Temperature Strength and Creep Behavior of Carborundum Alpha Silicon Carbide Fibers," *Ceramic Trans.* Vol. 38, pp. 655–666.

Glimcher, M. J. (1984), "Recent Studies of the Mineral Phase in Bone and Its Possible Linkage to the Organic Matrix By Protein-Bound Phosphate Bonds," in *Phil Trans. R. Soc. Lond. B.*, Vol. 304, pp. 479–508.

Hashin, Z. (1979), "Analysis of Properties of Fiber Composites with Anisotropic Constituents," *J. Appl. Mech.* Vol. 46, p. 543.

Hashin, Z. (1965), "On Elastic Behavior of Fibre-Reinforced Materials of Arbitrary Transverse Phase Geometry," *J. Mech. Phys. Solids*, Vol. 13, p. 119.

Highsmith, A. L. (1984), "Damage Induced Stress Redistribution in Composite Laminates," Ph.D. Dissertation, College of Engineering, Virginia Polytechnic and State University, Blacksburg.

Highsmith, A. L., and Reifsnider, K. L. (1982), "Stiffness-Reduction Mechanisms in Composite Laminates," in *Damage in Composites*, ASTM STP 775, K. L. Reifsnider, Ed., American Society for Testing and Materials, Philadelphia, pp. 103–117.

Hill, R. (1964), "Theory of Mechanical Properties of Fiber Strengthened Materials: I. Elastic Behavior," *J. Mech. Phys. Solids*, Vol. 12, p. 199.

Hilmas, G. E., Holmes, J. W., Ghatt, R. T., and DiCarlo, J. A. (1993), "Tensile Creep Behavior and Damage Accumulation in a SiC-Fiber/RBSN-Matrix Composite," *Ceramic Transactions*, Vol. 38, pp. 291–304.

Hiltner, A., Orberg, J., and Baer, E. (1983), "Organization of Collagen Fibers in the Intestine," in *Connective Tissue Res.*, Vol. 11, p. 285.

Holmes, J. W., and Cho, C. (1992), "Experimental Observations of Frictional Heating in Fiber-Reinforced Ceramics," *J. Am. Ceramic Soc.*, Vol. 75, No. 4, pp. 929–938.

Horton, P. E., and Whitehead, R. (1988), *Damage Tolerance of Composites*, Vols. I and II, AFWAL, TR-87-3030, Air Force Wright Aeronautical Laboratories, Akron, OH.

Jamison, R. D. (1986), "On the Interrelationship between Fiber Fracture and Ply Cracking in Graphite/Epoxy Laminates," in *Composite Materials: Fatigue and Fracture*, ASTM STP 907, American Society for Testing and Materials, Philadelphia, pp. 252–273.

Jamison, R. D., and Reifsnider, K. L. (1982), "Assessment of Microdamage Development during Tensile Loading of Graphite/Epoxy Laminates," Final Report, Contract No. DAAG29-82-K-0190, U.S. Army Research Office, Research Triangle Park.

Knauth, P. (1974), *The Emergence of Man: The Metalsmiths*, D. M. Brown, Series Ed., Time-Life Books, New York.

Lee, S. S., and Stinchcomb, W. W. (1996), "Damage Mechanisms and Fracture Modes in Nicalon/CAS-II Laminates," in *Fracture of Composites*, E. Armanios, Ed., Transtec Publications, New York, pp. 227–256.

Lesko, J. J. (1994), "Interphase Properties and Their Effects on the Compression Mechanics of Polymeric Composites," Ph.D. Dissertation, Virginia Polytechnic Institute and State University, Blacksburg.

Liao, K. (1994), "Tensile and Uniaxial/Multiaxial Fatigue Behavior of Ceramic Matrix Composites at Ambient and Elevated Temperatures," Ph.D. Dissertation, Virginia Polytechnic Institute and State University, Blacksburg.

Lorenzo, L., and Hahn, H. T. (1986), "Fatigue Failure Mechanisms in Unidirectional Composites," in *Composite Materials: Fatigue and Fracture*, H. Thomas Hahn, Ed., ASTM STP 907, American Society for Testing and Materials, Philadelphia, pp. 210–232.

McNicol, L. A., and Strahlman, E., Eds. (1989), *Corneal Biomechanics and Wound Healing*, National Eye Institute, National Institute of Health, Bethesda, MD.

Mori, T., and Tanaka, K. (1973), "Average Stress in Matrix and Average Elastic Energy of Materials with Misfitting Inclusions," *ACTA Metal.*, Vol. 21, p. 571.

O'Brien, T. K. (1982), *Damage in Composite Materials: Basic Mechanisms, Accumulation, Tolerance, and Characterization*, ASTM 775, Reifsnider, K. L., Ed., American Society for Testing and Materials, Philadelphia, pp. 140–167.

Piggott, M. R. (1980), *Load Bearing Fibre Composites*, Pergamon, New York.

Razvan, A. (1992), "Fiber Fracture in Continuous-Fibre-Reinforced Composite Materials During Cyclic Loading," Ph.D. Dissertation, Virginia Polytechnic Institute and State University, Blacksburg.

Razvan, A., Bakis, C. E., and Reifsnider, K. L. (1990), "S.E.M. Investigation of Fiber Fracture in Composite Laminates," *Mater. Characterization*, Vol. 24, No. 2, pp. 179–190.

Razvan, A., and Reifsnider, K. L. (1991), "Fiber Fracture and Strength Degradation in Unidirectional Graphite/Epoxy Composite Materials," *Theoret. Appl. Fracture Mech.*, Vol. 16, No. 1, pp. 81–89.

Reifsnider, K. L., Ed. (1982), *Damage in Composite Materials: Basic Mechanisms, Accumulation, Tolerance and Characterization*, ASTM 775, American Society for Testing and Materials, Philadelphia.

Reifsnider, K. L., Ed. (1991), *Fatigue of Composite Materials*, Elsevier Science, New York.

Reifsnider, K. L., Henneke, E. G., and Stinchcomb, W. W. (1979), *Effect Property Relationships in Composite Materials*, AFML-TR-76-81, Air Force Materials Laboratory, Wright-Patterson AFB, OH.

Russel, W. B. (1973), "On the Effective Moduli at Composite Materials: Effect of Fiber Length and Geometry at Dilute Concentrations," *Z. Angew. Math. Phys.*, Vol. 24, p. 581.

Sharpe, L. H. (1972), "The Interphase in Adhesion," *J. Adhesion*, Vol. 4, No. 1, pp. 51–64.

Stinchcomb, W. W., and Bakis, C. E. (1991), "Fatigue Behavior of Composite Laminates," in *Fatigue of Composite Materials*, K. L. Reifsnider, Ed., Elsevier, New York, pp. 105–180.

Subramanian, S., Reifsnider, K. L., and Stinchcomb, W. W. (1995a), "A Cumulative Damage Model to Predict the Fatigue Life of Composite Laminates Including the Effect of Fiber-Matrix Interphase," *Int. J. Fatigue*, Vol. 17, No. 5, pp. 343–351.

Subramanian, S., Reifsnider, K. L., and Stinchcomb, W. W. (1995b), "Tensile Strength of Unidirectional Composites: The Role of Efficiency and Strength of Fiber-Matrix Interface," *J. Composites Technol. Res.*, Vol. 17, No. 4, pp. 289–300.

CHAPTER 2

ENGINEERING CONCEPTS OF STRENGTH

Thesis: Scalar functions of stress and principal material strengths can be constructed to represent the state of the material for each distinct failure mode under current or progressive applied conditions.

We will develop this thesis using the following approach. First, we will identify the physical factors that influence the definition and measurement of *intrinsic* material strengths, or *principal strengths*, as we will call them, for isotropic and anisotropic materials. This will ensure a consistent nomenclature, avoid subsequent inconsistencies in mathematical operations, and establish a general foundation for modeling. Then, we will turn our attention to the question of strength concepts for multiaxial loading, so that we can associate material strengths with the multiaxial stress states that occur in "real" engineering components. Finally, we will generalize the concepts to provide a basis for discussing the accumulation of damage and the degradation of strength during service.

2.1 FACTORS THAT DETERMINE COMPOSITE MATERIAL STRENGTH

How do we answer the question, "how strong is this material?" The general concept of strength is deceptively simple; we think in terms of how much mechanical force it takes to break it. Beyond this simple idea, providing a well-defined engineering answer requires a rather careful account of several important mechanics and materials factors. We will consider the ones listed below:

1. How many material parameters (material properties) are needed to describe strength? Is one enough?
2. Does the shape of the specimens influence the values obtained?
3. Does the size of the specimens influence the values obtained?
4. Do the loading conditions (e.g., method of gripping, temperature, loading rate) influence the values obtained?
5. Do we have to determine a unique set of "strengths" for each failure mode?

2.1.1 Principal-Strength Components

In Chapter 1, we alluded to the idea of principal strengths as a minimum set of values needed to define the intrinsic strength of a material. As an example, we indicated that a sheet of continuous-fiber-reinforced composite material with all of the fibers aligned required five such quantities to define its strength for in-plane loading only. Figure 2.1 illustrates this idea. But how do we determine how many principal strengths are needed in general? (We will formulate all of our models in terms of these values, to minimize complexity and to assure uniqueness of our definitions of strength.)

The first consideration is isotropy or anisotropy. Assume that the material of interest is isotropic; that is, that the *strength* of the material is the same when determined from specimens taken from a larger sample of the material at any and all orientations relative to any reference system. We have italicized strength because

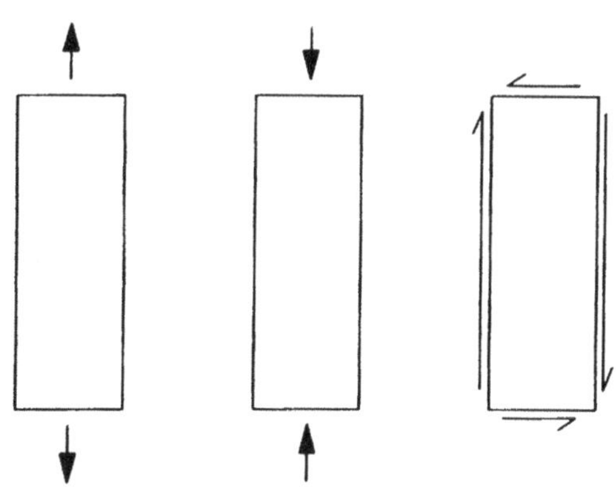

Figure 2.1 Simple engineering loading methods typically used to define (*a*) tension strength, (*b*) compression strength, and (*c*) shear strength for in-plane loading of coupon specimens.

we usually think of isotropic materials in terms of their stiffness values, and we often assume that if a material has isotropic stiffness, then it also has isotropic strength. This assumption may not be true. We must be careful to define isotropy in terms of the material parameter that we wish to characterize, which is strength in the present case.

If we are sure that the material is isotropic, then how many quantities are needed, and how do we specify them? Strength is an engineering definition, but we will assume that it is defined in such a way that it can be measured uniquely; that is, we will treat it as an independent, external variable (or array of variables). If this premise is true, then we ask, "what independent experiments can be done?" Figure 2.1 suggests some common choices. It is possible to apply a force (or stress) in *only* one direction (Fig. 2.1*a*) if we are careful not to constrain the resulting strain in any other direction. This *uniaxial* loading is the most universal answer to the last question. It is also apparent that we can "push" or "pull" along only one axis to measure a *tensile strength* or a *compressive strength*. We will take these two values as principal strengths for an isotropic material.

Figure 2.1 also suggests that we can conduct an independent experiment by applying a shear force (and stress) to a material element, such as a plate or a rod (or tube). So we could claim that the *shear strength* of the material is also a principal strength. Then we have three principal material strength values for an isotropic material; we will call them X_t, X_c, and S. Are they really independent? If the physical failure mode for all of these quantities (or any two) is identical, they are not independent. We will discuss this topic further under item 5 in our topic list, but, for example, if we are testing a very ductile material and the tensile and compressive failures occur by shear (which is independent of the sign of the applied loading), then the tensile strength and compressive strength will be the same and the shear strength will be equal to half of that value. (This is easily shown from a Mohr circle plot; see Exercise 2.1). This assumption is often made, but it is not a general "truth." If the failure mode is different in tension and compression (as it is likely to be for a brittle material like glass) and the failure mode in shear is different from them both, then all three of the principal strengths are independent, and our models must depend on all of them, explicitly.

The generality to be made here is that *the number of principal strengths needed to uniquely describe the strength of a material (isotropic or anisotropic) is determined by the number of unique failure modes*. There are a number of subtleties involved in this axiom, and we will review several of them as we further develop the subject in this chapter. In the case of isotropic materials, if the failure modes in tension, compression, and shear really are different physical processes, then we need three principal strengths to describe the material, even though only two elastic stiffness or compliance material constants are needed to fully describe the elastic response. Although this contrast has been noted in the literature (Hill, 1983; Marin, 1962; Paul, 1961), the fundamental issues associated with it are usually not discussed explicitly; instead, a phenomenological failure criterion is adopted, as we will discuss below. Nevertheless, for anisotropic materials, an explicit account of these issues is essential and will provide the basis for a consistent model of strength.

2.1.2 Effect of Specimen Geometry

The applied loading types shown in Figure 2.1 are "ideal" in the sense that they are the *result* that we hope to get from an experimental effort to measure the values of material strength. A unidirectional stress, needed to measure tensile and compression strength, for example, can, in principle, be created by simply gripping the ends of a coupon of material and pulling on it, as suggested in Figures 2.1 and 2.2*a*. However, the act of gripping the material may cause a complex stress state, and fracture may occur in or near the gripping area. The strength measured in such an experiment is not the principal tensile strength of the material under uniaxial stress, as we defined it. A common solution to this problem is to reduce the cross section of the specimen in the gage section, as shown in Figure 2.2*b*. This creates a region of the specimen in which the stresses are higher and "forces" failure to occur in that area. If the stress concentrations caused by the "shoulders" on the specimen do not cause failure due to complex stress, such a shape may be successful.

The choice of specimen geometry to assure the success of experimental efforts to measure the principal strengths of materials is an art and a science. Efforts to define the many details essential to the success of such experiments are very widespread, and the literature in this area is extensive. Handbooks and design guides provide collections of results, and standard test methods and procedures, developed by organizations such as the American Society for Testing and Materials, American Society for Mechanical Engineers, and the Society for Engineering Stress Analysis should be consulted for this purpose (Faupel and Fisher, 1981; *ASTM Book of Standards*).

Applying a *pure shear* stress to a specimen is also difficult. For example, it is very difficult to apply the shear stresses shown in Figure 2.1 to a flat-plate specimen. A more common arrangement is shown in Figure 2.2*c*. Torsion of a tube (especially a thin tube) produces a state of shear stress that is usually sufficiently well controlled for the experimental determination of shear strength if the gripping regions are

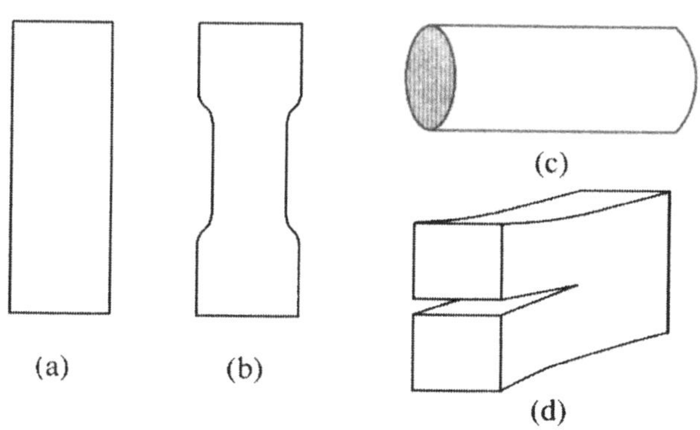

Figure 2.2 Typical specimen geometries used to measure strength under more general conditions.

carefully designed. Although tubes can be difficult to make and gripping them is challenging, they are growing in popularity as test specimens.

Of course, many other test geometries are possible. Brittle fracture is sometimes induced by notched specimens of various geometries, such as the *compact tension specimen* sketched in Figure 2.2*d*. For the materials discussed in this book, self-similar crack propagation through the specimen thickness over large in-plane distances is not a typical failure mode for composites of interest to engineering applications. Hence, test specimen geometries of this type are not commonly used.

Size effects are a universal concern to the characterization of materials of all types, including composites. Fundamentally, the most common physical basis for this effect is the fact that material spatial nonuniformities and defects or flaws are statistically distributed in any and all volume elements of the global volume of the material. If one considers the strength of a large-volume element, then the chance of this element having a large or serious flaw, even though that flaw is statistically rare, becomes significant. Therefore, such large volumes will have, on the average, a lower strength than smaller elements, which have a lesser chance of having such a large flaw. This concept is often described with a probability of failure having the form

$$P = 1 - \exp\left[-\frac{V_e}{V_0}\left(\frac{\sigma}{\sigma_0}\right)^{\alpha}\right]$$

where V_0 is the volume of the representative volume element being described, V_e is the volume of an element under test, σ is the applied stress level, and σ_0 and α are material constants. This type of size dependence is widely observed and discussed in the literature, especially for brittle materials in which the defect or flaw distribution discussed above is a controlling feature. It has also been used for composite materials of various types (Batdorf, 1985; Curtin, 1991; Lissart and Lamon, 1997).

There are at least two other types of size effects that are important for composites. The first concerns the number of internal elements in the composite. We will learn many of the details of this effect in Chapter 4 when micromechanical relationships for strength are developed. For now, we will conduct a thought experiment. First, we must break a large number of individual fibers and determine the average strength of this sample. Second, we take an identical number of fibers, grip them together in a bundle, and determine the break strength of that bundle. The strength of the bundle will, in general, be less than the average strength of the individual fibers (except for some very special cases). One reason for this result is that the first fibers to break in the bundle influence the conditions in the rest of the fibers. They release their load before the strength of the bundle is measured; that is, many *weak fibers*; fail before the bundle fails. But the area of the broken fibers is not discounted when the strength of the bundle is calculated. This is one (of several) interactions that are possible between the fibers in the bundle, as we will see in Chapter 4. The result is that the *bundle strength* of composites with a larger number of interacting constituent elements can be expected to have a strength that is both less than the average of those elements and dependent on the number of elements in the global volume.

The second important additional size effect in composites is associated with surfaces. We will see in later chapters that there are many *edge effects* in composites, especially in composite laminates. The internal stresses and deformations in regions near an edge are usually complex and may introduce theoretical singularities in stress or strain. In addition, the behavior of brittle materials is often greatly influenced by surface effects such as roughness and defects. Everyone knows that a small crack on the edge of a sheet of glass can quickly propagate to cause failure. In ductile materials, such as common metals, surface flaws are one of the most common sources of the initiation of fatigue failure (Suresh, 1991). So it is not surprising that edges, surface condition, and surface area influence the strength and life of composites in a manner influenced by surface-to-volume ratio.

However, we are left with one final observation of engineering importance, which we can pursue with a specific question. Should we use a lot of small fibers in a composite (with a lot of surface area) or large fibers (with a lot of volume per unit surface area)? As usual in engineering, there is obviously a balance between several size effects that produce contrasting or reversed result. This premise is one of the motivations to fully understand the nature of these effects and to find models that are sensitive to the material parameters that control them.

2.1.3 Influence of Loading Conditions

We have already mentioned the influence of the manner in which specimens are loaded by the gripping devices. In general, because continuous-fiber-reinforced composites often have high stiffness and low strain to failure, it is difficult to introduce load into the material with grips, fixtures, or joints. Consequently, characterizing these materials requires special care.

But several other factors are also important, to a degree not common to the characterization of other engineering materials such as homogeneous metals. One of these factors is strain rate.

Figure 2.3 illustrates some typical features of the dependence for fiber-controlled behavior. The dependence of stiffness on strain rate is often assumed to be linear. The data in Figure 2.3, which show behavior that is often seen in composites, indicate that the stiffness varies roughly linearly with strain rate over certain ranges of the independent variable, but the slope of that linear variation may be discontinuous and may be quite different in different strain range regions. Figure 2.3 also indicates that the changes may be significantly large, of the order of 50% for some of the data in that example. Finally, Figure 2.3 illustrates that the composite behavior is similar in the general sense to the fibers used to make them. In the present context, all of these general features can be regarded as justification for the requirement that the rate of strain be recorded and controlled as an important loading condition during the characterization of composite materials, even for fiber-controlled response. There is a large body of literature on this subject (see, e.g., Sierakowski, 1997).

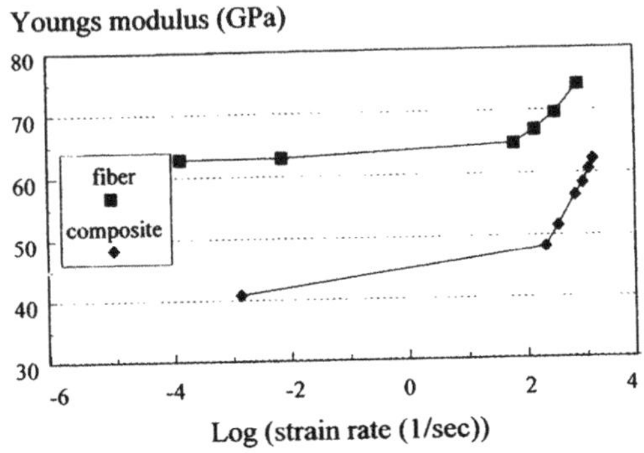

Figure 2.3 Effect of rate of strain during tensile elongation on measured stiffness of glass fibers and glass epoxy composite (Hartwig, 1978; Xia et al., 1995).

We could also discuss the influence of the rate of loading on the resulting properties when the behavior is matrix controlled. This behavior is usually discussed in terms of viscoelastic or viscoelastic–plastic representations of these materials; loading rate has a strong influence on the viscoelastic response. However, loading rate also influences strength and fatigue behavior (Gates, 1991; Miyano et al., 1994; Rotem, 1993). The dependence of polymer matrix behavior on rate and temperature and the attendant viscoelastic representations are well-developed subjects in the literature. Summaries of particular interest to the present discussion can be found in Dillard (1991) and Christensen (1971).

Environmental moisture or the presence of other chemical species during a characterization test may also substantially alter the properties of a composite material. The diffusion of water vapor into polymer composites, for example, is widely described and can be a substantial influence on matrix-controlled properties. We will discuss in detail the influence of such moisture on stiffness and strength in later chapters. However, Figure 2.4 illustrates the nature of the influence of moisture on the tensile strength in the direction perpendicular to the fibers for an AS4/3501 carbon–epoxy lamina (data from Browning, et al., 1976). One can see that moisture can reduce the strength of the material in that direction by 50% or more. Moreover, the data show that the combination of moisture and temperature is even more severe. Hence, controlling test conditions, especially such things as moisture content, is essential to the recovery of consistent material data. Figure 2.5 shows another interesting aspect of this issue. Although the influence of moisture and temperature on matrix-dominated properties appears to have easily identifiable roots in the effects of those environments on the matrix materials, it also happens that fiber-dominated properties may be significantly influenced. Figure 2.5 shows that temperature can change the tensile strength of unidirectional polymer matrix composites in the *fiber* direction by as much as 40% or more (Kerr and Haskins, 1987). This result is indeed surprising and will be discussed in our chapter on the micromechanics of strength. The data in Figure 2.5 also show that this influence is not a "simple degradation," as the strength is a highly nonlinear function of temperature.

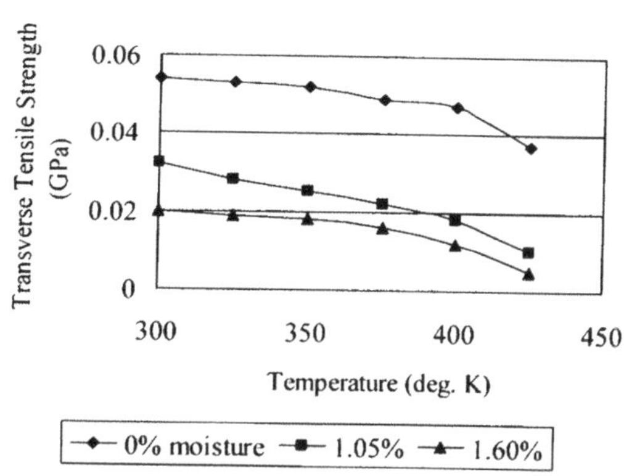

Figure 2.4 Variation of tensile strength of unidirectional carbon–epoxy lamina in direction perpendicular to axis of fibers as function of moisture and temperature.

2.1.4 Influence of Failure Modes

In Chapter 1, we discussed the idea of damage modes and failure modes, from the physical standpoint. The discussion centered on the premise that, for the materials considered here, damage, in various physical forms, occurs in a *widely distributed manner*, and can, in some cases, *change the properties and performance of the material greatly without causing fracture* (separation into two or more parts). In many respects, this idea is the central theme of this book.

In the present context, this fact has considerable influence on the manner in which our models are formulated, and the definitions we use for the parameters involved. We will regard strength as rupture or fracture strength, i.e., separation into parts.

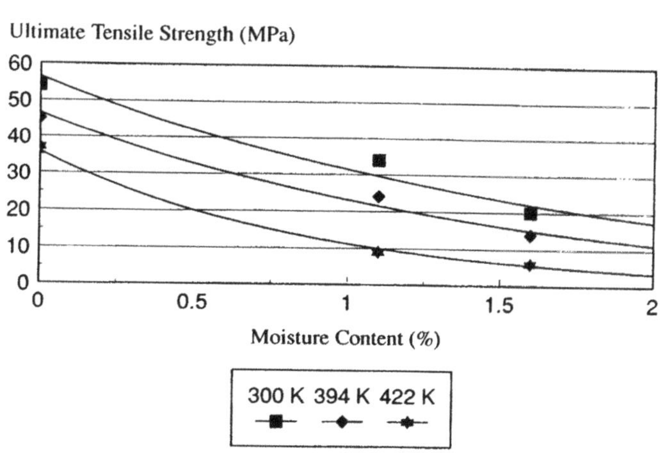

Figure 2.5 Variation of fiber direction tensile strength of unidirectional glass–epoxy specimen as function of temperature.

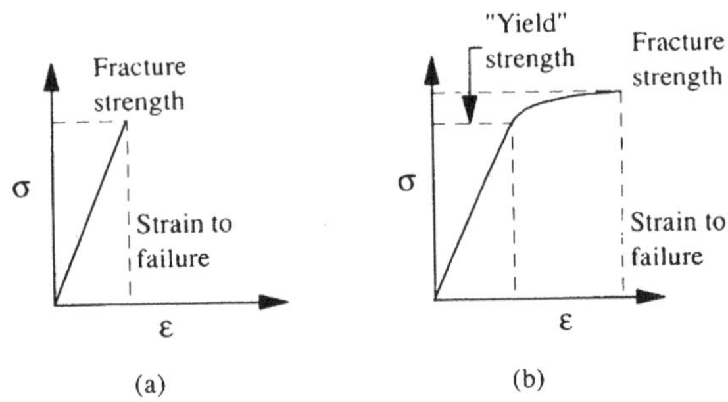

Figure 2.6 Stress–strain response for (*a*) *brittle* composite material and (*b*) *ductile* composite material.

Strength will be defined by the fracture event and process. Figure 2.6 illustrates the first consequence of this point. Many of the continuous fiber-reinforced composite materials in use are quite brittle, at least under uniaxial quasi-static loading. In many cases, their stress–strain diagrams have the general appearance of Figure 2.6*a*. But some materials, even under quasi-static loading, have substantial nonlinearity and stiffness reduction, even under the uniaxial loading commonly used to define the principal strengths of the material, as suggested in Figure 2.6*b*. If the matrix is a metal, there may be a *yield strength* value beyond which the material response is nonlinear. For ceramic composites, there is often a "knee" in the stress–strain curve, sometimes at stresses that are of the order of one-third of the fracture value, caused by matrix cracking. After that point, the response is often fairly linear until quite near the fracture load. Also, for laminates having off-axis plies, those plies may sustain damage such as matrix cracking and cause substantial nonlinear response before fracture. We will consider these *damage modes* in considerable detail in later chapters, but for the present discussion, strength will mean fracture strength, a measurable engineering quantity.

What do we mean by *failure mode*, and why do we have to be so careful to observe and define it precisely? To understand this point, two examples will be given. The first concerns an effort to measure the shear strength of a continuous-fiber-reinforced lamina, from which a specimen has been cut so that the fibers have the orientation shown in Figure 2.7*a*. When a shear force is applied to such a coupon, as in the top diagram, the resulting normal stresses in a coordinate system at 45° to the first are shown in the bottom diagram of Figure 2.7*a*. If the sign of the applied shear stress is reversed, as in Figure 2.7*b*, then the sign of the normal stresses also reverse, as shown. However, physically, the effect of the stresses on the material (and, therefore, the resulting failure mode) will be the same. Then, a pure shear test with the fiber orientation shown in Figures 2.7*a,b* is a valid test of stiffness and strength, independent of the direction of the applied shear stress. Consider the choice of fiber orientation shown in Figures 2.7*c,d*. In Figure 2.7*c*, the resulting normal tensile stress is along the fiber direction, and the resulting normal compression stress acts in a direction perpendicular to the fibers. If the direction of the applied shear stress is reversed, as in Figure 2.7*d*, the resulting

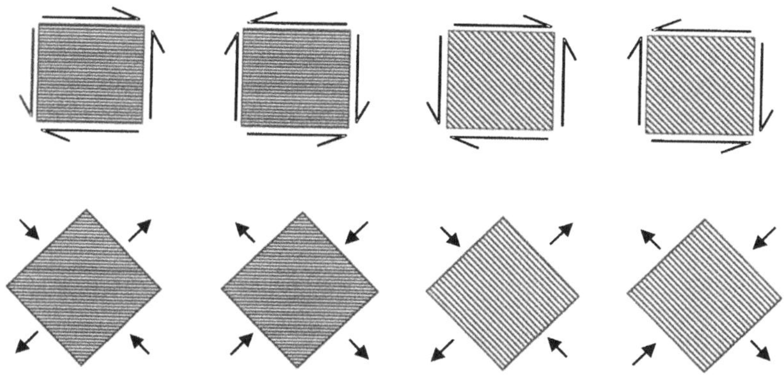

Figure 2.7 Possible shear test configurations for a unidirectional coupon, producing *(a,b)* valid identical result and *(c,d)* invalid, different result.

tensile normal stress is perpendicular to the fibers and the compression stress is in the direction of the fibers. One can imagine that the failure mode of the composite is different in these two cases, especially since many composites have a very low strength in tension in the direction perpendicular to the fibers compared to the strength in the fiber direction. Hence, the pure shear test shown in Figures 2.7*c,d* is not a valid test for the principal shear strength of the composite.

There is a second, more rigorous illustration of the importance of determining the failure mode. Consider a test of compressive strength in the fiber direction, conducted as suggested by Figure 2.8. The figure indicates that there is uniaxial compression loading along the 2 axis of an element of unidirectional continuous-fiber-reinforced material. Under this loading, the strain in the direction of the fibers (the 1 axis) can be written as

$$\varepsilon_1 = S_{11}\sigma_1 + S_{12}\sigma_2 + S_{16}\sigma_6 \tag{2.1}$$

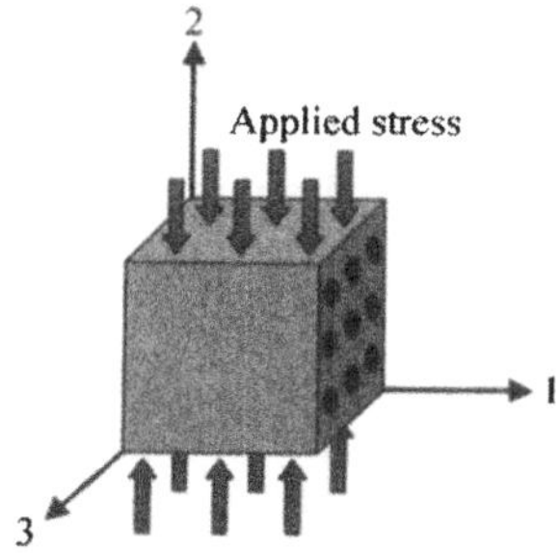

Figure 2.8 Compressive loading of unidirectional element in 2 direction, transverse to fiber axis.

where the compliance tensor of the material is S and the indices have their usual meaning. We assume that we have made a prior measurement of the tensile strength of the material in the 1 direction to obtain a value that we will call X_t. Then, the fracture *strain* in the 1 direction is simply equal to $S_{11}X_t$. Then using (2.1), we could ask, "what applied stress along the 2 axis will cause failure in the 1 direction?" Apparently, the value of that stress is

$$\sigma_2 = \frac{S_{11}}{S_{12}} X_t - \frac{S_{16}}{S_{12}} \sigma_6 \tag{2.2}$$

Now we ask the fundamental question, "can we measure the compressive strength in the 2 direction, independently?" From (2.2), if the applied stress σ_2 reaches the value given in that equation before fracture occurs due to a different failure mode (such as matrix cracking driven by a limiting strain in the 2 direction), then the compressive loading in the 2 direction will produce the same failure mode as tensile loading in the 1 direction. The value of strength measured by this experiment, which we will call Y_c, will not be independent of X_t. Hence, if we can measure an independent principal strength in compression in the 2 direction, we should require that

$$\sigma_2 = \frac{S_{11}}{S_{12}} X_t - \frac{S_{16}}{S_{12}} \sigma_6 \le -Y_c \tag{2.3}$$

remembering that the applied stress is a negative number. We can rewrite this expression in a more convenient form as

$$\frac{S_{16}}{S_{12}} \frac{\sigma_6}{X_t} - \frac{S_{11}}{S_{12}} \ge \frac{Y_c}{X_t} \tag{2.4}$$

Based on the definition of the compliance S_{12}, that quantity is known to have a negative value. In this case, the smallest number on the left of (2.4) (the *least upper bound*) will occur when the shear stress is equal to its largest value, namely, the shear strength, which we will call S (with no subscripts, in deference to the literature). The final condition is

$$\frac{S_{16}}{S_{12}} \frac{S}{X_t} - \frac{S_{11}}{S_{12}} \ge \frac{Y_c}{X_t} \tag{2.5}$$

Hence, if we define principal material strengths in terms of the principal material directions of the composite (which is the approach used to define unique stiffness values), the principal strength values must satisfy (2.5) and five additional conditions of that type for a two-dimensional anisotropic plate that obeys (Wu, 1974) Hooke's law. If they do not, some of the principal strengths are not unique, and the material direction chosen may not be appropriate. As far as the analytical representation of strength is concerned, failure to satisfy one of these conditions represents an inconsistency.

To examine this inconsistency, let us return to (2.1), where this argument began. The assumption is that stress and strain are uniquely related by a generalized stress–strain relationship such as

$$\sigma_{ij} = C_{ijkl}\varepsilon_{kl} \tag{2.6}$$

If the C_{ijkl} stiffness tensor components (and the contracted inverse of that array, the compliance components S_{ij} discussed above) are elastic quantities, then the strength values discussed above represent the *first departure* from elastic response, as discussed in Figure 2.6. However, we are concerned with the *final* fracture strength, as definitions of our strength array, X_{ij}, and we will further claim in Chapter 3 that these values may be changing with time as the result of exposure to applied environments.

The question remains of how we can construct a consistent failure criterion. More than one answer exists, depending on the nature of the unloading response of the material. We consider the material behavior shown in Figure 2.6 and define two distinct possibilities for the nature of the stress–strain behavior when the load is removed, as suggested in Figure 2.9. If damage occurs and the stiffness of the material is reduced (to the value E' in Fig. 2.9*a*) but the strain on the material is fully recovered when the applied load is removed, then we have essentially linear elastic response with a stiffness of E' upon reloading, even for the final loading to failure to measure X_{ij}. Surprisingly, many composite materials respond in this way. Many engineering composite materials are made with fibers in many directions, either as plies of a laminate or as tows in a weave. Matrix cracking, for example, that is parallel to fibers in one direction is "spanned" by fibers in another direction. If the fibers remain intact, they are still elastic and return to their original shape upon unloading, closing the cracks and bringing the material back to the origin of its stress–strain curve. For these materials, we will use the *failure function* and *failure surface* concepts at any given instant of time, as defined by the instantaneous (possibly reduced) stiffness values and the corresponding generalized stress–strain relationship.

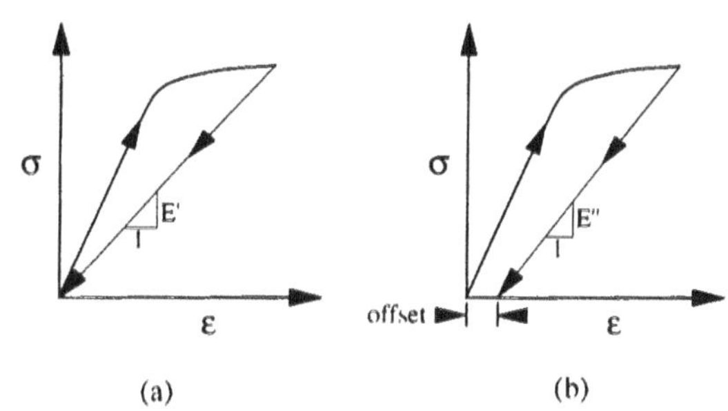

Figure 2.9 Loading and unloading curves for composite materials having (*a*) elastic response and (*b*) elastic–plastic response.

For materials that do not return to their original shape, as shown in Figure 2.9*b*, the interpretations are less rigorous. We will still require that, at any instant of time, there is a single-valued relationship between stress and strain. We will also require that the loading of the material be proportional, that is, that all components of stress increase in proportion to one another to failure when the strength of the material is measured to define X_{ij}. Finally, we will require that our criteria and modeling philosophy not depend, explicitly, on any of the unloading behavior of the material.

2.2 STRENGTH UNDER MULTIAXIAL LOADING

The fundamental problem of estimating strength in the presence of multiaxial stress has been widely considered. Literally dozens of approaches to this problem have been postulated. Review articles by Sandu (1978) and Soni (1983) discuss theories that account for anisotropy of stiffness and strength. Most of these theories can be expressed in terms of the principal material strengths that we discussed in earlier sections. However, the manner in which they are formulated reflects very different approaches to the general question of how strong a material is.

2.2.1 Failure Surfaces

A convenient method of presenting an answer to the question of strength is to construct a surface in stress space that defines the boundaries of the allowable stress states for a given material with specific strength. This failure surface can then be easily used by designers with no additional interpretation. Another advantage of this scheme is that such surfaces are ideally suited to the *effective stress* concepts that are so popular for the description of yielding and plasticity of isotropic, ductile materials. Popular answers to the question of the strength of homogeneous, isotropic materials are typically given in one of the following forms:

1. This material will fail when the maximum principal normal stress reaches a critical value, which we will call the *strength of the material* (a *maximum-stress criterion*).
2. This material will fail when the maximum principal normal strain reaches a critical value, which we will call the strength of the material (a *maximum-strain criterion* usually associated with St. Venant).
3. This material will fail when the maximum shear stress reaches a critical value (an approach usually attributed to Tresca and Guest).
4. This material will fail when the total strain energy reaches a critical value (often attributed to Beltrami) or when the distortional energy reaches a critical value (usually discussed in association with Huber, Von Mises, and Hencky).

These (and many other) approaches are widely discussed in texts on homogeneous materials (Collins, 1981; Dowling, 1993; Faupel and Fisher, 1981; Juvinall,

1967). We will defer to those descriptions for the details of those approaches. For our present purposes, we will note only a few features that are pertinent to our development.

The first such feature is demonstrated by the Tresca and Von Mises–Hencky criteria. Figure 2.10 shows a plot of these two criteria for a two-dimensional stress state in a coordinate system defined by the two principal stresses in that plane. The entirety of the failure surface is defined by a single material constant in both cases, the tensile yield or fracture strength. The Von Mises–Hencky criterion, for example, is usually stated in the form

$$2\sigma_{yd}^2 = (\sigma_1 - \sigma_2)^2 + (\sigma_1 - \sigma_3)^2 + (\sigma_2 - \sigma_3)^2 \tag{2.7}$$

where σ_{yd} is commonly evaluated from a single test, usually a tensile test. Hence, in this case, the *effective stress* [the octahedral shear stress on the right-hand side of (2.7)] is compared to only one material constant to establish a *critical value* on the premise that the criterion must be correct for all stress states so any convenient one can be used to evaluate the associated material constant. This assumption is not correct if the material has different failure modes under the application of different stress states. Hence, while these approaches have obvious utility for ductile, homogeneous, isotropic materials such as common metals in many cases, they have, in general, very limited use for inhomogeneous, anisotropic composites.

Figure 2.11 demonstrates a second category of criteria for which a failure envelope based on maximum stress, maximum strain, and combinations of shear and normal stress is constructed. The essential difference in these approaches is that different failure modes (and different strengths under different stress states) are explicitly admitted and accounted for. The maximum-stress criterion accounts for

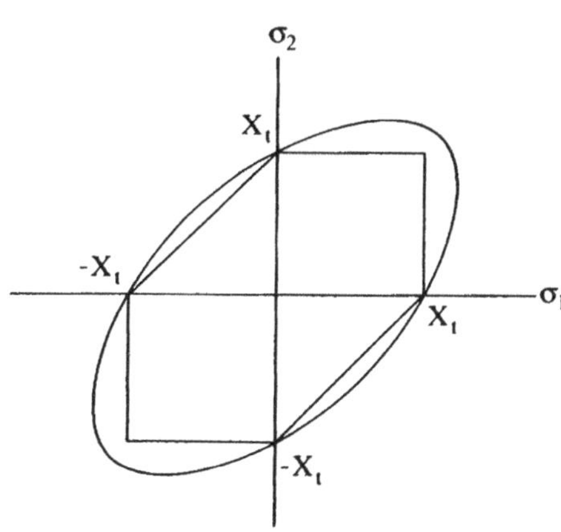

Figure 2.10 Failure envelopes for Tresca criterion (polygon) and Von Mises–Hencky criterion (ellipse).

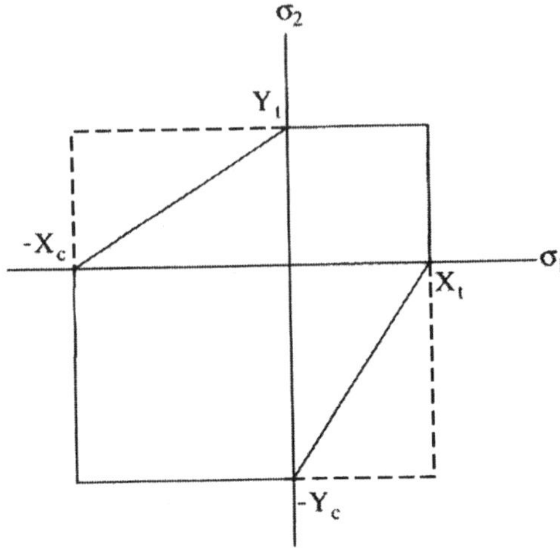

Figure 2.11 Failure envelopes for Coulomb–Mohr–Duguet criterion (solid polygon) and maximum-stress criterion (rectangle formed by solid and dotted lines).

the fact that the strength in tension may be quite different from the strength of the same material in compression. The same is true of the Coulomb–Mohr–Duguet criterion (also shown in Fig. 2.11) and a variety of modifications of that criterion that are widely used for brittle (isotropic) materials. Such criteria typically represent measured strength values under multiaxial stress better than single-parameter models, such as the ones discussed above, especially for brittle materials like ceramics or glasses (Paul, 1961).

For the composite materials considered here, the need for multiparameter approaches is clear. We adopt the paradigm that *failure criteria for anisotropic, inhomogeneous composite materials must incorporate a sufficient number of material parameters to correctly represent the failure modes observed in the material under multiaxial applied stress states*. This paradigm must also be associated with an important observation. The fact that a material is isotropic and homogeneous does not mean that only one material parameter is necessary to define its strength, as we saw in the above discussion. If more than one failure mode is observed, more than one material constant is needed.

2.2.2 Polynomial Criteria for Composite Materials

One way of introducing multiple material constants into a failure criterion for composite materials is to postulate a polynomial in the active stresses, in which each term is multiplied by a coefficient that reflects the material resistance to the stress component or components in the term. Examples of such a polynomial have been formulated by Tsai and Wu (1971) and Tsai and Hahn (1980). They talk about a *failure potential* that may have the general form

$$f(\sigma_{ij}) = F_{ij}\sigma_{ij} + F_{ijkl}\sigma_{ij}\sigma_{kl} + F_{ijklmn}\sigma_{ij}\sigma_{kl}\sigma_{mn} + \cdots \tag{2.8}$$

in which F_{ij}, F_{ijkl}, F_{ijklmn}, ... are tensor arrays of material properties and each term is a scalar. Tsai and Wu (1971) discuss the admissibility of the surfaces that are represented by this very general form, some 17 different types in all. By requiring that admissible surfaces have only one root for any radial loading path (to eliminate imaginary surfaces, surfaces with two sheets, and hyperbolic paraboloids), they are left with only real ellipsoids and elliptic paraboloids as possible surfaces. These restrictions introduce constraints on the values of the coefficients, such as

$$F_{ii}F_{jj} - F_{ij}^2 \geq 0 \quad \text{(no sum on repeated indices here)} \tag{2.9}$$

Although the material parameter arrays are usually evaluated (and discussed) in the material coordinate system (defined by the fiber directions for continuous-fiber-reinforced materials, as discussed earlier), it should be noted that the *principal directions* of these arrays may not be coincident with the *principal material directions* defined by the fiber directions, as we have defined them. The principal directions and values of the arrays can be found in the usual manner. For the second-order array, for example, the usual eigenvalue and eigenvector problem is solved using the equations

$$F_{ij}n_i - \lambda\delta_{ij}n_i = 0 \tag{2.10}$$

For each of the three roots of these equations, there will be a direction vector, n, of the eigenvector that may or may not coincide with the material directions. Hence, the material strengths defined in this limited context may or may not coincide with the engineering principal material strengths defined in a more general sense that always coincide with the material directions defined by the direction(s) of the fibers in the material.

If we continue to restrict our attention to the 1,2 plane of a fiber-reinforced material and use the usual contracted notation, then (2.8) can be stated as

$$F_1\sigma_1 + F_2\sigma_2 + 2F_{12}\sigma_1\sigma_2 + F_{11}\sigma_1^2 + F_{22}\sigma_2^2 + F_{66}\sigma_6^2 = 1 \tag{2.11}$$

where all terms that were linear in the shear-stress component σ_6 have been dropped, since the criterion cannot depend on the sign of the shear stress. Using tension, compression, and shear tests and one combined stress test, all of the constants in

Equation (2.11) can be determined. Alternatively, if we use the minimum value for F_{12} suggested by Tsai and Hahn (1980), the resulting expression becomes

$$\sigma_1\left(\frac{1}{X_t}-\frac{1}{Xc}\right)+\sigma_2\left(\frac{1}{Y_t}-\frac{1}{Y_c}\right)+\sigma_1^2\frac{1}{X_tX_c}+\sigma_2^2\frac{1}{Y_tY_c}-\sigma_1\sigma_2\frac{1}{\sqrt{X_tX_cY_tY_c}}=1 \quad (2.12)$$

which is an ellipse in the 1,2 plane. To examine the nature of this failure surface, data from Tsai and Hahn (1975) were averaged (for each test condition) and used to determine the coefficients in (2.12) and to compare to various combinations of biaxial applied stress, as shown in Figure 2.12. First, in (2.12), if only one stress is applied until failure (say, a tensile stress in the 1 direction), then the equation produces an identity at the failure stress; that is, the ellipse crosses the axes of the material stress axes at exactly the correct values of these stresses, as also borne out by the plot in Figure 2.12. The rotation of the semimajor and semiminor axes of the ellipse is controlled by the value of F_{12}. For the present data, this value is very small for this case, since the ellipse is nearly horizontal (as shown by the position of the fiducial marks at either end of the semimajor axes). Finally, for the data used, the "fit" of the representation appears to be reasonably good, with the possible exception of the data in the third quadrant for which the stress in the 1 direction (parallel to the fibers) is compressive, and just a little compressive stress is applied perpendicular to the fibers, in the 2 direction. For that case, the criterion appears to overestimate the apparent strength of the material for this example.

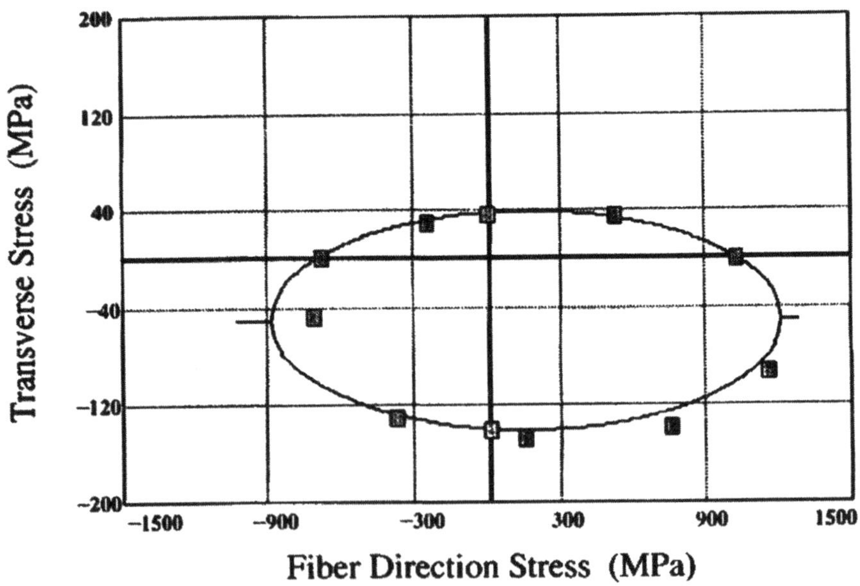

Figure 2.12 Failure envelope for Tsai–Wu criterion fit to data from Tsai and Hahn (1980).

Several features of the criterion shown in (2.12) are significant in the context of our discussion. First, the general form of the equation is a comparison of stress in a given direction to strength in that direction. We will expand on this idea in subsequent sections, but the physical significance of this simple concept is very great. Anisotropic materials may be very much stronger in one direction than in another, so that the stress required to cause failure may be very much smaller in that direction than in another. Hence, criteria based *only* on effective stress, distortional energy, principal stress, and principal strain are inherently *unsuited* for the representation of the strength of anisotropic, inhomogeneous materials. Finding the magnitude and direction of the maximum normal stress, for example, is of little value if a smaller value in another direction is larger than the strength in that direction and causes failure. Since the principal stress directions are not, in general, the same as the principal strain directions, these concepts are not generally useful. The fundamental question for these materials is not what the stresses are, but rather, what the stress components are compared to the strength components in comparable directions.

Hill (1983) has postulated a number of forms with these general characteristics. For orthotropic materials, for example, he suggests the representation

$$\frac{\sigma_1^2}{X_1^2} - \left(\frac{1}{X_1^2} + \frac{1}{X_2^2} - \frac{1}{X_3^2}\right)\sigma_1\sigma_2 + \frac{\sigma_2^2}{X_2^2} + \frac{\sigma_6^2}{X_6^2} = 1 \tag{2.13}$$

in which the X_i terms are the material strengths in the corresponding material directions. Of course, if the tensile strength is different from the compressive strength in a given direction, then we must ask the direction of the stress before choosing the value of the denominators to use. In this sense, the slope of the failure surface represented by (2.13) is discontinuous. This brings us to another general class of criteria.

2.2.3 Criteria for Specific Failure Modes

The material parameters that appear in failure criteria for composites, polynomial criteria, or others can be chosen to correspond to the strength of the material associated with specific failure modes, which may be induced by a combination of stresses over a range of values. This simple idea introduces a lot of flexibility into the representation of the strength of inhomogeneous, anisotropic materials and is the most popular approach in use at this point. The simplest example of such a representation is the maximum-stress criterion, in which the maximum values of tension and compression may represent very different failure modes (and usually do) and the corresponding values of those strengths will be quite different. Figure 2.13 shows such a failure surface for the present example data. While the general features of the representation and the simplicity are appealing, the agreement with this data set is not good in the "corners" of the surface. The slope of this surface, of course, is not continuous.

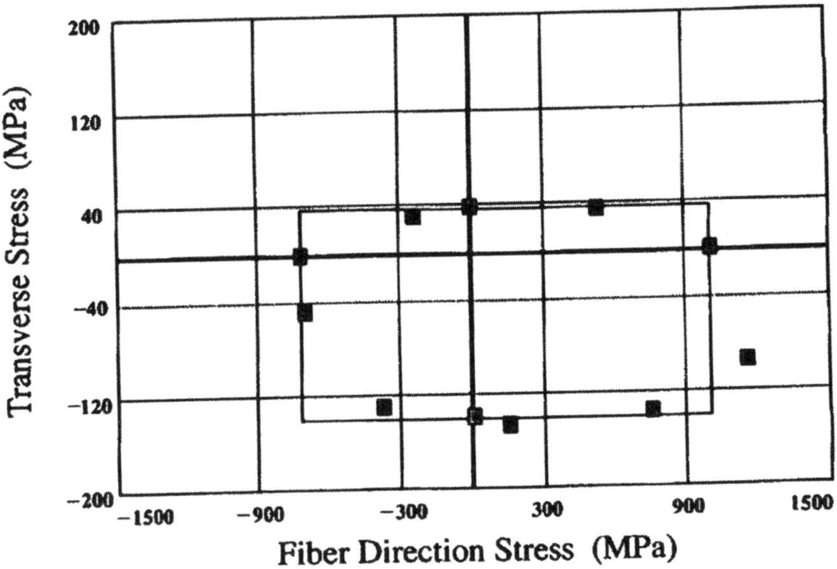

Figure 2.13 Failure envelope for maximum-stress criterion fit to data from Tsai and Hahn (1975).

A modification of (2.13) has been suggested by Tsai and Hahn (1975), on the grounds that fibrous composites should have the same strength when pulled (or pushed) in any direction that is perpendicular to the fiber direction since there is no geometric distinction between any of those directions. Therefore, they reasoned, X_2 and X_3 should be the same in (2.13), and the equation should reduce to the form

$$\frac{\sigma_1^2}{X_1^2} - \frac{1}{X_1^2}\sigma_1\sigma_2 + \frac{\sigma_2^2}{X_2^2} + \frac{\sigma_6^2}{X_6^2} = 1 \tag{2.14}$$

In this form, we can also choose the compression or tension value of X_i to match the sign of the corresponding stress for σ_1 or σ_2. But there will be some question about how to choose the denominator of the cross term in (2.14). Strictly speaking, the cross term is not rigorous, as written, since a transformation of the coordinates, exchanging the 1 direction for the 2 direction, does not leave the criterion unchanged, as it must to ensure frame indifference. By such coordinate transformations we might choose either X_1 or X_2 for the denominator. Moreover, physically, the assumption that $X_2 = X_3$ is not generally true, since the manufacturing process of nearly all "real" composites renders them different in those directions. Nevertheless, the Tsai–Hill criterion, as it is generally called, given in (2.14) is widely used because of simplicity and general utility. Figure 2.14 shows the representation of the data for the present example achieved by (2.14). In this case, the strengths in the denominators, X_{ij}, have been chosen to match the sign of the stress in the numera-

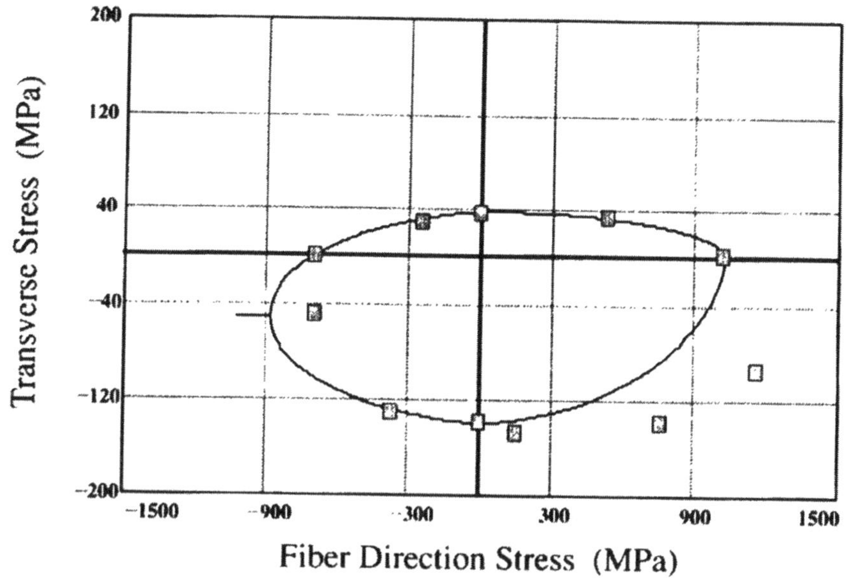

Figure 2.14 Failure envelope for Tsai–Hill criterion fit to data from Tsai and Hahn (1975).

tors, so that the slope of the ellipse is not continuous across the material axes when the sign of the stress changes.

It is also clear from Figure 2.14 that the axes of the ellipse are tilted relative to the material axes, as a result of the cross term in (2.14). In general, the cross term, or *interaction term*, as it is called, is very influential in this representation and is the subject of much discussion in the literature. For example, if the sign of the coefficient of the cross term in (2.14) is reversed, the representation shown in Figure 2.15 is obtained. For the data set shown, the "benefit" of having a biaxial stress state occurs in the fourth quadrant, when the stress in the fiber direction is tensile and the stress perpendicular to the fibers is compressive. The difference between the representation in Figures 2.14 and 2.15 is striking; the importance of the interaction of the stresses in these materials is often quite significant.

A further demonstration of this fact is realized by consideration of the maximum-strain criterion. First, we consider the first quadrant of stress in the material system. Then, the critical strain components, in terms of the compliance tensor components S_{ij} of the material (in contracted notation), can be written as

$$\varepsilon_{1t} = S_{11}X_t \qquad \varepsilon_{2t} = S_{22}Y_t \tag{2.15}$$

along the 1 and 2 axes, respectively. Then we can also write the strain along those two directions (for the case of no shear stress for this example) as

$$\varepsilon_1 = S_{11}\sigma_1 + S_{12}\sigma_2 \qquad \varepsilon_2 = S_{22}\sigma_2 + S_{12}\sigma_1 \tag{2.16}$$

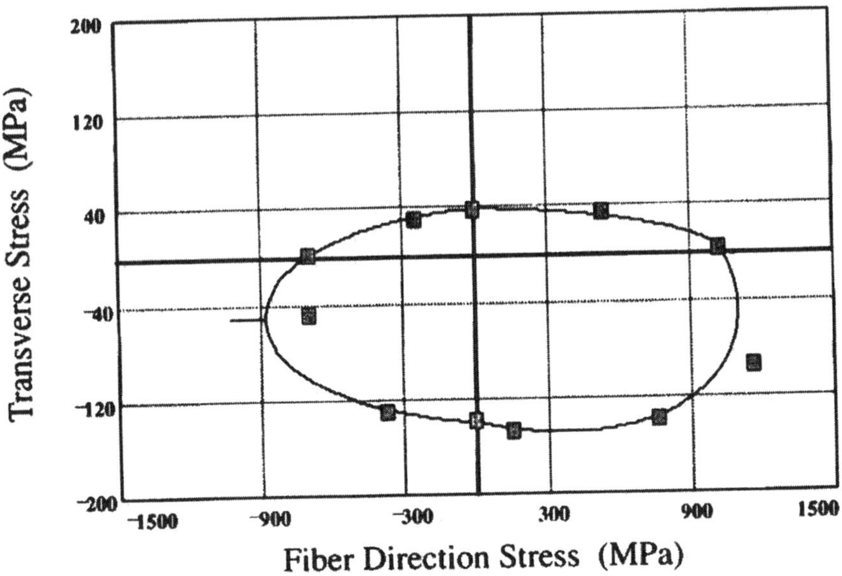

Figure 2.15 Failure envelope for Tsai–Wu criterion fit to data from Tsai and Hahn (1975), with sign of interaction term reversed for demonstration purposes.

The requirement that these values be less than or equal to the limiting values shown in (2.15) results in the expressions

$$\sigma_1 + \frac{S_{12}}{S_{11}}\sigma_2 \leq X_t \qquad \sigma_2 + \frac{S_{12}}{S_{22}}\sigma_1 \leq Y_t \tag{2.17}$$

where S_{12} is essentially a Poisson effect parameter and generally has a negative value. Consequently, these equations suggest that the material is "stronger" than X_t, for example, when both a positive stress in the 1 direction and a positive stress perpendicular to the fibers in the 2 direction are applied. The physical "logic" of this prediction is clear, as suggested by Figure 2.16. The premise of this criterion is that the maximum *strain* controls failure of the material ("regardless" of the level of stress). In the 1 direction, for example, the strain to failure is given by the first of (2.15). If uniaxial stress is applied, that strain occurs for an applied σ_1 of X_t. However, if we apply a tensile stress σ_2 in the direction perpendicular to the fiber direction, then the Poisson effect will result in less strain in the 1 direction for a corresponding level of applied stress in this direction; that is, the strain for a stress level of X_t will not reach the limiting value if σ_2 is positive. Depending on the ratio of S_{12}/S_{11}, the *stress interaction* will give us a greater strength for the combined stress state than for the uniaxial cases.

Figure 2.17 shows an example of the maximum-strain criterion for the example data we have been considering. The *strengthening* effect is very apparent in the first and third quadrants. The data shown seem to follow the increase for the third

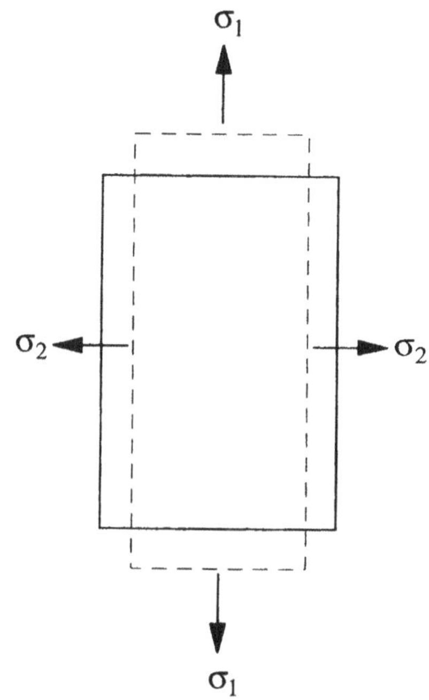

Figure 2.16 Deformation of coupon under biaxial loading.

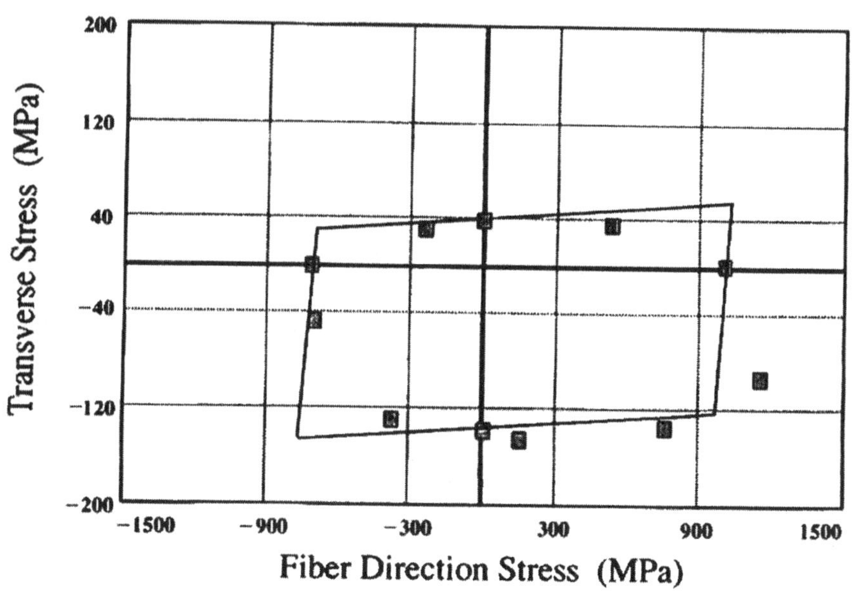

Figure 2.17 Failure envelope for maximum-strain criterion fit to data from Tsai and Hahn (1975).

quadrant but do not seem to agree well in the fourth quadrant. Also, the slopes of the lines that describe the failure surface are small; that is, the "interaction" coefficients are modest, in general, for this criterion. There is much evidence to support the use of the maximum-strain criterion for continuous-fiber-reinforced composites made from brittle constituents. This popular criterion for industrial designers also reflects well the differences between the tensile and compressive behavior of the materials.

The idea of constructing failure criteria that reflect specific failure modes can be greatly extended. Cuntze (1994) has postulated an *action plane strength criterion* that is based only on the stresses that are active in the *fiber parallel action plane* and the corresponding action plane strengths. The criterion is piecewise smooth in the domains of the prevailing fracture, with a discontinuous slope. The distinctive feature of this approach is that it is based on stresses and strengths in the (unknown) action plane of the fracture event, which may not be coincident with the global or material axes. The criterion is primarily concerned with the initiation of matrix cracking and rests on the Coulomb–Mohr concept that the *friction* caused by the normal stress component acting perpendicular to the action plane influences the initiation of fracture, as suggested by Figure 2.17. Cuntze's criteria take the general form

$$\left(\frac{\tau_{nt}}{R_{\perp\perp} - \mu_{\perp\perp}\sigma_n}\right)^2 + \left(\frac{\tau_{nl}}{R_{\perp\parallel} - \mu_{\perp\parallel}\sigma_n}\right)^2 = 1 \tag{2.18}$$

where τ is the shear stress on the active plane, n is the direction vector perpendicular to the plane, l and t are direction vectors in the plane, R are the material strengths in tension across the plane (the $\perp\perp$ component) and in shear (the $\perp\parallel$ component), σ_n is the normal stress across the active plane, and μ is the *frictional coefficient* (Fig. 2.18).

A variety of criteria, such as the one just discussed, are concerned with the initiation of damage, but the final strength of the material is determined by the final damage event. Segmented failure criteria for specific damage modes can be postulated and used for essentially all failure modes (and damage modes). An example of such an application is provided by Socie and Wang (1994). These authors produced fiber direction tension and compression failure (and avoided out-of-plane kinking) by applying combined hoop compression and torsion to tubes. The resulting data were well represented by maximum-stress and maximum-strain theories, as shown in Figure 2.19.

The material was a NEMA/ASTM G-10 grade woven composite laminate, epoxy resin reinforced with E-glass plain woven fabric. The crimp angle of both the fill and warp yarns was less than 10°. Figure 2.19 shows that a careful observation of failure mode can produce consistent interpretations, even with this simple criterion. The tubular and coupon specimens they tested had different failure modes under compressive stress in the warp direction. This was well represented by the maximum-stress criterion. Interestingly, the strain-to-failure plot (Fig. 2.19) does not separate the delamination failure modes well. Socie and Wang were also able to

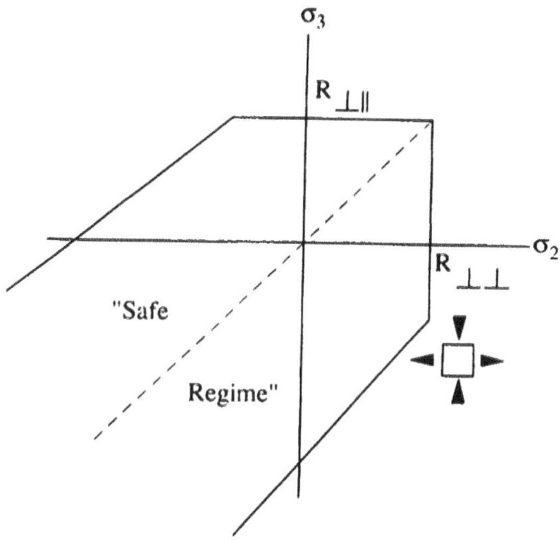

Figure 2.18 Failure envelope segment illustrating Equation (2.18) (Reprinted with permission from Cuntze, 1994).

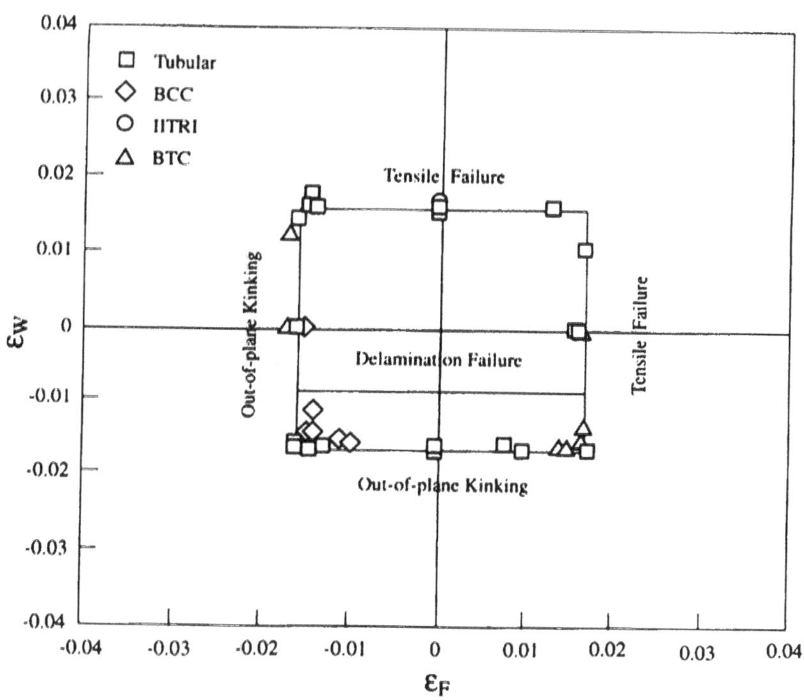

Figure 2.19 Comparison of experimental failure stress data with segmented maximum-strain criterion. (Data are for plain-weave glass–epoxy samples; loading is referred to fill (F) and warp (W) directions.) (Reprinted with permission from Socie and Wang, 1994.)

separate failure modes associated with a combination of matrix cracking, interface debonding, fiber pull-out, fiber buckling, and tensile fiber fracture under in-plane shear loading.

2.3 FAILURE FUNCTIONS FOR DAMAGE ACCUMULATION

The historical failure criteria discussed above are designed to address the strength of materials under monotonically increasing quasi-static loading. The subject of this book is damage accumulation. How can the failure criteria be used to describe this type of behavior? The approach to be followed in the remainder of this book is based on the following interpretations of the idea of a classical failure criterion.

Our descriptions will be based on a failure function, generally written in the form

$$\mathrm{Fa}\left(\frac{\sigma_{ij}}{X_{ij}}\right) \tag{2.19}$$

where the stress components in the numerator of the fraction, σ_{ij}, match the direction and sense of the material strength parameters, X_{ij} in the denominator (the indices ij are not summed). This form is assumed to be a generalization of relationships such as (2.13), in which the stress in a given direction is compared to the strength of the material in that direction. However, we will require the following interpretations of this expression:

1. The stress and strength components used in Fa will be "local," in the sense that they represent the state of stress and state of material that control the *final* material failure event that defines strength, with the definition that is appropriate for a given failure mode.
2. The failure function will be defined for a specific failure mode; that is, when we talk of Fa, we will be discussing the controlling failure mode. Each failure mode will have a different Fa, possibly a different form, and certainly a different value for a given set of applied conditions.
3. The local stress state and material state, represented by σ_{ij} and X_{ij}, respectively, will be assumed to be functions of time in some general sense; that is, the local stress state and material state will be assumed to be functions of the history of the applied conditions and the response of the material.

A summary of these features is shown in Figure 2.20. We will expand on the first requirement in Chapter 3. For inhomogeneous materials, we could decide to apply any criterion at any one of various dimensional levels in the material: fiber–matrix–interface (or interphase), ply, laminate, or structural component. For various situations, one can imagine all of these selections being useful and appropriate. For the present development, we will apply Fa to local conditions, with some latitude in our definition of local. We will attempt to set the problem and description at the smallest

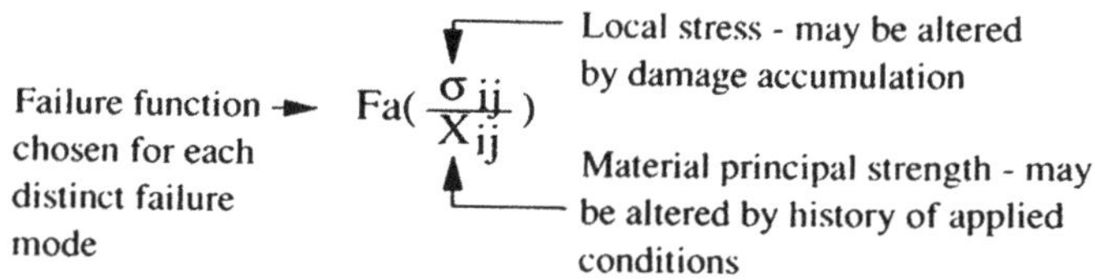

Figure 2.20 Interpretation of failure function for damage accumulation.

dimensional scale we need to capture the mechanism of failure, with the admission that the absence of data in some situations may force us to use a larger scale than we would wish. It is widely known, for example, that predictions of laminate strength made by using the failure criteria we discussed above at the ply level are better than those made by attempting to assess the strength of the total laminate as an anisotropic material. Currently, use of ply-level analysis of laminates is nearly universal in the industrial world. The primary reason for this approach is that damage events such as matrix cracking *precede* the final failure event, and they may substantially change the stresses in the *last ply* to fail. Of course, it is possible that one would want to predict the initial onset of *any* damage (like *first-ply failure*) for design purposes, but most designs concerned with reliability, safety, damage tolerance, and durability are based on the fracture or separation strength of the material or component, the primary focus of our discussion here.

In fact, when the data are available to support physically consistent formulations of micromechanical representations (at least down to the fiber–interphase–matrix level), we will use that definition of local. (See Chapter 4.) Another reason for this focus on the local details is that for many brittle composite materials currently in use, the initiation of damage and final failure is a highly local event, as we read in Chapter 1.

Following Figure 2.20, we will also assume that a given Fa is defined for a given failure mode. In the case of tensile failure controlled by stresses in the fiber direction, for example, we might choose

$$\text{Fa} = \frac{\sigma_{\text{fiber direction}}}{X_t} \tag{2.20}$$

where X_t is the tensile strength of the material in the fiber direction. The normal stress in the fiber direction might be a ply stress, perhaps in the 0° plies. When calculated in that way, if the local stress is updated as damage develops in other plies, the final strength of a laminate can be estimated with good accuracy, as we will show in Chapter 3. This very simple example of a failure function is remarkably useful for "simple" failure modes in fiber-dominated materials. For other situations, one of the criteria discussed in Figures 2.10–2.17 may be more appropriate, or a variety of other criteria that have been advanced in the literature may be better representations of the observed failure mode. The critical issue is that failure modes should be well established and a proper failure function Fa should be chosen.

Figure 2.20 also suggests that the arguments in Fa may be changing with time. The manner in which those changes alter the prediction of (remaining) strength is the subject of Chapter 3. Here, we note that the material parameters and local stresses change with time, indicating that the failure surfaces that we discussed using Figures 2.10–2.17 are also changing with time. We must remind ourselves of the requirement, mentioned earlier, that we have to assume that the "damaged" material, at any point in the loading history, has a single-valued stress–strain relationship if we are to use the philosophy discussed above to estimate instantaneous strength.

Finally, with the requirements stated in Figure 2.20 and discussed above, we introduce another complication into our mechanics problem. If the stress and strength arrays that appear in Fa are local quantities, we are faced with solving a global–local problem each time we wish to use Fa to make a strength assessment. If σ_{ij} and X_{ij} are ply-level quantities, for example, we must use classical laminate analysis, elasticity, or numerical methods to relate the global applied loadings to the local resulting stresses. While this exercise is familiar, establishing a local X_{ij} is not. Obtaining the data needed to determine how internal elements of material are influenced by such external applied conditions as temperature, moisture, or other environments can be very difficult. Since we are ultimately concerned with durability and damage tolerance in the presence of combined effects, this subject will receive considerable attention in later sections of this book.

REFERENCES

ASTM Book of Standards, American Society for Testing and Materials, Philadelphia.

Batdorf, S. B. (1985), "Probabilistic Methods in the Mechanics of Solids and Structures," in *Proceedings of the IUTAM Symposium*, S. Eggwertz and N. C. Lind, Eds., Springer, Berlin, pp. 299–305.

Browning, C. E., Husman, G. E., and Whitney, J. M. (1976), "Moisture Absorption and Desorption of Composite Materials," *J. Composite Mater.*, Vol. 10, p. 2.

Christensen, R. M. (1971), *Theory of Viscoelasticity*, Academic, New York.

Collins, J. A. (1981), *Failure of Materials in Mechanical Design*, Wiley, New York.

Cuntze, R. G. (1994), "Evaluation and Application of a New, Physically Based 2D/3D Inter-Fibre Fracture (IFF) Strength Criterion," in *Proc. I., Symposium on Advanced Materials for Lightweight Structures*, H. Bergman and S. Stavrinitis, Eds., *European Space Agency*, Noordwijk, The Netherlands, March 22–25, 1994.

Curtin, W. A. (1991), *J. Am. Ceramic Soc.*, Vol. 74, p. 2837.

Dillard, D. A. (1990), "Viscoelastic Behavior of Laminated Composite Materials," in *Fatigue of Composite Materials*, K. L. Reifsnider, Ed., Elsevier Science, New York.

Dowling, N. E. (1993), *Mechanical Behavior of Materials*, Prentice-Hall, Englewood Cliffs, NJ.

Faupel, J. H., and Fisher, F. E. (1981), *Engineering Design; a Synthesis of Stress Analysis and Materials Engineering*, Wiley-Interscience, New York.

Gates, T. S. (1991), "Rate Dependent Stress/Strain Behavior of Advanced Polymer Matrix Composites," in *High Temperature Constitutive Modeling—Theory and Application*, MD-

Vol. 26/AMD-Vol. 121, A. D. Freed and K. P. Walker, Eds., American Society of Mechanical Engineers, New York.

Hartwig, G. (1978), "Mechanical and Electrical Low Temperature Properties of High Polymers," in *Nonmetallic Materials and Composites at Low Temperatures*, A. F. Clark, R. P. Reed, and G. Hartwig, Eds., Plenum, New York, 1978, pp. 33–50.

Hill, R. (1983), *The Mathematical Theory of Plasticity*, Oxford University Press, London.

Hill, R. (1948), "A Theory of Yielding and Plastic Flow of Anisotropic Metals," *Proc. Roy. Soc. Series A*, Vol. 193, no. 1033, pp. 281–297.

Juvinall, R. C. (1967), *Stress, Strain, and Strength*, McGraw-Hill, New York.

Kerr, J. R., and Haskins, J. F. (1987), "Time-Temperature-Stress Capabilities of Composite Materials for Advanced Supersonic Technology Application," NASA Contractor Report 178272, NASA Langley Research Center, Hampton, VA.

Lissart, N., and Lamon, J. (1997), "Damage and Failure in Ceramic Matrix Minicomposites: Experimental Study and Model," *Acta Mater.*, Vol. 45, No. 3, pp. 1025–1044.

Marin, J. (1962), *Mechanical Behavior of Engineering Materials*, Prentice-Hall, Englewood Cliffs, NJ.

Miyano, Y., McMurray, M. K., Enyama, J., and Nakada, M. (1994), "Loading Rate and Temperature Dependence on Flexural Fatigue Behavior of a Satin Woven CFRP Laminate," *J. Composite Mater.*, Vol. 28, No. 13, pp. 1250–1260.

Paul, B. (1961), "A Modification of the Coulomb–Mohr Theory of Fracture," *J. Appl. Mechanics Trans. ASME Ser. E*, Vol. 28, No. 2, pp. 259–268.

Rotem, A. (1993), "Load Frequency Effect on the Fatigue Strength of Isotropic Laminates," *Composites Sci. Technol.*, Vol. 46, pp. 129–138.

Sandhu, R. S., Gallo, R. L. and Sendeckyj, G. P. (1982), "Initiation and Accumulation of Damage in Composite Laminates," *Composite Materials: Testing and Design, 6th Conf.*, STP 787, I. M. Danial, Ed., American Society for Testing and Materials, Philadelphia, pp. 163–182.

Sierakowski, R. J., and Chaturvedi, J. K. (1997), *Dynamic Loading and Characterization of Fiber-Reinforced Composites*, Wiley, New York.

Socie, D., and Wang, J. (1994), "Failure Strength and Mechanisms of a Woven Composite Laminate Under Multiaxial In-Plane Loading," in *Durability and Damage Tolerance*, AD-Vol. 43, A. K. Noor and K. L. Reifsnider, Eds., American Society of Mechanical Engineers, New York.

Soni, S. R. (1983), "A Comparative Study of Failure Envelopes in Composite Laminates," *Journal of Reinforced Plastics Composites*, Vol. 2, p. 34.

Suresh, S. (1991), *Fatigure of Materials*, Cambridge University Press, Cambridge, U.K.

Tsui, S. W., and Hahn, H. T. (1980), *Introduction to Composite Materials*, Technomic Publishing.

Wu, E. M. (1974), "Phenomenological Anisotropic Failure Criterion," in *Composite Materials*, Vol. 2, Mechanics of Composite Materials, G. P. Sendeckyj, Ed., Academic Press, New York, pp. 353–431.

Xia, Y., Rao, S. and Yang, B. (1995), "A Statistical Model and Experimental Study of the Strain Rate Dependance of the Strength f Fibres," Composite Science and Technology, Vol. 53, pp. 499–504

EXERCISES

2.1 The following data (collected from the general literature listed in the references) represent a series of biaxial test results for the strength of unidirectional coupons of a continuous-fiber-reinforced composite (values of the stress are in megapascals):

Stress in Fiber Direction	Stress Perpendicular to Fibers
0	42.0
46.5	42.0
125.4	42.0
534.2	32.7
1049.8	0
1128.8	−4.7
1277.4	−58.4
1240.3	−93.4
1161.3	−133.1
966.2	−116.8
789.7	−137.8
599.2	−135.5
520.3	−121.5
432.0	−147.1
367.0	−126.1
269.4	−128.5
144.0	−149.5
0	−142.5
−60.4	−147.1
−153.3	−137.8
−380.9	−116.8
−738.6	−44.4
−701.4	0
−445.9	28.0
−269.4	35.0
−51.1	39.7

(a) Plot these data on a diagram showing the failure surface (in the form of Figures 2.12–2.15) for this material. Discuss the strength of this composite, especially the comparison of strength in tension and compression and in directions parallel to the fibers compared to directions perpendicular to the fibers.

(b) Using the above data, determine the material parameters in the maximum-stress, Tsai–Wu (2.12), Hill (2.13), Tsai–Hill (2.14), and maximum-strain (2.15–2.17) failure criteria discussed in the text. Construct failure envelopes to compare to the data, and compare the relative utility and accuracy of the criteria. Also, compare those plots to the results for the different data sets shown in Figures 2.12–2.15 and 2.17. Based on this limited comparison, what criterion or criteria would you be most likely to use for a situation in which safety was the overriding design requirement? Which criterion or criteria would you most likely use if economy were the overriding design driver? (see Soni, 1983.)

2.2 Using three of the criteria for which material parameters were determined in Exercise 2.1(b), determine the fracture value of the torque applied to a cylinder made from unidirectional material of this type with the fibers aligned with the axis of revolution of the element. The dimensions of the cylinder are shown below. The thickness of the unidirectional lamina from which it is constructed is 4 mm.

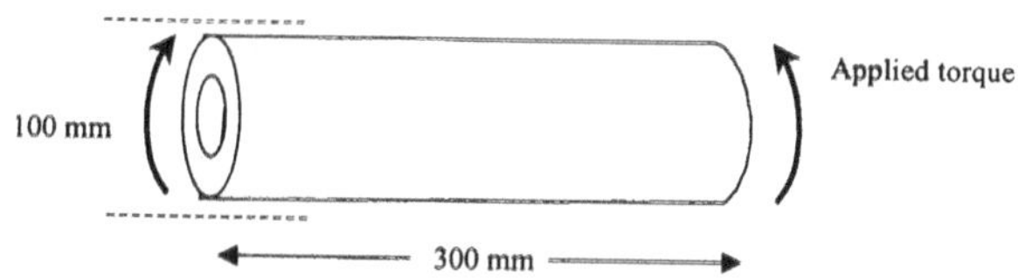

Discuss the variations in the predicted design failure torque associated with the different criteria used. Are they significant? If you had to design this cylinder using these results, what value(s) for the failure torque would you specify, and why would you make that selection?

2.3 Derive one of the companion expressions to (2.5). For example, begin by assuming that a tensile stress is applied in the 2 direction. Or begin with any other uniaxial or shear loading that might be used in an attempt to measure one of the engineering principal material strength values, as discussed in the text.

2.4 Construct the failure envelope illustrated in Equations 2.15–2.17 for several cases of non-zero-applied shear stress. What effect does the applied shear stress have on the failure envelope?

CHAPTER 3

STRENGTH EVOLUTION

Thesis: Using remaining strength as a damage metric, the evolution of global strength during long-term exposure to applied conditions can be represented in terms of changes in the local stresses and principal material strengths in an appropriate failure function (applied at the local level) to predict remaining strength and life as a function of material characteristics and the history of exposure.

3.1 NATURE OF THE PROBLEM

In Chapter 1 we observed that there are many processes that can cause changes in the properties and performance of continuous-fiber-reinforced composites under applied mechanical, thermal, and environmental conditions. In Chapter 2 we examined some typical approaches to the representation of the strength of such a composite material under quasi-static loading conditions, that is, under a single application of the applied conditions to cause fracture over a short period of time (sometimes called the fast fracture of the material, under mechanical loading, for example). In this chapter, we will combine those two concepts to establish a fundamental approach to the description, representation, and prediction of the changes in the initial (quasi-static) strength as a function of damage accumulation due to a variety of degradation processes.

3.1.1 Basic Behavior

We begin by examining the basic behavior we wish to model. Some of the physical features of the problem are represented in Figure 3.1. The data in that figure were

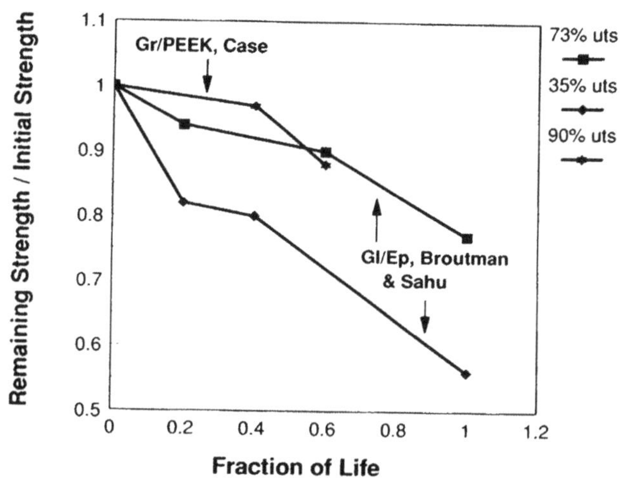

Figure 3.1 Strength reduction during fatigue loading of composite laminates (glass–epoxy data from Broutman and Sahu, 1972).

recorded during the room-temperature cyclic loading of laminates made from two types of material. The glass-reinforced epoxy laminates of Broutman and Sahu (1972) were cross-ply laminates with seven plies having a 0° orientation and six plies having a 90° orientation. The carbon fiber-reinforced PEEK material was a quasi-isotropic laminate tested in our laboratory. Several features of the data are typical. First, there is a small early reduction in remaining strength, followed by a period of small reduction, followed by a sharp drop in strength as the end of life is approached. As we will show in a later chapter, this behavior is sometimes exaggerated, in the sense that there is very little change in strength until quite near the end of life, often called *sudden-death behavior*.

A second feature is the magnitude of the changes for fatigue loading at various fractions of the ultimate strength of the laminates. An important point associated with these magnitudes is obvious but often overlooked; *the largest changes in strength occur for the lowest loading levels, since the strength must be reduced to a lower level before failure occurs (defining life) at that low stress level. Hence, the greatest damage occurs for the lowest stress level, in that sense*. For many engineering purposes, this situation is the condition of greatest interest, since most designs are conservative and involve low stress levels. If we wish to generate a model of this behavior, we must deal with the long-life *large* strength reduction case (with low applied stress levels) as well as the short-life *small* strength reduction situation more typical of laboratory tests. From Figure 3.1, one can see that the reduction of strength for an applied stress amplitude of 35% of static ultimate strength is larger than that of the higher stress levels, even in the early part of the life.

Figure 3.2 shows several other basic features of the problem that occur when the material is notched (a center hole, in this case). The laminates represented by the

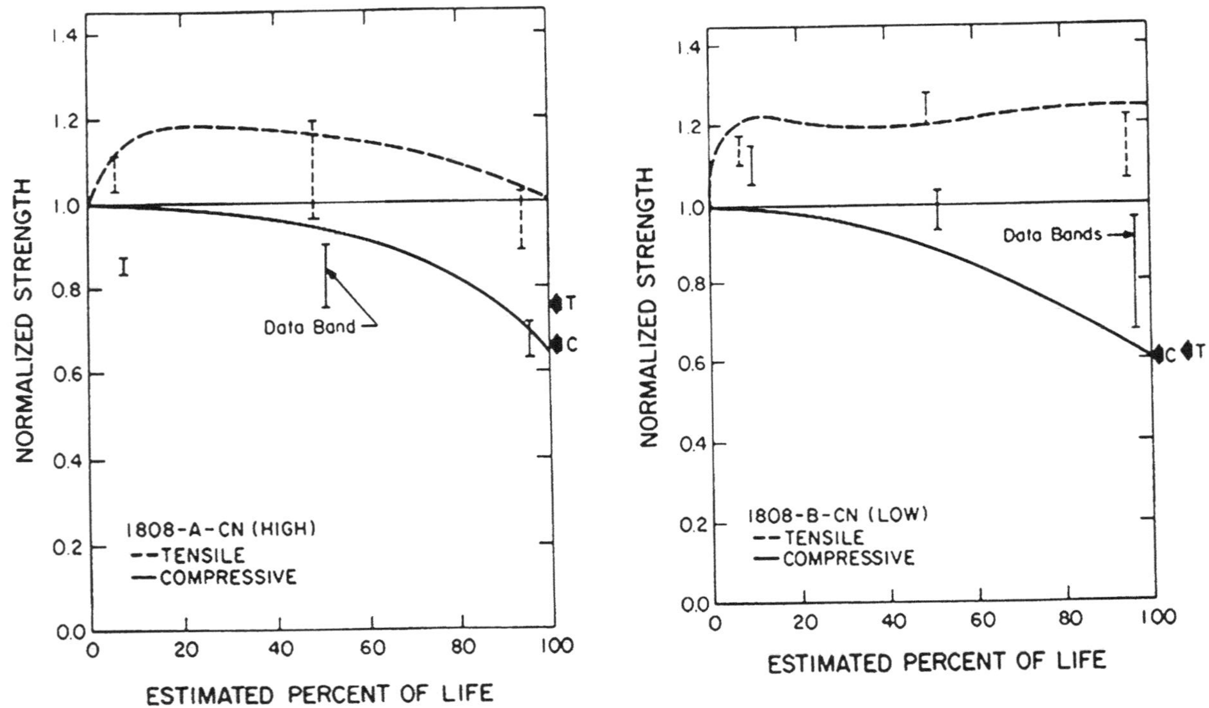

Figure 3.2 Residual strength in tension and compression during fatigue loading of (*a*) $(0, 90, +45, -45)_S$ carbon–epoxy laminates and (*b*) $(0, +45, -45, 0)_S$ laminates as function of fraction of life.

data were quasi-isotropic, with carbon-reinforced epoxy material constituents. In this case, the comparison between high and low load levels is not as simple as for the unnotched material discussed in Figure 3.1. (The curves in those figures are predicted results that will be discussed in a later section.) Figure 3.2 shows data for fully reversed cyclic loading at a high stress level (short life). A striking feature is the increase in remaining strength in tension during the early part of life, when the development of damage at the local level near the notch relaxes the stress concentration and "strengthens" the specimen. The material is not getting stronger, but the coupon (which is an engineering component in the strict sense) is stronger because the notch produces a smaller stress concentration as the damage develops. Of course, as damage continues to develop, the degradation in the remaining ligaments of material reduces the global strength. However, the laminate fails in compression, as Figure 3.2 shows; the tensile strength is never reduced very much below the initial value of the laminate. The compressive behavior is more normal, in the sense that it is monotonically decreasing, and shows basic features like those seen in Figure 3.1 (in which the graphite/PEEK data were shown for compressive loading). There is an early decrease followed by a quiescent period of small reduction and a sharp reduction near the end of life.

For a low-load level, the data are different. Figure 3.2 shows that the tensile strength is never reduced below the initial strength of the laminate. The laminate fails in compression, but there is an early rise in strength, rather than a drop, as seen for the high-stress level. In a later chapter, we will see that all of these details are quite typical of various laminates and can be traced to the basic nature of the damage development details at the local level. For the purpose of our discussion here and as a general conceptual statement of the problem, we will use the behavior summarized in Figure 3.3 as a representation of the physical situation of interest.

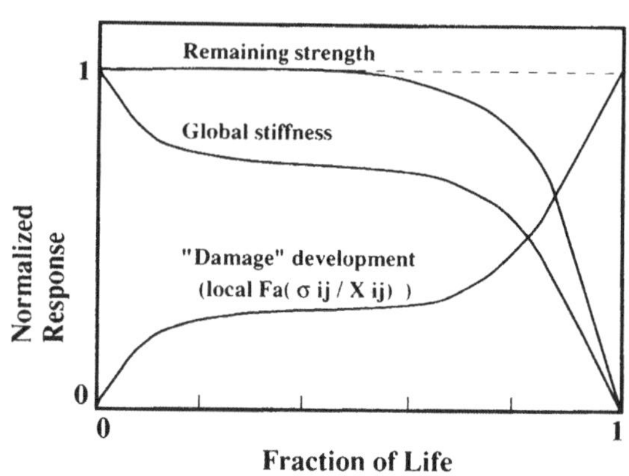

Figure 3.3 Typical variations of remaining strength, global stiffness, and amount of damage as function of fraction of life for composite laminates.

3.1.2 Phenomenological Features

Although it is deceptively simple, Figure 3.3 summarizes many of the basic phenomenological features of the physical problems we wish to address in this book and the approach we will take to modeling them. Our primary concern is the determination of remaining strength (which we will call Fr throughout the text). This strength changes as a function of the history of applied conditions, that is, with the experience of the specimen or component during its life. From an engineering standpoint, we are also interested in the change in stiffness since, in many cases, it is an indicator of the manner in which various degradation processes are reducing strength. Other material parameters may also be of interest (e.g., changes in conductivity, changes in dissipation). Stiffness may decrease or increase, depending on the type of internal process that alters it, a point we will develop more fully in a later chapter. And most importantly, the development of damage, which we will define as any process that alters strength, will be followed by attempting to determine how the failure function Fa *defined at the local level* (by the controlling failure mode) changes. Perhaps the most difficult part of this problem is to find the relationship between the time history of local failure function and the global value of remaining strength. This chapter will provide the basic framework for relating all of the variables and parameters represented in Figure 3.3. We will use the physics, chemistry, and mechanics of the degradation processes to combine their effects to predict remaining strength and life at the global level.

3.2 PROGRESSIVE FAILURE

As the data in Chapter 1 and Figures 3.1 and 3.2 show, strength is degraded by a progressive process during the life of the component. In homogeneous materials, this progressive process is often the self-similar growth of a single crack, which eventually reaches a critical length and becomes unstable, causing fracture. However, in composite materials, this “simple” mechanism is somewhat rare. More often, combinations of the damage modes discussed earlier *accumulate* and interact to cause changes in the material state and local stress states in the material. These changes bring about failure defined by a limiting value of Fa, the appropriate local failure criterion for the controlling failure mode. Perhaps the most common progressive damage mode is microcracking. We have seen the physical evidence for this in Chapter 1. Now we will examine this progressive damage mode, *first under quasi-static loading*, to develop an understanding of how to use our failure function concepts at the local level and why this is a good approach to the problem.

Figure 3.4 shows a multiaxial laminate that includes plies having orientations of 0° and 90° under uniaxial (quasi-static) mechanical loading. As the level of loading increases, microcracks typically form in the matrix material (for brittle matrix materials like many polymer and ceramic composites). (For cyclic loading, these matrix cracks also form in ductile matrix materials, including metal matrix composites.) We will show, below, that it is possible to use the failure criteria discussed above, *applied*

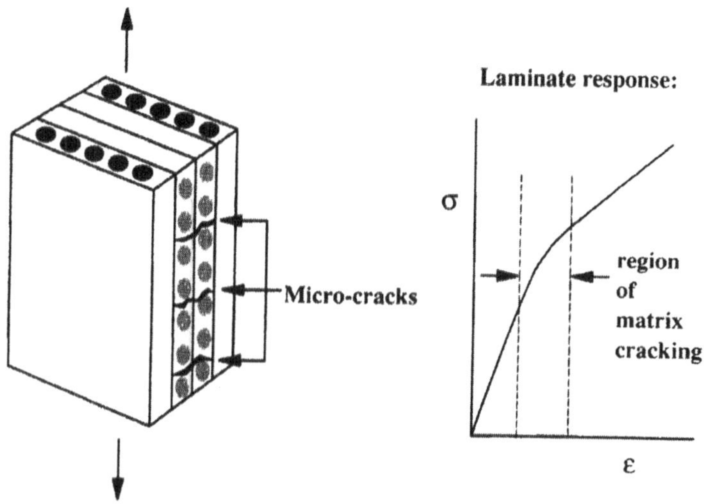

Figure 3.4 Matrix cracking in cross-ply laminate (left) and resulting change in global stiffness (right) for quasi-static, uniaxial loading.

at the ply level, to estimate the level of applied loading that will initiate matrix cracking. In a later chapter, we will show that it is also possible to calculate the amount of reduction of stiffness in the plies that break as a function of applied loading and, in that manner, to calculate the correct (observed) change in the global stiffness of the specimen or component. At this point, we will only be concerned with understanding the internal (local) process that accompanies this progressive cracking.

The first observation is that all of the cracks do not form at exactly the same stress level. The stiffness of the plies that crack is not reduced instantaneously to some constant lower value when the level of applied stress causes cracks to begin to form. Hence, as shown in Figure 3.4, the global stiffness will be reduced over some finite range of applied loading. At the local level, this progression of cracking is driven by spatial variations in matrix strength and numerous variations in local geometry due to manufacturing, as emphasized in Figure 1.3. For quasi-static loading, the range of applied stress over which the crack density varies may be quite large, as shown in Figure 1.26. Furthermore, the density of cracks reaches a constant saturation level as a function of loading, as we will discuss in a later chapter on the subject of stiffness change. For long-term cyclic loading, in general, the cracks form the saturation state (characteristic damage state, or CDS, as it is often called) at all stress amplitudes above the level required for the initiation of matrix cracking (Reifsnider and Highsmith, 1981).

But how can we use this local information to calculate the strength of a multiaxial laminate under quasi-static loading? The key to answering this question is illustrated in Figure 3.5. This figure shows a schematic of the stress in the composite and in the 0° and 90° plies of the composite as a function of global strain, that is, the stress–

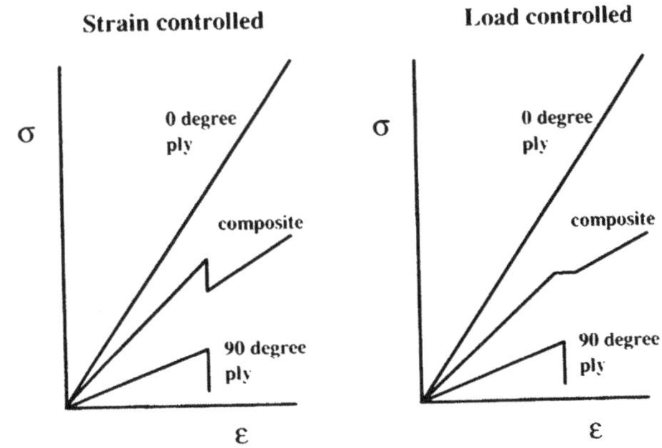

Figure 3.5 Stress–strain relationships for (*a*) strain controlled loading and (*b*) load-controlled loading of cross-ply laminates.

strain curve of the composite, the 0° plies (that do not have cracks), and the 90° plies in which cracks form. We will assume that the strain in all of the plies is the same in the direction of uniaxial loading (at this point), a common assumption. In this case, the stress in the two plies is proportional to their relative stiffness, in the same sense as the rule of mixtures discussed earlier. Figure 3.5 shows the local situation if the composite is subjected to a controlled strain. If matrix cracking occurs as a single event, that is, if all the cracks were to form at once, there would be a drop in the load on the composite (for that fixed strain), and the subsequent slope of the composite stress–strain curve would be reduced beyond that point. If the load (or engineering stress) is controlled at the global level and all cracking occurs at one level, then there would be an increment of added strain in the composite when the cracks form, as shown in Figure 3.5. In a complex engineering component in which there is nonuniform stress and large structural constraints, either of these extreme cases may be found. What is usually observed is the smooth transition shown in Figure 3.4, which is a combination of the controlled strain and controlled stress cases that allows the microcracks to form as a progressive process.

3.2.1 Critical Elements

For our question of strength, however, the most important point is that for a given level of applied stress, the amount of the total *load* carried by the 0° plies increases as the stiffness of the off-axis plies is reduced by microcracking. Figure 3.5 shows that the stress in the fibers rises faster compared to the rise in stress in the composite after the onset of microcracking. This well-known phenomenon introduces three important concepts that we will use throughout the rest of the book:

1. During the *failure process* that controls the strength (and remaining strength) of a laminate or a composite system, *subcritical elements* of the material exist

at the local level. Their failure only serves to redistribute stress at that level. We can also identify *critical elements* of the composite whose failure causes and defines the fracture of the composite at the global level. Obviously, failure of a critical element occurs last in the sequence of events that defines the failure process.

2. Stress redistribution in subcritical elements (and other events, as we will see) can greatly influence the environment (e.g., level of stress) in which the critical elements continue to sustain the loading.
3. Failure function concepts (or criteria) can be applied at the local level, where the physical events that define the failure process occur, to estimate the consequences of the events and to calculate the strength of the resulting material or component.

To illustrate these concepts, at least in a very simple form, we will consider their application to the problem of the estimation of the strength of laminates under quasi-static loading, including the effect of matrix crack formation in the off-axis plies. Then we will broaden the scope of our discussion to include other modes of damage and failure. Finally we will introduce explicit time (or cycles) into our discussion and learn how to extend our approach to the question of *remaining strength* after long-term exposure to mechanical, thermal, and environmental applied conditions.

3.2.2 Progressive Failure of a Laminate under Quasi-Static Loading

For the purpose of illustrating the concepts listed above, we consider a laminate having only two ply orientations, a $[0,\ 50,\ -50]_S$ laminate with ply thicknesses of 0.125 mm. The lamina properties of the material used in the laminate are given in Table 3.1, where E_i is Young's modulus G_{12} is the shear modulus, and P_{12} is the Poisson's ratio in the material coordinate system and X_i, ε_{it}, and α_{ij} are the corresponding values of material principal strengths, strains to failure, and thermal expansion coefficients, respectively. Subscripts t and c refer to tension or compression loading. We will not include moisture-driven strains in this example, but they would be entered in exactly the same way as the thermal terms. We will further assume that the laminate is cured at a stress-free temperature of 190°C and cooled to room temperature (25°C) for use. Following our discussion in Chapter 2, we first determine the thermal resultants and stresses from the equations

$$[N^T] = \int \bar{Q}\alpha\, \Delta T\, dz \qquad [\sigma]_k = [\bar{Q}]_k[\varepsilon - \alpha\, \Delta T]_k \tag{3.1}$$

TABLE 3.1 Lamina Properties for Example Analysis

$E_1 = 39$ GPa	$E_2 = 4.8$ GPa	$\nu_{12} = 0.3$	$G_{12} = 6.0$ GPa	
$X_t = 1130$ MPa	$X_c = 620$ MPa	$Y_t = 90$ MPa	$Y_c = 600$ MPa	$S = 83$ MPa
$\varepsilon_{1t} = 28{,}000$	$\varepsilon_{1c} = 15{,}897$	$\varepsilon_{2t} = 18{,}750$	$\varepsilon_{2c} = 125{,}000$	$\varepsilon_{12} = 3{,}833$
$\alpha_{11} = -1.8 \times 10^{-6}/°C$	$\alpha_{22} = 27 \times 10^{-6}/°C$			

TABLE 3.2 Thermal Stresses at Room Temperature after Cool-Down

Stress (MPa/ply)	0°	50°	−50°
σ_1	−10	−22.8	−22.8
σ_2	17.8	18.9	18.9
σ_6	0	−0.34	0.34

where the subscript k indicates a specific ply. For the properties given above, the thermal stresses in the plies due to cooling the laminate down to room temperature, only, are found to be those listed in Table 3.2.

We notice that there is a rather large tensile stress perpendicular to the fibers in the off-axis (50°) plies. The unconstrained contraction of the plies in that direction during cooling would be much greater than is allowed by the constraint of the 0° plies. We then determine what applied resultant N_x in the fiber direction of the laminate will cause the off-axis plies to fail (by matrix cracking in this case). Using a maximum-stress criterion, that value is determined to be $N_x = 174{,}000\,\mathrm{N/m}$. The stresses in the plies for that loading turn out to have the values shown in Table 3.3.

It is interesting to note that the off-axis plies did not fail in tension; for this laminate and loading, they failed due to the fact that the shear stress reached the allowable value of 83 MPa. This onset of matrix cracking is typically called first-ply failure and is sometimes used as a design limit. As a result of this failure, the stresses carried by the 50° plies in shear and tension perpendicular to the fibers must be reduced (since we know from experiments that the cracks run along the fiber directions in this case), but we do not know, a priori, to what level the stress is reduced. The specific reduction can be estimated, and we will address this calculation in a later chapter. For now, we make an engineering assumption that the stiffness in the directions controlled by the matrix (E_2 and G_{12}) is reduced to unit value. That reduction changes the global stiffness of the laminate to

$$E_x = 13.1 \quad (21.9) \quad E_y = 7.07 \quad (14.8) \quad \nu_{xy} = 0.683 \quad (0.27) \quad G_{xy} = 6.75 \quad (8.81)$$

TABLEe 3.3 Ply Stresses for Axial Loading of $N_x = 174{,}000\,\mathrm{N/m}$

Stress (MPa/ply)	0°	50°	−50°
σ_1	403	89.9	89.9
σ_2	19.1	47.1	47.1
σ_6	0	−83	83

in gigapascals (with the previous undamaged laminate values in parentheses). As discussed in Figure 3.5, the composite stiffness in the loading direction decreases. But the other stiffness values also change significantly, and Poisson's ratio actually increases, reflecting a relaxation of constraint on contraction in the direction perpendicular to the loading direction.

Another change influences the results noted above. The thermal coefficients of expansion/contraction are altered by crack formation. We have assumed that α_2 (perpendicular to the fibers) in the 50° plies is reduced to zero by the matrix cracking. Hence, the matrix contraction upon laminate cooling that would normally occur in that ply is relaxed as well. This changes the thermal resultants associated with cooling, alone, as shown below:

$$N_x^T = 4.49 \quad (-3.84) \qquad N_y^T = 0.748 \quad (-8.14) \qquad N_{xy} = 0 \quad (0)$$

in units of newtons per millimeter. The laminate stiffness changes and changes in the stresses in the plies of the laminates shown above are not simple linear variations associated with a reduction of area, as we might have expected from our rule-of-mixtures analogy. The manner in which the local damage changes the material *constituent* properties (at the ply level for this example) greatly influences the resultant local stresses and the resulting global properties. In our case, the changes in ply-level stiffness and thermal expansion coefficients are combined by laminate analysis to produce changes in local stresses and global properties that are not just *reduction-in-area effects*. This essential lesson motivates many features of the present approach to the more general problem, described later in this chapter. Also, the first-ply failure stress level is influenced by the thermal stress components. The applied resultant for first-ply failure for this example would have been 182 N/mm instead of 174 N/mm if the thermal stresses were not included in the calculation.

Continuing our example, if the 50° plies are broken, then the applied loading can be increased to the point where the stress in the 0° plies is equal to the tensile strength of these plies. Since these plies are the last plies to fail in tensile loading, they define the strength of the laminate, which is $N_x = 282$ N/mm for this example. This value is determined from the laminate analysis with the reduced ply properties noted above by determining the applied resultant that causes the fiber direction stress in the 0° plies to reach the limit given in Table 3.1. We can check this result at the engineering level. Since the 50° plies are broken, we may assume that the applied load is held by the 0° plies alone (which is only true if the stiffness of the 50° plies is actually zero). If this is true, the stress in the 0° plies should be about three times the applied level. Then

$$\sigma_1^0 = \frac{282 \times 10^3}{0.00075} \times 3 = 1128 \tag{3.2}$$

in megapascals (where we have divided by the ply thickness to correct the units), which compares closely with the 1130-MPa value, the axial strength limit of the 0° plies given in Table 3.1. If we had not reduced the stiffness in the 50° plies to allow for internal stress redistribution into the 0° plies as matrix cracking occurs, we would have estimated the strength to be about 488 N/mm, a much greater number than the correct value of 282 N/mm that we estimated above.

For a laminate such as $[0, +45, -45, 90]_S$, there is the possibility that the progressive failure may involve cracking of the 90° plies and subsequent cracking of the 45° plies before the final failure load of the laminate is reached. However, the problem must be solved sequentially, by reducing ply properties, increasing the applied load, determining the stresses on each ply for each load level, and using these stresses in a failure criterion (or failure function, as we will call it) to determine the sequence of damage development. For the quasi-isotropic laminate mentioned here, with thermal stresses due to processing of the laminate, only the 90° plies crack before the 0° plies become overloaded. The estimated strength of the laminate, based on independently measured ply properties and the method discussed above, is about 509 MPa compared to observed values that range between 496 and 530 MPa. If the stress redistribution is ignored, the predicted value is about 572 MPa, an error of about 12%. In this case, the amount is relatively small since the stiffness of the 90° plies is not very large.

This method of reducing local properties as damage accumulates in a laminate under quasi-static loading is long-standing and widely used in the applications community. It is sometimes called discount theory because the ply properties are discounted as damage progressively develops. As such, it represents the most rudimentary progressive damage method and serves as a well-established starting point for the development of a more general approach based on these same general concepts. The discount concept provides a familiar basis for the three essential concepts (listed in Section 3.2.1) that we will use in broadening this analysis to other damage and failure modes and to the development of progressive damage under long-term conditions. A recap of these concepts follows:

1. For damage accumulation during progressive damage development, careful experimental investigations generally allow the identification of a critical element for a given failure mode. The failure of the critical element defines the global strength of the laminate or material component. (It was the 0° ply in the examples above; other failure modes may define other critical elements.) (See Reifsnider 1991a, 1992; Reifsnider and Stinchcomb, 1986.)
2. Other *subcritical* local material elements (at the ply level or constituent level) may fail during the damage development process. Their failure does not cause global failure, but they do, in general, cause internal stress redistribution that can significantly change the stress state of the critical elements. Consequently, the remaining global properties are affected as well as the stress-related degradation rates throughout the material.
3. This redistribution process can be followed using mechanics analysis, and the remaining strength of the critical element (and of the global component that it controls) can be determined by a correct assessment of a failure criterion or failure function *applied to the critical element at the local level.*

4. Since failure of the critical element defines failure at the global level, we can use the normalized failure function Fa at both levels.

3.2.3 Changes in Local Failure Function

We have adopted the basic premise that the global strength of a fibrous composite system can be described in terms of a local failure function in a critical element that controls the global failure event for a given failure mode. We have also addressed the question of how to determine the manner in which this local failure function changes due to internal stress redistribution. In the next section we will extend this question to include the question of how we can relate those changes to global remaining strength over long-term applications of applied conditions. We remind ourselves that the failure function has the general form

$$\mathrm{Fa} = \mathrm{Fa}\left(\frac{\sigma_{ij}}{X_{ij}}\right) \tag{3.3}$$

where Fa is the local failure criterion that describes the controlling failure mode, σ_{ij} are the local stresses in the critical element, and X_{ij} are the principal material strengths, as described earlier. We have examined how the fiber direction stresses σ_{11} in the 0° plies of a laminate changed with matrix cracking in the off-axis plies. But there may be many damage modes that alter the stress state and alter it in a more general way.

Figure 3.6 shows some of the most common damage modes that are known to cause changes in the local stresses. We have discussed matrix cracking in the context of uniaxial loading of a laminate in which microcracks develop in the off-axis plies. These types of cracks are indicated by A in Figure 3.6. Although the cracks shown

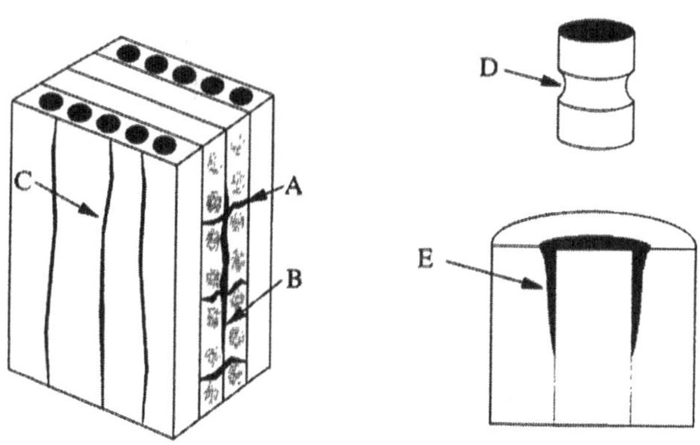

Figure 3.6 Transverse matrix cracking (*a*), delamination (*b*), longitudinal matrix cracking (*c*), fiber degradation (*d*), and debonding of fibers (*e*) in fibrous composite.

are in the 90° plies, they occur in plies having various angles to the 0° direction, as discussed in our example, above. Such cracks may also occur in the 0° plies, usually because off-axis plies constrain the Poisson contraction of the 0° plies during loading or during cool-down in the manufacturing process. Such cracks are indicated by C in Figure 3.6. All of these cracking modes redistribute stress at the local level, since the stiffness of the plies in the direction perpendicular to the crack planes and the shear stiffness in the cracked plies are significantly reduced. Hence, these changes must be calculated to obtain the correct stresses in the remaining plies as inputs to our local failure function in those plies (and, ultimately, in the critical element).

Also shown in Figure 3.6 is delamination, indicated as B and discussed earlier. The stiffness change and internal stress redistribution caused by delamination can be very large, illustrated by the following simple example. Consider a quasi-isotropic laminate with a stacking sequence of $[0, +45, -45, 90]_S$ made from carbon fiber-reinforced epoxy. The lamina stiffness values for this material are given in Table 3.4.

Using the lamina properties, we can use classical laminate analysis (CLT) to calculate the stiffness of laminates having various lamina stacking sequences. Table 3.4 also shows some of those calculations for sublaminates of the quasi-isotropic laminate, that is, for sublaminates that could be formed by delaminating various parts of the $[0, \pm 45, 90]_S$ laminate. In Chapter 5, we will study methods of estimating the stiffness changes in laminates that undergo partial delamination (along with various other stiffness change mechanisms). For our example, we simply consider the difference in stiffness of the quasi-isotropic laminate and the stiffness we would measure if we were to clamp and load various combinations of sublaminates that were *completely* delaminated. For example, suppose that our quasi-isotropic laminate completely separates into two $[0, \pm 45]$ laminates and a $[90]_S$ laminate. If we grip the ends of these three laminates and apply a uniaxial load, what would their collective stiffness be?

Since the sublaminates are uniaxially loaded, we will use a simple rule of mixtures to estimate that collective stiffness. We will assume that

$$E_x([0, \pm 45//90]_s) = 0.75(E_x([0, \pm 45])) + 0.25(E_x([90])) \tag{3.4}$$

since 75% of the bundle formed by the two $[0, \pm 45]$ laminates and the $[90]_S$ laminate has the stiffness of the first of the sublaminates and 25% has the stiffness of the second sublaminate, as shown. This simple calculation gives a fully delami-

TABLE 3.4 Laminate Stiffness Values Used for Delamination Example

Laminate Property (MPa)	[0] (lamina)	$N[\pm 45]_S$	$[0, \pm 45]_S$	$[0, \pm 45, 90]_S$
E_x	143.3	14.47	58.56	54.43
E_y	10	14.47	24.8	54.43
G_{xy}	4.13	37	26	20.6
ν_{xy}	0.31	0.813	0.76	0.322

nated stiffness of 46.44 MPa, a significant reduction from the 54.43 MPa stiffness of the undelaminated laminate. Hence, the stiffness change for this delamination is nearly 15%. If the quasi-isotropic laminate delaminates into three sublaminates resulting in two [0] laminates, two [±45] laminates, and one $[90]_S$ laminate, the stiffness change is about 16%. In each case (as we will show in Chapter 5), these stiffness changes are matched by increases in the fraction of the global load carried by the critical elements, that is, by the 0° fibers in the example discussed above. Consequently, they result in changes in the stresses that appear in the numerator of the failure function that we will use to calculate remaining strength, in exactly the same manner as the changes in the example calculation of strength we discussed earlier for matrix cracking.

Many other damage modes can cause these kinds of stiffness changes and increases in the local stress in the critical elements. If chemical degradation reduces the local diameter of the fibers, caused by oxidation of exposed fibers in high-temperature air, for example, then the stiffness of the fibers and composite can be affected, and the local stresses can be greatly altered (D in Fig. 3.6). If the fibers debond from the matrix material or local inelastic behavior occurs between the fiber and the matrix (a typical situation in the neighborhood of a fiber fracture, for example, as sketched by E in Fig. 3.6), then stiffness changes also occur, and the local stress is redistributed. In fact, a myriad of local damage events can cause local stress redistribution that will change the stress in the failure function in the critical element. We will examine the major types of changes in Chapter 5. The lesson we must learn, at this point, is that the local stresses in the critical element, and therefore in the failure function that we will use to estimate current composite global strength, will change as a function of time (or cycles of loading) when progressive damage development and accumulation occur.

But it is also essential to note that the value of the local failure function may also change because of changes in the principal-strength values in the denominator of this function. As we noted, the X_{ij} strength values are engineering definitions, but we will use them at the local level. The intrinsic strengths of the material also may change with time. This is a typical situation for materials serving in high temperature or in chemically aggressive environments, for example. Chapter 6 will be devoted to this topic. For now, we note that either a local stress increase or an intrinsic material strength decrease can drive the failure function to larger values, and these increases (or decreases) are typically functions of the applied conditions and the history of these conditions.

3.3 FAILURE MODES

As noted in Chapter 1, although the number of damage modes in composite materials may be large, the number of distinct failure modes is typically small. For our present discussions, failure will be defined by separation of the material into two or more parts. In later chapters, we will see that we can use the models and representations that we develop for a broader definition of failure modes, but for our present development of the equations, we will use separation as failure. It is also possible to talk about matrix-controlled failure. However, for the present, we will concern ourselves with composites and situations in which the fibers control failure. This is not a strict limitation, but it will simplify the presentation of the logic that drives the models and methodology that will be developed. With these notations, we address the principal subject of this book: How can we calculate and predict the remaining strength and life of fibrous composite materials in the presence of damage accumulation and interaction, as we have described those topics?

3.4 REMAINING STRENGTH UNDER LONG-TERM LOADING

In Chapter 1 and in earlier sections of this chapter, we have learned that fibrous composite materials fail by the (statistical) accumulation of damage in a progressive process that reduces stiffness and strength and defines the life of the material or component. Now we address the problem of how to calculate the remaining strength of such materials at various points in this process.

Figure 3.7 illustrates the damage accumulation process to be addressed and introduces time (or cycles) as an explicit variable in our problem. In general, in the early stages of life, damage initiation dominates the damage development process. For fiber-dominated composites, fiber fracture and matrix cracking are common results of the process. Then, again in general terms, damage accumulates and begins to interact over a portion of the life that is often a major fraction of the total. In the final stages, damage interaction creates locally intense concentrations that ultimately cause failure. It should be emphasized that, as we have shown in earlier physical examples, this process is generally widely distributed and repeated throughout the composite. We will assume, on the average, that all elements of the

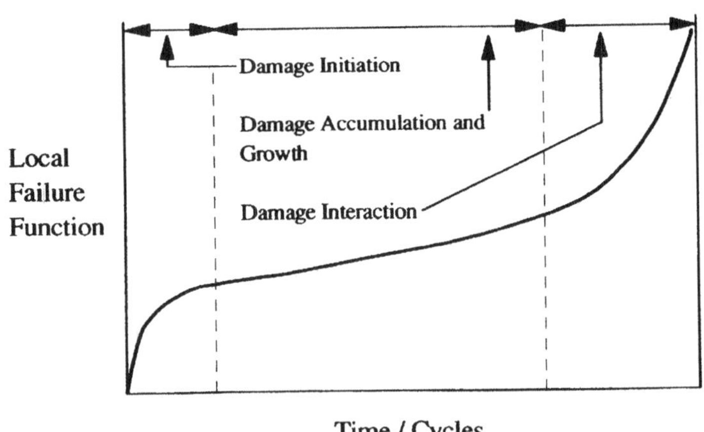

Figure 3.7 Damage accumulation rate in typical fibrous composite laminate during fatigue loading.

material that have the same loading and environmental history develop the same damage state. For now, we will construct a deterministic model of remaining strength.

3.4.1 Damage Metrics

We are concerned with long-term behavior that involves extended exposure of a component to applied conditions that may include mechanical, thermal, and chemical environments. The object of our philosophy and our calculations is strength, or, more precisely, the remaining strength after some period of service. The first element of our approach to this subject is to select remaining strength as a *damage metric*.

Figure 3.8 shows the essence of this argument. It is assumed that a service history that produces a given remaining strength results in damage that is "equivalent" to that produced by a different service history that results in the same remaining strength value (assuming the same initial strength). The most important result of this assumption is the resulting damage accumulation rule implied by the definition. As shown in Figure 3.8, if n_1 cycles of loading are applied at cyclic amplitude Fa^1, then the equivalent number of cycles at some other amplitude, Fa^2, is found by locating the cycles at the second level that produces the same strength reduction, and the remaining cycles to failure is easily determined. This summation is nonlinear unless the remaining strength curves are straight lines. Hence, it is quite different from familiar linear damage fraction rules (such as *Miner's rule*). Moreover, the result of "summing damage" by summing changes in remaining strength is affected by the sequence of the increments of applied conditions. The idea of defining equivalent damage by equivalent remaining strength is the central concept of the present approach and is the element of philosophy that enables us to combine the effects of various damage modes and their interactions.

We will attempt to calculate these remaining strength values, Fr (in terms of the same failure function chosen for the critical element). First, we will determine the failure mode of the material for a given loading condition. Next, we will set and solve a boundary-value problem based on a representative volume that defines that

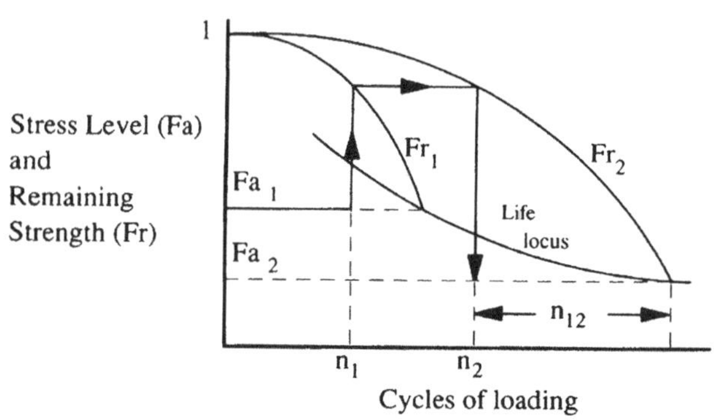

Figure 3.8 Concept of using remaining strength as a damage metric.

failure mode; that is, we will solve a separate problem for each distinct failure mode. All of the stress and strength quantities that we will use in our failure function will be *local* values in the critical element for that failure mode. The boundary-value problem to be solved for a given failure mode is identified by the critical element in the representative volume. The critical element, in turn, is defined physically by the fact that global fracture occurs when the critical element fails. Hence, if we define the problem in the manner described, our local calculations will define the global behavior; that is, the local–global transition is achieved by the critical-element concept and the associated definition of the boundary-value problem that is used to provide inputs to the failure function.

3.4.2 Damage Accumulation Kinetics

Neither thermodynamics nor statistical mechanics are applicable to systems that are not in equilibrium. Many nonequilibrium processes are, however, of great importance to engineering problems. Classical examples include chemical reactions and solidification of supercooled liquids. In the present context, these processes include micro- (or sub-micro-) defect formation and accumulation, which is generally dissipative and spontaneous.

Kinetic theory is uniquely suited to the description and interpretation of the speeds or rates associated with nonequilibrium processes. Typically, kinetic theory is used to derive relationships such as the equation of state of a material. The most common example is the equation of state for an ideal gas, which is easily derived from a simple consideration of the kinetic energy of the molecules in a control volume. However, we will be concerned with the kinetic theory of solids, and we will take a broad interpretation of the purview of kinetic theory. In particular, we will address the general question of the rate of damage initiation and accumulation and try to examine the question of how to construct appropriate equations of state for material volumes in which such damage processes are active to define the *state of the material*. The most familiar discussion of this aspect of kinetic theory is generally concerned with deformation and recovery processes in homogeneous materials, especially metals. Although we will use these foundations in the general sense, our focus will be on the description of processes that may, alternatively, be defined by sequences of discrete events such as microcrack formation, debonding, constituent–boundary separation, and constituent fracture in inhomogeneous systems. In contrast to the robust literature on the micromechanics of kinetic theory in metals, the subject of kinetic theory is in the early formative stages for composite systems.

The second concept we need for our discussion is the construct of free energy, the metric for the driving force of spontaneous processes. A typical form is the Gibbs free energy, defined as $F = E + PV - TS$, where F is the Gibbs free energy, P is pressure, V is volume, T is absolute temperature, and S is entropy. It is easily shown that the change in free energy is negative for a spontaneous process and that a spontaneous process can occur when the free-energy change is negative. Since changes in free energy define the driving force for spontaneous processes, we will use these variations to establish the rate equations associated with the damage development processes that control remaining strength and life of fibrous composite systems.

Consider the general accumulation of discrete local damage events (defined as observables such as single- or multiple-fiber fractures, or microcrack formation) and assume that the energy involved in their formation is also discrete. We assume that the damage events are being formed by a process (such as fatigue loading or sustained loading with time) that supplies input energy at a steady rate, that is, that the available energy per unit time is fixed. We also assume that the damage events can exchange energy by means of the elastic response of the continuum in which they exist. Then we address the question of how these discrete events are populated by a given energy input rate.

To illustrate the fundamental nature of our argument, we will first postulate a simple example and then generalize our argument in the limit. For this purpose, we consider four damage events that have discrete energy values of zero, e, $2e$, $3e$, or $4e$, with a total system energy (per unit time) of $3e$. Ultimately, we will let the value of e go to zero in the limit as the number of damage events goes to infinity to create continuous variables for the rate process of damage accumulation. At present, we ask, what are all the possible divisions of the available energy among the states? Table 3.5 gives those possibilities.

We assume that all divisions of the total energy occur with the same probability. Table 3.5 shows that there are three possible arrangements (or divisions) of the available energy among the damage events. Then the probability that a given division will occur is the total number of duplicate divisions divided by the total number of divisions, including all duplicates. Finally, the probability of populating a given damage event with a given energy is the sum of the number of events having that energy in each division multiplied by the probability of that division, which are the (unnormalized) fractions on the bottom line of Table 3.5. If one plots these numbers as a function of energy e, one obtains a perfect fit to an exponential having the form

$$P(e) = A \exp\left(-\frac{e}{e_{\text{av}}}\right) \tag{3.5}$$

TABLE 3.5 Occupation of Damage States as Function of Energy/Rate Needed to Create That Damage State

	0	e	$2e$	$3e$	$4e$	Number of Duplicates	Probability of State
Energy division							
1	3			1		4	/20
2	2	1	1			12	12/20
3	1	3				4	/20
Probability of event	40/20	24/20	12/20	4/20	0/20	20 (total)	

which has the form of a Boltzmann distribution. Although space does not allow us to present the details, the rigorous derivation of this form from the conditional probability that n articles can occupy q cells, given by

$$P_i = \frac{q_i!}{n_i!(q_i - n_i)!} \tag{3.6}$$

is well known. In (3.5), e_{av} is the average energy of all damage events. Equation (3.5) assumes that we have taken our discrete energy levels to the infinitesimal limit of a continuous variable.

We now introduce the familiar damage mechanics *continuity* Ψ, defined in the usual way with a value of 1 when the state of the material is undamaged and 0 when the material is completely damaged (Kachanov, 1986). As an essential feature of the present model, we interpret this continuity parameter as the normalized probability of survival of our material and, with that definition, refer to it as the state of the material. Then we argue that this variable is, in fact, the left-hand side of (3.5); that is, we can write

$$\Psi = A \exp\left[-\frac{e}{e_{av}}\right] \tag{3.7}$$

which now represents the probability of survival of the material as a function of the occupied energy level of the damage states, e. Finally, we postulate that the occupied energy is proportional to the total time over which energy is supplied to the system (at our steady rate) and specify that

$$\Psi = A \exp[-\eta\tau^j] \quad \text{where } \tau = \frac{t}{\hat{\tau}} \tag{3.8}$$

in which t is a time variable (such as sidereal time), j is a material parameter, and $\hat{\tau}$ is a characteristic (or average) time associated with the process. Characteristic times of damage processes have been discussed by Christensen (1981) and Miyano et al. (1994). They could be, for example, a stress rupture life, a fatigue life, or a stress corrosion life in the context of the present discussion. Taking the natural logarithm and the variation of (3.8), one obtains the rate equation for the change in material state due to damage accumulation as a function of generalized time, shown below:

$$\frac{\delta\Psi}{\delta\tau} = -\eta\Psi j\tau^{j-1} \tag{3.9}$$

Then we introduce a second postulate: The continuity of the material can be set equal to the quantity 1 – Fa, where Fa is the failure function that applies to a specific controlling failure mode. In general, Fa has the form suggested by Hill (1948) and others; that is, it is a function of (local) stress components divided by the corresponding material strength components. We could then write

$$\Psi = 1 - \text{Fa}\left(\frac{\sigma_{ij}}{X_{ij}}\right) \tag{3.10}$$

Although the continuity is almost certainly a more complex function of Fa, this simple first approximation is correct in the limit; when the applied load is null, the probability of survival is unity, and when the appropriate failure function reaches unity, the probability of survival is null and fracture is predicted. Finally, since we are presently considering a damage process with progressive time, we postulate that the remaining strength Fr is given by

$$\text{Fr} = 1 - \Delta\,\text{Fa}\left(\frac{\sigma_{ij}(\tau)}{X_{ij}(\tau)}\right) \tag{3.11}$$

where it is understood that the changes in the failure function are being induced by the damage process. Then combining (3.9)–(3.11) and taking η to be a unit value as a first approximation, we obtain an expression for the remaining strength of our material system in the form

$$\text{Fr} = 1 - \int_0^{\tau_1} (1 - \text{Fa})\left(\frac{\sigma_{ij}(\tau)}{X_{ij}(\tau)}\right) j\tau^{j-1}\,d\tau \tag{3.12}$$

This basic form of the strength evolution equation has been used for more than a dozen years by us, with many industrial partners, to solve various applied problems in which time, temperature, and cyclic loading are explicit influences (Reifsnider, 1991a,b, 1992, 1997; Reifsnider and Stinchcomb, 1986). Before looking at an example application of this equation, we note some recent refinements of the form of Equation (3.12) that have proven to be very useful. Suppose that fatigue is the dominant process for the determination of remaining strength and assume that the characteristic "time" to failure can be represented by N, the fatigue life in cycles. Then for fiber-controlled failure we argue that we can express that life in terms of the stress in the direction of the fibers of the most heavily loaded plies and write a one-dimensional equation of the form

$$\frac{\sigma_f(\tau)}{X_t(\tau)} = A + B\{\log[N(\tau)]\}^p \tag{3.13}$$

where σ_f is the fiber direction stress (which may be changing with time due to, e.g., stress redistribution caused by matrix cracking in other plies), X_t is the unidirectional tensile strength in the direction of the fibers (which may be decreasing due to, e.g., oxidation of the fibers), and A, B, and p are material constants. Then we introduce the frequency of cyclic loading, f, so that the number of cycles of loading, n, is given by $n = ft$. Then (3.12) can be stated in the form

$$\mathrm{Fr}(t_1) = 1 - \int_0^{t_1} (1 - \mathrm{Fa})\left(\frac{\sigma_{ij}(t)}{X_{ij}(t)}\right) j \left(\frac{ft}{N(t)}\right)^{j-1} d\left(\frac{ft}{N(t)}\right) \tag{3.14}$$

where all of the variables are evaluated at the local level, in the critical element, as we have discussed. We will show that these integral forms are able to combine the effects of time dependent processes like creep, stress rupture, viscoelastic stress redistribution, and corrosion with cycle-dependent effects such as fatigue or thermo-mechanical damage. Combining time- and cycle-dependent processes may be further facilitated by expressing the frequency of cyclic loading, explicitly, in terms of the applied cyclic frequency f_0 and the time to rupture in a stress rupture test (under constant applied loading with otherwise the same conditions), t_r, in the form

$$f = f_0(1 - R) + \frac{0.5}{t_r} R \tag{3.15}$$

where R is the usual stress ratio in fatigue. Hence, if the minimum and maximum stress is the same ($R = 1$), there is a constant loading and one recovers the correct result that there is only half a cycle of loading during the life, while a minimum stress of zero ($R = 0$) produces the correct cyclic loading frequency f_0. In Chapter 8, we will discuss an alternative (more complicated) approach to this last approximation in Equation (3.15).

Now our problem has been reduced to calculating inputs to (3.12) and (3.14). We can make a clear connection to the phenomenological behavior we discussed earlier. The local stiffness changes will act to change the tensor stiffness values used to calculate the local stress fields in the critical element. If we can determine the rate of the different stiffness changes, we can calculate the correct local stresses as a function of cycles and time as inputs into (3.14). In fact, we can calculate the stresses based on representations at various levels of the microstructure of the composite system. If we wish to use stiffness evolution information at the ply level, we can use laminated plate theory, or if we have information about the stiffness changes in the constituents (such as the viscoelastic flow of the matrix), we can use micromechanics to calculate the redistribution of local stress in a very direct way.

Similar comments apply to the changes in material state at the local level. We can develop inputs to (3.12) and (3.14) at the ply level or use micromechanics to represent the principal material strengths in terms of the properties, geometry, and arrangement of the constituents and the interphase regions between them. This latter approach has many advantages. Many of the processes that control behavior such as creep, creep rupture, and fatigue degradation are highly nonlinear in their dependence on the applied conditions and history. A major question for any representation is how to correctly combine the effects of these phenomena when they are happening at the same time. Using micromechanics has the following advantage. Changes in the properties (and geometry and arrangement, if appropriate) of the constituents and the interphase regions between them directly enter the micromechanical models of principal strengths and the calculations of local stress states as continuous functions of time (or cycles) in (3.5)–(3.14). This allows their collective effect to be summed incrementally, by integration to the present time of evaluation. Hence, the physics and mechanics of the quantities defined in (3.9) correctly combine the effects without "artificial" algorithms. However, great care must be taken to ensure that the rate equations that are measured in the laboratory (and all of the constitutive behavior, for that matter) under combined conditions are determined in such a way that the influence of independent external variables such as temperature and stress (or strain) are properly isolated.

3.5 FEATURES OF STRENGTH EVOLUTION INTEGRAL

In the remainder of this chapter, we will investigate the nature of the residual strength integral and the inputs that are needed. The fundamental behavior of the integral can be demonstrated by a few simple examples. For these examples, we will assume that fatigue is the dominant process for the determination of remaining strength and assume that the characteristic time to failure can be represented by N, the fatigue life in cycles. Then we restrict our consideration to one-dimensional (uniaxial) loading of a specimen with uniform stress (such as a coupon with uniform cross section) and assume that the fatigue behavior of the material involved can be written in the form

$$\frac{S}{X} = A + B(\log N)^p \tag{3.16}$$

where S is the uniaxial applied stress: A, B, and p are material constants; X is the ultimate intrinsic strength in the direction of loading; and N is the life of the member. We introduce the frequency of cyclic loading, f, so that the number of cycles of loading $n = ft$, where t is sidereal time, and write the residual strength integral as

$$\mathrm{Fr}(t_1) = 1 - \int_0^{t_1} \left(1 - \frac{S}{X(t)}\right) j \left(\frac{ft}{N(t)}\right)^{j-1} d\left(\frac{ft}{N(t)}\right) \tag{3.17}$$

where we have introduced the intrinsic material strength X as a function of time so that we can discuss failure modes such as creep rupture. Since $X = X(t)$, from (3.16), $N = N(t)$ as well, that is,

$$N(t) = 10^{\{[S/X(t)-A]/B\}^{1/p}} \tag{3.18}$$

For our illustration, we begin with the simplest possible case, when only fatigue behavior of a coupon is considered. For demonstration, we consider a material with $X = 100$ (in arbitrary units, all quantities in the integral are normalized),

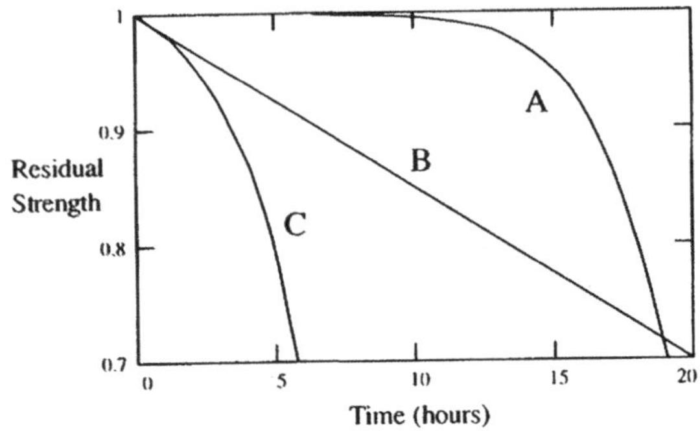

Figure 3.9 Example outputs from (3.17).

$A = 1$, $B = -0.06$, and $p = 1$. We consider the applied situation in which $S = 70$; that is, we are to model the change in strength when we apply a stress amplitude that is 70% of the ultimate strength. For this case, X is taken to be constant. Then, (3.17) shows that the life of the material is 100,000 cycles for this loading. For our example, we will assume that the specimen is being cycled at 1.388 Hz, so that the time for failure in fatigue (alone) will be 20 h. If we take $j = 1$ in (3.17) and everything else is constant (i.e., independent of time), we should get the same result given by (3.18), since (3.17) could be integrated directly. Integrating (3.17) numerically (with any standard method) produces the result shown in curve B of Figure 3.9. We notice that the "life," defined by the reduction of the residual strength to the level of applied loading (0.7 of the initial value in this example), is 20 h, as specified by the inputs.

Now we consider two complementary cases. First, we greatly lower the frequency to 0.0025 Hz so that only a small number of cycles of loading occur in 20 h. Then we introduce a reduction in the intrinsic strength of the critical element as a function of time to simulate a stress rupture condition, that is, a time-dependent failure. For our calculation, we change the critical-element strength according to the equation

$$X(t) = 100(1 - 0.015t \qquad 3.19)$$

so that the initial strength of 100 is reduced to the applied level of 70 by stress rupture effects over the time of 20 h. With those changes in (3.17) and (3.18), the integral produces the predicted strength reduction shown by curve A in Figure 3.9. Since we have only a few cycles of loading over the 20-h life and the cyclic life is still 100,000 cycles for this level of loading, the reduction in global strength is due only to the reduction in the critical-element strength, as represented by (3.18). Of course, the shape of curve A is affected by other variables, as we will show below, but the life predicted by the integral equation is consistent with the inputs, the point to be made by this example.

If we only had one type of degradation due to damage accumulation, using an equation such as (3.17) would not be necessary. The value of the equation and this approach is the ability to combine a variety of effects in a manner consistent with the physics and mechanics of the situation. If we combine only the two effects just described, the cyclic degradation (due to fatigue effects on the critical element), and the stress rupture effects as described, the integral produces curve C in Figure 3.9. The combined effect of fatigue and stress rupture degradation is greater than either of them alone, and the prediction is consistent with this idea. But is the large difference predicted observed? We will see in a later chapter that comparisons with data suggest that the general (but not exclusive!) answer to this question is yes. But we should remember that the details of the present example were entirely hypothetical and we intentionally chose a fatigue life (and loading frequency) and stress rupture life that produced exactly the same life (of 20 h) alone, something that is quite unlikely in any "real" situation but very useful for our consistency check.

We have also used a simple definition of failure so far by assuming that global failure occurs when the value of Fr equals the failure function Fa in the critical element, an idea that is the conceptual equivalent of saying that it breaks when the applied stress is the same as the remaining strength. Of course, our applied stress is a local failure function in the critical element that compares the local multiaxial stress to the full complement of intrinsic material strengths, X_{ij}, and Fr should be viewed as the remaining global strength under the conditions that define the inputs to Fa. In particular, a given failure mode will define the critical element and the appropriate failure criterion Fa.

The fatigue life representations that we have discussed have been cast in the form of cycles of life for a given value of Fa, that is, cycles of life for a constant value of applied conditions, evaluated by the failure function appropriate for the failure mode under consideration. Several aspects of the classical representations of fatigue involving stress amplitude and mean stress are mute in this representation. First of all, the maximum value of Fa has been used for the S–N relationships rather than the amplitude of Fa. This reflects the fact that maximum stress or strain correlates more reliably with life than stress or strain amplitude for many brittle materials, like continuous-fiber advanced composites. If the maximum value of the failure function is used, classical questions of mean stress and R ratio do not enter the analysis.

However, if the failure mode involves local shear stresses, such as in the initiation and growth of delamination from the edge of a multiaxial laminate, then the proper failure function may be the total range of Fa, that is, $\mathrm{Fa}_{\mathrm{max}} - \mathrm{Fa}_{\mathrm{min}}$. This situation is especially common for matrix-controlled failure modes and ductile matrix materials. However, since advanced composite materials usually fail by the accumulation of distributed damage and rarely fail by self-similar propagation of a single crack, questions of crack closure and retardation rarely enter the analysis.

The fundamental principle here is that questions of the effect of minimum stress, maximum stress, mean stress, or corresponding strain values are part of the question of what failure mode controls the life of the material or component and what failure function Fa describes that controlling failure mode. That information must come from careful characterization and must be determined before the form is Fa is set.

Representations of S–N behavior then become representations of Fa–N behavior, as discussed in the foregoing paragraphs.

There is one other characteristic of (3.17) that we need to understand before we undertake the task of generating inputs for it. We mentioned that the parameter j is a material parameter but did not associate it with physical behavior in our earlier discussion. To make the identity of j somewhat more clear, we return to the rate equation that we assumed in our derivation of (3.17). We recall that the form of the rate equation was

$$\frac{\partial\psi}{\partial\tau} = \psi j \tau^{j-1} \tag{3.20}$$

Clearly, if $j = 1$, there is no explicit time dependence in the rate equation. But j may also be greater or less than 1. Figure 3.10 shows the essence of the effect of those two different conditions. If $j > 1$, then Equation (3.20) suggests that the rate of degradation is increasing as a function of (generalized) time; that is, one would expect the residual strength to decrease with an increasing rate, and that is the effect shown in Figure 3.10. This behavior, sometimes called sudden death, is rather common for composite materials and other brittle systems, as we will see. However, if $j < 1$, then (3.20) suggests that the rate of degradation is greatest at the beginning and decreases as a function of time. The lower curve in Figure 3.10 shows that result from the integral. This early damage case is also observed, especially in the uniaxial fatigue of certain unidirectional composites. The two values of j used to generate the curves in Figure 3.10 are 1.4 and 0.4, respectively. It is clear that the *shape* of the resulting remaining strength curve is very sensitive to the value of j and that the actual value of j is determined by the basic nature of the time dependence of the rate that controls the life of the critical element.

It is difficult to determine j independently, beyond comparing the shape of measured remaining strength plots. Future research will address this point as well as many other aspects of the formulation of the remaining strength equation. It should be noted that, like any other strength equation, the form of the rate equation

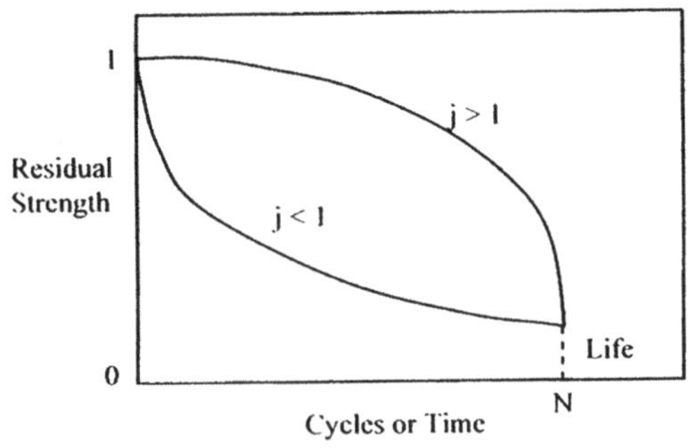

Figure 3.10 Effect of value of material parameter j on predicted remaining strength locus from (3.17).

and the resulting integral is a postulate and is not unique, nor should it be. Other choices driven by logic and physical reasoning are possible and may better represent physical behavior.

A final characteristic of the evolution integral should be noted before more specific features are addressed. The fundamental assumption associated with this approach is that strength is degraded as a function of time or cycles as damage accumulates. This assumption accounts for at least two basic issues.

First, what is the effect of an *overload*, that is, a momentary increase in the amplitude or maximum value of cyclic applied stress? This is a typical service event, in which an uncharacteristically high load is applied, similar to an automobile hitting a pothole. If damage accumulation controls strength, one would expect such an overload to subtract a corresponding increment of remaining strength, almost like a step function. Figure 3.11 shows that the integral predicts just such behavior. For the example shown, a cyclic load with maximum value of 70% of the ultimate strength (in tension) is applied with a constant amplitude, except for a sharp spike overload to about 72.5%, as shown in Figure 3.11*a*. The resulting predicted strength reduction is shown in Figure 3.11*b*. The step drop in remaining strength associated with the overload is clearly shown. It is also interesting to note that while all subsequent values of remaining strength are reduced, the length of the remaining life is not affected as much as the strength values; it stays nearly the same (close to one million cycles for the 70% constant-amplitude case).

The second issue is that damage accumulation assumes that strength is monotonically decreasing. Conceivably, the remaining global strength of a component may increase at some point during the life. We will see several examples of such behavior later, especially for notched components. In that situation, some additions to the model must be employed to follow the physical situation. We will deal with this problem in a later chapter using a simple mechanistic scheme. With this correc-

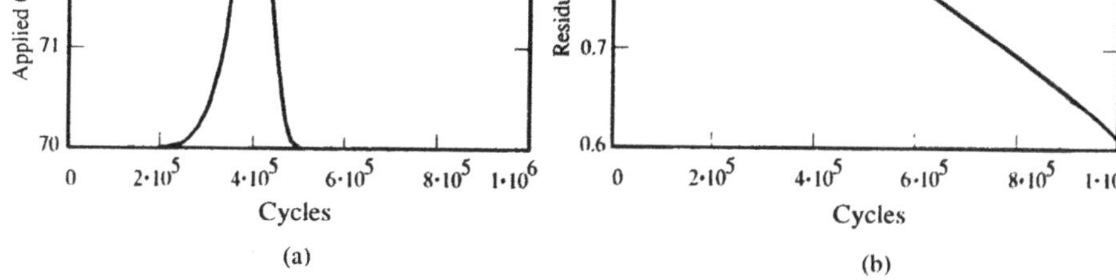

Figure 3.11 (*a*) Spike in applied loading and (*b*) corresponding drop in predicted remaining strength.

tion, the integral method follows essentially all behavior types observed (to date) in the laboratory.

The equations and approach presented above are used only because of the body of physical data that support them, developed over a period of about 18 years, and the familiarity of the writers with the origins, nature, and utility of the forms (Reifsnider, 1991b, 1992; Reifsnider and Stinchcomb, 1983). With these qualifications, we proceed with our effort to generate inputs to Equation (3.12) (or various other statements of that form), essentially the purpose of the rest of this book. However, the subjects that we will discuss are of general interest and will form a rational description of the field of damage accumulation in fibrous composites, notwithstanding their use in the equations above. We will use the equations to codify and combine the information and the associated understandings and representations and to predict remaining strength and life for engineering applications.

3.6 SUMMARY OF APPROACH

We can summarize our approach using Figures 3.12 and 3.13. Figure 3.12 is the flow diagram of the methodology. We begin by identifying a unique and well-defined failure mode by careful laboratory testing. Since the present method is mechanistic, this step is critical to the success of the representation. We discussed damage and failure modes earlier. For many fibrous composite systems used for engineering components, the materials are designed and constructed to achieve a high degree of fiber influence or control of the damage and failure modes; that is, fiber-controlled materials. The basic nature of the fiber influence will be the subject of some of our discussion in Chapter 4. But in some situations, the matrix may control failure or a variety of other failure modes may exist. While we have tacitly assumed that failure

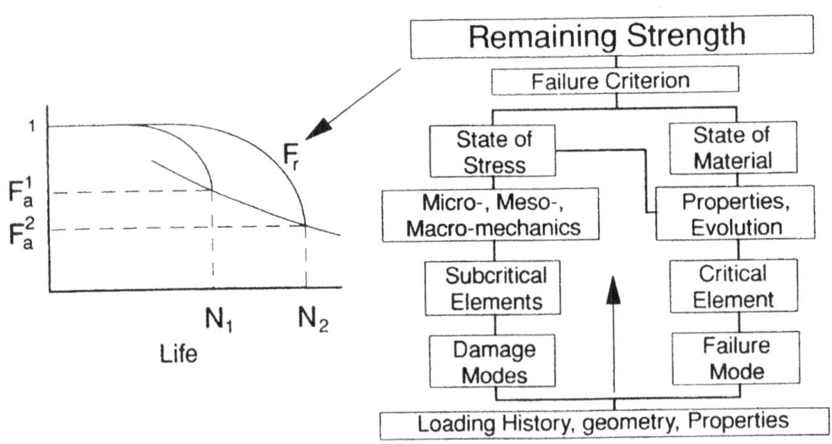

Figure 3.12 Flow diagram of present methodology for predicting remaining strength and life of composite systems.

$$Fr(n/N) = 1 - \int_0^{n/N_1} \left[1 - Fa\left(\frac{\sigma(\tau)_{ij}}{X(\tau)_{ij}} \right) \right] j\left(\frac{t}{\tau}\right)^{j-1} d\left(\frac{t}{\tau}\right)$$

Chapter:
4 - Micromechanical models of strength and stiffness - $X(\tau)_{ij}$, $E(\tau)_{ij}$
5 - Stiffness Reduction: physical processes, stress redistribution - $\sigma(\tau)$, $N(\tau)$
6 - Strength Reduction: physical processes, stress rupture - $X(\tau)$, $N(\tau)$
7 - Non-uniform stress states: stress concentrations, edge and boundary layer stresses - $Fa(\tau)$
8 - Application of the concepts: prediction of remaining strength and life - $Fr(\tau)$

Figure 3.13 Outline of how present subject will be addressed in remainder of book.

is separation of the material into parts, failure may be alternatively defined in terms of critical stiffness change, allowable creep strain, or other limiting behavior. So it is necessary to determine the criterion clearly for the application involved and to determine the physical nature of limiting event(s) as a first step in generating a representation of remaining strength and life using this approach.

Once the failure mode is determined, the critical element can be specified. It may be a critical number of fiber fractures in tension or a microbuckling region in compression. If ply-level representations are used, the critical element may be a controlling ply that has a low angle of orientation to a principal loading direction. We will provide several examples throughout the book. The critical element is the object of our stress analysis and is used to specify the state of the material, in the sense that continuum constitutive theory is used to describe it. For the predominantly brittle materials considered in this book, the constitutive equations used to describe the behavior of the critical element may be linear in some cases. But we will find that the effects of temperature and environment often introduce highly nonlinear effects into the constitutive equations that describe stiffness, strength, conductivity, thermal expansion, and diffusion. Hence, even though the critical element remains intact until the failure of the component, the constitutive equations used to describe its behavior may be highly nonlinear. In many cases, they will be tensor equations reflecting the anisotropy of the response.

The next step, in Figure 3.12, is the determination of the damage modes that control and influence the local conditions in the critical element. These are the subcritical elements in the sense that their failure does not cause failure of the global component. Examples of these elements are easily identified and often include matrix cracking, delamination, debonding, or constituent separation or relative motion. It is then possible to determine the problem to be solved by defining a representative volume, as explained earlier. From the standpoint of the mechanics analysis to be performed, specifying the representative volume sets the boundary-value problem for us. For the present approach, we will use the imminent-failure condition to set the problem; that is, we define our boundary-value problem on the basis of the geometry and boundary conditions at the end of life, rather than at the beginning of life in the undamaged condition, as is more typically done.

Following along in Figure 3.12, we then determine the changes in stress state and material state associated with the processes that control damage and degradation. Experimentally, this is the most challenging part of the problem. It is necessary to isolate each contributing process and determine the rate equation for these processes as a function of the extensive variables, a demanding requirement in many cases. For example, it may be necessary to determine the rate of matrix cracking as a function of applied cyclic load level, or the degradation of strength of the fibers as a function of oxygen content and temperature, or the anisotropic diffusion rate of a corrosive environmental element, or a variety of other rate equations. Of course, this can be difficult for any mechanistic approach to this problem. When data are not available, it may be necessary to use phenomenological representations that correlate a few existing data points or to rely on heuristic postulates for a general form of the rate equations. We will see examples of these schemes in later chapters.

The format for this discussion is outlined in Figure 3.13. Chapter 4 will discuss micromechanical concepts and models associated with intrinsic material strengths, or principal material strengths, as we will call them. Special attention will be given in that chapter to the manner in which these strengths depend on the strength, geometry, and arrangement of the constituents of the composite. Since these characteristics may be altered by the development of damage, these principal strengths are functions of generalized time, as suggested by the equation. Chapter 5 will be concerned with stiffness reduction, caused by degradation processes or by explicit time dependence (such as viscoelastic/plastic behavior). This stiffness reduction generally occurs at the microlevel, so that the internal distribution of stress (especially in the critical element) becomes a function of time, and this, in turn, will alter the characteristic time to failure [$N(t)$ in Fig. 3.13]. Chapter 6 addresses the reduction of material strength as a function of the degradation processes. Since these changes also alter the characteristic time to failure [e.g., according to (3.16) for fatigue-controlled failure], N becomes a function of time, $N(t)$. Chapter 7 addresses the effects of stress concentrations, such as holes, edges, and joints. As suggested by Figure 3.13, the form of Fa(t) is altered by such nonuniform stresses. Finally, Chapter 8 will address the application of the concepts to various engineering situations and problems and will present several examples of such solutions.

REFERENCES

Broutman, L. J., and Sahu, S. (1972), A New Theory to Predict Cumulative Fatigue Damage in Fiberglass Reinforced Plastics, in *Composite Materials: Testing and Design* (2nd conf.), H. T. Corten, Ed., ASTM STP 497, American Society for Testing and Materials, Philadelphia, pp. 170–188.

Christensen, R. M. (1981), "Lifetime Predictions for Polymers and Composites under Constant Load," *J. Rheol.*, Vol. 25, No. 5, pp. 517–528.

Hill, R. (1948), "A Theory of Yielding and Plastic Flow of Anisotropic Metals," *Proc. Roy. Soc.*, Series 14, Vol. 193, pp. 281–297.

Kachonov, L. M. (1986), *Introduction to Continuum Damage Mechanics*, Martinus Nijhoff, Boston.

Miyano, Y., McMurray, M. K., Enyama, J., Nakada, M. (1994), "Loading Rate and Temperature Dependence on Flexural Fatigue Behavior of a Satin Woven CFRP Laminate," *Journal of Composite Materials*, Vol. 28, no. 13, pp. 1250–1260.

Reifsnider, K. L. (1991a), "Performance Simulation of Polymer-Based Composite Systems," in *Durability of Polymer-Based Composite Systems for Structural Applications*, A. H. Cardon and G. Verchery, Eds., Elsevier Applied Science, New York, pp. 3–26.

Reifsnider, K. L., Ed. (1991b), *Fatigue of Composite Materials*, Elsevier Science, London.

Reifsnider, K. L. (1992), "Use of Mechanistic Life Prediction Methods for the Design of Damage Tolerant Composite Material Systems," in *Cyclic Deformation, Fracture and Nondestructive Evaluation of Advanced Materials*, ASTM STP 1157, M. R. Mitchell and O. Buck, Eds., American Society for Testing and Materials, Philadelphia, pp. 205–223.

Reifsnider, K. L. (1997), "Durability and Damage Tolerance: Testing, Simulation, and Other Virtual Realities," in *Composite Materials: Testing and Design*, Vol. 13, ASTM STP 1242, S. J. Hooper, Ed., American Society for Testing and Materials, pp. 45–59.

Reifsnider, K. L., and Highsmith, A. L. (1981), "Characteristic Damage States: A New Approach to Representing Fatigue Damage in Composite Laminates," in *Materials: Experimentation and Design in Fatigue*, F. Sherratt and J. B. Sturgeon, Eds., Westbury House, Surrey, pp. 246–260.

Reifsnider, K. L., and Stinchcomb, W. W. (1983), "Cumulative Damage Model for Advanced Composite Materials," Phase II Final Report to Air Force Materials Laboratory, Wright Patterson AFB (see also General Dynamics Report No. FZM-7149, contract no. F33615-81-C-5049, Fort Worth, TX, 1983).

Reifsnider, K. L., and Stinchcomb, W. W. (1986), "A Critical Element Model of the Residual Strength and Life of Fatigue-Loaded Composite Coupons," in *Composite Materials: Fatigue and Fracture*, ASTM STP 907, H. T. Hahn, Ed., American Society for Testing and Materials, Philadelphia.

EXERCISES

3.1 Construct a progressive damage code as follows. Use the lamina properties given in Table 3.1.

(a) Starting with a classical laminate code, construct a laminate analysis that allows progressive step increases in the applied resultants and changes in ply-level stiffness values at each loading step.

(b) Add to that code a routine that iteratively changes matrix-controlled stiffness values in each ply (E_2 and G_{12}) at a given applied load level by an amount that just keeps the normal stress in the direction perpendicular to the fibers constant as a function of increasing strain above a given applied load level. This will have the effect of keeping the matrix-controlled stresses in the plies constant after matrix cracking begins at the ply-level value that initiates such cracking. The progressively reduced ply stiffness represents a progressively increasing crack density with increased applied loading.

(c) Using the ply properties given in Table 3.1, apply the code to a $[0,\ 90_3]_S$ laminate. Plot the calculated stiffness of the 90° plies as a function of applied uniaxial, in-plane loading, N_x, for values that extend well above the matrix crack initiation level.

(d) Plot the stress–strain curve of the laminate for that range of applied values.

(e) Plot a comparison stress–strain curve for that laminate, generated by reducing the stiffness of the 90° plies in the direction of loading to 10^{-6} times the original value for all applied loads that produce a local ply stress above the ply-level matrix failure stress.

(f) Discuss the shape of the plots determined in (c) and (d) on the basis of the physical behavior that is being modeled and identify advantages and disadvantages of the two approaches.

3.2 Using the data and information given in the text, construct a strength evolution equation and verify the results shown in Figures 3.9 and 3.10.

3.3 Using the analysis constructed in Exercise 3.2, conduct a test of sequence effects. Use the parameters quoted in the text for curves A and B in Figure 3.9. Use the following loading increments (or others of your choosing). (Assume the following values for material parameters: $j = 1$, $X^0 = 100$, $A = 1$, $B = -0.06$, $p = 1$, $F_a^0 = 0.7$). Apply sequence 1 (defined below), and compare the remaining strength value at the end of that sequence with the value obtained by applying sequence 2 (the reverse combination of loadings).

First sequence: Fatigue, alone, for 10 h (as in curve B of Fig. 3.9) followed by strength reduction alone for 10 h. Remember to use 1.388 Hz for the frequency and to hold the material strength constant for the first increment and to use 0.028 Hz with the strength reduction equation quoted in the text for the second increment. Also remember to use the reduced strength from the first increment as the starting value for the strength reduction in the second increment.

Second sequence: Reverse the two increments used in the first sequence; that is, in this case, apply strength reduction alone for 10 h followed by fatigue alone for 10 h.

(a) What is the estimated remaining strength after each of the two sequences?

(b) Discuss the comparative results on the basis of the physical behavior being simulated.

(c) What makes one sequence "more damaging" than the other?

3.4 Repeat Exercise 3.3 for different values of the material parameter j: $j = 1.4$ and $j = 0.8$. What physical variations are represented by the different values of j?

3.5 Construct a basic strength evolution code, to be expanded and improved in subsequent chapters. The following steps are suggestions only and will result in a rudimentary model. The reader may wish to construct a more comprehensive and robust model by substituting more elaborate representations for each element listed below.

(a) Construct a laminate code (as used in Chapter 2) that has all physical inputs as a function of generalized time. A simple method of doing this might be to set:

$$P^k(t) = P_0^k + P_1^k t + P_2^k t^2 + P_3^k t^3 + P_4^k t^4$$

where $P^k(t)$ is any k property such as stiffness $[E_{ij}(t)]$, strength $[X_{ij}(t)]$, thermal expansion coefficient $[\alpha_{ij}(t)]$, and moisture coefficient $[\beta_{ij}(t)]$. Generalized time may need to have different definitions in each $P^k(t)$ expression. Typical choices are linear time t; logarithmic time [log(time, min) or log(time, h)]; or normalized cycles such as n/N, where n is the current cycle count and N is the fatigue life under current instantaneous conditions.

(b) Construct a failure function module that evaluates several classical failure criteria in the material coordinate system of each ply as a function of $P^k(t)$, that is, which determines the instantaneous value of $\text{Fa} = \text{Fa}(P^k(t))$.

(c) Create a time loop so that all ply-level values of $\text{Fa}(t)$ can be calculated as time is incremented and $P^k(t)$ changes.

(d) Create a strength evolution module using Equation (3.17) (or a form that you prefer) that calculates $\text{Fr}(t)$ for selected plies (critical elements) as a function of generalized time.

(e) Validate your code by running the examples given in the text and verified in Exercise 3.2.

3.6 Upgrade your basic strength evolution code using more general or more complete representations and models. Examples of improvements to be considered include the following:

(a) Associate your $P^k(t)$ equations with specific engineering behavior, such as matrix cracking, fatigue, stress rupture, environmental degradation, and viscoelastic response. Customize the form of the equations used for each according to accepted representations of those phenomena in the literature.

(b) Integrate your analysis with a more general stress analysis, such as a closed-form analysis for a notched coupon, an appropriate solution for an elasticity problem, or a finite-element solution for a complex body or component.

(c) Construct pre- and postprocessors for your code that facilitate inputs and the interpretation of outputs.

CHAPTER 4

MICROMECHANICAL MODELS OF COMPOSITE STIFFNESS AND STRENGTH

Thesis: Principal composite material strengths and stiffness can be represented by micromechanical models. These models provide a mechanistic means of representing the collective action and interaction of the constituents.

The failure of composite materials in the presence of general loading conditions involves a complex interaction between the fiber and the matrix. Traditional composite failure theories (such as those described in Chapter 3) require principal material stiffness and strength values as inputs. These inputs may be determined in two ways: either by direct experimental measurement or by calculation through the use of models. Direct experimental characterization tends to be time consuming and expensive and as a result does not lend itself well to "what if" or trade-off studies. Micromechanical models are powerful tools in these types of studies and can also help us to understand how and why composites fail. As a result, the purpose of this chapter is to develop micromechanical models for each of the principal composite stiffness and strength components. As it happens, the subject of micromechanical stiffness models is well developed and rather easily accessible. However, micromechanical strength models are less mature, so we will consider that subject in more detail. We will first consider the estimation of the fiber direction tensile strength of unidirectional, continuous-fiber, composites. In doing so, we will begin with simple models and then consider more complicated models of composite behavior.

4.1 AXIAL TENSILE STRENGTH OF UNIDIRECTIONAL COMPOSITES

The introductory description of Weibull statistics that follows is not meant to be complete but will serve our purposes as we examine tensile strength models. If we apply Weibull statistics to a single fiber of length l, the cumulative number of defects that can fail at stress σ is given by

$$\Phi(l, \sigma) = \frac{l}{l_0}\left(\frac{\sigma}{\sigma_0}\right)^m \tag{4.1}$$

We may interpret the parameters l_0 and σ_0 by recognizing that σ_0 is the stress required to cause one failure, on average, in a fiber of length l_0. The parameter m, the Weibull modulus, describes the amount of variability (the scatter) in the strength values about the average (large values of m correspond to small amounts of scatter in the data). We may relate Φ to the measured (experimental) probability of failure by

$$\begin{aligned} P_f(l, \sigma) &= 1 - \exp[\Phi(l, \sigma)] \\ &= 1 - \exp\left[-\frac{l}{l_0}\left(\frac{\sigma}{\sigma_0}\right)^m\right] \end{aligned} \tag{4.2}$$

Values of $P_f(l, \sigma)$ are illustrated in Figure 4.1 for various values of m at a fixed length l as a function of σ/σ_0.

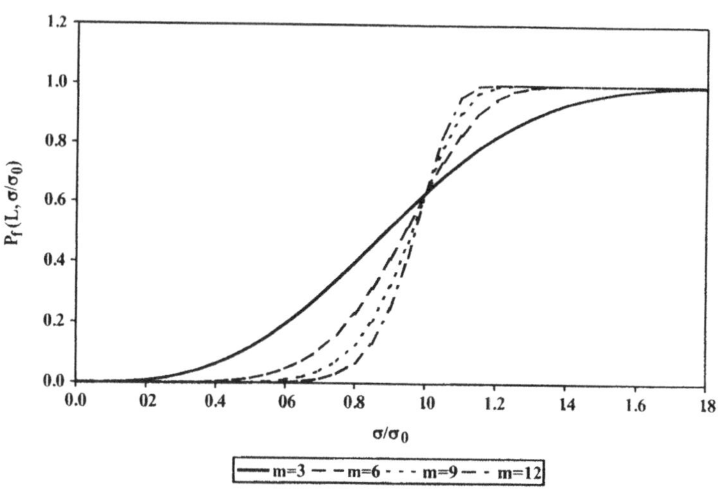

Figure 4.1 Weibull cumulative probability of failure.

The corresponding reliability is given by

$$\begin{aligned} R(l, \sigma) &= 1 - P_f(l, \sigma) \\ &= \exp\left[-\frac{l}{l_0}\left(\frac{\sigma}{\sigma_0}\right)^m\right] \end{aligned} \tag{4.3}$$

Because Equations (4.2) and (4.3) depend upon the length l, values for σ_0 and m must always be quoted (or determined) for a reference value of l_0. Two different methods are available for us to determine m and σ_0. Unfortunately, each of these involves testing large numbers of fibers. In the first method, many fibers are tested at a fixed gage length. The resulting strength values are fit to a Weibull distribution. This process is repeated at other gage lengths, so that new values for m and σ_0 are obtained. The values for σ_0 are converted to one reference length by using the relation

$$\sigma_0 = \sigma_{02}\left(\frac{l_{02}}{l_0}\right)^{1/m} \tag{4.4}$$

Ideally these values should be identical. In practice, however, the values are different. A common procedure is then to average the resulting values for m and σ_0. However, simulations suggest that a better method is available. In this method, one proceeds as before, determining σ_0 at different gage lengths, only in this case the values are not converted to a common reference length. Rather, a linear regression is performed on $\log(\sigma_0)$ versus l_0. From the form of Equation (4.4), we can see that such a plot has a slope of $-1/m$. The resulting line may also be used to determine σ_0.

4.1.1 Bundle Strength

Our first effort to estimate the axial tensile strength of a unidirectional composite will be a bundle strength model. In this model, we disregard the effect of the matrix entirely. We begin with a bundle of fibers of length l whose strains to failure may be represented in terms of the Weibull parameters ε_0 and m (the reason for choosing the strain to failure rather than the strength will become clear). The resulting reliability may be written as

$$R(\varepsilon) = \exp\left[-\frac{l}{l_0}\left(\frac{\varepsilon}{\varepsilon_0}\right)^m\right] \tag{4.5}$$

The stress carried by the bundle of fibers at a given strain ε is then given by

$$\sigma = E_f \varepsilon R(\varepsilon \qquad 4.6)$$

which has the physical interpretation of being the stress carried by an individual unbroken fiber multiplied by the fraction of unbroken fibers.

To estimate the strength of the bundle, we find the maximum of Equation (4.6) by setting $d\sigma/d\varepsilon = 0$ and solving for ε. This critical value of ε is given by

$$\varepsilon_c = \varepsilon_0 \left(\frac{l_0}{ml} \right)^{1/m} \tag{4.7}$$

Substituting this value into Equation (4.6), we obtain the maximum fiber stress,

$$\sigma_{fc} = E_f \varepsilon_0 \left(\frac{l_0}{mle} \right)^{1/m} \tag{4.8}$$

The corresponding composite stress is then given by

$$X_t = V_f E_f \varepsilon_0 \left(\frac{l_0}{mle} \right)^{1/m} \tag{4.9}$$

where V_f is the composite fiber volume fraction. An interesting observation can be made if we consider the average strength of the fibers in the bundle. This average strength for fibers having a length l is defined by

$$\bar{\sigma}_f = E_f \varepsilon_0 \left(\frac{l_0}{l} \right)^{1/m} \Gamma\left(\frac{m+1}{m} \right) \tag{4.10}$$

where $\Gamma(z)$ is the Euler gamma function. If we take the ratio of the maximum fiber stress given by Equation (4.8) to the average fiber strength given by Equation (4.10), we obtain

$$\frac{\sigma_{fc}}{\bar{\sigma}_f} = \frac{E_f \varepsilon_0 (l_0/l)^{1/m} (1/me)^{1/m}}{E_f \varepsilon_0 (l_0/l)^{1/m} \Gamma[(m+1)/m]} = \frac{1}{(me)^{1/m} \Gamma[(m+1)/m]} \tag{4.11}$$

A plot of (4.11) is presented in Figure 4.2.

We note that, for all cases, the strength of the bundle is less than the average fiber strength. As explained in Chapter 1, the load carried by each fiber that breaks under increasing applied load is redistributed to the other fibers. Unless all the fibers have the same strength (the limiting case in which m approaches infinity), as the weakest fibers in the distribution break, they increase the load on the remaining fibers. Typical values for the Weibull modulus of fibers used as reinforcements in composite materials range from 3 to 10. For such cases, the critical bundle stress ranges from 0.556 to 0.755 times the average fiber strength.

In the development of this bundle strength model, we have completely disregarded the role of the matrix in transferring load back into a broken fiber. A widely used technique for determining the nature of this load transfer is the shear-lag analysis. An analysis of this type was used by Cox (1952) and Rosen (1964). In

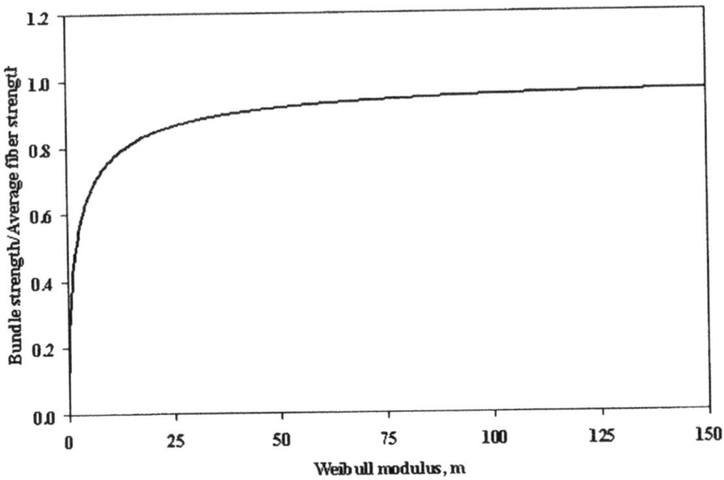

Figure 4.2 Comparison of bundle strength to coverage fiber strength.

this analysis, we may consider the representative volume shown in Figure 4.3. The assumptions used in the analysis are as follows. First, we assume that the fiber supports normal stresses only (no shear stresses). Second, we assume that the matrix supports shear stresses only (no normal stress). With these assumptions, the equations of axial equilibrium become

$$2\pi r_f \tau + \pi r_f^2 \frac{d\sigma_f}{dx} = 0 \quad \text{(fiber} \tag{4.12)}$$

$$\pi r_f^2 \sigma_f + \pi (r_c - r_m)\sigma_c = \pi r_c^2 \sigma \quad \text{(composite)} \tag{4.13}$$

where $\bar{\sigma}$ is the average stress applied to the composite away from the break, σ_f is the stress in the fiber, σ_c is the stress in the composite in the vicinity of the break, and τ is the shear stress along the fiber. Next we claim that the shear strain in the matrix is given by

$$\gamma_m = \frac{u_c - u_f}{r_m - r_f} \tag{4.14}$$

so that

$$(r_m - r_f)\frac{d^2\gamma_m}{dx^2} = \frac{d^2 u_c}{dx^2} - \frac{d^2 u_f}{dx^2} \tag{4.15}$$

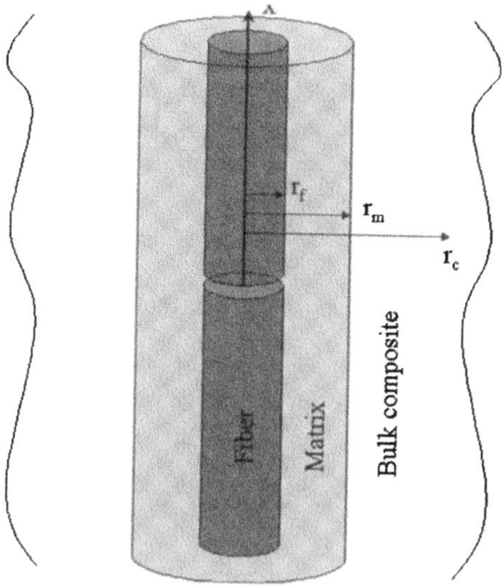

Figure 4.3 Representative volume for shear-lag analysis.

or

$$\frac{r_m - r_f}{G_m}\frac{d^2\tau}{dx^2} = \frac{1}{E_c}\frac{d\sigma_c}{dx} - \frac{1}{E_f}\frac{d\sigma_f}{dx} \tag{4.16}$$

where G_m is the matrix shear modulus, E_c is the composite modulus, and E_f is the fiber modulus.

If we differentiate (4.13) and make use of the fact that the value of $\bar{\sigma}$, a constant, does not depend upon the axial coordinate x, we have

$$\pi r_f^2\frac{d\sigma_f}{dx} + \pi(r_c^2 - r_m^2)\frac{d\sigma_c}{dx} = 0 \tag{4.17}$$

Combining (4.17) and (4.12), we obtain

$$2\pi r_f\tau - \pi(r_c^2 - r_m^2)\frac{d\sigma_c}{dx} = 0 \tag{4.18}$$

so that if we substitute (4.18) into (4.16) and (4.17) and simplify the resulting expression, we obtain the governing equation for the shear stress τ:

$$\frac{d^2\tau}{dx^2} - \eta^2\tau = 0 \tag{4.19}$$

In (4.19), η is given by

$$\begin{aligned} \eta^2 &= \frac{2G_m}{E_f(r_m - r_f)r_f}\left[1 + \frac{E_f}{E_c}\left(\frac{r_m^2}{r_c^2 - r_m^2}\right)\right] \\ &\approx \frac{2G_m}{E_f(r_m - r_f)r_f} \end{aligned} \tag{4.20}$$

assuming that $r_c \gg r_m$. Making use of the fact that, for our representative volume,

$$\frac{r_f}{r_m} = \sqrt{V_f} \tag{4.21}$$

we may simplify (4.20) to

$$\eta = \frac{1}{r_f}\sqrt{\frac{2G_m\sqrt{V_f}}{E_f(1 - V_f)}} \tag{4.22}$$

The appropriate boundary conditions for the problem are

(a) $\lim_{x\to\infty} \tau(x) = 0$,
(b) $\lim_{x\to\infty} \sigma_f(x) = \bar{\sigma}(E_f/E_c)$, and
(c) $\sigma_f(0) = 0$.

The solution for that fiber stress that results from the solution to (4.19) that satisfies the boundary conditions is given by

$$\sigma_f(x) = \frac{\bar{\sigma}E_f}{E_c}(1 - e^{-\eta x} \tag{4.23)}$$

If we define an efficiency parameter ϕ so that

$$\phi = \frac{\sigma_f(\Delta)}{\lim\limits_{x\to\infty} \sigma_f(x)} = 1 - e^{-\eta\Delta} \tag{4.24}$$

where Δ is half the *ineffective length* (the region over which the fiber does not carry the full far-field load), then

$$\frac{\Delta}{r_f} = \sqrt{\frac{E_f}{2G_m}}\left(\frac{1 - \sqrt{V_f}}{\sqrt{V_f}}\right)\ln\left(\frac{1}{1 - \phi}\right) \tag{4.25}$$

If we denote the ineffective length by δ, then we have

$$\frac{\delta}{r_f} = \sqrt{\frac{2E_f}{G_m}\left(\frac{1-\sqrt{V_f}}{\sqrt{V_f}}\right)} \ln\left(\frac{1}{1-\phi}\right) \tag{4.26}$$

Thus, the ineffective length has been determined in terms of the constituent properties (E_f and G_m) as well as the geometry (r_f and V_f).

We can next attempt to estimate the strength of the composite by using a rule of mixtures for a fiber bundle that has a length δ, so that we have

$$X_t = V_f X_b + V_m \bar{\sigma}_m \tag{4.27}$$

where X_b is the strength of a fiber bundle having a length δ given by (4.26), V_m is the matrix volume fraction, and $\bar{\sigma}_m$ is the average matrix stress. Using our work from (4.11), the strength of such a composite may be estimated as

$$X_t = \frac{1}{(me)^{1/m}\Gamma[(m+1)/m]}\left(\frac{l}{\delta}\right)^{1/m} V_f \bar{X}_f + V_m \bar{\sigma}_m \tag{4.28}$$

where $\bar{X}_f$ is the average strength of the fibers in our composite of length l. However, in such an approach, we have not considered the overstresses (stress concentrations) that result on the fibers adjacent to a broken fiber within its ineffective length. Our next analysis will describe an attempt to incorporate these overstresses into the tensile strength model.

4.1.2 Batdorf Tensile Strength

Batdorf (1982a,b) considers a composite containing N fibers, each of length l, which are held together by a matrix. Damage in the composite due to loading is assumed to consist solely of breaks in the fibers. There will be single isolated breaks (*singlets*), pairs of breaks (*doublets*), three adjacent breaks (*triplets*), and (in general) i adjacent breaks (*i-plets*). Each i-plet is surrounded by n_i nearest neighbors, each of which is subjected to a maximum stress concentration of c_i in the plane of the break. This stress concentration acts over an axial distance of δ_i (the ineffective length). We first assume that the fiber failure conforms to a two-parameter Weibull representation. Therefore, when a stress $\hat{\sigma}$ is applied to a fiber of length l, the probability of failure P_f is given by

$$P_f(\sigma) = 1 - \exp\left[-\frac{l}{l_0}\left(\frac{\hat{\sigma}}{\sigma_0}\right)^m\right] \tag{4.29}$$

where, as in the bundle strength model, σ_0 is the Weibull characteristic value, m is the Weibull modulus, and l_0 is the reference length. For the case in which $P_f \ll 1$, Batdorf approximates (4.29) by

$$P_f \approx \frac{l}{l_0}\left(\frac{\hat{\sigma}}{\sigma_0}\right)^m \tag{4.30}$$

The number of singlets, Q_1, may then be determined by multiplying the probability of failure given by (4.30) so that

$$Q_1 = NP_f = N\frac{l}{l_0}\left(\frac{\hat{\sigma}}{\sigma_0}\right)^m \tag{4.31}$$

Now we look in the vicinity of the broken fiber. The load that was originally carried by the broken fiber is now supported by the adjacent fibers. As a first approximation, we will assume that only the n_1 fibers immediately adjacent to the broken fiber experience an overstress, or stress concentration, given by

$$\sigma(z) = \hat{\sigma}\left[c_1 + \frac{2|z|}{\delta_1}(1 - c_1)\right] \tag{4.32}$$

where z is the axial distance away from the plane of the broken fiber, c_1 is the stress concentration on the adjacent fibers in the plane of the break, and δ_1 is the ineffective length, the distance over which stress is transferred back into the broken fiber by shear stresses in the surrounding matrix. Next we must determine the reliability of a fiber having a stress variation of the type given by the linear form in (4.32). To do so, we divide our fiber up into $N_{\text{increments}}$ pieces having length Δz_i ($1 \le i \le N_{\text{increments}}$) in the interval $-\frac{1}{2}\delta \le z \le \frac{1}{2}\delta$. The reliability of the ith increment is given by

$$R_i = \exp\left[-\frac{\Delta z_i}{l_0}\left(\frac{\sigma_i}{\sigma_0}\right)^m\right] \tag{4.33}$$

so that the total reliability is given by the products of the individual reliabilities, or

$$\begin{aligned} R &= \prod_{i=1}^{N_{\text{increments}}} \exp\left[-\frac{\Delta z_i}{l_0}\left(\frac{\sigma_i}{\sigma_0}\right)^m\right] \\ &= \exp\left[-\sum_{i=1}^{N_{\text{increments}}} \frac{\Delta z_i}{l_0}\left(\frac{\sigma_i}{\sigma_0}\right)^m\right] \end{aligned} \tag{4.34}$$

where σ_i is the stress in the ith increment. Now, if we take the limit as the number of intervals goes to infinity (or as Δz_i goes to zero), we obtain

$$R = \exp\left[-\int_{-\delta_1/2}^{\delta_1/2}\left(\frac{\sigma(z)}{\sigma_0}\right)^m \frac{dz}{l_0}\right] \tag{4.35}$$

For the linear variation in stress given by Equation (4.32), Equation (4.35) reduces to

$$R = \exp\left[-\frac{\delta_1}{l_0}\left(\frac{\hat{\sigma}}{\sigma_0}\right)^m \frac{c_1^{m+1}-1}{(m+1)(c_1-1)}\right] \tag{4.36}$$

so that the probability of failure within the interval $-\frac{1}{2}\delta \leq z \leq \frac{1}{2}\delta$ is given by

$$\begin{aligned} P_f &= 1 - \exp\left[-\frac{\delta_1}{l_0}\left(\frac{\hat{\sigma}}{\sigma_0}\right)^m \frac{c_1^{m+1}-1}{(m+1)(c_1-1)}\right] \\ &\approx \frac{\delta_1}{l_0}\left(\frac{\hat{\sigma}}{\sigma_0}\right)^m \frac{c_1^{m+1}-1}{(m+1)(c_1-1)} \end{aligned} \tag{4.37}$$

or, in Batdorf's notation,

$$P_f \approx \frac{\lambda_1}{l_0}\left(c_1\frac{\hat{\sigma}}{\sigma_0}\right)^m \tag{4.38}$$

where

$$\lambda_1 = \frac{c_1^{m+1}-1}{c_1^m(m+1)(c_1-1)}\delta_1 \tag{4.39}$$

Because there are n_1 nearest neighbors to each singlet, the probability that a singlet becomes a doublet is given by

$$P_{1\to 2} = n_1\frac{\lambda_1}{l_0}\left(c_1\frac{\hat{\sigma}}{\sigma_0}\right)^m \tag{4.40}$$

The number of singlets, Q_1, is given by Equation (4.31). Therefore, we may estimate the number of doublets by

$$Q_2 = Q_1 n_1 \frac{\lambda_i}{l_0}\left(c_i\frac{\sigma}{\sigma_0}\right)^m \tag{4.41}$$

In general, the number of i-plets may be given by

$$Q_{i+1} = Q_i n_i \frac{\lambda_i}{l_0}\left(c_i \frac{\sigma}{\sigma_0}\right)^m \tag{4.42}$$

or equivalently,

$$Q_i = N\frac{L}{l_0}\left(\frac{\sigma}{\sigma_0}\right)^{im} \prod_{j=1}^{i-1} c_j^m n_j \frac{\lambda_j}{l_0} \tag{4.43}$$

Examining the form of Equation (4.43), it is readily apparent that a log-log plot of Q_i versus σ is a straight line having slope im. A representative schematic diagram of Equation (4.43) is shown in Figure 4.4 for several i-plets as a function of the applied stress level. This envelope of intersection points has a special significance in the Batdorf formulation. Over the stress range in which an i-plet lies on the envelope, it is unstable [as soon as it is created it will immediately become an $(i+1)$-plet, which will immediately become an $(i+2)$-plet until composite failure occurs]. The failure stress is given by the lowest stress at which any unstable i-plet is present. This is the stress at which the envelope intersects the horizontal line $Q_i = 1$ (or $\ln Q_i = 0$).

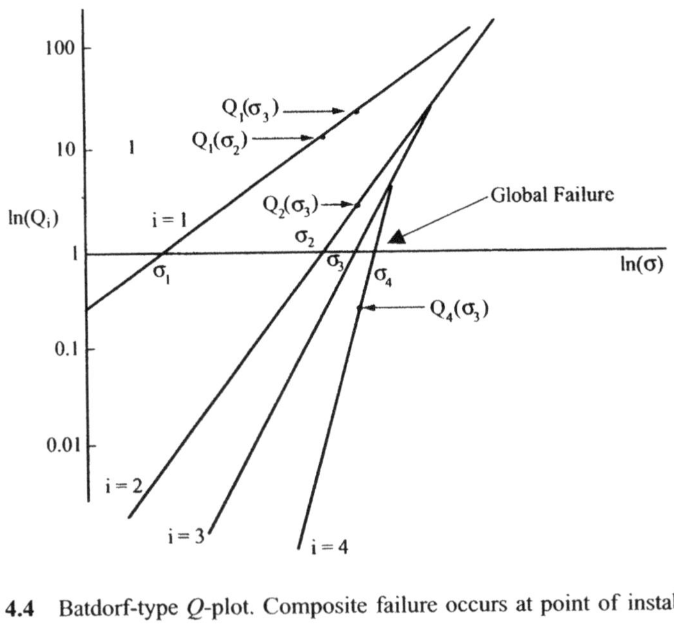

Figure 4.4 Batdorf-type Q-plot. Composite failure occurs at point of instability.

4.1.3 Stress Concentrations and Ineffective Lengths

Although the shear-lag approach described previously does give us insight into a number of factors that affect the stress redistribution around a fiber break, it does not give us the stress concentrations on the adjacent fibers. The most widely quoted values for stress concentrations are those obtained by Hedgepeth (1961) and Hedgepeth and Van Dyke (1967). This approach may be illustrated for the two-dimensional case of a series of aligned fibers shown in Figure 4.5, although solutions have been obtained for three-dimensional arrangements of fibers (Hedgepeth and Van Dyke, 1967).

Each of the fibers has a displacement and a corresponding load so that

$$P_n = EA\frac{dU_n}{dx} \tag{4.44}$$

By using an approach similar to that employed in the shear-lag analysis, we arrive at the equilibrium equation for the nth fiber given as

$$EA\frac{d^2U_n}{dx^2} = -\frac{Gh}{d}(U_{n+1} + U_{n-1} - 2U_n) \tag{4.45}$$

where h is the thickness. For the case in which r fibers are broken in the plane $x = 0$, the boundary conditions are given by

$$\begin{aligned} P_n(0) &= 0 \qquad 0 \le n \le r-1 \\ U_n(0) &= 0 \qquad \text{all others} \\ \lim_{x\to\infty} P_n(x) &= p_\infty \end{aligned}$$

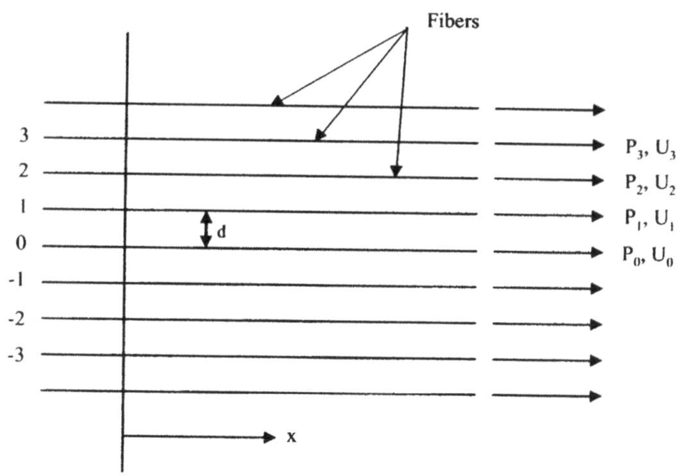

Figure 4.5 Two-dimensional geometry considered by Hedgepeth (1961).

We may rewrite Equation (4.45) in nondimensional form as

$$\frac{d^2 u_n}{d\xi^2} + (u_{n+1} + u_{n-1} - 2u_n) = 0 \tag{4.47}$$

where

$$\begin{aligned} P_n &= p_n p_\infty \\ U_n &= u_n p_\infty \sqrt{\frac{d}{EAGh}} \\ x &= \sqrt{\frac{EAd}{Gh}}\xi \end{aligned} \tag{4.48}$$

Hedgepeth first considers an auxiliary problem in which a single fiber ($n = 0$) is broken and displaced one unit, while a zero displacement condition at $\xi = 0$ is maintained in the remaining fibers. The resulting elemental displacements are represented by $V_n(\xi)$ and the stresses by $L_n(\xi)$. These influence functions are superimposed in the general case to yield

$$\begin{aligned} p_n(\xi) &= 1 + \sum_{m=0}^{r-1} L_{n-m}(\xi) u_m(0) \\ u_n(\xi) &= \xi + \sum_{m=0}^{r-1} V_{n-m}(\xi) u_m(0) \end{aligned} \tag{4.49}$$

The influence functions are determined from the solution of the auxiliary problem having boundary conditions

$$\begin{aligned} V_0(0) &= 1 \\ V_n(0) &= 0 \qquad n \neq 0 \\ \lim_{x \to \infty} \frac{dV_n(x)}{d\xi} &= 0 \end{aligned} \tag{4.50}$$

If we employ the discrete Fourier transform so that

$$\begin{aligned} \bar{V}(\xi, \theta) &= \sum_{n=-\infty}^{\infty} V_n(\xi) e^{-in\theta} \\ V_n(\xi) &= \frac{1}{2\pi} \int_{-\pi}^{\pi} \bar{V}(\xi, \theta) e^{-in\theta} \, d\theta \end{aligned} \tag{4.51}$$

applying Equation (4.51) to Equation (4.47) [and replacing $u_n(\xi)$ with $V_n(\xi)$], the governing equation becomes

$$
\begin{aligned}
\sum_{n=-\infty}^{\infty} \frac{d^2 V_n}{d\xi^2} e^{-in\theta} + \sum_{n=-\infty}^{\infty} V_{n+1} e^{-in\theta} + \sum_{n=-\infty}^{\infty} V_{n-1} e^{-in\theta} - 2 \sum_{n=-\infty}^{\infty} V_n e^{-in\theta} &= 0 \\
\frac{d^2 \bar{V}}{d\xi^2} + e^{i\theta} \sum_{n=-\infty}^{\infty} V_{n+1} e^{-i(n+1)\theta} + e^{-i\theta} \sum_{n=-\infty}^{\infty} V_{n+1} e^{-i(n-1)\theta} - 2\bar{V} &= 0 \\
\frac{d^2 \bar{V}}{d\xi^2} + (e^{i\theta} + e^{-i\theta} - 2)\bar{V} &= 0 \\
\frac{d^2 \bar{V}}{d\xi^2} - 4 \sin^2\left(\frac{\theta}{2}\right)\bar{V} &= 0
\end{aligned}
\tag{4.52}
$$

whose solution subject to the boundary conditions given in Equation (4.46) is

$$
\bar{V}(\xi, \theta) = e^{-2\sin(\theta/2)|\xi|} \tag{4.53}
$$

Applying the inverse transformation, we arrive at the influence functions

$$
\begin{aligned}
V_n(\xi) &= \operatorname{sgn}(\xi) \frac{1}{\pi} \int_0^{\pi} \cos(n\theta) e^{-2|\xi|\sin(\theta/2)}\, d\theta \\
L_n(\xi) &= -\frac{2}{\pi} \int_0^{\pi} \cos(n\theta) \sin\frac{\theta}{2} e^{-2|\xi|\sin(\theta/2)}\, d\theta
\end{aligned}
\tag{4.54}
$$

so that

$$
\begin{aligned}
L_n(0) &= -\frac{2}{\pi} \int_0^{\pi} \cos(n\theta)\sin(\tfrac{1}{2}\theta)\, d\theta \\
&= \frac{1}{\pi} \int_0^{\pi} [\sin(n - \tfrac{1}{2})\theta - \sin(n + \tfrac{1}{2})\theta] \\
&= \frac{1}{\pi} \left[-\frac{\cos(n - \frac{1}{2})\theta}{n - \frac{1}{2}} + \frac{\cos(n + \frac{1}{2}\theta}{n + \frac{1}{2}} \right]_0^{\pi} \\
&= \frac{4}{\pi(4n^2 - 1)}
\end{aligned}
\tag{4.55}
$$

Suppose that we break one fiber ($n = 0$) so that $r = 1$. In this case,

$$\begin{aligned} p_0(0) &= 0 = 1 + L_0(0)u_0(0) \\ 0 &= 1 + \left(-\frac{4}{\pi}\right)u_0(0) \Rightarrow u_0(0) = \tfrac{1}{4}\pi \end{aligned} \tag{4.56}$$

The stress concentration on the adjacent fibers is then given by

$$\begin{aligned} C_1 &= p_1(0) = 1 + L_1(0)u_0(0) \\ &= 1 + \left(\frac{4}{3\pi}\right)\left(\frac{\pi}{4}\right) \\ &= \tfrac{4}{3} \end{aligned} \tag{4.57}$$

This process may be repeated for any number of breaks to yield the well-known result that

$$C_N = \frac{2^{2N+1}[(N+1)!]^2}{(2N+2)!} \tag{4.58}$$

The stress distributions on the broken fiber and the nearest unbroken fibers for the case of a single break are shown in Figure 4.6. As the number of adjacent breaks increases, both the stress concentration and ineffective length increase. An important feature of this result is that the stress concentrations are *independent* of the constituent properties as well as the fiber volume fraction (although the resulting ineffec-

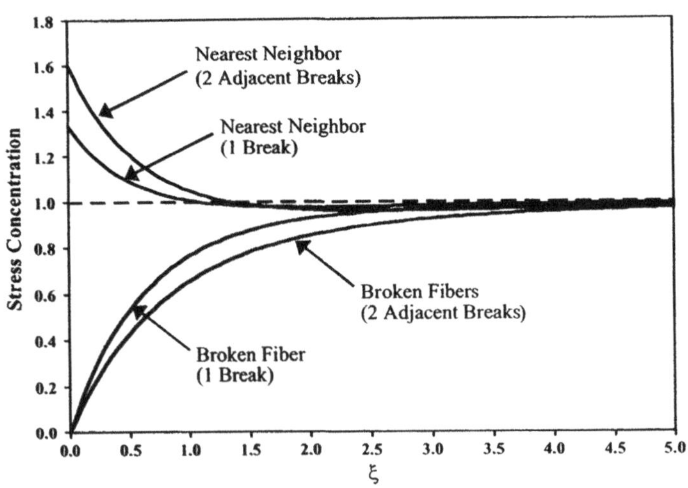

Figure 4.6 Results from Hedgepeth (1961) analysis for single break and two adjacent breaks.

tive lengths do depend upon these properties). Carman et al. (1993b) and Case et al. (1995) used a model composite system to obtain an experimental comparison of this result. A limitation of the Hedgepeth and Van Dyke (1967) analysis is that it can become computationally intensive, particularly for larger numbers of adjacent broken fibers. Also, it is difficult to include yielding at the interface between the fiber and matrix. Therefore, a simpler model is desirable. For our purposes, we will examine a model similar to the shear-lag model developed by Gao and Reifsnider (1992).

Gao and Reifsnider begin by assuming there is a central core of i broken fibers, as shown in Figure 4.6.

The broken fibers are flanked by unbroken fibers. A homogeneous "effective" material that is regarded as being strained uniformly throughout flanks the unbroken fibers. It is assumed that the core of i broken fibers can be approximated by a homogeneous material with circular cross section whose Young's modulus is obtained by a rule of mixtures. The radius of this core is given by $r_0 - d$. The radius r_0 is determined from the equivalency of areas so that the area of the equivalent broken core is equal to the total areas of i concentric cylinders of radius $r_f + d$,

$$\pi r_0^2 = i\pi(r_f + d)^2 = iA_f + iA_m \tag{4.59}$$

where r_f is the radius of the fibers and d is half of the width of the matrix region (also the thickness of the matrix in the concentric cylinder, as shown in Figure 4.7.

The fiber area is given by

$$A_f = \pi r_f^2 \tag{4.60}$$

and the area of the matrix is given by

$$A_m = \pi(r_f + d)^2 - \pi r_f^2 \tag{4.61}$$

The number of neighboring unbroken fibers, denoted n_i, is dependent upon the number of broken fibers, i. It is assumed that the fibers carry only axial stress while the matrix supports shear only. The half-length of the region of matrix and interfacial yielding is given by a. In the region of matrix and interfacial yielding, the equilibrium equations are written as

$$\begin{gathered} E\pi(r_0 - d)^2 \frac{d^2U_0}{dx^2} - 2\pi r_0 \eta \tau_0 = 0 \\ n_i A_f E_f \frac{d^2U_1}{dx^2} + 2\pi(r_0 + 2d + 2r_f)\frac{G_m}{2d}(U_2 - U_1) + 2\pi r_0 \eta \tau_0 = 0 \\ 0 \le x \le a \end{gathered} \tag{4.62}$$

where U_0, U_1, and U_2 are displacements of the broken core, neighboring unbroken fiber, and average composite, respectively; E_f and G_m are the Young's modulus of

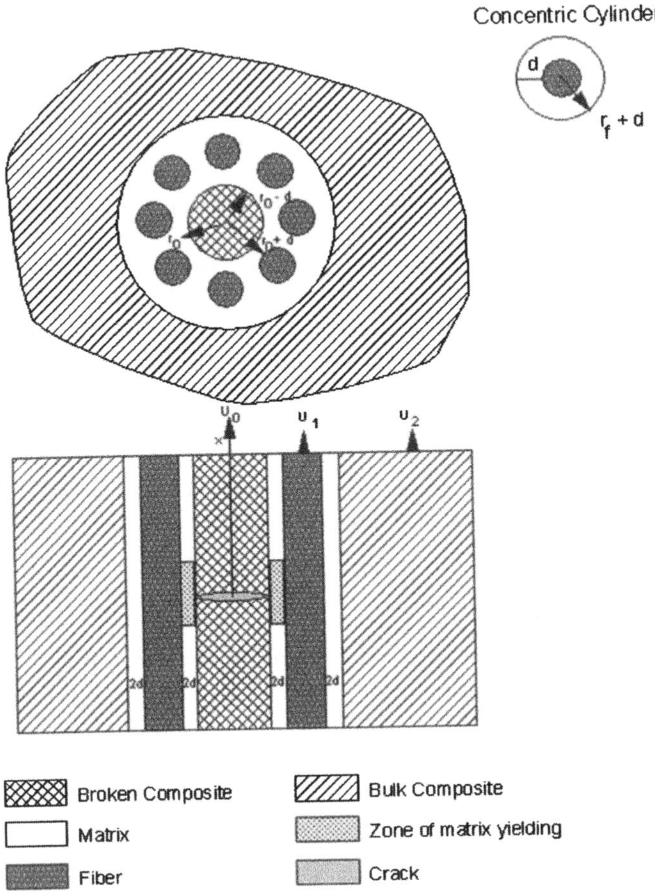

Figure 4.7 Micromechanical model used by Gao and Reifsnider (1992) to describe fiber fracture in unidirectional composites.

the fiber and the shear modulus of the matrix; A_f is the area of the fiber; τ_0 is the yielding stress of the matrix and interface; and η is a shear parameter. This shear parameter defines the shear stress in the inelastic region. When η is equal to zero, there is no shear stress transfer between broken fibers and their neighbors in that region. This would be representative of complete fiber–matrix debonding or matrix cracking in the region. When $\eta = 1$, the shear stress in the inelastic region is equal to τ_0 (an elastic, perfectly-plastic material behavior). The Young's modulus of the broken core, E, is determined by the rule of mixtures, so that

$$E = \frac{iA_f E_f + [iA_m - \pi(r_0^2 - (r_0 - d)^2)]E_m}{\pi(r_0 - d)^2} \tag{4.63}$$

where E_m is the Young's modulus of the matrix. Hence,

$$\begin{aligned} E\pi(r_0 - d)^2 &= iA_f E_f + \{iA_m - \pi[r_0^2 - (r_0 - d)^2]\}E_m \\ &= iA_f E_f \quad 1 + \frac{iA_m - \pi[r_0 - (r_0 - d)^2]}{iA_f E_f} E_m \\ &= iA_f E_f \beta \end{aligned} \tag{4.64}$$

where

$$\beta = \left\{ 1 + \frac{iA_m - \pi[r_0^2 - (r_0 - d)^2]}{iA_f E_f} E_m \right\} \tag{4.65}$$

Using Equation (4.63), we may rewrite Equation (4.61) as

$$\begin{aligned} iA_f E_f \beta \frac{d^2 U_0}{dx^2} - 2\pi r_0 \eta \tau_0 = 0 \\ n_i A_f E_f \frac{d^2 U_1}{dx^2} + 2\pi(r_0 + 2d + 2r_f)\frac{G_m}{2d}(U_2 - U_1) + 2\pi r_0 \eta \tau_0 = 0 \end{aligned} \tag{4.66}$$

Similarly, for the region where no interfacial yielding has occurred, we have

$$\begin{aligned} E\pi(r_0 - d)^2 \frac{d^2 U_0}{dx^2} + 2\pi r_0 \frac{G_m}{2d}(U_1 - U_0) = 0 \\ n_i A_f E_f \frac{d^2 U_1}{dx^2} + 2\pi(r_0 + 2d + 2r_f)\frac{G_m}{2d}(U_2 - U_1) - 2\pi r_0 \frac{G_m}{2d}(U_1 - U_0) = 0 \\ a \le x < \infty \end{aligned} \tag{4.67}$$

which can be rewritten as

$$\begin{aligned} iA_f E_f \beta \frac{d^2 U_0}{dx^2} + 2\pi r_0 \frac{G_m}{2d}(U_1 - U_0) = 0 \\ n_i A_f E_f \frac{d^2 U_1}{dx^2} + 2\pi(r_0 + 2d + 2r_f)\frac{G_m}{2d}(U_2 - U_1) - 2\pi r_0 \frac{G_m}{2d}(U_1 - U_0) = 0 \end{aligned} \tag{4.68}$$

Gao and Reifsnider (1992) assume that the strain in the average composite is constant, so that

$$U_2 = \frac{\sigma}{E} x \tag{4.69}$$

where σ is the remote fiber stress. If we next define

$$U_0 = \sigma u_0 \sqrt{\frac{2\,dA_f}{E_f G_m r_0}} \qquad U_1 = \sigma u_1 \sqrt{\frac{2\,dA_f}{E_f G_m r_0}}$$
$$x = \xi \sqrt{\frac{2A_f E_f d}{G_m r_0}} \qquad \tau_0 = \bar{\tau}_0 \sigma \sqrt{\frac{A_f G_m}{2\,dE_f\,r_0}} \qquad a = \alpha \sqrt{\frac{2A_f E_f d}{G_m r_0}} \tag{4.70}$$

we can rewrite Equations (4.66) and (4.68) as $(0 \le \xi \le \alpha)$

$$\frac{d^2 u_0}{d\xi^2} - \frac{2\pi}{i\beta}\eta\bar{\tau}_0 = 0$$
$$\frac{d^2 u_1}{d\xi^2} - \phi t u_1 + \phi(\eta\bar{\tau}_0 + t\xi) = 0 \tag{4.71}$$

and $(\alpha \le \xi < \infty)$

$$\frac{d^2 u_0}{d\xi^2} + \frac{2\pi}{i\beta}(u_1 - u_0) = 0$$
$$\frac{d^2 u_1}{d\xi^2} - \phi(1+t)u_1 + \phi u_0 = -\phi t \xi \tag{4.72}$$

where

$$t = \frac{r_0 + 2d + 2r_f}{r_0} \tag{4.73}$$

$$\phi = \frac{2\pi}{n_i} \tag{4.74}$$

Equations (4.71) and (4.72) must be solved subject to the boundary conditions

$$\frac{du_0(0)}{d\xi} = u_1(0) = 0 \qquad \frac{du_0(\infty)}{d\xi} = \frac{du_1(\infty)}{d\xi} = 1 \tag{4.75}$$

The solution of Equation (4.71) is given by

$$u_0 = \frac{\pi\eta\bar{\tau}_0}{i\beta}\xi^2 + A_1$$
$$u_1 = \frac{\eta\bar{\tau}_0}{t} + \xi + A_2 \exp(\xi\sqrt{t\phi}) - \left(A_2 + \frac{\eta\bar{\tau}_0}{t}\right)\exp(-\xi\sqrt{t\phi}) \tag{4.76}$$

where A_1 and A_2 are constants to be determined from continuity conditions. For Equation (4.71), the solution is given by

$$\begin{aligned} u_0 &= \xi + B_1\left(1 + t - \frac{\lambda_1}{\phi}\right) \exp(-\sqrt{\lambda_1}\xi) + B_2\left(1 + t - \frac{\lambda_2}{\phi}\right) \exp(-\sqrt{\lambda_2}\xi) \\ u_1 &= \xi + B_1 \exp(-\sqrt{\lambda_1}\xi) + B_2 \exp(-\sqrt{\lambda_2}\xi) \end{aligned} \tag{4.77}$$

where B_1 and B_2 are constants and

$$\begin{aligned} \lambda_1 &= n_i + i\beta(1+t) + \sqrt{n_i^2 + 2i\beta n_i(1-t) + (i\beta)^2(1+t)^2} \quad \frac{\pi}{i\beta n_i} \\ \lambda_2 &= n_i + i\beta(1+t) - \sqrt{n_i^2 + 2i\beta n_i(1-t) + (i\beta)^2(1+t)^2} \quad \frac{\pi}{i\beta n_i} \end{aligned} \tag{4.78}$$

The stress concentrations on the unbroken fibers, $C_i(\xi)$, and the dimensionless shear stress, $\bar{\tau}(\xi)$, are expressed in terms of the displacements as

$$C_i(\xi) = \frac{du_1(\xi)}{d\xi} \qquad \bar{\tau}(\xi) = u_0 - u_1 \tag{4.79}$$

The coefficients A_1, A_2, B_1, B_2, and α are determined from the continuity from the displacement and stress continuity conditions at $\xi = \alpha$ and

$$u_0(\alpha) - u_1(\alpha) = \bar{\tau}(\alpha) = \bar{\tau}_0 \tag{4.80}$$

The stress concentrations determined from Equation (4.79) may be used to determine the ineffective length for Batdorf's model (1982a,b) by defining the ineffective length to be the distance required for the stress on the adjacent fibers to return to its far-field value. To illustrate the effectiveness of this modeling approach, we may compare the results for the Gao and Reifsnider (1992) model to those from the Hedgepeth and Van Dyke (1967) analysis. Although the stress concentrations for the Hedgepeth and Van Dyke model are independent of the fiber and matrix properties as well as the volume fraction, the results from the Gao and Reifsnider model do depend weakly on these values. For illustration purposes, we will use values typical of polymer matrix composites, so that

$$\mathrm{E} = 235\ \mathrm{GPa} \qquad G_{\mathrm{m}} = 1.074\ \mathrm{GPa} \qquad V_f = 0.6$$

and the Gao and Reifsnider analysis will be strictly elastic (so that $a = 0$). The results of such a comparison for a square packed array are shown in Figures 4.8 and Figure 4.9.

For the Gao and Reifsnider analysis, each of the adjacent fibers has the same stress concentration and ineffective length, while for the Hedgepeth and Van Dyke analysis, the stress concentrations and ineffective lengths depend upon the location

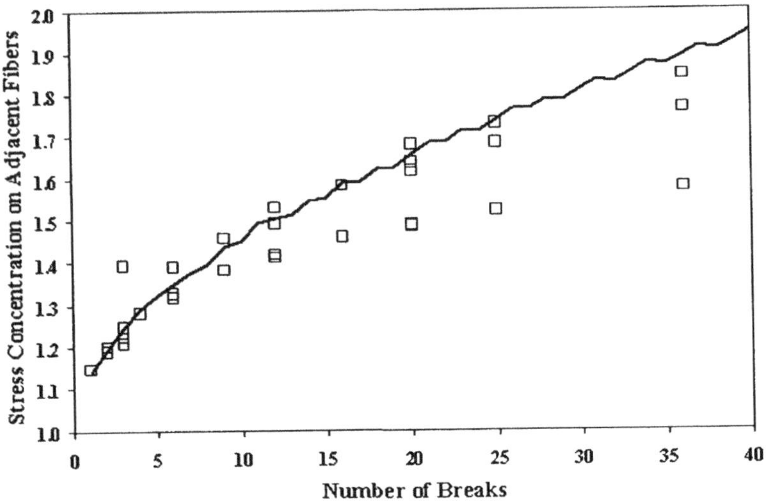

Figure 4.8 Comparison of stress concentrations from Gao and Reifsnider (1992) model (solid line) to those from Hedgepeth and Van Dyke (1967) analysis (open squares) for square-packed array.

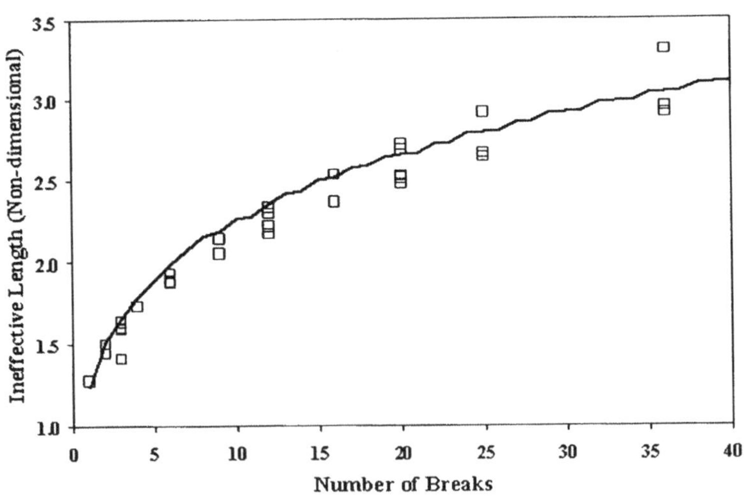

Figure 4.9 Comparison of ineffective lengths from Gao and Reifsnider (1992) model (solid line) to those from Hedgepeth and Van Dyke (1967) analysis (open squares) for square-packed array.

of the adjacent fibers with respect to the broken fibers. However, the overall agreement is quite good, particularly up to approximately 12 broken fibers. Similar results may also be obtained for a hexagonal-packed array. Once these stress concentrations and ineffective lengths have been determined, we may apply the Batdorf analysis to estimate the tensile strength of the composite. As an example, we will consider composites reinforced with AS-4 fibers. Wimolkiatisak and Bell (1989) have studied the strength of these fibers using the single-fiber fragmentation test. Using their results, we may estimate the Weibull characteristic strength value σ_0 to be 5.25 GPa at a reference length l_0 of 1 mm. The Weibull modulus m is estimated to be 10.65. To apply the Batdorf analysis (1982a,b), we also require the length of the composite, l. A typical gage length for a laboratory specimen is 150 mm, so we will use this value in our initial estimate of the tensile strength. Also, a typical tensile specimen geometry is 12.5 × 1.0 mm. The number of fibers in the composite, N, may be estimated as

$$N = \frac{A_{\text{composite}} V_f}{A_{\text{fiber}}} \tag{4.81}$$

where V_f is the fiber volume fraction and $A_{\text{composite}}$ and A_{fiber} are the composite cross-sectional area and the area of a single fiber, respectively. For AS-4 fibers, the fiber diameter is roughly 7 μm. Therefore, the number of fibers present in a composite with a 60% fiber volume fraction is

$$N = \frac{(12.5\ \text{mm})(1.0\ \text{mm})(0.6)}{\pi(0.007\ \text{mm})^2/4} \approx 195{,}000 \tag{4.82}$$

The required values for the stress concentrations and ineffective lengths are determined using the Gao and Reifsnider (1992) analysis for elastic behavior (no yielding region). These values, along with the number of adjacent fibers, are given in Table 4.1. Applying the Batdorf (1982a,b) analysis yields a *fiber* stress at failure of 4.01 GPa. This critical fiber stress is reached when the stress in the composite reaches 2.41 GPa—a value that corresponds to the tensile strength (X_t) of the composite. For comparison purposes, typical values for AS-4 fiber based composites from the literature are 1890 MPa for AS-4/Epon828 composites (Madhukar and Drzal, 1991), 2090 MPa for AS-4/PEEK composites (Vure, 1993), and 1830 MPa for AS-4/J2 composites (Chang et al., 1994). Other researchers have attempted to improve upon the shear-lag and related analysis methods. In his analysis, McCartney (1993) developed an approximate solution to the problem of multiple concentric cylinders. The analysis is approximate in the sense that the axial stress in each of the cylinders is assumed to be constant in the radial direction. By smearing the nearest neighboring fibers into a cylinder, this multiple concentric cylinder model may be used to estimate the stress concentrations and the ineffective lengths surrounding isolated fiber fractures. Using this approach, McCartney was able to determine that the maximum stress concentration experienced by the neighboring fibers (to a single break) was 1.1694. However, because the stresses decay rapidly away from the crack

TABLE 4.1 Stress Concentrations and Ineffective Lengths for Fiber Fracture Analysis

i	c_i	δ_i (μm)	n_i
1	1.104	27.7	6
2	1.155	32.5	8
3	1.204	35.1	9
4	1.243	37.3	10
5	1.276	39.3	11
6	1.304	41.1	12
7	1.349	41.8	12
8	1.349	44.3	14
9	1.388	44.8	14
10	1.405	46.2	15
11	1.441	46.7	15
12	1.453	47.9	16
13	1.487	48.3	16
14	1.497	49.5	17
15	1.529	49.9	17
16	1.536	51.0	18
17	1.566	51.3	18
18	1.572	52.4	19
19	1.600	52.7	19
20	1.604	53.7	20

tip, the (area) average stress concentration over the adjacent fibers was 1.0671. The maximum value is greater than that predicted by the Hedgepeth and Van Dyke (1967) analysis (1.104), although the average value is less. A similar analysis has been developed by Case and Reifsnider (1996). In this analysis, as in the McCartney (1993) analysis, a smearing technique was employed to model the fibers adjacent to the broken fiber. The resulting axisymmetric domain allows the development of an exact elasticity solution in the form of an integral equation that may be solved numerically. Two major limitations exist in this solution. The first limitation is a result of the solution technique employed and requires that the crack be contained within a single domain (i.e., the fiber) so those cracks that extend into the matrix cannot be modeled. The second limitation is a result of the displacement potentials employed in the solution and requires that isotropic constituents only be modeled. As a result, stress analyses were conducted on glass fiber composites. Results were obtained that were similar in character to those obtained by McCartney. In an effort to improve upon the limitations of the elasticity solution, Caliskan (1996) and Case (1996) applied the axisymmetric analysis of Pagano (1991) to the fiber fracture problem. Again, a smearing technique was employed to model a hexagonal array. To determine the stress concentrations due to multiple broken fibers, mixed boundary conditions were applied to multiple rings of fibers. Thus, for the smeared hexagonal array model, it was possible to determine the stress concentrations due to

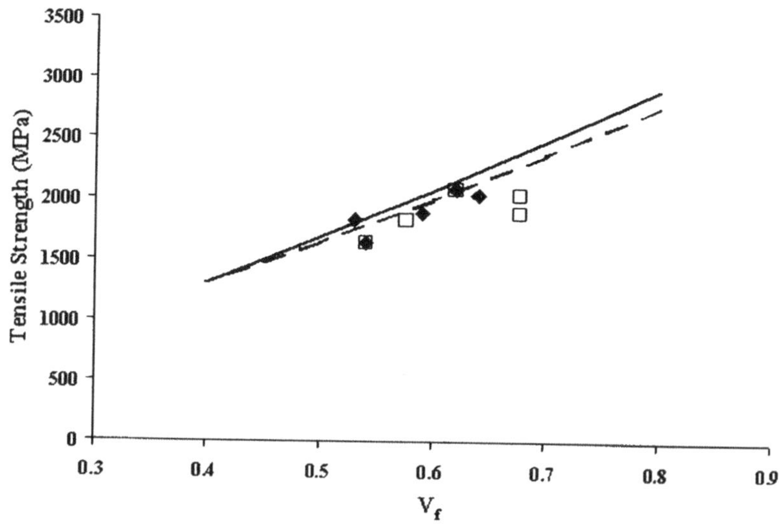

Figure 4.10 Comparison of predicted and observed tensile strengths for AS-4 based composites. Solid line corresponds to average stress concentration; dashed line to maximum stress concentration.

$1, 7, 19, \ldots$ adjacent breaks. Because of the radial decay of the stresses, once again two different measures of the stress concentration were obtained: the maximum stress concentration and the average stress concentration. These values were used to calculate the strength for a number of AS-4 graphite-based composites. The resulting predictions, as well as the experimental values, are summarized in Figure 4.10.

As an alternative to the Batdorf approach, we may construct an analytical expression by arguing that unstable fiber fracture occurs when the chance of breaking n fibers together (i.e., forming an n-plet) is the same as the chance of forming an $(n+1)$-plet. If that is so, the number of $(n+1)$-plets would be exactly equal to the number of n-plets times the number of adjacent fibers around the n-plet, namely n_n fibers. Analytically this is stated as

$$Q_{n+1} = Q_n n_n \tag{4.83}$$

Then, using the Batdorf recursion relationships, (4.83) can be written as

$$Q_i n_i \frac{\lambda_i}{l_0}\left(c_i \frac{\sigma_c}{\sigma_0}\right)^m = Q_i n_i \tag{4.84}$$

where σ_c represents the critical stress. After simplification in (4.84), we obtain

$$\sigma_c = \frac{\sigma_0}{c_i} \frac{1}{(\lambda_i/l_0)^{1/m}} \tag{4.85}$$

Substituting our previous result for λ_i found in (4.39) into (4.85), we obtain

$$\sigma_c = \frac{\sigma_0}{\left[\frac{\delta_i}{l_0}\left(\frac{c_i^{m+1}-1}{c_i-1}\right)\right]^{1/m}} \tag{4.86}$$

Using the expansion of $c_i^{m+1} - 1$, we obtain

$$\sigma_c = \frac{\sigma_0}{\left[\frac{\delta_i}{l_0}\frac{1}{m+1}(c_i^m + c_i^{m-1} + \cdots 1)\right]^{1/m}} \tag{4.87}$$

The expression for the ineffective length δ_i in (4.87) for the case in which the fiber–matrix interfacial shear stress is constant is given by the relation

$$\delta_i = 2\left(\frac{DE_f\varepsilon}{4\tau_0}\right) \tag{4.88}$$

where D is the diameter of n broken fibers and ε and τ_0 represent the fiber strain to failure and fiber–matrix shear strength, respectively. The expression for the critical stress then becomes

$$\sigma_c = \frac{\sigma_0}{\left[\frac{DE_f\varepsilon}{2\tau_0 l_0(m+1)}(c_i^m + c_i^{m-1} + \cdots 1)\right]^{1/m}} \tag{4.89}$$

Collecting terms, we obtain

$$\sigma_c = \sigma_0^{(m-1)/m}\left(\frac{2\tau_0 l_0}{D}\right)^{1/m}\frac{(m+1)^{1/m}}{(c_i^m + c_i^{m-1} + \cdots 1)^{1/m}} \tag{4.90}$$

where the stress concentration factors c_i remain unknown at this point.

Figure 4.11 shows the basic concept for a model of fiber fracture that does incorporate local load sharing. As the diagram suggests, the stress in a broken fiber builds back up to the undisturbed level by shear transfer from the surrounding matrix, composite, and interphase region. That rate of buildup is directly proportional to the stress concentration in the next nearest fibers. If the buildup occurs over a short distance, the stress concentration in the neighboring fibers is great and they tend to break, causing very brittle composite behavior. If the buildup occurs over a large distance (e.g., there is a very low fiber–matrix coupling by the interphase region), the strength of each fiber is almost completely lost when the first local fiber break in that fiber occurs (at the lowest strength location in that fiber). Hence, a correct model of this situation would predict that there is an optimum

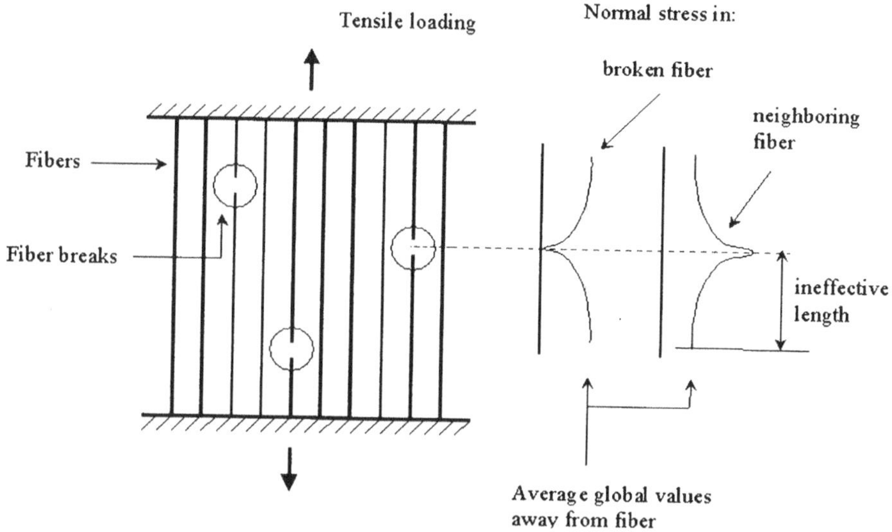

Figure 4.11 Local load-sharing concept for composite strength model based on fiber strength distribution.

strength for the interface/interphase that provides an intermediate condition to those two extremes.

Gao and Reifsnider (1992) and subsequently Reifsnider and Case (1997) have constructed such a model. Figure 4.12 shows a typical result of predictions with such a model, as in Equation (4.35). The most striking aspect of the predictions is that there is, in fact, an optimum in the strength of the composite as a function, in this case, of the interfacial strength between the fibers and matrix. Also shown in Figure 4.12 are observations for a unidirectional polymer composite in which the interface was systematically varied in strength. It is remarkable that the observed strength of the unidirectional composite shows the same large change in value as the general range predicted by the model. Changes of as much as 34% in unidirectional strength have been demonstrated in polymer composites by the authors. This is most certainly not a typical textbook result and is not suggested by concepts such as the rule of mixtures for unidirectional strength. Details of this prediction and the observations can be found in Reifsnider and Case (1997).

4.1.4 Ceramic Matrix Composites

While the Batdorf analysis (1982a,b) combined with the stress concentrations of Hedgepeth and Van Dyke (1967) or Gao and Reifsnider (1992) provide useful insights into the fiber direction strength of polymer matrix composite materials, these models are generally not appropriate for ceramic matrix composites. In these materials, fiber–matrix interface is usually designed to be weak; that is, the fibers are not bonded strongly (or not bonded at all in some cases) to the matrix. As a

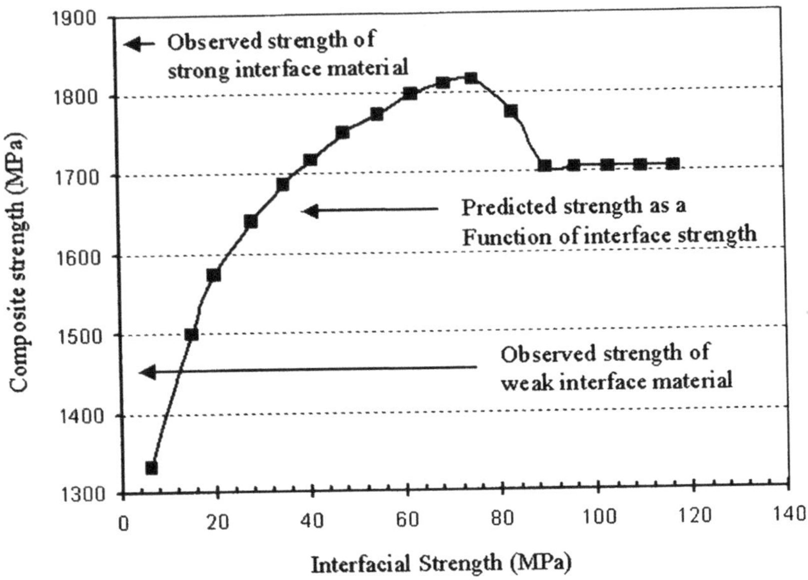

Figure 4.12 Predicted and observed composite strength as function of interface strength in carbon–PEEK system.

result, when fiber breaks occur, the load is transferred more or less uniformly to all unbroken fibers within the broken fiber plane. The behavior of these materials in which this *global load-sharing* behavior occurs has been considered in detail by Curtin (1991). In his analysis, Curtin considers a ceramic matrix of modulus E_m uniaxially reinforced with a volume fraction f of cylindrical fibers of radius r, modulus E_f, characteristic tensile strength σ_0 at a reference length L_0, and Weibull modulus m describing both the distribution of strengths about σ_0 and the scaling of strength with fiber length L. Until the point of matrix cracking, the response of the composite is linear, with the modulus given by

$$E = f E_f + (1 - f)E_m \tag{4.91}$$

After matrix cracking, the load-bearing capacity of the composite is the same as that of the fibers in any matrix crack plane. If we consider a single matrix crack plane at applied load σ, the average stress per fiber is σ/f. This load is partially carried by fibers with no breaks within $\pm l_f$ (the fiber slip length) of the matrix crack. The remainder of the load is carried by fibers that are broken within a slip length of the crack plane. The sliding length l_f is given by

$$l_f = \frac{r\sigma_f(l_f)}{2\tau} \tag{4.92}$$

where $\sigma_f(l_f)$ is the fiber stress previously existing at l_f away from the break. Following Curtin, we denote the stress carried by the unbroken fibers as T and the fraction of fibers broken within $\pm l_f$ by $q(l_f)$, so that we have

$$\frac{\sigma}{f} = T[1 - q(l_f)] + \frac{2\tau}{r}\langle L(T)\rangle q(l_f \qquad 4.93)$$

The first term in Equation (4.93) represents the load carried by the fraction $1 - q$ of unbroken fibers and the second term is the average load carried by the q broken fibers because of shear stress τ acting over the average pullout length $\langle L(T)\rangle$. Equation (4.93) may be simplified by recognizing that

$$l_f = \frac{rT}{2\tau} \qquad (4.94)$$

so that we have

$$\frac{\sigma}{f} = T\left[1 - \left(1 - \frac{\langle L(T)\rangle}{l_f}\right)q(l_f)\right] \qquad (4.95)$$

To make the tensile strength predictions, Curtin makes two further approximations. The first is to assume that at the value of T that maximizes Equation (4.95), the probability of finding two breaks on the same fiber within $\pm l_f$ of the matrix crack is negligible. As a result, the average pullout length is given by $\langle L(T)\rangle = \frac{1}{2}l_f$ so that Equation (4.95) reduces to

$$\frac{\sigma}{f} = T(1 - \tfrac{1}{2}q(l_f)) \qquad (4.96)$$

The second approximation is to approximate $q(l_f)$ by the fraction of breaks occurring in a gage length $2l_f$. This approximation gives

$$q(l_f) \approx \frac{2l_f}{L_0}\left(\frac{T}{\sigma_0}\right)^m \qquad (4.97)$$

Substituting Equation (4.94) into Equation (4.97), the result into Equation (4.96), and maximizing with respect to T, we arrive at

$$\sigma_u = f\left(\frac{\sigma_0^m \tau L_0}{r}\right)^{1/(m+1)}\left(\frac{2}{m+2}\right)^{1/(m+1)}\frac{m+1}{m+2} \qquad (4.98)$$

which is the expression for the tensile strength of the ceramic matrix composite material.

4.2 COMPRESSION STRENGTH

Compressive loading of continuous-fiber composites in the direction of the fibers can produce several different failure modes, sometimes in sequence. Some general features of this process are suggested by Figure 4.13. The stress–strain diagram is generally quite linear until a maximum is reached but upon close examination may be slightly nonlinear close to the failure loads. Failure is often quite abrupt, and "unloading" portions of the curve may not be apparent. However, as suggested by Figure 4.13, the compressive failure process can be quite complex and may involve combinations of the following:

1. *Fiber Crushing.* If the fibers are initially very straight and well aligned (A in Figure 4.13), and if those fibers have some resistance to bending (e.g., if they have a large diameter), and if the matrix and surrounding composite are stiff, then the fibers and surrounding material may crush in the manner of a brittle material, as suggested by B in Figure 4.13. Such failures have been observed in large-diameter boron fiber-reinforced epoxy composites, for example. This is a very strong failure mode, an ideal case in may ways, but difficult to obtain for "soft" matrix composites and small fiber diameters or for low-volume-fraction materials (if they are still fiber controlled). The compressive strength for this mode of failure is generally estimated well by the rule of mixtures, such that $X_c = V_f X_c^f + V_m X_c^m$ where X_c^f is the compressive strength of the fibers alone and X_c^m is the compressive strength of the matrix alone.
2. *Simple Shear.* If the fibers are misaligned relative to the loading axis (as shown by C in Figure 4.13) then axial loading may simply produce a shearing action, generally along the fiber or in the matrix between the fibers. This

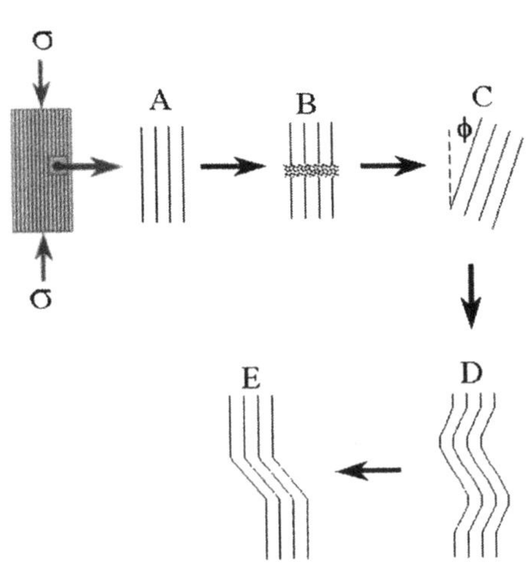

Figure 4.13 Failure processes for uniaxial compressive loading of unidirectional continuous-fiber composites.

general model has been discussed by Argon (1972) and will be described as part of more general equations derived below.

3. *Microbuckling.* If the matrix and surrounding composite do not provide sufficient support to prevent the fibers from buckling, a failure mode like the one shown by D in Figure 4.13 may occur, in which a "wavy" region forms in the region of instability, or the undulations of fibers may be due to manufacturing.
4. *Shear Band Formation.* If the fibers fail due to bending during the microbuckling process, they usually do so in a band of failure, as suggested by E in Figure 4.13. Subsequent deformation occurs primarily within that band, by a shearing motion. We will also describe such a model below.

As we will see, each of these failure modes can be described by a reasonable model. But the application of the models is not always straightforward. Combinations of the modes often occur. It is not unusual for a slight misalignment to contribute to the initiation of microbuckling and subsequent shear band formation, the sequence A–C–D–E in Figure 4.13. Then, the question becomes "which of the failure modes (and models) correspond to the highest point in the stress–strain curve?" If the fibers have a very small diameter, microbuckling may occur at a very low load compared to the subsequent load required to shear the material in a shear band. If the fibers are small but the volume fraction is large and the matrix if very stiff, microbuckling may require the largest load in the process, corresponding to the composite strength. And if that load is very large, that mode may not even occur and the load corresponding to one of the other failure modes may constitute the composite strength. Careful experimental observation of the microdetails of the failure modes and of manufacturing details such as the initial fiber alignment or curvature must be used to define which of the modes should be used and in what sequence. If definitive experimental data are available, models can predict the compressive strength with sufficient accuracy for most engineering applications. We will outline three of the major approaches to modeling uniaxial compression strength for unidirectional composite materials below.

4.2.1 Undulating Fiber Models

For fibers with small diameter, the orientation of the fibers may vary from point to point along their length. This "undulation," depicted in Figure 4.14, may be caused by manufacturing or be the result of buckling of the fibers under axial compressive load. A common representation of this behavior is usually ascribed to Rosen (1964).

Following that development, we consider an element of a fiber of length dx and width w_f in the deformed geometry shown in Figure 4.14. The equilibrium equation for the bending moments yields

$$\frac{dM}{dx} - V + \sigma A \frac{dv}{dx} = 0 \tag{4.99}$$

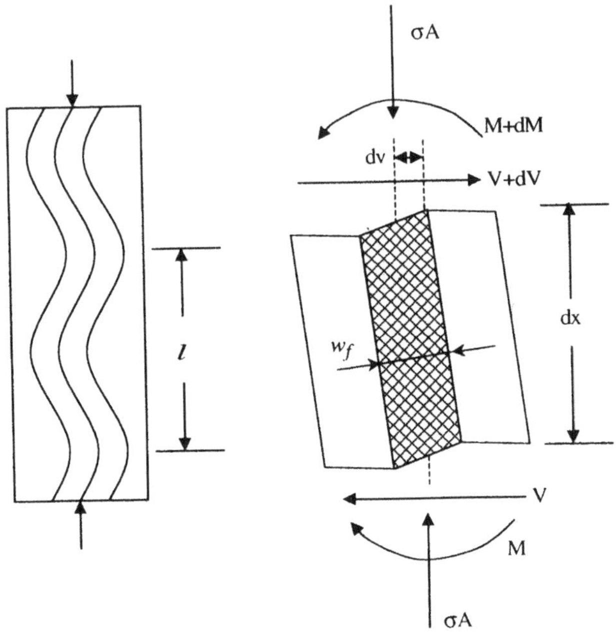

Figure 4.14 Local deformation associated with fiber bending and matrix shearing.

The bending moment is related to the fiber curvature κ by

$$M = E_f I_f \kappa \tag{4.100}$$

where E_f and I_f are the fiber modulus of elasticity and area moment of inertia, respectively.

If we assume linear behavior, we can approximate the curvature of the fiber by the second derivative of the transverse displacement. Then the bending moment becomes

$$M = E_f I_f \frac{d^2}{dx^2}(v - v_0 \qquad 4.101)$$

where v represents the transverse displacement and v_0 defines the initial shape of the fiber (the initial fiber waviness). Assuming that the axial displacement is much lower than the transverse displacement, the shear deformation can be approximated by

$$\gamma = \frac{d}{dx}(v - v_0 \qquad 4.102)$$

Then the shear force in the matrix is given by

$$V = AG_{LT} \frac{d}{dx}(v - v_0 \qquad 4.103)$$

where G_{LT} is the shear modulus of the composite. We next need a description of the initial waviness and the deformed shape of the fibers. If we assume the initial waviness may be described as

$$v_0 = f_0 \sin\left[\frac{\pi x}{l}\right] \tag{4.104}$$

and that the deformed shape is

$$v = f \sin\left[\frac{\pi x}{l}\right] \tag{4.105}$$

then the composite strength may be estimated by

$$\sigma_c = \left(G_{LT} + \frac{\pi^2 E_f I_f}{A l^2}\right)\left(1 - \frac{f_0}{f_c}\right) \tag{4.106}$$

where f_c is the critical bending curvature and l is the length of the fiber.

It is generally assumed that the second term in (4.106) is small compared with G_{LT} so that the compression strength is given by

$$\sigma_c = G_{LT}\left(1 - \frac{f_0}{f_c}\right) \tag{4.107}$$

There are now two critical cases that must be considered. In the first case, local shear stress controls failure, so that

$$\tau_c = G_{LT}\frac{\pi}{l}(f_c - f_0) \Rightarrow f_c^{\text{shear}} = \frac{\tau}{G_{LT}}\frac{l}{\pi} + f_0 \tag{4.108}$$

where τ_c is the shear strength. In the second critical case, failure is controlled by the bending strength of the fibers. In this case

$$X_c^f = E_f r_f \frac{\pi^2}{l^2}(f_c - f_0) \Rightarrow f_c^{\text{bending}} = \frac{X_c^f}{E_f}\frac{l^2}{\pi^2 r_f} \tag{4.109}$$

Substituting (4.108) and (4.109) into (4.107), we obtain two estimates of the composite compressive strength for shear and fiber bending controlled failure given by

$$\sigma_c^{\text{shear}} = G_{LT}\frac{1}{1+\pi\left(\frac{f_0}{l}\right)\frac{G_{LT}}{\tau_c}} \tag{4.110}$$

$$\sigma_c^{\text{bending}} = G_{LT}\frac{1}{1+\pi^2\left(\frac{f_0}{2l^2}\right)\frac{w_f E_f}{X_{cf}}} \tag{4.111}$$

To estimate the value of G_{LT}, let us evaluate the strain energy associated with fiber bending. Assuming for simplicity that $v_0 = 0$, the bending energy associated with the fibers alone is

$$U_b = \int_0^l \int_A \frac{\sigma^2}{2E_f}\, dA\, dx = \int_0^l \frac{M^2}{2E_f I_f}\, dx = \int_0^l \frac{E_f I_f}{2}\left(\frac{d^2 v}{dx^2}\right)^2 dx \tag{4.112}$$

Taking the lateral displacement to have the form $v = Bx^3$, where B is a constant, we obtain

$$U_b = 18B^2 E_f I_f \frac{l^3}{3} \tag{4.113}$$

For the energy stored by shear we can write

$$U_s = \int_0^l \int_A \frac{\tau_f^2}{2G_f}\, dA\, dx = A_f \int_0^l \frac{1}{2G_f A_f^2}\left(\frac{dM}{dx}\right)^2 dx = A_f \int_0^l \frac{E_f^2 I_f^2}{2G_f A_f^2}\left(\frac{d^3 v}{dx^3}\right)^2 dx \tag{4.114}$$

$$U_s = 18B^2 l \frac{E_f^2 I_f^2}{A_f G_f} \tag{4.115}$$

Setting the total energy equal to the applied work, we obtain

$$U = U_b + U_s = SBl^3 \tag{4.116}$$

so that we may define the shear modulus as

$$G_{LT} = \frac{\tau}{\gamma} = \frac{S}{(h+2c)(v/l)} = \frac{S}{(h+2c)Bl^2} \tag{4.117}$$

where h and $2c$ are defined in Figure 4.15. Referring to Figure 4.15, we can write

$$V_f = \frac{h}{h+2c} \qquad I_f = \frac{h^3}{12} \qquad A_f = h \tag{4.118}$$

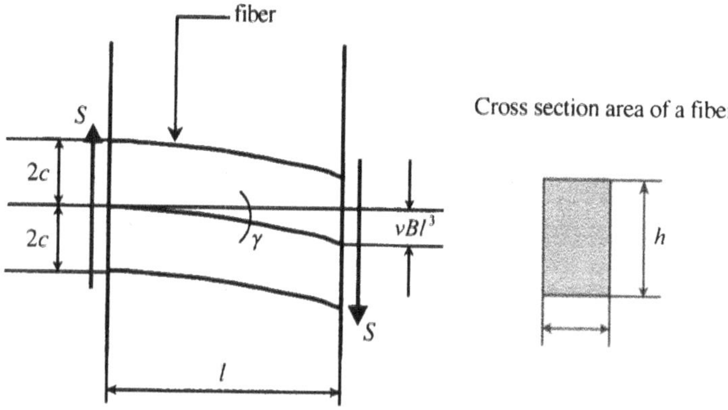

Figure 4.15 Shearing of fibers.

Introducing (4.118) in (4.117), the in-plane shear stiffness for this failure mode can be estimated as

$$G_{LT} = \frac{E_f V_f}{2} \left(\frac{h}{l}\right)^2 \left[1 + \left(\frac{h}{l}\right)^2 \frac{E_f}{G_f} \frac{1}{4}\right] \tag{4.119}$$

4.2.2 Straight-Fiber Models

At the beginning of this chapter, we stated that if the fibers are straight, large in diameter, and perfectly aligned with a uniaxial loading axis, then it is possible that the fibers may fail in a crushing mode, especially if the matrix (and composite) is very stiff so that buckling or bending of the fibers is strongly resisted. This is a very strong mode of failure, and the rule of mixtures works well for an estimation of that strength. Boron fibers (of the order of 40 μm in diameter) with high volume fraction can be used to make a composite that fails in this way, for example, and some compression-loaded structures like the landing gear of certain aircraft are made that way. However, it is very difficult to create such a composite with perfect alignment, and large-diameter fiber composites are hard to manufacture and to use. Therefore, even if the fibers are initially straight and remain straight when loaded, there is usually a likelihood that the fibers will have some finite misalignment with the loading axis. This is especially true when multiaxial loading is present. Let us suppose that due to an applied planar stress, the fibers form a sheared region, as shown in Figure 4.16. The angle γ represents the shear angle and α_0 is the initial misalignment, as shown in the figure. The stresses in the two coordinates systems shown in the figure are related by the equation

$$\sigma'_{ij} = a_{ik} a_{jl} a_{kl} \tag{4.120}$$

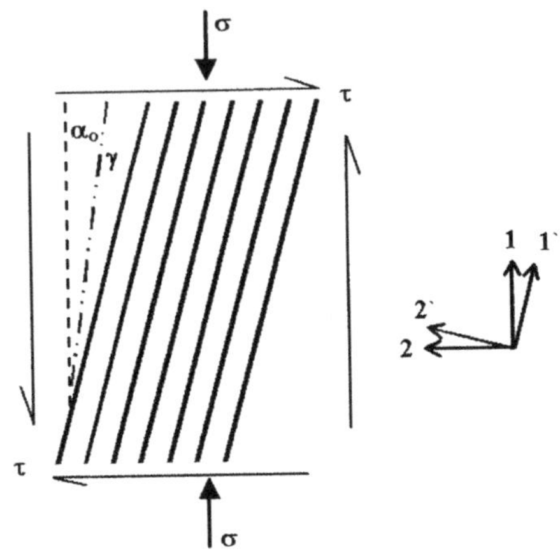

Figure 4.16 Fibers misaligned with loading axis by initial angle α_0 sheared through shear angle γ.

where $a_{ij} = \cos(x_i,\ x_j)$. As a result, the shear stress is given by

$$\tau' = -\sigma \cos\alpha \ \sin\alpha + \tau(\cos^2\alpha - \sin^2\alpha) \tag{4.121}$$

and the normal stress by

$$\sigma' = \sigma\cos^2\alpha + 2\tau\ \sin\alpha\cos\alpha \tag{4.122}$$

If α is small, then

$$\sigma' = \sigma - 2(\alpha_0 + \gamma)\tau \qquad \tau' = \tau + (\alpha_0 + \gamma)\sigma \tag{4.123}$$

and we can write

$$\sigma = \frac{\tau' - \tau}{\alpha_0 + \gamma} = \sigma(\gamma \qquad \text{4.124)}$$

We are interested in the maximum value of σ. To find this value, we consider the equation

$$\frac{d\sigma}{d\gamma} = 0 \Rightarrow \frac{d\tau'}{d\gamma}\frac{1}{\alpha_0 + \gamma} - \frac{\tau' - \tau}{(\alpha + \gamma)^2} = 0 \tag{4.125}$$

which leads to

$$\frac{d\tau'}{d\gamma} = \frac{\tau' - \tau}{\alpha_0 + \gamma} \tag{4.126}$$

If there is no global shear, then $\tau = 0$ and

$$\sigma_c = \frac{\tau_c}{\alpha_0 + \gamma_f} \tag{4.127}$$

is the compressive strength, where γ_f is the critical shear strain of the local material (the matrix or interface). This is the classical form given by Argon (1972). It is interesting to note that this expression is essentially identical to (4.110), since

$$\sigma_c = \frac{G_{LT}}{1 + (\pi f_0/l)(G_{LT}/\tau_c)} = \frac{\tau_c}{\gamma_f(1 + \alpha_0/\gamma_f)} = \frac{\tau_c}{\gamma_f + \alpha_0} \tag{4.128}$$

where $\tau_c = G_{LT}\gamma_f$ and $\alpha_0 = \pi f_0/l$ is the local slope in the axial direction of the deformed fiber.

4.2.3 Microbuckling Models

Referring to D in Figure 4.13, if the fibers are initially straight and well aligned but the matrix and surrounding composite are relatively compliant, local regions of fiber instability may form, a mode of failure called *microbuckling*. The microbuckling analysis of the fibers in a composite material is based on the following assumptions:

two-dimensional deformation,
no shear in the fiber,
no matrix extension or compression, and
normal stress in the fiber that produces no mechanical work during the deformation of the fiber.

Under these assumptions, and with reference to Figure 4.17, we can write

$$\tau_{xy} = G_m \gamma_{xy} \qquad \gamma_{xy} = \left(\frac{c + r}{c}\right)\frac{dv}{dx}$$

where G_m is the matrix shear modulus, r the fiber radius, and $2c$ the spacing between two consecutive fibers. The mechanical work done by the axial force is given by

$$P_{\text{work}} = \int_0^l P(ds - dx) = P\int_0^l \left[\sqrt{1 + \left(\frac{dv}{dx}\right)^2} - 1\right] dx = \frac{P}{2}\int_0^l \left(\frac{dv}{dx}\right)^2 dx \tag{4.129}$$

Similarly the work done by the shear force is

$$S_{\text{work}} = \int_0^l S\, dv = \int_0^l S\frac{dv}{dx}\, dx \tag{4.130}$$

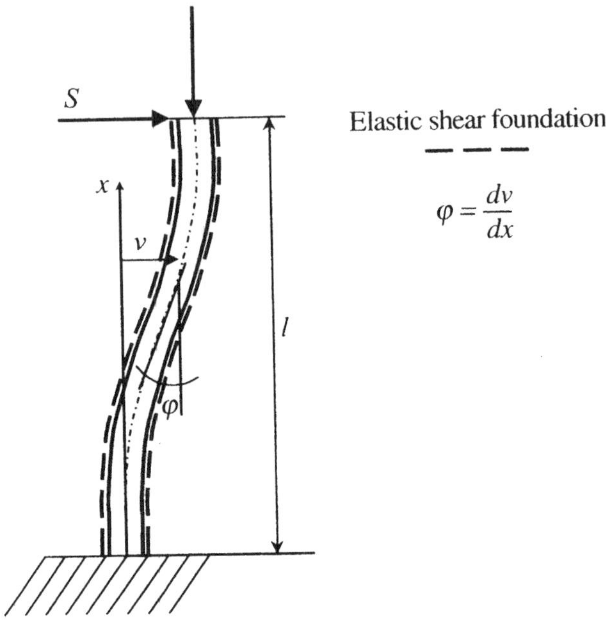

Figure 4.17 Fiber buckling model.

Then the total potential energy has the form

$$U = \frac{1}{2}\int_0^l E_f I_f \left(\frac{d^2 v}{dx^2}\right) dx + \frac{1}{2}\int_0^l V_m \tau_m \gamma_m \, dx - \frac{P}{2}\int_0^l \left(\frac{dv}{dx}\right)^2 dx - \int_0^l S\frac{dv}{dx}\, dx \quad (4.131)$$

where V_m represents the volume of matrix corresponding to one fiber. To derive the governing equation of the problem as well as the boundary conditions, we use the principle of virtual work, such that

$$\delta U = 0 \quad (4.132)$$

This leads to the equations

$$\frac{d^2}{dx^2}\left(E_f I_f \frac{d^2 v}{dx^2}\right) + \left[P - V_m G_m \left(1 + \frac{r}{c}\right)^2\right]\frac{d^2 v}{dx^2} = 0 \quad (4.133)$$

$$\frac{d^3 v}{dx^3} = -\frac{S}{E_f I_f} \quad (4.134)$$

with the boundary conditions

$$v = 0 \qquad \frac{dv}{dx} = 0 \quad \text{at} \quad x = 0 \tag{4.135}$$

$$\frac{d^2 v}{dx^2} = 0 \qquad \frac{d^3 v}{dx^3} = -\frac{S}{E_f I_f} \quad \text{at} \quad x = l \tag{4.136}$$

The solution of (4.133) and (4.134) with the boundary conditions (4.135) and (4.136) depends on the sign of the expression

$$A = P - V_m G_m \left(1 + \frac{r}{c}\right)^2 \tag{4.137}$$

For $A < 0$

$$v = -\frac{S}{E_f I_f B_1^3} \quad \sinh(B_1 x) - B_1 x - \frac{\cosh(B_1 l) - 1}{\sinh(B_1 l)} [\cosh(B_1 x) - 1] \tag{4.138}$$

and for $A > 0$

$$v = \frac{S}{E_f I_f B_2^3} \quad \sin(B_2 x) - B_2 x + \frac{\cos(B_2 l) - 1}{\sin(B_2 l)} [\cos(B_2 x) - 1] \tag{4.139}$$

where

$$B_1^2 = \left[V_m G_m \left(1 + \frac{r}{c}\right)^2 - P\right] \frac{1}{E_f I_f} \tag{4.140}$$

$$B_2^2 = \left[P - V_m G_m \left(1 + \frac{r}{c}\right)^2\right] \frac{1}{E_f I_f} \tag{4.141}$$

Using

$$\sigma = \frac{Mr}{I_f} \qquad M = E_f I_f \frac{d^2 v}{dx^2} \tag{4.142}$$

we obtain

$$\sigma_f^{\max} = \frac{P}{A_f} + \frac{Sr}{I_f B_1}\left[\frac{\cosh(B_1 l) - 1}{\sinh(B_1 l)}\right] \quad \text{for } A < 0 \tag{4.143}$$

$$\sigma_f^{\max} = \frac{P}{A_f} + \frac{Slr}{2I_f} \quad \text{for } A = 0 \tag{4.144}$$

$$\sigma_f^{\max} = \frac{P}{A_f} - \frac{Sr}{I_f B_2}\left[\frac{\cos(B_2 l) - 1}{\sin(B_2 l)}\right] \quad \text{for } A > 0 \tag{4.145}$$

If fibers fail, then $\sigma_f^{\max} = X_c^f$. If the matrix failure controls global failure, then

$$\tau_m = G_m \gamma_m = G_m\left(1 + \frac{r}{c}\right)\frac{dv}{dx} \tag{4.146}$$

and using the solution corresponding to $A < 0$, the matrix shear stress is

$$\tau_m = \frac{SG_m}{E_f I_f B_1^2}\left[1 + \frac{\cosh(B_1 l) - 1}{2\cosh(B_1 l/2)} - \cosh\left(\frac{B_1 l}{2}\right)\right]\left(\frac{1}{V_f}\right)^{1/4}\left(\frac{r}{c}\right) \tag{4.147}$$

The critical values of applied shear S and axial force P are found from the minimum of these conditions. If the models are applied at the ply level in a laminate, P is the in-plane fiber direction force resultant and S is the in-plane shear force.

4.2.4 Kink-Band Shear Models

If the fibers in the local instability discussed in the previous section fail in bending, then a local region of shear forms a "kink band" or "shear band," as shown by E in Figure 4.13. This problem has been extensively analyzed by Budiansky and Fleck (1993). Following that development, let us consider a kinked fiber region, as shown in Figure 4.18, under a far-field normal and shear stress denoted by σ^∞ and τ^∞. Equilibrium requires that

$$\sigma \cdot n = \sigma^\infty \cdot n \tag{4.148}$$

With respect to the system defined by ($\hat{e}_1$, $\hat{e}_2$), where $\hat{e}_1$ and $\hat{e}_2$ represent the unit vectors associated with the far-field stress, we can write

$$\sigma^\infty = -\sigma^\infty \hat{e}_1\hat{e}_1 + \tau^\infty(\hat{e}_1\hat{e}_2 + \hat{e}_2\hat{e}_1) \tag{4.149}$$

Similarly by defining a local system ($\hat{\varepsilon}_1$, $\hat{\varepsilon}_2$) with $\hat{\varepsilon}_1$ and $\hat{\varepsilon}_2$ being the unit vectors shown in Figure 4.18,

$$\sigma = \sigma_L \hat{\varepsilon}_1\hat{\varepsilon}_2 + \sigma_T \hat{\varepsilon}_2\hat{\varepsilon}_2 + \tau(\hat{\varepsilon}_1\hat{\varepsilon}_2 + \hat{\varepsilon}_2\hat{\varepsilon}_1) \tag{4.150}$$

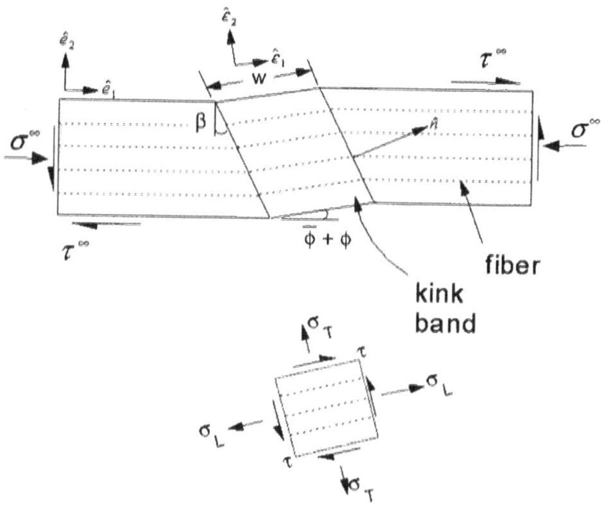

Figure 4.18 Geometry for kink-band shear model.

The vectors $\hat{e}_1$ and $\hat{e}_2$ are related to $\hat{\varepsilon}_1$ and $\hat{\varepsilon}_2$ by the geometric relations

$$\hat{e}_1 = \hat{\varepsilon}_1 \cos(\phi + \bar{\phi}) - \hat{\varepsilon}_2 \sin(\phi + \bar{\phi}) \tag{4.151}$$

$$\hat{e}_2 = \hat{\varepsilon}_1 \sin(\phi + \bar{\phi}) + \hat{\varepsilon}_2 \cos(\phi + \bar{\phi}) \tag{4.152}$$

The normal $\hat{n}$ can be expressed as

$$\hat{n} = \hat{e}_1 \cos\beta + \hat{e}_2 \sin\beta \tag{4.153}$$

Introducing (4.149)–(4.153) in (4.148), we obtain two scalar equations:

$$-\sigma^{\infty}\cos\beta\cos(\varphi + \bar{\varphi}) + \tau^{\infty}\sin(\beta + \varphi + \bar{\varphi}) = \sigma_L \cos(\beta - \varphi - \bar{\varphi}) + \tau\sin(\beta - \varphi - \bar{\varphi}) \tag{4.154}$$

$$\sigma^{\infty}\cos\beta\sin(\varphi + \bar{\varphi}) + \tau^{\infty}\cos(\beta + \varphi + \bar{\varphi}) = \sigma_T \sin(\beta - \varphi - \bar{\varphi}) + \tau\cos(\beta - \varphi - \bar{\varphi}) \tag{4.155}$$

For small $\phi + \bar{\phi}$, the linearization of (4.155) yields

$$(\varphi + \bar{\varphi})[\sigma^{\infty}\cos\beta - 2\tau^{\infty}\sin\beta] = (\tau - \tau^{\infty})[\cos\beta + (\varphi + \bar{\varphi})\sin\beta] + \sigma_T[\sin\beta - (\varphi + \bar{\varphi})\cos\beta] \tag{4.156}$$

By neglecting $\sigma_T(\varphi + \bar{\varphi})$ in (4.156), we obtain

$$\sigma^\infty - 2\tau^\infty \tan\beta = \frac{\tau - \tau^\infty + \sigma_T \tan\beta}{\phi + \bar{\phi}} \tag{4.157}$$

and if $\beta = 0$, $\tau^\infty = 0$, then

$$\sigma^\infty = \frac{\tau}{\phi + \bar{\phi}} \tag{4.158}$$

If $\phi \to \gamma$ (local shear), then by assuming a plastic deformation in the shear zone

$$\frac{\sigma^\infty}{G} = \frac{\tau/\tau_y}{\gamma/\gamma_y + \bar{\varphi}/\gamma_y} \tag{4.159}$$

For deformation plasticity theory with a nonlinear constitutive law of the Ramberg–Osgood type,

$$\gamma = \frac{\tau}{G}\left[1 + \frac{3}{7}\left(\frac{\tau}{\tau_Y}\right)^{n-1}\right] \tag{4.160}$$

Introducing (4.160) in (4.159) and maximizing the new expression with respect to τ/τ_Y, we obtain

$$\frac{\sigma_c}{G} = \frac{1}{1 + n(3/7)^{1/n}[(\sigma/\gamma_Y)/(n-1)]^{(n-1)/n}} \tag{4.161}$$

where $G \approx G_m$. We note that the far-field shear stress τ^∞ does have an influence on the critical stress. Writing

$$\tau^\infty = G\gamma^\infty = e\sigma^\infty \tag{4.162}$$

where e is a parameter, the normal stress in the transverse direction can be written as

$$\sigma_T = E_{CT} e_T \tag{4.163}$$

Taking the ratio

$$\frac{\sigma_c}{G} = \frac{1 + (E_T/G)\tan^2\beta}{1 - 2e\tan\beta} \tag{4.164}$$

we obtain a function β. The minimum value of (4.164) occurs for a specific value of β. For polymer composites, that value is typically between about 10° and 30° and can be verified by experiment. A variety of other situations are discussed by Budiansky and Fleck (1993).

4.2.5 Elastic Foundation Models

Xu and Reifsnider (1993) introduced two additional features into the microbuckling models that significantly influence the predictions. One is modeling of the support of the fiber as a beam on an elastic foundation and an evaluation of that foundation stiffness to obtain

$$k = \frac{\pi E_2^c b}{1 + \nu_{12}^c} \tag{4.165}$$

where E_2^c is the transverse Young's modulus of the composite, ν_{12}^c is the in-plane Poisson's ratio of the composite, and b is the fiber diameter. The second is fiber-matrix slippage. With these additions (and a few other minor changes) to the formulation noted in Section 4.2.2, Xu and Reifsnider obtain the explicit form composite compressive strength

$$\begin{aligned}\sigma_c = G_m &\left[V_f + \frac{E_m}{E_f}(1 - V_f) \right] \\ &\times \left[2(1 + \nu_m) \sqrt{\frac{\pi\sqrt{\pi}\eta r_f}{3\dfrac{E_m}{E_f}\left(V_f\dfrac{E_m}{E_f} + 1 - V_f\right)[1 + V_f\nu_f + \nu_m(1 - V_f)]}} \right. \\ &\left. + 1 - \xi - \frac{\sin(\pi\xi)}{2\pi} \right]\end{aligned} \tag{4.166}$$

where $\xi = 2s/l$ is the fractional length of matrix slippage, s is the local fiber–matrix slippage length, and η is used to represent the fiber–matrix bonding conditions. If the fiber is completely separated from the matrix on one side, $\eta = 1$; if the fiber is well bonded on both sides, $\eta = 2$; if the fiber is partly bonded to the matrix "foundation," $0 < \eta < 2$. Figure 4.19 illustrates a comparison of the predictions from Equation (4.166) with experimental data for two materials. The predictions for the carbon–epoxy system are typical of high strength and stiffness reinforcement in a relatively soft matrix. The predictions shown assume no fiber–matrix slipping and strong bonding of the fibers, that is, $\eta = 2$, so there are no adjustable constants in the predictive model. Agreement is typically within a few percent of experimental results. However, if the failure mode is not microbuckling, other models are needed. For the E-glass reinforced epoxy system shown in Figure 4.19, it can be

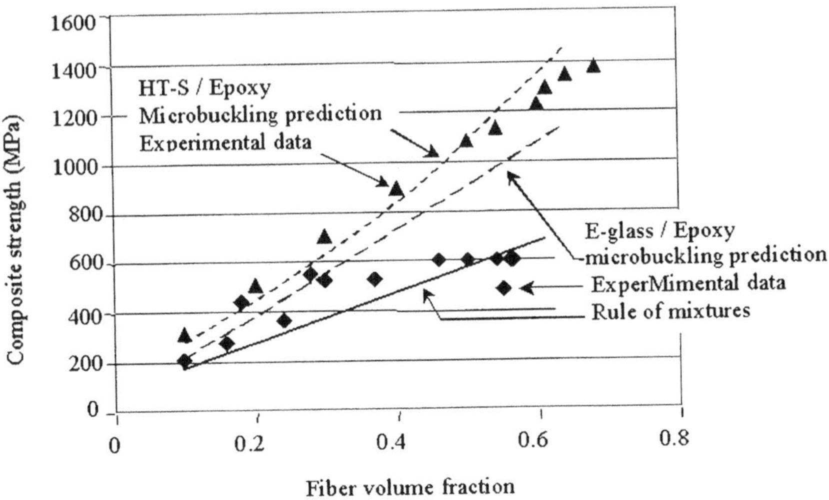

Figure 4.19 Comparison of predicted uniaxial compression strength of uniaxial continuous-fiber composites with experimental data.

seen that the strength begins to drop below the predicted results for microbuckling above a fiber volume fraction of about 30%. In this case, a simple rule-of-mixtures prediction, also shown in the figure, seems to follow the data. This could only have been anticipated by careful determination of the failure mode in the laboratory.

Having a closed-form solution that is explicit in material parameters, like expression (4.166), enables parameter studies that can guide the material system designer. Figure 4.20 shows the dependence on several parameters for a carbon–epoxy system.

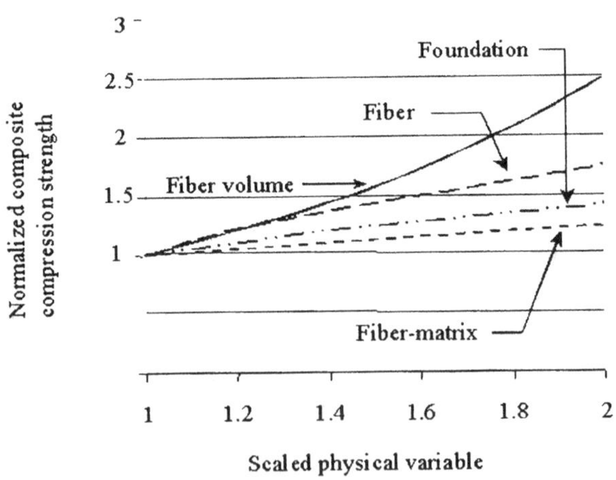

Figure 4.20 Effect of manufacturing parameters on uniaxial composite compressive strength.

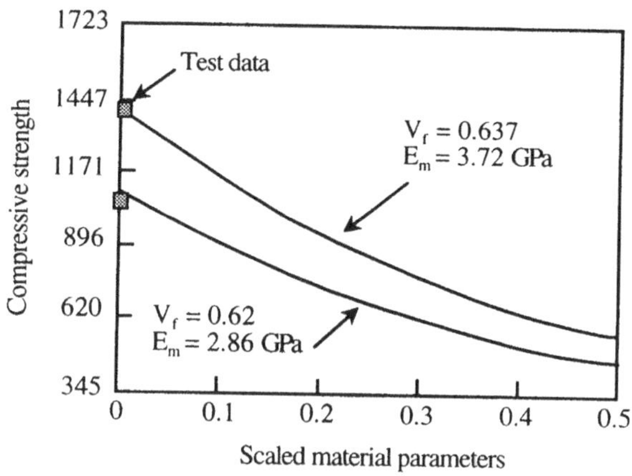

Figure 4.21 Effect of fiber–matrix slipping on uniaxial compressive strength of composite materials.

To achieve a realistic comparison, the material parameters were all scaled to the interval 1–2 on the abscissa. All strengths were normalized by the initial value for the minimum abscissa value. For volume fraction, the physical values ranged from 0.32 to 0.72. The dependence on volume fraction is seen to be nonlinear, with increasing "gain" as the fraction increases. Volume fraction also had the greatest effect on the strength of all parameters considered. Fiber diameter was scaled from 3 to 20 μm. The variation is quite nonlinear for low values but nearly linear after about 6 μm. Hence, the greatest gains for this system are achieved by using fibers greater in diameter than about 6 μm. Fiber diameter is the second most important material parameter. Stiffness of the matrix was varied between 689 and 6338 kPa. The influence is significant, but third in importance, and relatively linear as a function of matrix stiffness. Fiber matrix bonding ranged between $0 < \eta < 2$. That caused a variation of about 25% in the predicted strength.

Fiber–matrix slipping was not included in Figure 4.20, since it was not easily scaled to the presentation. Figure 4.21 shows that variation. The results suggest that fiber–matrix slipping over half the length of the fiber (in the microbuckling region) can reduce the strength by more than 50%. Figures 4.20 and 4.21 show that compressive strength is remarkably sensitive not only to the constituent properties but also to how the composite system is made.

4.3 TRANSVERSE STRENGTH AND SHEAR STRENGTH

4.3.1 Transverse Loading

Generally, the *first-ply failure* of a composite lamina is due to transverse tension. The typically low transverse strength (and corresponding strain to failure) is due to stress

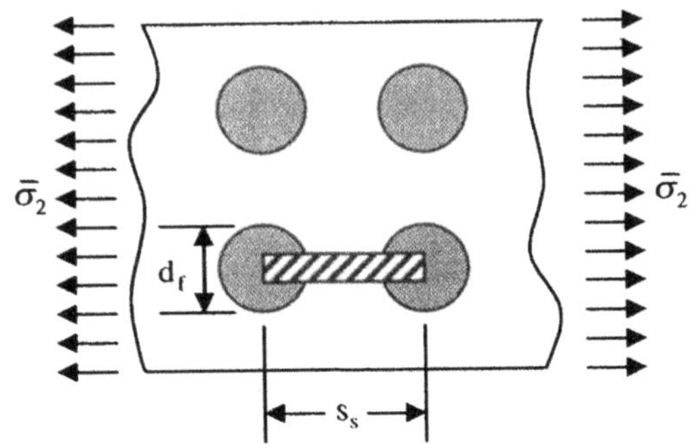

Figure 4.22 Representative volume element for mechanics of materials model for transverse strength.

concentrations in the matrix around the fibers. A simple way to analyze such a case has been developed by Kies (1962) based upon a mechanics of materials approach, obtained by analyzing a representative volume element, illustrated in Figure 4.22. For the shaded strip shown in Figure 4.22, the total elongation is found by summing the deformations in the fiber and in the matrix:

$$\delta = \delta_f + \delta_m = s_s\bar{\varepsilon}_2 = d_f\varepsilon_f + (s_s - d_f)\varepsilon_m \tag{4.167}$$

Now we assume that the stresses in the composite, the matrix, and the fiber are equal (an isostress assumption). In this case, we have

$$\varepsilon_f = \frac{\bar{\sigma}_2}{E_2^f} \qquad \varepsilon_m = \frac{\bar{\sigma}_2}{E_2^m} \qquad \frac{\varepsilon_f}{\varepsilon_m} = \frac{E_2^m}{E_2^f} \tag{4.168}$$

Substituting Equation (4.168) into Equation (4.167), we obtain

$$s_s\bar{\varepsilon}_2 = \left(d_f\frac{E_2^m}{E_2^f} + s_s - d_f\right)\varepsilon_m \tag{4.169}$$

Then the strain concentration (which is equivalent to the stress concentration for elastic materials) is given by

$$K_t = \frac{\varepsilon_m}{\bar{\varepsilon}_2} = \frac{1}{(d_f/s_s)(E_2^m/E_2^f - 1) + 1} \tag{4.170}$$

Failure is assumed to occur when the stress in the matrix is equal to the matrix strength X_t^m for tensile loading. The corresponding composite strength is then given by

$$Y_t = \frac{E_2 X_t^m}{E_2^m K_t} \tag{4.171}$$

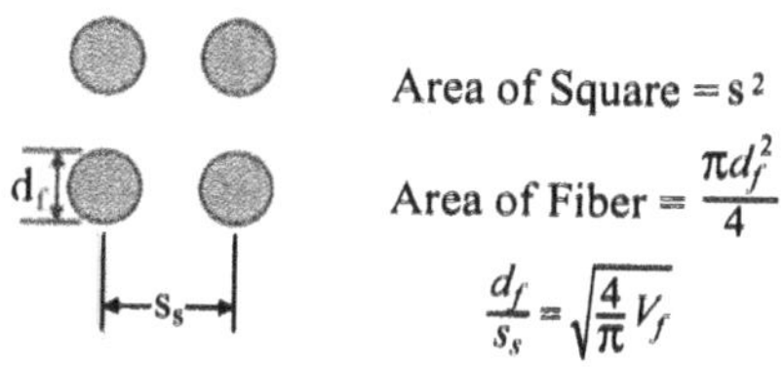

Figure 4.23 Determination of s_s and d_f for square-packed fiber array.

where E_2 is the composite transverse stiffness, X_t^m is the tensile strength of the matrix, and E_2^m is the transverse stiffness of the matrix. Now all we need is to determine the dimension s_s in terms of the fiber diameter d_f and the fiber volume fraction V_f. Figure 4.23 illustrates how this can be accomplished.

We need not be restricted to square packing, however. If, instead of Figure 4.23, we had started with the hexagonally packed arrangement shown in the Figure 4.24, we would have arrived at the same result (with s_s replaced by s_h).

Other researchers have attempted to use more complicated stress analyses in order to understand the transverse strength of composite materials. In particular, the concentric cylinder model has received much attention (e.g. Arnold and Wilt, 1992; Carman et al., 1993a; Ghosn and Lerch, 1989; Pagano and Tandon, 1988). In their work, Carman et al. (1993a) suggested a means by which specific fiber coatings could be used to minimize the stress concentrations in composites subjected to transverse loading caused by the stiffness mismatch between the fiber and the matrix. To illustrate this effect, we may consider the case of an AS-4/Epon 828 composite (a graphite–epoxy system) subjected to a transverse stress. The composite has a fiber volume fraction of 60%. If we first consider the case in which no interphase region is present and look at the maximum principal stress in the matrix, we find that the maximum stress occurs at the location where the normal to the fiber is parallel to the applied load (at $\theta = 0°$), as shown in Figure 4.25.

If we next consider the case in which an extremely compliant interphase region with a Young's modulus that is 1% of the matrix modulus exists, we find that the maximum principal stress in the matrix occurs at the location where the normal to the fiber is perpendicular to the applied loading direction (at $\theta = 90°$), as shown in Figure 4.25. By selecting an interphase region that has the appropriate modulus (the

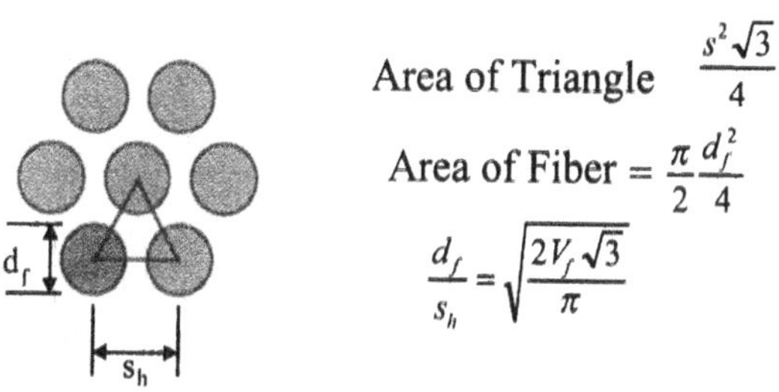

Figure 4.24 Determination of s_f and d_f for hexagonally packed fiber array.

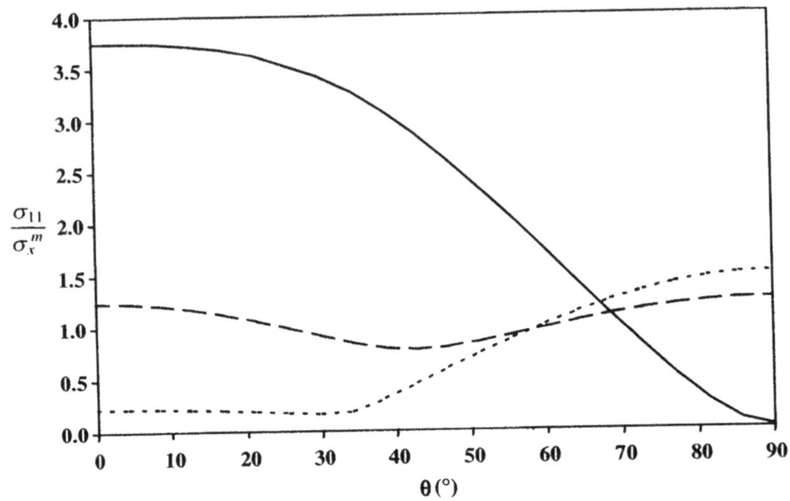

Figure 4.25 Maximum principal-stress variation with angular position for 60% fiber volume fraction AS-4–Epon 828 composite having various interphase properties.

"optimum" modulus), it is possible to make the maximum principal stresses at $\theta = 0°$ and $\theta = 90°$ equal to one another. In such a case, failure is predicted to initiate simultaneously at the two locations (if a maximum-stress-failure criterion is used). For the composite system under consideration, this optimum interphase modulus has a value of 0.211 GPa (approximately 7% of the matrix modulus). The resulting principal-stress variation as a function of angular position for this optimum case is also shown in Figure 4.25. To attempt to validate these findings experimentally, Case and Carman (1994) made use of a model composite system in which the fibers were 3-mm-diameter glass rods. Two different fiber volume fractions were considered: 25 and 44%. Two extremes in interphase conditions were considered. In the first extreme, no coating was applied to the fibers during the specimen preparation. In the second extreme, vacuum grease was applied to the fibers to simulate an extremely compliant interphase material. Tests were then conducted to determine the transverse strength and strain to failure of these model composite systems. Observations of the failure surfaces, as shown in Figure 4.26, highlight the movement of the failure location. For the composites in which no vacuum grease was applied, failure occurs at $\theta = 0°$. For the composites in which the vacuum grease was applied, failure occurs at $\theta = 90°$, consistent with the model predictions. In addition, Case et al. attempted to predict the normalized strains to failure of these model composite systems (relative to the 25% fiber volume fraction with no coating). These results are summarized in Figure 4.27. These investigations of model composites as well as results obtained by other researchers on real composite systems highlight the importance of considering the interphase region in any micromechanical modeling effort.

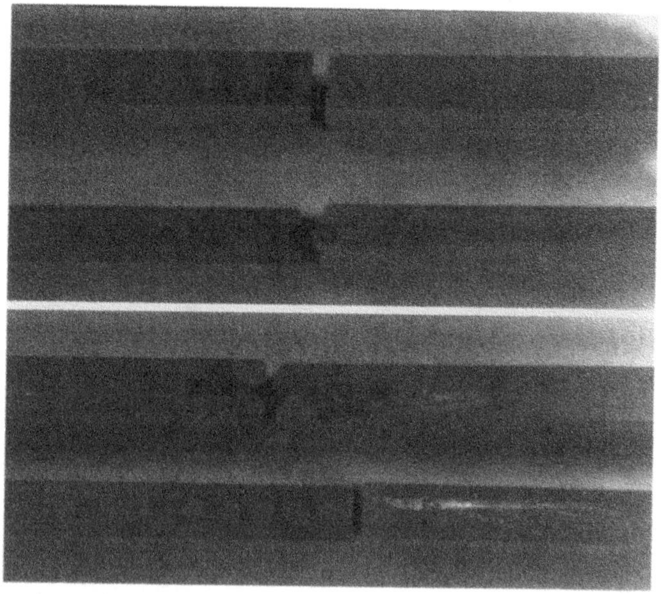

Figure 4.26 Comparison of failure locations for model composite systems. Specimens are, from the top, glass–epoxy (25% fiber volume) with vacuum grease, glass–epoxy (25% fiber volume) with no coating, glass–epoxy (44% fiber volume) with vacuum grease, and glass–epoxy (44% fiber volume) with no coating.

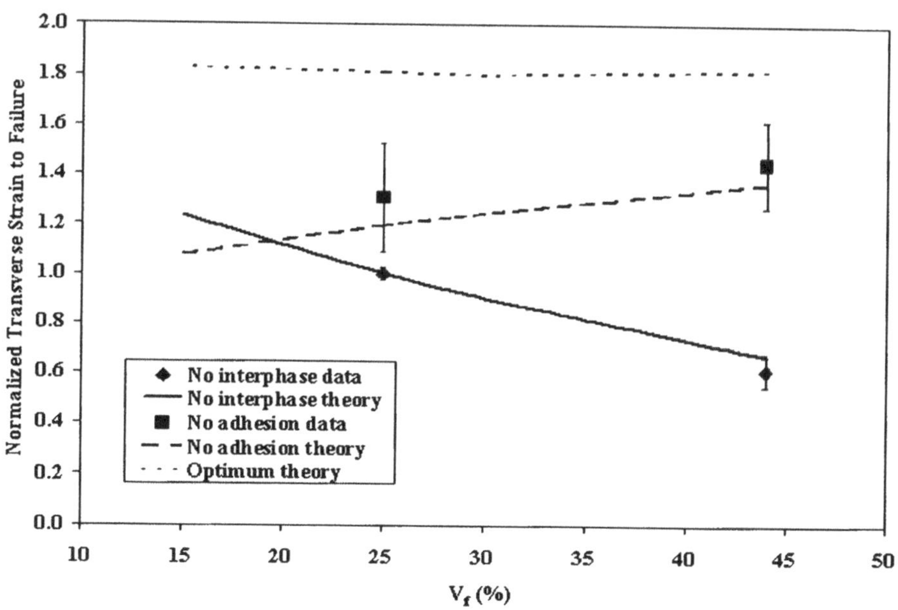

Figure 4.27 Comparison of predicted and measured strain-to-failure results for glass–epoxy model composites.

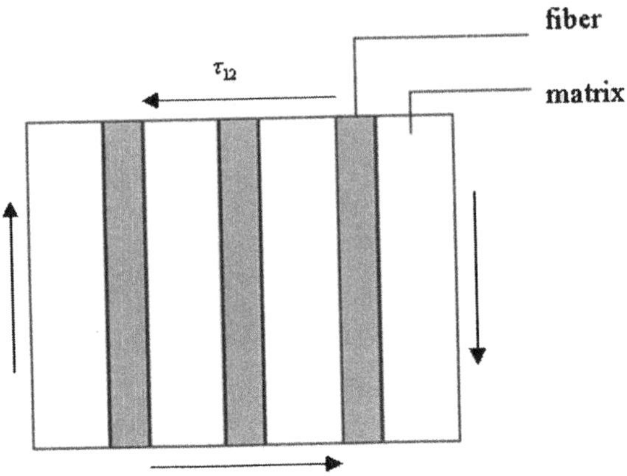

Figure 4.28 Far-field shear loading.

4.3.2 Shear Loading

Micromechanical models for shear loading conditions generally follow much the same approach as those used for transverse loading. The in-plane shear strength S may also be estimated by the Kies (1962) approach outlined in the previous section. In this case, referring to Figure 4.28, the expression analogous to (4.167) is given by

$$s\bar{\gamma}_{12} = d_f\gamma_f + (s_s - d_f)\gamma_m \tag{4.172}$$

so that the stress concentration factor is given by

$$K_t = \frac{\gamma_m}{\bar{\gamma}_{12}} = \frac{1}{(d_f/s_s)(G_{12}^m/G_{12}^f - 1) + 1} \tag{4.173}$$

The strength S is then given by

$$S = \frac{G_{12}S_m}{G_{12}^m K_t} \tag{4.174}$$

where G_{12} is the composite shear stiffness, S_m is the matrix shear strength, and G_{12}^m is the matrix shear stiffness. Carman et al. (1993a) have suggested that we may also follow a similar approach to that used for the composite subjected to transverse loading to determine the optimum interphase modulus (the one that minimizes the stress concentration in the matrix). To do so, we consider a unidirectional composite material subjected to an axial shear strain and use a cylindrical representative volume element as shown in Figure 4.29. Based on the assumption that the representative

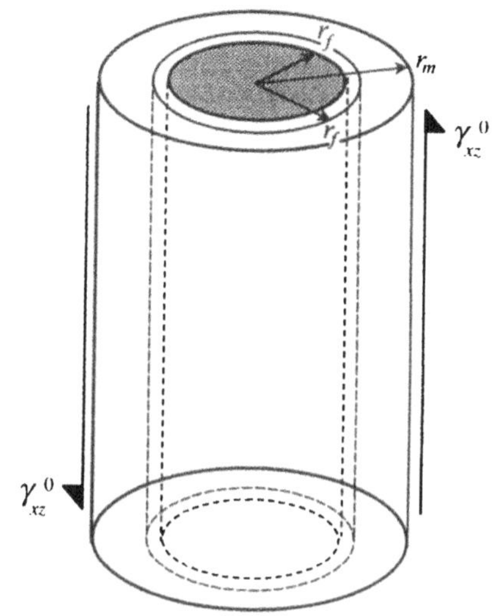

Figure 4.29 Representative volume element for concentric cylinder analysis.

volume element experiences the same strain as the composite, the following boundary conditions are applied:

$$u_r^m(r_m, \theta) = u_\theta^m(r_m, \theta) = 0 \qquad u_z^m(r_m, \theta) = \gamma_{xz}^c r_m \cos\theta \tag{4.175}$$

where u is the displacement field, r_m is the radius of the matrix defining the boundary of the element, and γ_{xz}^c is the applied shear strain. Throughout this section, the superscripts f, i, and m refer to the fiber, the interphase, and the matrix, respectively. For this problem, the relevant equilibrium equation is

$$\frac{\partial \sigma_{rz}}{\partial r} + \frac{\sigma_{rz}}{r} + \frac{1}{r}\frac{\sigma_{\theta z}}{\partial \theta} = 0 \tag{4.176}$$

Because of the form of the boundary conditions given in (4.175), we assume displacements in the form of

$$u_z^n(r, \theta) = f^n(r, \theta) \qquad u_r^n(r, \theta) = 0 \qquad u_\theta^n(r, \theta) = 0 \tag{4.177}$$

where $n = f,\ i,\ m$, so that the only nonzero strain components are

$$\gamma_{rz}^n = \frac{\partial u_z^n}{\partial r} \qquad \gamma_{\theta z}^n = \frac{1}{r}\frac{\partial u_z^n}{\partial \theta} \tag{4.178}$$

with the corresponding stresses given by

$$\sigma_{rz}^n = G_{12}^n \gamma_{rz}^n = G_{12}^n \frac{\partial u_z^n}{\partial r} \qquad \sigma_{\theta z}^n = G_{12}^n \gamma_{\theta z}^n = G_{12}^n \frac{1}{r}\frac{\partial u^n}{\partial \theta} \tag{4.179}$$

Combining (4.179) and (4.176), we find the pertinent governing equation is

$$u^n_{z,rr} + \frac{1}{r}u^n_{z,r} + \frac{1}{r^2}u^n_{z,\theta\theta} = 0 \tag{4.180}$$

Following Pagano and Tandon (1988), we can easily solve (4.180) subject to the boundary conditions presented in (4.175). The solution is then given by

$$u^n_z(r, \theta, z) = \left[\frac{F^n_1}{r} + F^n_2 r\right]\cos\theta \tag{4.181}$$

where F^n_1 and F^n_2 ($n = f,\ i,\ m$) are constants. The nonzero stress components are given by

$$\sigma^n_{rz} = G^n_{12}\left[-\frac{F^n_1}{r^2} + F^n_2\right]\cos\theta \qquad \sigma^n_{\theta z} = G^n_{12}\left[-\frac{F^n_1}{r^2} - F^n_2\right]\sin\theta \tag{4.182}$$

where G^n_{12} is the shear modulus of the nth phase. The constants in (4.181) and (4.182) can be evaluated by applying the boundary conditions in (4.175), demanding continuity of the traction and displacement at each interface, and requiring that the stresses be bounded (i.e., $F^f_1 = 0$). For the three-cylinder representative volume element under consideration, the following system of equations is obtained:

$$\begin{aligned}
\frac{F^m_1}{r^2_m} + F^m_2 &= \gamma^c_{xz} \\
\frac{F^m_1}{r^2_i} + F^m_2 - \frac{F^i_1}{r^2_i} - F^i_2 &= 0 \\
\frac{F^i_1}{r^2_i} + F^i_2 - F^f_2 &= 0 \\
-G^m_{12}\frac{F^m_1}{r^2_i} + G^m_{12}F^m_2 + G^i_{12}\frac{F^i_1}{r^2_i} - G^i_{12}F^i_2 &= 0 \\
-G^i_{12}\frac{F^i_1}{r^2_f} + G^i_{12}F^i_2 - G^f_{12}F^f_2 &= 0
\end{aligned} \tag{4.183}$$

where r_i and r_f are the fiber and interphase radii, respectively. If the interphase shear stiffness is known, the five unknown constants F^n_i can be determined by solving the system of linear equations. However, if we treat the interphase shear modulus as an unknown, a total of six unknowns exist, so that we require an additional condition. This final condition is given as

$$\sigma^m_{rz}(r_i,\ 0^\circ,\ z) = -\sigma^m_{\theta z}(r_i,\ 90^\circ,\ z) \tag{4.184}$$

To solve the resulting system of equations, we note that (4.184) implies that $F_1^m = 0$. If we use this result in the first of (4.183), we immediately learn that $F_2^m = \gamma_{xz}^c$. This result suggests that the stress and displacement fields in the matrix are identical to that in a homogeneous matrix material subjected to the same shear strain—this was in fact the motivation for (4.184). With further manipulations of (4.183), we can develop the following closed-form expression for the optimum interphase shear modulus:

$$(G_{12}^i) + \left[\frac{(-G_{12}^m + G_{12}^f)(1 + (r_i^2/r_f^2))}{(r_i^2/r_f^2) - 1} \right] G_{12}^i - G_{12}^f G_{12}^m = 0 \tag{4.185}$$

We note that the optimum interphase is not a function of the matrix radius (i.e., fiber volume fraction) and is solely dependent upon mechanical properties of the constituents and the relative ratio of the interphase-to-fiber radii. Eskandari et al. (1996) have suggested an important application of such an approach: the increase in composite compression strength that results by decreasing the stress concentration factor in the matrix. If we reexamine our undulating fiber model for compression strength and assume that the shear strength τ_c is controlled by shear failure in the matrix, then the compression strength is given by

$$\sigma_c = G_{LT} \frac{1}{1 + \pi(f_0/l)(G_{12}^m K/\tau_c^m)} \tag{4.186}$$

where G_{12}^m is the matrix shear modulus, τ_c^m is the matrix shear strength, and K is the stress concentration factor in the matrix given by

$$K = \max \left| \frac{(-F_1^m/r_i^2) + F_2^m}{\gamma_{xz}^c} \right|, \left| \frac{(-F_1^m/r_i^2) - F_2^m}{\gamma_{xz}^c} \right| \tag{4.187}$$

where the two terms result from the possibility of failure occurring at $\theta = 0°$ or $\theta = 90°$.

4.3.3 Gradient Interphase Properties

In our previous discussions, we have considered the interphase to have uniform properties as a function of position. However, this interphase region is often the product of processing conditions during the composite manufacturing. In particular, when fiber coatings or sizings are used, interdiffusion of the sizing with the matrix may result. As a direct result, the interphase may have spatially nonuniform properties. As an example of this influence on the micromechanical modeling, we may consider the special case of a composite subjected to longitudinal shear loading. The fiber and matrix are assumed to be homogeneous, while the interphase is assumed to have a shear modulus that varies as

$$G_{12}^{i}(r) = G_{12}^{0}\left(\frac{r}{r_f}\right)^{Q_y} \tag{4.188}$$

Our solution procedure in the fiber and the matrix are the same as in the previous section, resulting in displacements given by (4.181) and stresses given by (4.182). For the interphase region, we assume displacements of the form

$$u_z^i(r, \theta) = r^{\lambda} \cos\theta \tag{4.189}$$

If we substitute (4.188) and (4.189) into (4.179) and the resulting expression into (4.176), we find the interphase displacement is

$$u_z^i(r, \theta) = F_1^i r^{\lambda_1} + F_2^i r^{\lambda_2} \tag{4.190}$$

where

$$\lambda_1, \lambda_2 = \frac{-Q_y \pm \sqrt{Q_y^2 + 4}}{2} \tag{4.191}$$

To illustrate the results for such a case, we will consider an example of an AS-4/ Epon 828 composite. The relevant properties are

$$G_{12}^f = 28.0 \text{ GPa} \qquad G_{12}^m = 1.075 \text{ GPa}$$
$$V_f = 0.6 \qquad V_i = 0.2$$

where V_f and V_i are the fiber and interphase volume fractions, respectively. For the interphase properties, we will consider three cases:

$$\begin{aligned} &\text{Case II:} && G_{12}^i(r) = 1.87795 r^5 \text{ GPa} \\ &\text{Case II:} && G_{12}^i(r) = 0.61537 r^{-5} \text{ GPa} \\ &\text{Case III:} && G_{12}^i(r) = 1.075 \text{ GPa} \end{aligned} \tag{4.192}$$

In each case, the variation of the interphase shear modulus is chosen so that it has a value equal to that of the matrix at the interphase–matrix boundary. The resulting variation in σ_{rz} for these three cases is shown in Figure 4.30. Jayaraman et al. (1994) have used a similar technique in conjunction with the Mori–Tanaka method to evaluate the stress state in composites subjected to general thermomechanical loading conditions.

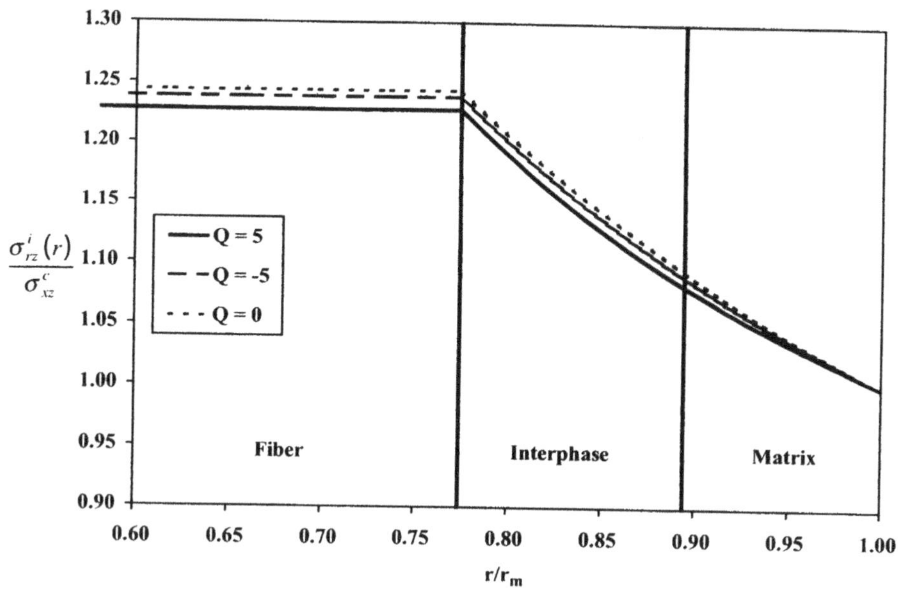

Figure 4.30 Variation in shear stress for composites with varying interphase properties.

REFERENCES

Argon, A. S. (1972), "Fracture of Composites," in *Treatise on Materials Science and Technology*, Vol. 1, Academic, New York, p. 79.

Arnold, S. M. and Wilt, T. E. (1992), "Influence of Engineered Interfaces on Residual Stresses and Mechanical Response in Metal Matrix Composites," NASA TN-105438.

Batdorf, S. B. (1982a), "Tensile Strength of Unidirectionally Reinforced Composites—I," *J. Reinforced Plastics Composites*, Vol. 1, pp. 153–164.

Batdorf, S. B. (1982b), "Tensile Strength of Unidirectionally Reinforced Composites—II," *J. Reinforced Plastics Composites*, Vol. 1, pp. 165–176.

Budiansky, B., and Fleck, N. A. (1993), "Compressive Failure of Fiber Composites," *J. Mech. Phys. Solids*, Vol. 41, No. 1, pp. 183–211.

Caliskan, A. G. (1996), "Micromechanics-Based Approach to Predict Strength and Stiffness of Composite Materials," M. S. Thesis, Virginia Polytechnic Institute and State University.

Carman, G. P., Averill, R. C., Reifsnider, K. L., and Reddy, J. N. (1993a), "Optimization of Fiber Coatings to Minimize Stress Concentrations in Composite Materials," *J. Composite Mater.*, Vol. 27, No. 6, pp. 589–612.

Carman, G. P., Lesko, J. J., and Reifsnider, K. L. (1993b), "Micromechanical Analysis of Fiber Fracture," in *Composite Materials: Fatigue and Fracture*, Vol. 4, ASTM STP 1156, W. W. Stinchcomb and N. E. Ashbaugh, Eds., American Society for Testing and Materials, Philadelphia, pp. 430–452.

Carman, G. P., Averill, R. C., Reifsnider, K. L., and Reddy, J. N. (1993c), "Optimization of Fiber Coatings to Minimize Stress Concentrations in Composite Materials," *Journal of Composite Materials*, Vol. 27(6), pp. 589–612.

Case, S. W. (1996), "Mechanics of Fiber-Controlled Behavior in Polymeric Composite Materials," Ph. D. Dissertation, Virginia Polytechnic Institute and State University.

Case, S. W., Carman, G. P., Lesko, J. J., Fajardo, A. B., and Reifsnider, K. L. (1995), "Fiber Fracture in Unidirectional Composites," *J. Composite Mater.*, Vol. 29, No. 2, pp. 208–228.

Case, S. W., and Carman, G. P. (1994), "Optimization of Fiber Coatings for Transverse Performance: An Experimental Study," *J. Composite Mater.*, Vol. 28, No. 15, pp. 1452–1466.

Case S. W., and Reifsnider K. L. (1996), "Micromechanical Analysis of Fiber Fracture in Unidirectional Composite Materials," *Int. J. Solids Structures*, Vol. 33, No. 26, pp. 3795–3812.

Chang, Y. S., Lesko, J. J., Case, S. W., Dillard, D. A., and Reifsnider, K. L. (1994), "Mechanical Properties of Thermoplastic Composites: The Interphase Effect," *J. Thermoplastic Composites*, Vol. 7, pp. 311–324.

Cox, H. L. (1952), "The Elasticity and Strength of Paper and Other Fibrous Materials," *British Journal of Applied Physics*, Vol. 3, pp. 72–79.

Curtin, W. A. (1991), "Theory of Mechanical Properties of Ceramic-Matrix Composites," *J. Am. Ceramic Soc.*, Vol. 74, No. 11, pp. 2837–2845.

Eskandari, S., Carman, G. P., and Case, S. W. (1996), "Evaluating the Influence of Fiber Coatings on the Compression Strength of a Unidirectional Polymer Composite," *J. Composite Mater.*, Vol. 30, No. 18, pp. 1958–1976.

Gao, Z., and Reifsnider, K. L. (1992), "Tensile Failure of Composites: Influence of Interface and Matrix Yielding," *J. Composites Technol. Res.*, Vol. 14, No. 4, pp. 201–210.

Ghosn, L. J., and Lerch, B. A. (1989), "Optimum Interface Properties for Metal Matrix Composites," NASA TN-102295.

Hedgepeth, J. M. (1961), "Stress Concentrations in Filamentary Structures," NASA TN D-882, National Aeronautics and Space Administration.

Hedgepeth, J. M., and Van Dyke, P. (1967), "Local Stress Concentrations in Imperfect Filamentary Composite Materials," *J. Composite Mater.*, Vol. 1, pp. 294–309.

Jayaraman, K., Gao, Z., and Reifsnider, K. L. (1994), "Interphase in Undirectional Fiber-Reinforced Epoxies: Effect on Local Stress Fields," *J. Composites Technol. Res.*, Vol. 16, No. 1, pp. 21–31.

Kies, J. A. (1962), "Maximum Strains in the Resin of Fiberglass Composite," U.S. Naval Research Laboratory Report 5752.

Madhukar, M. S., and Drzal, L. T. (1991), "Fiber-Matrix Adhesion and Its Effect on Composite Mechanical Properties: II. Longitudinal (0°) and Transverse (90°) Tensile and Flexure Behavior of Graphite/Epoxy Composites," *J. Composite Mater.*, Vol. 25, pp. 958–991.

McCartney, L. N. (1993), "Stress Transfer Mechanics for Multiple Perfectly Bonded Concentric Cylinder Models of Unidirectional Composites," National Physical Laboratory Report DMM(A)129.

Pagano, N. J. (1991), "Axisymmetric Micromechanical Stress Fields in Composites," in *Local Mechanics Concepts for Composite Material Systems*, J. N. Reddy and K. L. Reifsnider, Eds., pp. 1–26.

Pagano, N. J. and Tandon, G. P. (1988), "Elastic Response of Multi-Directional Coated-Fiber Composites," *Composites Science and Technology*, Vol. 31, pp. 273–293.

Reifsnider, K. L., and Case, S. W. (1997), "Strength-Based Predictive Methods for Large-Scale Combined Degradation Effects," AIAA-98-268, American Institute of Aeronautics and Astronautics.

Rosen, B. W. (1964), "Tensile Failure of Fibrous Composites," *AIAA J.*, Vol. 2, pp. 1985–1991.

Vure, N. R. S. (1993), "Effect of Cooling Rate and Stacking Sequence on the Fatigue Behavior of Notched Quasi-Isotropic APC-2 Laminates," M.S. Thesis, Virginia Polytechnic Institute and State University.

Wimolkiatisak, A. S., and Bell, J. P. (1989), "Interfacial Shear Strength and Failure Modes of Interphase-Modified Graphite-Epoxy Composites," *Polym. Composites*, Vol. 10, No. 3, pp. 162–172.

Xu, Y. L., and Reifsnider, K. L. (1993), "Micromechanical Modeling of Composite Compressive Strength," *J. Composite Mater.*, Vol. 27, No. 6, pp. 572–587.

EXERCISES

4.1 It is found by experimental observation of a series of tensile specimens that the fraction of fibers that are not broken in any volume of a unidirectional continuous-fiber composite is given by the probability function

$$P(\varepsilon) = \exp\left[-\frac{L}{L_0}\left(\frac{\varepsilon}{\varepsilon_0}\right)^{\alpha}\right]$$

where ε is the applied strain, ε_0 is the characteristic strain of the distribution, α is the shape factor of the distribution function that represents the surviving fraction of fibers, and L is the length of the composite under test and L_0 the length of the fibers from which the probability function was determined. If only the fibers support load in tension, the nominal stress on such a tensile specimen might be written as $\sigma = \varepsilon E_f P(\varepsilon)$, where E_f is the stiffness of the fibers. If that is so, show the following:

(a) An estimate of the tensile strength of such a composite can be written as

$$X_c \cong \frac{\varepsilon_0 E_f V_f}{[\alpha(L/L_0)e]^{1/\alpha}}$$

where e is the Napierian log base and V_f is the volume fraction of the fibers.

(b) The ratio of the "bundle strength" estimate of strength to the average strength of an equal number of fibers follows the relationship

$$\left(\frac{X_c}{X_f}\right) \cong \frac{1}{(\alpha e)^{1/\alpha}\Gamma(1 + 1/\alpha)}$$

so that the strength of a bundle of fibers together is less than the average strength of all of the individual fibers by an amount that is highly dependent on the statistical spread of the individual fiber strengths.

4.2 If the tensile stress near the end of a broken fiber in a composite is caused only by the maximum shear stress on the fiber surface, τ (due to matrix shear failure or fiber–matrix slipping), show that the ratio of the distance over which the tensile stress builds up from a null value to the undisturbed value in the rest of the fiber (the ineffective length) can be given by

$$\frac{\delta}{\text{fiber} \times \text{diameter}} = \frac{\sigma_{\text{fiber-ultimate}}}{4\tau}$$

4.3 Suppose that single-fiber tensile tests have been conducted on AS-4 fibers with the nominal properties given below:

$$E_1 = 235 \text{ GPa} \qquad \nu_{12} = 0.25 \qquad \alpha_1 = -0.55 \times 10^{-6}/^\circ\text{C}$$

$$E_2 = 17 \text{ GPa} \qquad \nu_{23} = 0.27 \qquad \alpha_2 = 5.6 \times 10^{-6}/^\circ\text{C}$$

$$G_{12} = 55 \text{ GPa}$$

The tensile tests were conducted at five different gage lengths. At each of the lengths, the resulting data were fit to a Weibull representation. The results are summarized in the following:

Gage Length (mm)	Weibull Reference Strength, σ_0 (GPa)
50	3.60
30	3.85
20	3.96
6	4.45
3	4.71

What are the Weibull parameters m and σ_0 for the AS-4 fibers, referenced to $L_0 = 1$ mm?

4.4 Use the bundle strength model to estimate the tensile strength X_t of an AS-4/"Generic" composite (fiber volume fraction 60%). You should use the values for m and σ_0 developed in (1). The length of the composite is 0.1524 m (6 in.). Generic matrix properties are

$$E_1 = E_2 = 2.90 \text{ GPa} \qquad \nu_{12} = \nu_{23} = 0.35 \qquad \alpha_1 = \alpha_2 = 4.8 \times 10^{-5}/^\circ\text{C}$$

4.5 The elastic shear-lag analysis results in Table 4.1 may be used to develop the stress concentrations and ineffective lengths due to multiple adjacent fiber fractures. Use these results in the Batdorf tensile strength model to estimate the tensile strength of the same composite considered in Exercise 4.2. Make (and discuss) reasonable estimates of any quantities that are not specified (such as the width/thickness of the composite). How does the strength change as the length is doubled?

Some values for the actual tensile strength of AS-4-based composites are as follows:

AS-4/Epon 828	1890 MPa
AS-4/PEEK	2090 MPa
AS-4/J2	1830 MPa

How do your bundle strength results and those based upon the Batdorf model compare with the experimental results? Do you have explanations for the discrepancies (if any are present)?

4.6 Using the $\underline{n} \cdot \underline{\underline{\sigma}} = \underline{n} \cdot \underline{\underline{\sigma}}^{\infty}$ identity, where σ^{∞} is the applied stress field in the remote coordinate system and σ is the stress field in the coordinate system of a kink band that forms in a unidirectional composite under compressive loading, show that

$$\sigma^{\infty} - 2\tau^{\infty} \tan(\beta) = \frac{\tau - \tau^{\infty} + \sigma_T \tan(\beta)}{\phi + \bar{\phi}}$$

where σ^{∞} and τ^{∞} are the remote normal and shear-stress components, σ_T and τ are the same components in the local (kinked) system, β is the kink-band angle, ϕ is the angle of local shear, and $\bar{\phi}$ is the initial fiber misalignment. (Refer to discussion in the text.) Some approximations are used by Budiansky and Fleck (1993) to get this expression.

4.7 On a single graph, plot estimates of the compressive strength of continuous-fiber composites as a function of fiber volume fraction for the various failure modes discussed in the text (e.g., undulating fibers, local shear, microbuckling, and fiber crushing). Compare the predictions and explain the different predicted values based on physical reasoning.

4.8 The following data are from Lo and Chim (1992):

Fiber Volume Fraction	Composite Strength (MPa)
0.65	848
0.68	938
0.71	1145
0.75	1076

Estimates of the constituent properties include the following:

$$E_f = 73.1 \text{ GPa} \qquad \nu_f = 0.22 \qquad E_m = 3.45 \text{ GPa} \qquad \nu_m = 0.35$$

Radius of fiber = 3.5 μm

Using these data and the logic (and models) discussed in Exercise 4.7, can you guess which failure mode controls the strength of this material system over this range of fiber volume fractions? What are the certainties and uncertainties associated with your guess?

CHAPTER 5

STIFFNESS EVOLUTION

Thesis: Pointwise tensor stiffness changes control local stress distributions and redistributions. Useful representations of these local stress states can be constructed from a precise knowledge of constituent, interface, and interphase behavior and geometry as a function of stress, time, cycles, and environments.

5.1 PROBLEM DEFINITION

Stiffness controls stress level for a given strain (or strain level for a given stress). In a component, the pointwise stiffness distribution controls stress distributions at the local level. As damage develops during service, local stress (or strain) distributions change with time and history of loading. These local stress (or strain) distributions influence the properties and performance of the structure and determine durability and damage tolerance. As mentioned earlier, brittle matrix composites, in general, have the distinctive feature of changing their stiffness during service, by large amounts in some cases (10–40% changes have been recorded). These local (and global) stiffness changes can be a dominant feature of the mechanics of durability and damage tolerance, and their description and modeling are a major subject of this text.

The foundation for this discussion is in Chapter 2, where we discussed the strength of laminates by considering the progressive failure of off-axis plies under uniaxial quasi-static loading. We found that, for many common composite systems, the stress in the matrix material of off-axis plies exceeds its strength before the laminate strength is reached, and matrix cracking occurs. We modeled this sequence by reducing the stiffness of the matrix to simulate the matrix cracks and found that the stress in the other plies was increased (in the loading direction) by the relaxation process; that is, there was a redistribution of stress at the local level.

Since the arguments in the strength evolution integral we are using to calculate global remaining strength are calculated at the local level, this redistribution is of direct concern to that calculation. We recall the form of that equation as

$$\mathrm{Fr} = 1 - \int \left[1 - \mathrm{Fa}\left(\frac{\sigma_{ij}(t)}{X_{ij}(t)} \right) \right] j \left(\frac{n}{N} \right)^{j-1} d\left(\frac{n}{N} \right) \tag{5.1}$$

The stress σ_{ij} that appears in the failure function (that represents conditions in the critical element) is directly altered by stiffness changes, and since we also use the representation

$$\mathrm{Fa}\left(\frac{\sigma_{ij}}{X_{ij}} \right) = A + B[\log(N)]^p \tag{5.2}$$

to determine the instantaneous life under current conditions of stress and strength in the critical element, that local stress also directly affects the current life, N, in (5.1). In Chapter 6, we will find that the rates of change of principal strengths $X_{ij}(t)$ can also be dependent on the local stress level, so that those values in (5.1) are also altered by stiffness changes, indirectly.

The subject of this chapter will be the stiffness change caused by sustained or cyclic (long-term) loading. The causes of these stiffness changes are numerous in continuous-fiber composites. The most pervasive (and widely discussed) is matrix cracking (as mentioned in Chapter 2). We will address this process first with a discussion that is based on the different embodiments of that damage mode in various material systems, including ply cracking, splitting, and delamination. We will then turn our attention to stiffness changes brought about by viscoelasticity, including the effects of moisture and temperature. Finally, we will address the subject of stiffness change caused by fiber fracture, an especially important subject for ceramic matrix composites.

5.2 STIFFNESS CHANGE DUE TO MATRIX CRACKING

5.2.1 Quasi-Static Loading

The most common cause of stiffness change in brittle matrix composites is microcracking of the matrix material. This progressive process is a function of increasing applied stress (or strain) level or a function of time or applied cycles (Reifsnider, 1982). To illustrate some of the fundamental aspects of this process, consider a laminate with two ply orientations, like the $[0, 90]_s$ laminate sketched in Figure 5.1, subjected to a uniaxial applied strain. For our conceptual discussion, we

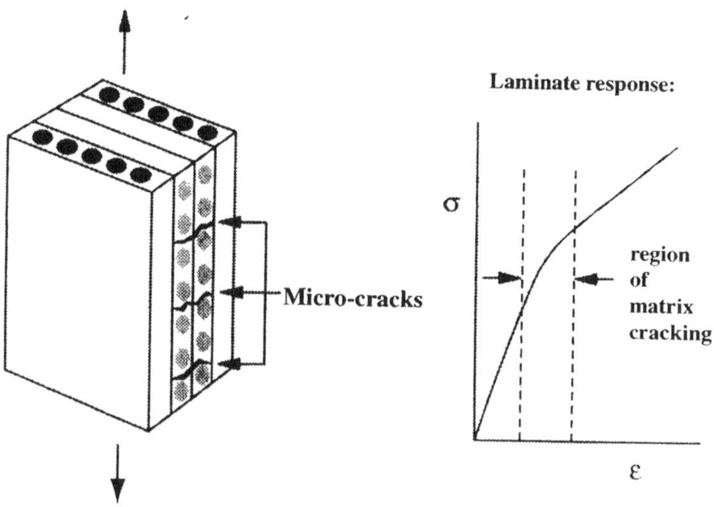

Figure 5.1 Matrix cracking under increasing applied loading in a $[0, 90]_s$ laminate.

make the simplifying assumption that the stiffness of the composite can be written in terms of the stiffness of the two ply types in the direction of loading using a rule of mixtures expression such as

$$E_c = v_1 E_1 + v_2 E_2 \tag{5.3}$$

Ply type 1 may be a 0° ply, and ply type 2 may be an off-axis ply with an orientation such as 45° or 90°. Of course, the correct value of the composite stiffness can be found using classical laminate theory; (5.3) is only a convenient approximation for our illustration. If we multiply (5.3) by the applied strain (which is the same for all plies), we would write

$$\sigma_c = v_1 \sigma_1 + v_2 \sigma_2 \tag{5.4}$$

where σ and σ_2 are stress in the direction of the applied loading in the 1 and 2 type plies, respectively.

Now we conduct a "thought experiment" wherein the applied composite stress increases quasi-statically to produce a stress–strain curve. As the composite stress increases, the stress in each ply increases in proportion to the volume fraction of the specific ply, according to (5.4). The result is a linear stress–strain law. However, let us assume that ply type 2 reaches a strain level that initiates matrix cracking. We will call the stress that causes such cracking to begin Y (in the direction of loading). Since the ply (or plies of that type) cannot support any more stress than that value, we assume that cracks nucleate at a rate whereby the compliance they add to that ply (as the global strain increases) is just enough to keep the stress in the cracked plies at that limit level. For example, we assume that for all strains above matrix crack

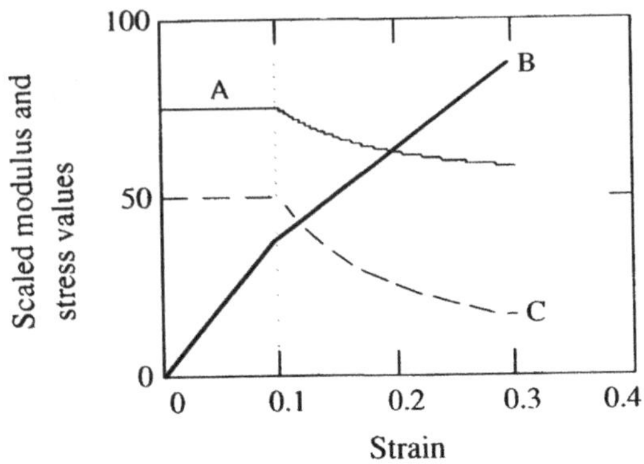

Figure 5.2 Stiffness of ply in which cracking is occurring (A) and resulting bilinear stress–strain curve of composite (B) for (hypothetical) composite shown in Figure 5.1.

initiation, $\sigma_2 = Y$ in the cracked ply(s). If we enter that value into (5.4) and divide through by the (uniform) applied strain level ε, it becomes

$$E_c = v_1 E_1 + v_2 \frac{Y}{\varepsilon} \tag{5.5}$$

for the composite stiffness for applied stress levels above the matrix crack initiation stress. This produces an interesting result. Figure 5.2 shows a plot of the stiffness of the composite as a function of strain and the resulting stress strain–curve for such a material. The composite stiffness is constant until the value of Y/ε drops below E_2, after which (5.5) controls, and the composite stiffness drops as a function of strain, reaching an asymptotic value of the first term in (5.5) as the second term is driven to zero by increasing strain values. However, the resulting stress–strain curve is remarkably bilinear, with a sharp transition point at the strain corresponding to the initiation of cracking of the second ply. Hence, even though the composite modulus is distinctly nonlinear, it is nonlinear in such a way that the stress–strain curve is linear, albeit *bi*linear. The reason is clear from (5.5), since multiplying that relationship by the strain ε yields a constant for the second term in that expression. The resulting bilinear stress–strain curve is a well-known experimental result for laminates that develop matrix cracking (cf. Highsmith and Reifsnider, 1982; Johnson, 1982; Reifsnider, 1982).

Figure 5.3 shows an example of this physical behavior for a graphite–epoxy laminate that has a large number of off-axis plies. The measured specimen stiffness is also shown. The behavior is quite similar to that shown in Figure 5.2. The specimen stiffness changes in a very nonlinear fashion, but the stress–strain curve is distinctly bilinear. For our discussion, the important point is that the reduction of the stiffness of the cracked plies as a function of strain above the cracking threshold introduces a progressively larger proportion of the applied load into the remaining uncracked ply as the load increases. This increase is critical to our efforts to calculate

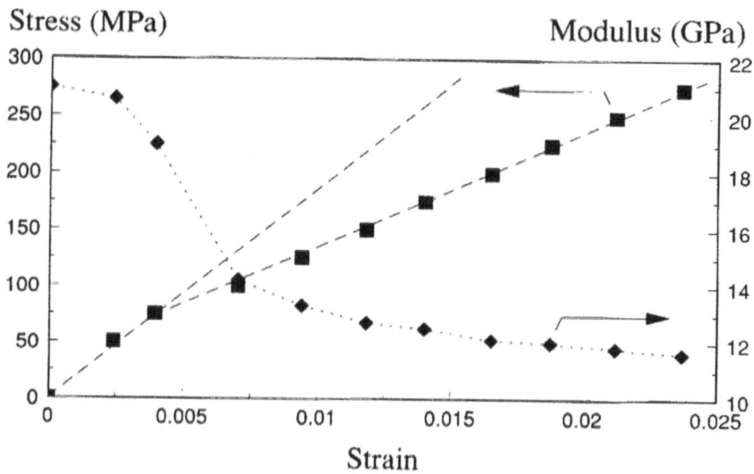

Figure 5.3 Stress–strain curve and specimen stiffness for $[0, 90_3]_S$ graphite–epoxy laminate showing bilinear stress–strain behavior caused by matrix crack initiation as function of increasing applied global stress.

remaining strength and life, controlled by the *last* ply failure, not by the first, second, and so on. Figure 5.4 shows these changes, as represented by the equations above, when (hypothetical) values of 100 for E_1 and 50 for E_2 are used as initial values in the equations above.

For these values of relative ply stiffness, the stress in the unbroken plies increases by 70% as the stiffness in the broken plies drops to about two-thirds of its initial value over the range of strains shown. The actual numbers in this example depend on the relative ply stiffness values and the volume fractions of the plies. But two very important points made by this example are general in nature:

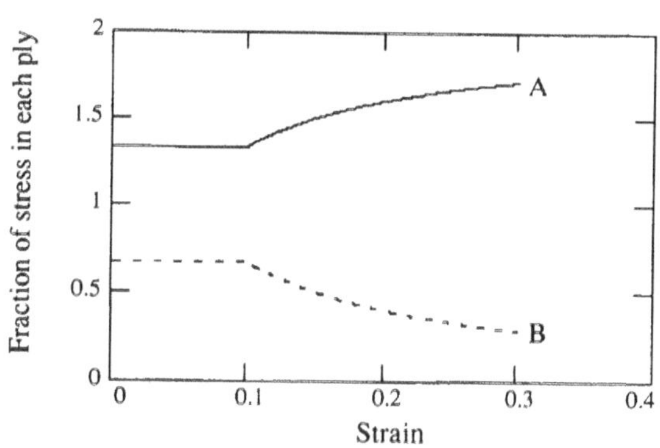

Figure 5.4 Fraction of total stress in unbroken ply (*a*) and broken ply (*b*) as function of applied stress. Matrix cracking reduces the stiffness of cracked plies to maintain constant stress in those plies.

1. On the basis of experimental data and the simple concepts noted above, stiffness change in a constituent or ply as a function of increasing stress occurs in such a way that the local stress in that constituent remains at or near the (constant) value that causes the initiation of damage in that constituent.
2. Stress redistribution occurs at the local level as a result of this progressive stiffness change in the constituents, generally having the effect of increasing the stress in the unbroken constituents.

Although these "rules" have been stated in a deterministic fashion, the statistical aspects of damage initiation can (and should) be considered, with only a change in the details of the statements. The consequence of these rules is that the correct local stress in unbroken material can be calculated as a function of applied global loading. As we have discussed, that local stress, ultimately in the critical element, controls the remaining strength and life of the laminate or composite material.

However, an important addition to this philosophy involves a limiting condition for matrix cracking, called *saturation cracking*, or the *characteristic damage state* (*CDS*) for matrix cracking by various authors. This phenomenon (observed in a great many physical systems from stratified rock layers to highways) has the effect, for example, of imposing an upper bound on the number of cracks that can form in a laminate such as the one shown in Figure 5.1. In fact, if one counts the number of cracks in a given off-axis ply as a function of the global applied load level, the number generally approaches an asymptotic value as the load approaches the fracture strength of the specimen. Figure 5.5 shows an example of such data for graphite–epoxy, quasi-isotropic laminates, in which the spacing of the cracks in the off-axis plies was measured (Masters and Reifsnider, 1982).

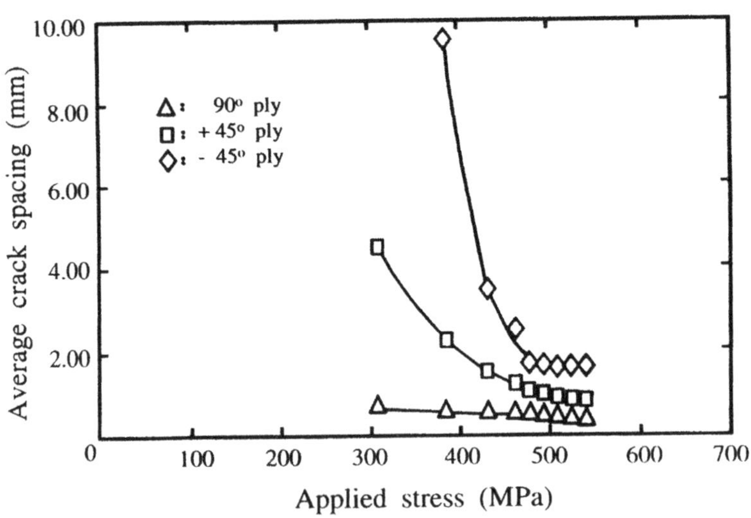

Figure 5.5 Crack spacing in 90°, +45°, and −45° plies as function of global applied stress level for quasi-isotropic laminate (Masters and Reifsnider, 1982).

Several features of Figure 5.5 are of general interest. Cracks begin to form in the weakest plies relative to the major normal stress in the matrix, the 90° plies for Figure 5.5. Then, cracks begin to initiate in the +45° plies, and finally in the −45° plies as loading continues. In all cases, the crack spacing approaches a saturation value, but it is interesting to note that the saturation value is not the same for the +45° and −45° plies of the $[0, 90, +45, -45]_s$ laminate (Masters, 1982); we will attempt to explain the reason for that fact below.

The primary physical consequence of the equilibrium crack density is a regular array of cracks in each of the cracked plies. Some physical evidence for such arrays is shown in Figure 5.6. If one looks carefully at the patterns in Figure 5.6, it becomes apparent that the spacing between the cracks has a characteristic value for each ply and laminate, and this spacing is fairly regular. (A statistical analysis of this observation has been presented by Masters, 1982.) The principal importance of this CDS, as we have named it, is that it is a laminate property. It depends only on the properties and thickness of the plies in the laminate and their stacking sequence. The CDS is not altered by extensive variables like stress level or temperature or by internal conditions like residual stress (Kriz and Stinchcomb, 1982). The analysis of this situation has attracted many investigators; references will be cited below. However, the physics and mechanics are of considerable interest to the remainder of our discussion, so we will use a simple shear-lag analysis to explain the physical behavior.

We consider the diagram of a crack in a laminate, shown in Figure 5.7. For this demonstration, we will consider the stresses in three plies, the one in which a crack has formed (a) and two neighboring plies (c and d). All three plies are assumed to have different ply orientations. The plane of the laminate is assumed to be parallel to the x, y plane in the figure, but the thickness of the laminate is not constrained. There may be many other plies on either side of plies c and d, but we will group their behavior as the "composite response." We will also describe a shear region b, in which all shear behavior is required to occur, for the purpose of the model. Hence, we have the usual shear-lag assumptions: The plies sustain only normal stresses, and only shear stresses occur in the shear regions b. In this case, however, these assumptions are motivated by a great number of laboratory observations that suggest there is usually a matrix-rich region in a laminated composite near the ply boundaries in which shear behavior is often concentrated. With these assumptions, simple equilibrium equations can be written for an equilibrium element, such as the one shown in Figure 5.7. In dimensional form, we could write

$$aE_{xa}\frac{d^2U_a}{dx^2} + \frac{G}{b}(U_c - U_a) + \frac{G}{b}(U_d - U_a) = 0 \tag{5.6}$$

$$cE_{xc}\frac{d^2U_c}{dx^2} + \frac{G}{b}(U_\infty - U_c) + \frac{G}{b}(U_a - U_c) = 0 \tag{5.7}$$

$$dE_{xd}\frac{d^2U_d}{dx^2} + \frac{G}{b}(U_\infty - U_d) + \frac{G}{b}(U_a - U_d) = 0 \tag{5.8}$$

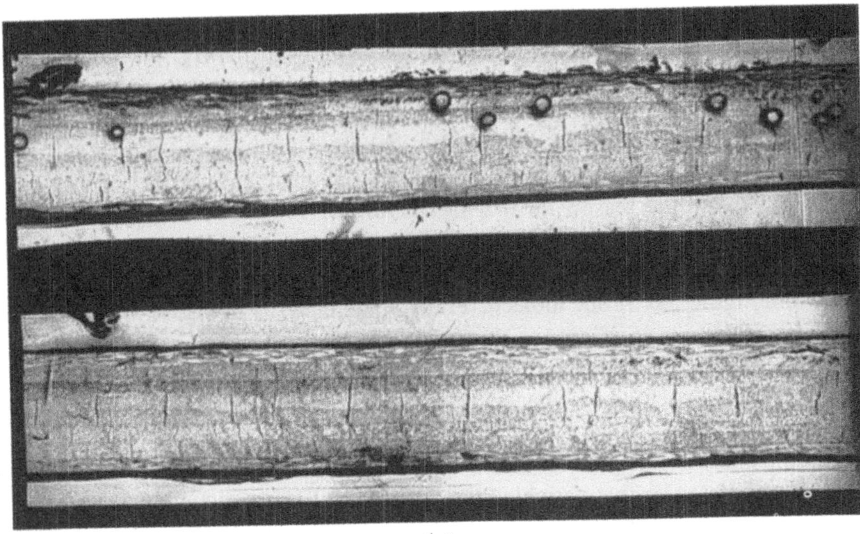

(a)

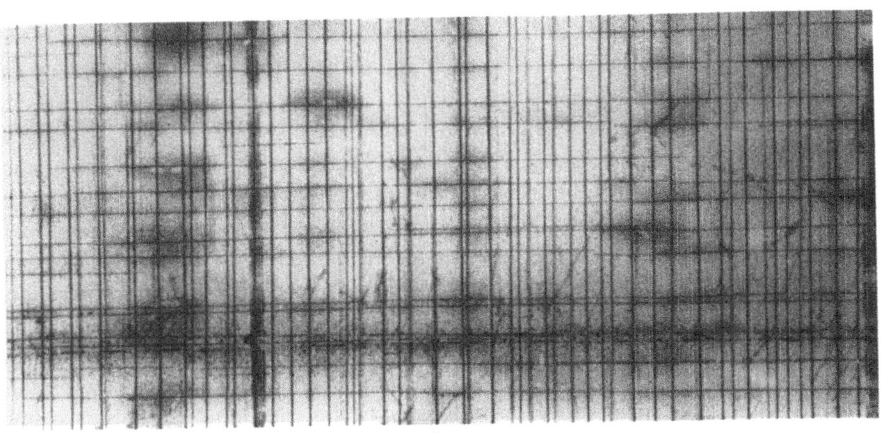

(b)

Figure 5.6 Regular arrays of cracks in laminates: (*a*) in off-axis plies of $[0, +45, -45, 90]_S$ graphite–epoxy laminate (shown by edge replica); (*b*) in 90° and 0° plies of $[0, 90_3]_S$ glass–epoxy laminate (shown in X-ray radiograph).

where, in addition to the variables defined by Figure 5.7, E_{xa}, E_{xc}, and E_{xd} are the elastic Young's modulus of the plies in the direction of principal loading; $U_{a,c,d}$ are the displacements (constant across the ply thickness) of those plies; and U_∞ is the (uniform) displacement of the surrounding composite at large distances from the local disturbance. By imposing the usual boundary conditions (no stress normal to the crack face, disturbance vanishing at infinity, symmetry about the crack plane), solutions to these one-dimensional equations can easily be found (cf. Reifsnider, 1977; Reifsnider and Highsmith, 1981; Reifsnider and Talug, 1980). Several useful

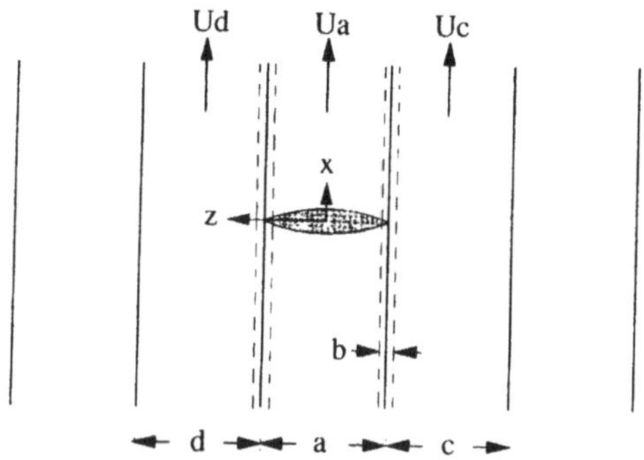

Figure 5.7 A crack in one ply of laminate, viewed on a section through the thickness of the laminate.

results can be obtained from such a simple analysis. The stress in the broken ply as a function of distance from the crack can be estimated, to produce a plot such as the one shown in Figure 5.8. This figure can be used to estimate the *ineffective length* of the ply created by the local crack, that is, the distance over which the local normal stress is reduced in the broken ply for a given laminate and material system. This concept is similar to the ineffective length discussed for fiber fracture. However, in this case, the local stress redistribution caused increased normal stress and shear stress in the neighboring plies, influencing the next event(s) in the sequence of damage during loading (or over time). Some of the associated physical features of this situation are of special interest in the present context for the purpose of estimating the consequence of matrix cracking on stiffness change and the attendant local stress redistribution.

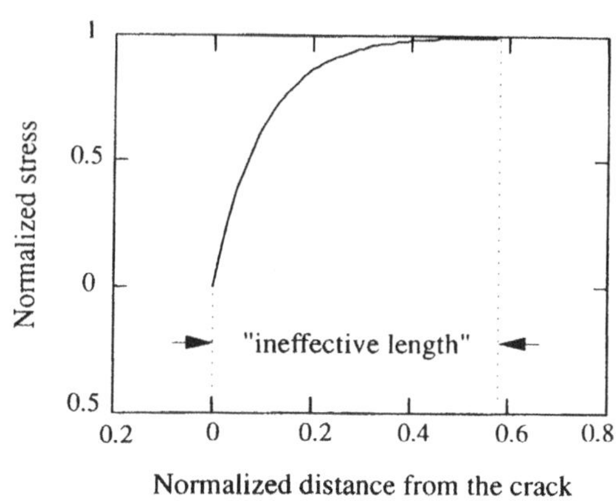

Figure 5.8 Normal stress in broken 90° plies of graphite–epoxy quasi-isotropic laminate as function of the normalized distance from crack face.

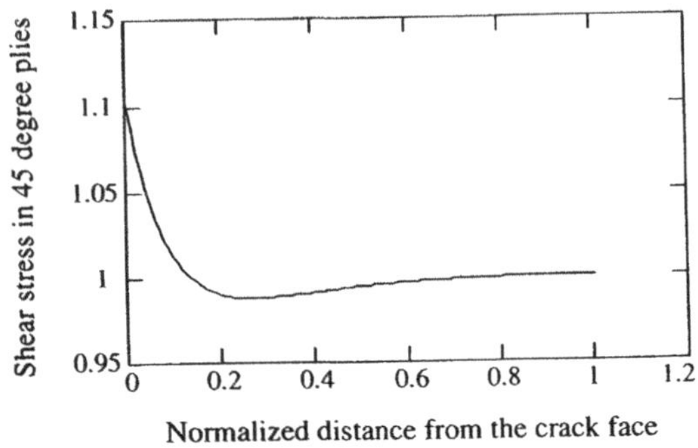

Figure 5.9 Plot of (normalized) shear stress in 45° plies of $[0, 90, +45, -45]_S$ graphite–epoxy laminate as function of the normalized distance from crack in 90° plies.

First, we examine the nature of the local shear-stress distribution. For the example used to generate the data shown in Figure 5.8 (from an AS4-5208 system), the shear stress distribution in the neighboring 45° ply of the cracked 90° ply in a $[0, 90, +45, -45]_s$ laminate is shown in Figure 5.9. As one approaches the ineffective length, the shear stress vanishes. Near the crack, the increase in the neighboring ply is substantial, and over a small region, the sign of the shear stress reverses as a function of distance from the crack. Note that we made an earlier assumption that the shear stress is restricted to a small region, with width b, near the boundary between the plies of the laminate, based on our observations of physical behavior such as the formation of local shear cracks. The consequences of that assumption are substantial. We will discuss them by considering an important application of the resulting shear calculations. Following Moore (1988), we consider the problem of predicting the occurrence of the first crack in a ply of a laminate. Moore suggested that such a prediction might be done by considering the change in strain energy caused by the introduction of a crack (calculated from the stress and strain distributions predicted by the model just described) and comparing it to the measured value for fracture of a specimen in which all plies have a 90° orientation under uniaxial loading. This analysis allows the comparison (and evaluation) of various models as well as the illustration of an important point in the present case. The analyses we will use for comparison include several concerned with the stiffness change of a cracked laminate (of direct interest to us here) (Garrett, et al., 1977; Hashin, 1985a,b; Highsmith, 1981; Kamimura, 1984), with the calculation of first ply failure (Bailey et al., 1979; Flaggs, 1985; Parvizi et al., 1978), and with progressive cracking (Bailey et al., 1979; Fukunaga et al., 1984; Garrett and Bailey, 1977; Kamimura, 1984; Nuismer and Tan, 1982; Parvizi and Bailey, 1978; Vasil'ev, et al., 1970).

A distinguishing factor among the models to be considered is the assumption concerning the nature of the shear distribution in the layers around the crack. As mentioned, the present model, in (5.6)–(5.8), assumes that the shear stress is concentrated in a narrow band of width b the dimensions of which are independent of the thickness of the adjacent plies but typically about one-tenth of the size of a single unidirectional ply. The normal stress is assumed to be constant across the layer, and the shear stress is assumed to be linear across the thickness b. Kamimura (1984) assumes that the shear stress is constant across half of the cracked and uncracked layers (i.e., the displacement is bilinear within the region of half of both layers). The authors cited previously make a variety of other assumptions. Moore uses data for T300/834 from Flaggs (1985) and Wang (as presented in Flaggs, 1985) and glass–epoxy data from Bailey et al. (1979) to compare the predicted strain for the onset of first-ply cracking as a function of the thickness of the plies in which the cracks form. An example of his calculations is shown in Figure 5.10.

The results in Figure 5.10 are typical for this situation. The strain levels for crack initiation are high for thin transverse layers and lower as the relative thickness of the cracking plies increases. In many of the models mentioned, the shear transfer region is a fraction of the ply dimension, so that the shear-stress distribution changes as the thickness of the ply changes. This feature appears to result in a predicted drop in initiation strain that is in excess of the present model, as suggested by Figure 5.10. The *Bailey model* in Figure 5.10 (and Figure 5.11) refers to the model presented by Bailey et al. (1979), Parvizi and Bailey (1978), and Garrett and Bailey (1977). In this comparison, and in others made by Moore (1988), the present approach with a fixed shear transfer layer appears to follow the observed data with significantly better fidelity than the other available models, which do not have this feature. Note that these calculations are directly dependent on the calculation of stiffness change as a

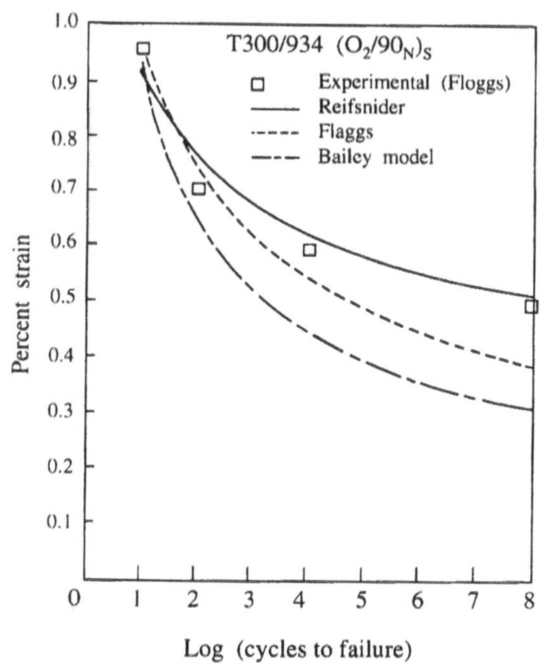

Figure 5.10 Comparison of predicted threshold strains for initiation of matrix cracking with experimental observations for several models (Moore, 1988; Flaggs, 1985).

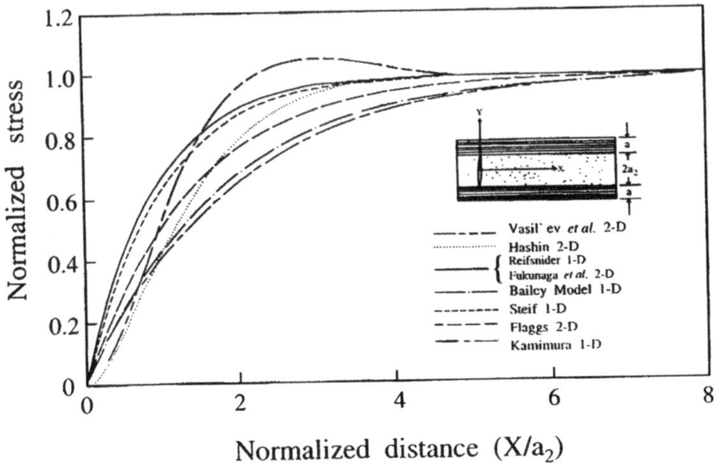

Figure 5.11 Comparison of seven models of local normal stress in cracked ply of $[0, 90_3]_S$ Kevlar–Epoxy laminate as function of distance from crack face (normalized by the half-thickness of broken ply).

function of matrix cracking, since the strain distributions are integrated to calculate a specimen deflection under a fixed load (i.e., the specimen stiffness) when the change in strain energy is calculated. Hence, we see again, as we noticed in Chapter 3 in the discussion of ply discount methods, that the global stiffness change is directly related to the local stress redistributions and the initiation of damage.

Now, we will consider the second fundamental question of the ineffective length predictions and the resulting stiffness change in the specimen for multiple cracking. Moore also compared the models discussed above with the present model and (for calibration) with the closed-form solution for a cross-ply laminate presented by Vasil'ev, et al. (1970). The results of Moore's (1988) comparison are shown in Figure 5.11. All of the approximate models considered underestimate the normal stress in the broken ply as a function of the distance from the crack face in the middle of the ineffective length. If one attempts to estimate the ineffective length by measuring where the normal stress just reaches the undisturbed value (i.e., reaches the normalized value of 1.0 in Figure 5.11), then the present model seems to give the shortest prediction of the approximate models, but still longer than the distance predicted by the closed-form solution. To relate these predictions to experimental data and to complete our discussion of the saturation crack spacing mentioned earlier, we consider the physical sequence of multiple matrix cracking.

We consider Figure 5.7, altered to represent the local stress distributions in the plies of a laminate when cracks form in one such ply, noted by a in Figure 5.12. The normal stress in the direction of primary loading is represented by the distributions sketched on the right-hand side of Figure 5.12, and the normal stress in the adjacent ply, ply b in the figure, is sketched on the left-hand side of the schematic. The familiar local redistribution length, or ineffective length δ, is noted in the figure. In the cracked ply, the normal stress rises from a null value at the crack face to the

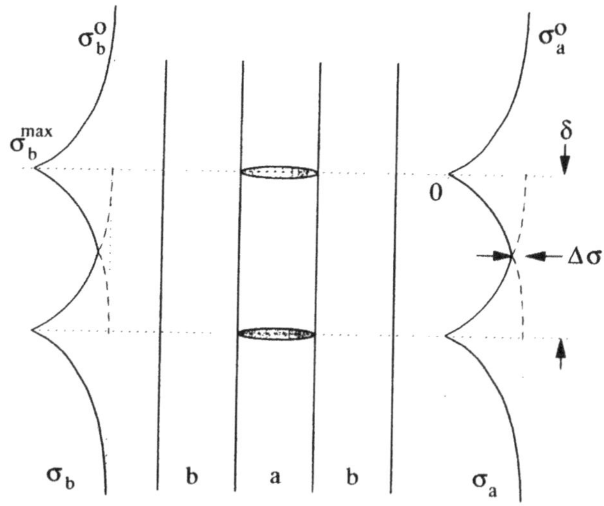

Figure 5.12 Stress distribution near multiple matrix cracks in laminate.

undisturbed value, σ_a^0 at a distance of δ from the crack face. If the remote loading increases to a level that causes σ_a^0 to reach the threshold for matrix cracking to begin, then we might expect to see the result shown in Figure 5.12. However, if σ_a^0 is just equal to the matrix cracking threshold, then the nearest position to the first such crack to appear where a second one can form is, in fact, equal to δ, since that distance is required for the stress to rise to the undisturbed value. When a crack forms at that minimum distance, the superposition of the local stress fields of the two adjacent cracks results in a reduction of the local stress over the ineffective length, as shown by the $\Delta\sigma$ in Figure 5.12. Hence, the global stress must increase by another increment (which is usually of substantial size in real situations) before another crack can form between the first and second one, if they form with a spacing of δ. This has the effect of "arresting" crack formation between any two such cracks and of resulting in the formation of an array of cracks with a (statistically) equal crack spacing, as suggested in Figures 5.5 and 5.6. Hence, we would expect the ineffective length discussed above to predict the minimum crack spacing in the plies of a laminate, which is the saturation spacing so widely observed, sometimes called the CDS for matrix cracking.

Many such predictions of saturation crack spacing have been made in the literature (cf. Bailey et al., 1979; Kriz and Stinchcomb, 1982; Masters, 1982; Moore, 1988; Reifsnider, 1977, 1980). For our present discussion, we will use (5.6)–(5.8) to estimate the minimum crack spacing for plies in several laminates. The results are shown in Table 5.1.

The agreement between predicted and observed results is generally quite good. The constraint offered by the laminate controls the spacing. The crack spacing in the 90° plies varies by over 400% in the table, depending on the laminate, for example. The stacking sequence can also be a major influence. The crack spacing in the 90° plies of the $[0, 90]_s$ laminate is nearly half the spacing in the $[90, 0]_s$ laminate, for

TABLE 5.1 Predicted and Observed Crack Spacing for Several Laminate Types

Specimen Type	Predicted Crack Spacing (mm)	Observed Crack spacing (mm)
Center plies of $[0, 90]_s$ laminate	0.882	1.087–0.532
Outside 90° plies of $[90, 0]_s$ laminate	1.66	1.73–0.775
Center 90° plies of $[0, +45, -45, 90]_s$ laminate	0.76	Static, 1.51–0.62 fatigue 1.44–0.47
Center 45° plies of $[0, 90, +45, -45]_s$ laminate	1.21	1.25–0.995
90° plies of $[0, 90, +45, -45]_s$ laminate	0.411	0.423–0.241
45° plies of $[0, 90, +45, -45]_s$ laminate	0.875	0.909–0.524
Outside 90° plies of $[90_2, 0, +45, -45]_s$ laminate	1.12	1.40–1.07
−45° plies of $[0, +45, -45, 90]_s$ laminate	0.879	0.96–0.990
90_2° plies of $[0, 90_2, +45, -45]_s$ laminate	0.701	0.60–0.460
+45° plies of $[0, 90_2, +45, -45]_s$ laminate	0.890	1.05–0.650

example. Hence, the simple physical arguments given in Figure 5.12 and the local stress analysis (even in simple one-dimensional shear-lag form) seem to capture the essence of the mechanics involved in the formation of the saturation spacing. Many additional calculations of this type have been done, with similar comparisons to experimental data (cf. Masters, 1982; Moore, 1988). An example of a predicted and observed edge-crack pattern using the calculations shown in Table 5.1 is shown in Figure 5.13 (Stinchcomb et al., 1981).

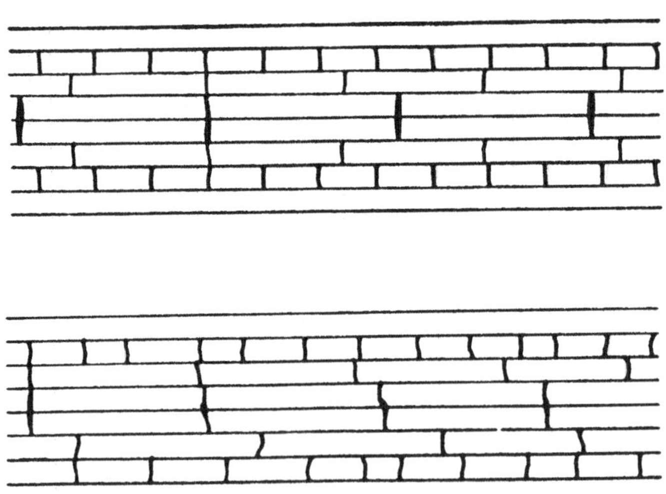

Figure 5.13 Predicted (top) and observed (bottom) edge crack pattern in graphite–epoxy quasi-isotropic laminate after fatigue loading, showing agreement between predicted characteristic crack spacing for saturation and observed crack patterns.

A second feature of Figure 5.12 is also important to our discussion. On the left of Figure 5.12, the normal stress in the next-nearest unbroken ply is also plotted schematically. As one would expect, the local redistribution of stress in the broken ply causes an increase in the stress (a stress concentration) in the neighboring plies. Also, if the distance over which the redistribution occurs (the ineffective length) is small, the local stress concentration is high, since the rate of shear transfer in the neighboring plies is high. Of course, if the ineffective length is very long, there is very little shear rate in the neighboring plies and the stress concentration is low. Figure 5.14 shows measured and predicted strain distributions in the 0° plies that constrain the 90° plies in a $[0, 90_3]_s$ glass–epoxy laminate. In (5.6)–(5.8), the ratio G/b is a material parameter. It is very difficult to measure, independently, the shear stiffness of the matrix-rich region or the exact dimensions of that shear region, b. Hence, that ratio becomes an adjustable parameter, in the sense that a value is determined for one comparison with experimental data for a given material system and that value is used to make all other predictions. Figure 5.14 shows that an intermediate value of G/b seems to fit the observed data well and that changes in that value do not have radical effects on the resulting local stress distribution over values of about one order of magnitude.

The stress concentrations shown in Figure 5.14 can cause additional damage, including matrix cracking in the adjacent plies, fiber fracture, and delamination. Figure 5.15 shows an example of fiber fractures (resolved with edge replication methods) caused by matrix cracks in an adjacent ply. We will discuss the stiffness changes and stress redistributions associated with those damage modes in a subsequent section.

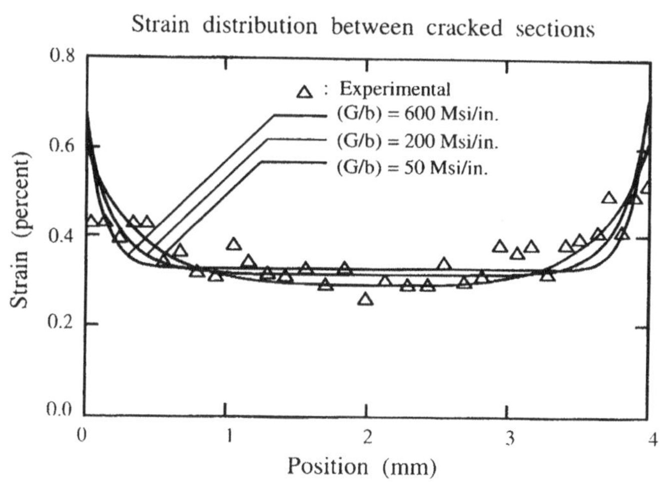

Figure 5.14 Observed and predicted strain distribution in 0° plies in regions near cracks in underlying 90° plies of $[0, 90_3]_S$ glass–epoxy laminate. Predictions are shown for several values of (only) parameter in shear-lag model used to do predictions (Reifsnider and Highsmith, 1981). Crack positions are on left and right extremes of plot.

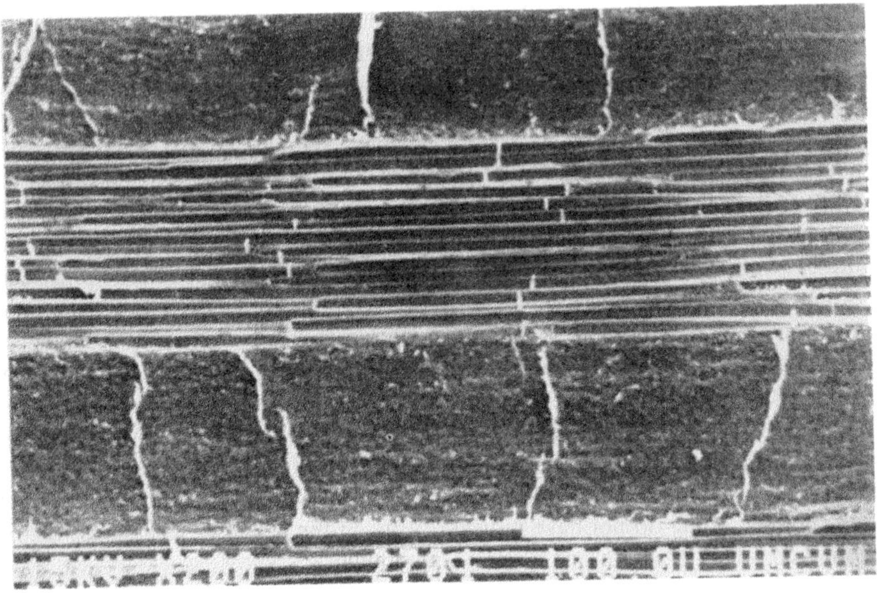

Figure 5.15 Fiber fracture patterns in 0° plies caused by matrix cracks in 45° plies of quasi-isotropic graphite–epoxy laminate.

Another representation of stiffness change associated with matrix cracking can be constructed using constitutive equations with internal variables, as described by Coleman and Gurtin (1967). Talreja (1990, 1994) introduces an internal variable associated with matrix cracking by defining a damage entity tensor having the form

$$d_{ij} = \int_S a_i n_j \, dS \tag{5.9}$$

where a_i is a vector placed at a point on the damage entity surface that represents a selected influence of the damage entity and n_j is a unit outward normal to the surface at that point. If only the influence of damage normal to the entity surface is considered, then Talreja obtains the result

$$d_{ij} = \int_S a_i n_j \, dS = \int_S \kappa \alpha n_i n_j \, dS = K t_c t_c W \tag{5.10}$$

where $\alpha = \frac{1}{2} t_c$ is half the length of a transverse crack (half of the thickness of the cracked ply) and W is the width of the specimen and κ and K are constants. Using these forms, Talreja (1990, 1994) constructs a polynomial form of the Helmholtz free energy for a damaged laminate undergoing in-plane strains using an irreducible integrity basis for the combination of the strain tensor and damage vector [kernel of Equation (5.9)] as independent variables. Talreja obtains the following explicit forms for the lamina stiffness in terms of the damage variables and the usual material stiffness tensor:

$$
\begin{aligned}
E_1 &= E_1^0 + 2\frac{Kt_c^2}{ts}[c_3 + c_7(\nu_{12}^0)^2 - c_{13}\nu_{12}^0] \\
E_2 &= E_2^0 + 2\frac{Kt_c^2}{ts}[c_7 + c_3(\nu_{12}^0)^2 - c_{13}\nu_{12}^0] \\
\nu_{12} &= \nu_{12}^0 + \frac{Kt_c^2}{t_s}\left[\frac{1 - \nu_{12}^0\nu_{21}^0}{E_2^0}\right](c_{13} - 2c_7\nu_{12}^0) \\
G_{12} &= G_{12}^0 + \frac{2Kt_c^2}{ts}c_{11}
\end{aligned}
\tag{5.11}
$$

where the superscript 0 indicates initial (undamaged) values, c_{ij} are the usual material stiffness tensor values, s is the average spacing between cracks, and c_j are material constants. Among the advantages of this approach is the explicit form for all of the in-plane tensor values of stiffness as a function of matrix crack spacing s and laminate thickness t. Figure 5.16 shows two examples of predicted stiffness change as a function of cycles of loading (using measured crack densities) taken from the data of Talreja (1985b). It is seen that the changes in the different moduli are quite distinct. The changes in some of the tensor values are much greater than those in others, and the relative magnitudes depend on the laminate type.

A myriad of internal variable, damage mechanics approaches to this subject have been proposed. The reader is referred to Kachanov (1986), Krajcinovic (1984), and Lee et al. (1989) as starting points. The sophistication of this approach is attractive.

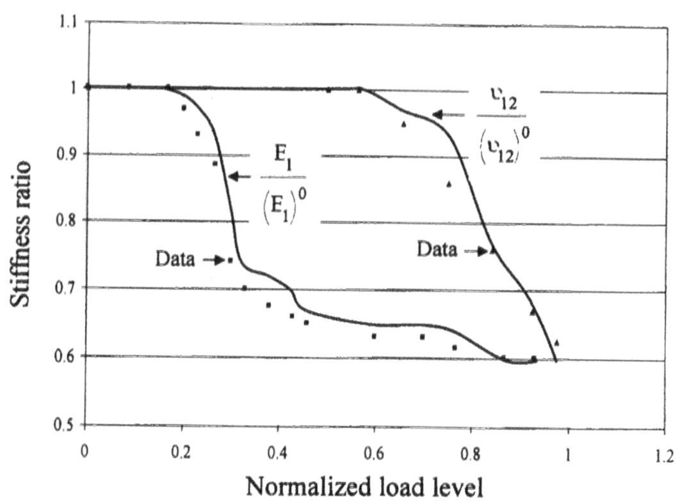

Figure 5.16 Predicted and observed stiffness changes with cyclic loading for cross-ply glass-reinforced epoxy laminate (Talreja, 1985b).

The most substantial challenge associated with the methodology is the physical interpretation and experimental determination of the large number of material constants that are inherent in the method.

5.2.2 Matrix Cracking under Cyclic Loading

Under typical applied service conditions, the CDS for matrix cracking, or saturation cracking as it is more commonly called in recent literature, is much more likely to occur under cyclic loading than for monotonically increasing or quasi-static loading. For quasi-static loading, depending on the statistical nature of local ply strength and local stress variations due to nonuniform stiffness (caused by such things as manufacturing flaws or variations in volume fraction of reinforcement), all regions of an off-axis ply may not reach saturation cracking before the laminate breaks under increasing load. Cyclic loading introduces the possibility of crack growth so that small starter cracks and other defects have more opportunity to contribute to a fully developed CDS that is more uniformly developed across the specimen.

Figures 5.17 and 5.18 show examples of the stiffness changes that are typically observed for cyclic loading of laminates. The stiffness changes observed in Figure 5.17 show three "regions," called regions I, II, and III in the figure. Most of the matrix cracking that we would observe under quasi-static loading to the same level of applied loading as the maximum cyclic stress (about 60% of the laminate strength, $R = -1$, for this case) occurs in region I, causing an initial sharp drop in stiffness, and a sharp increase in stress in the unbroken plies, internally, as we discussed above. In region II, the cracks that have nucleated grow and interact to form saturation spacing, as we will show below. The stiffness change (and internal

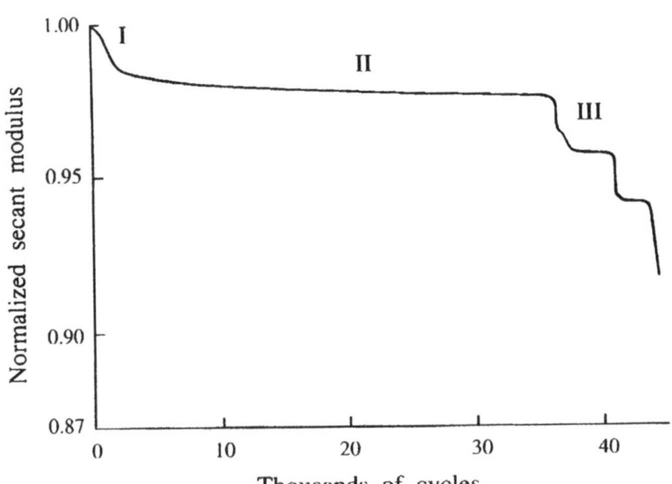

Figure 5.17 Typical global stiffness reduction observed in $[0, 90_2]_S$ graphite–epoxy laminate during cyclic loading to failure.

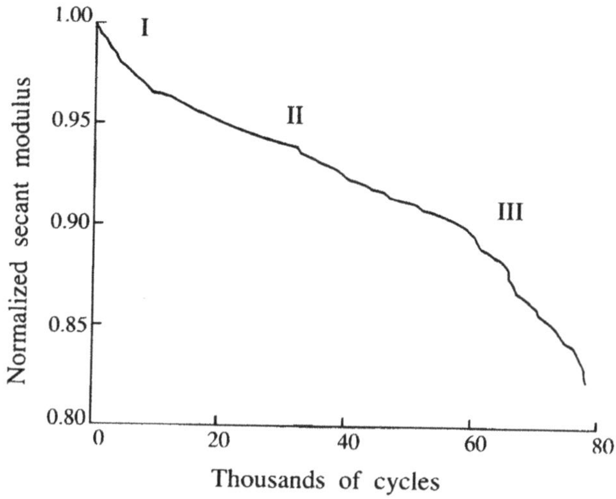

Figure 5.18 Typical global stiffness reduction observed in $[0, 90, +45, -45]_S$ graphite–epoxy laminate during cyclic loading to failure.

stress redistribution, which is our concern here) occurs at a low level. In region III, a significant increase in damage rate and stiffness change occurs, usually because of damage interaction, localization, and growth. For the cross-ply specimen type shown in Figure 5.16, region III is driven primarily by delamination initiation and growth.

For a quasi-isotropic laminate, the situation is somewhat different but fundamentally the same, as shown in Figure 5.18. Region I is marked by a sharp stiffness drop followed by a more modest stiffness change rate and ending in an accelerated stiffness change leading to the final failure. In this case, not only the 90° form characteristic crack patterns but also the 45° plies saturate. (See Figure 5.6*a*.) Region II shows a significant level of matrix crack initiation in the 45° plies, as well as growth of cracks in all of the off-axis plies.

In many cases, for a given material system, the shape of the stiffness change curve is quite similar for different laminate types when the data are normalized by the life of the laminate. (In Figures 5.17 and 5.18 the life differed by a factor of 2.) Figure 5.19 shows an example of this feature for a T300/914C system (Schulte and Stinchcomb, 1984).

Schulte and Stinchcomb report that the characteristic crack patterns develop in the first region of Figure 5.19, cracks couple and grow to cause some delamination initiation in the second region, and delamination and fiber fracture seem to be the main features of damage development in the final region of stiffness change. In the early stages of stiffness change (regions I and II), the global stiffness change is a remarkably good indicator of the rate of matrix cracking in the off-axis plies. Figure 5.20 shows the correlation for a cross-ply laminate. The rate of matrix cracking in each of the plies in Figure 5.20 as a function of the cycles of loading is shown in Figure 5.21, along with the global stiffness change. Cracks form first and more

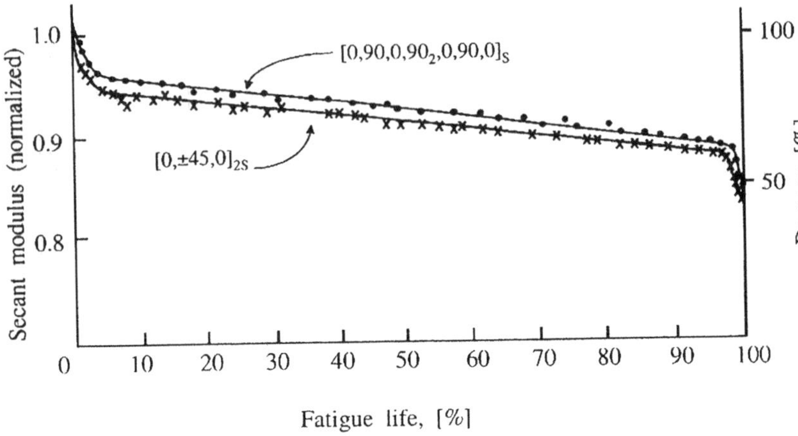

Figure 5.19 Secant stiffness change during fully reversed fatigue loading of graphite–epoxy laminates normalized by life of each laminate type.

densely in the 90° plies and reach an equilibrium saturation spacing at about midlife for this loading level. Crack development for the 45° plies is slower and does not quite reach the stable value (for which no more stiffness change occurs) over the life of the specimen, in this case. The global secant modulus follows these events well, with the exception of a sharp step near the end of life that is due to the initiation of delamination along the edge of the coupon specimens in this case.

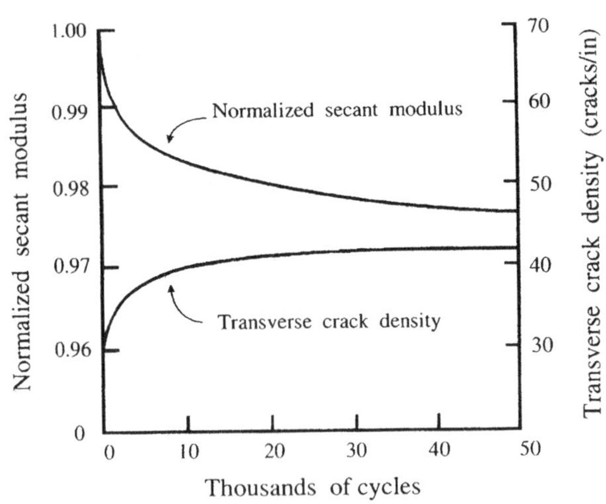

Figure 5.20 Stiffness reduction and crack development for $[0, 90_2]_S$ laminate in first region of damage development.

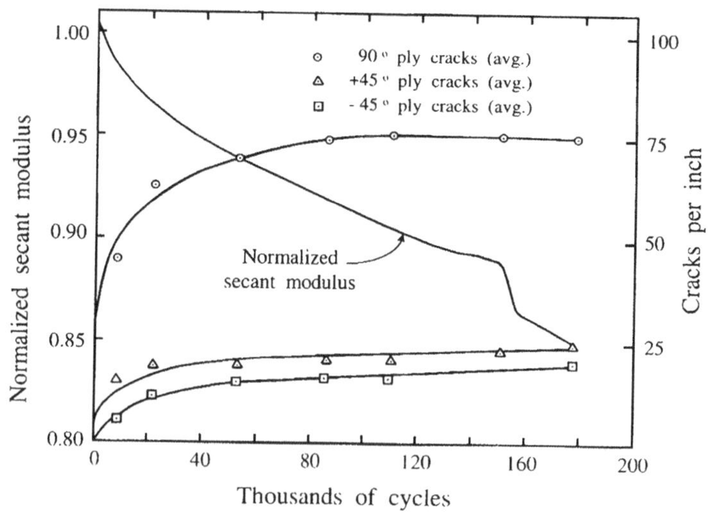

Figure 5.21 Stiffness reduction and crack development in $[0, 90, +45, -45]_S$ graphite–epoxy laminate.

An important feature of these results is shown in Figure 5.22. The saturation crack spacing produced by quasi-static loading is the same as that produced by cyclic loading if the saturation state is fully developed. This critical feature tells us that we can use the CDS calculations such as those discussed above to estimate the maximum stiffness change (and the maximum internal stress redistribution due to matrix cracking), independent of the mode of loading. Looking at the parameters

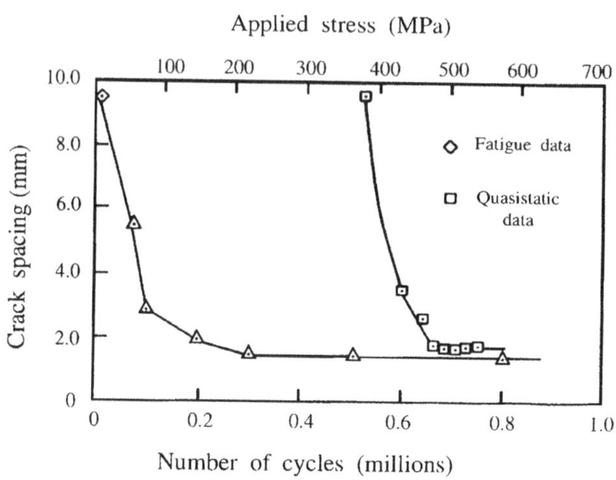

Figure 5.22 Crack spacing in 90° plies of quasi-isotropic laminate as function of quasi-static loading level and cyclic loading to 65% of quasi-static strength ($R = -1$).

in (5.5)–(5.8), the analysis and experiments show that the CDS for matrix cracking is independent of external variables such as temperature and moisture, unless they change the stiffness of the laminate (and thereby influence the rate of stress buildup discussed in Figure 5.11) (Kriz and Stinchcomb, 1982). Hence, this simple calculation can help us greatly to estimate the changes of the local stress in the critical element as the subcritical elements around it crack and redistribute their stress into the surrounding material.

5.2.3 Additional Analysis Methods

We have discussed the calculation of global stiffness change as a function of matrix cracking in an earlier section using a very simple one-dimensional shear-lag approach with some distinctive features based on physical observations. A variety of more complete analyses have been reported that provide us with a fairly comprehensive approach to this subject. A few of these methods will be noted here as starting points for the reader. In our earlier discussion, we noted that the shear-lag method, as proposed in (5.6)–(5.8) and the attending text, produced useful estimates of the saturation crack spacing observed in a range of plies and laminates, including laminates with off-axis plies at various angles (cf. Table 5.1). Highsmith (1982) used this method to estimate the stiffness change as a function of crack density to obtain results such as those shown in Figure 5.23. The fundamental behavior is followed well by the simple analysis, but a difference as high as 15% between predicted and observed results was observed for the different laminate types considered. The laminate analysis result shown in Figure 5.23 is obtained by reducing the stiffness of the cracked plies to zero to produce a lower bound on the laminate stiffness (which appears to be the case, in comparison to experimental data). Of course, we know from our earlier discussion that the stiffness in the cracked plies seems to reduce in such a way that the average stress in the ply is close to the breaking stress

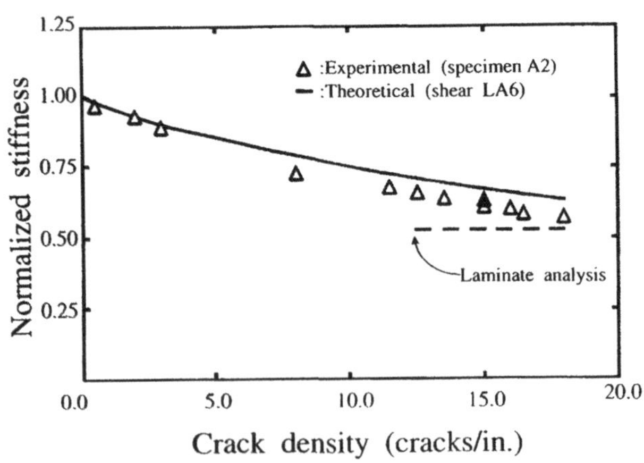

Figure 5.23 Predicted and observed stiffness change as function of matrix cracking using shear-lag method (Highsmith and Reifsnider, 1982).

of the ply; that is, the stiffness does not relax completely for the entire ply when cracks appear.

This is not surprising in the sense that a one-dimensional model cannot represent the variations of the local stress field across the specimen (or ply) thickness well; each ply has only one displacement variable in the analysis described. Highsmith and Reifsnider (1982) showed that the local stresses obtained from a full-field solution to a finite-difference formulation of the problem of local cracking were generally different from the shear-lag result (sometimes by significant amounts) but were usually similar at small distances from the crack of the order of one ply thickness (when compared to the stress along the centerline of the broken ply, for example) (Talug, 1978). An example calculation is shown in Figure 5.24.

The shear-lag results are "too stiff" near the crack face. It predicts stresses that are too high in that region, a result that we also saw in Figure 5.11 in comparison to the analysis of Hashin. Hashin (1985a,b) used the principle of minimum complementary potential energy to derive approximate stress and strain fields in a cracked cross-ply laminate and used the results to estimate global stiffness change. Good comparisons with results were obtained.

A variety of other approaches are available. Laws et al. (1983) and Laws and Dvorak (1985) used an *effective medium* to simulate the cracked ply and self-consistent methods applied to an infinite homogeneous material to calculate global stiffness changes. A variety of damage mechanics methods have been developed that address global stiffness change due to matrix cracking. Allen et al. (1987a,b) and Lee et al. (1989) have developed thermomechanical constitutive equations for laminates with matrix cracking and related the stiffness changes to local strain energy release in that context. Talreja (1985a, 1986, 1994) proposes an internal variable approach. An advantage of that method is its ability to represent a wide range of damage types, including matrix cracking or a variety of other damage modes. For

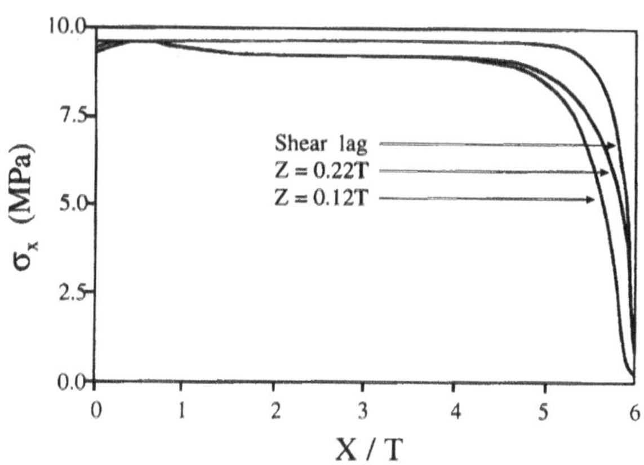

Figure 5.24 Normal stress distribution in cracked 90° ply of $[0, +45, -45, 90]_S$ graphite–epoxy laminate predicted by finite-difference method (Talug, 1978) and by present shear-lag analysis.

matrix cracking, Talreja chooses a vectorial representation of crack-induced damage, which carries information about the crack orientations and characteristic dimensions.

Gudmundson et al. (1991) and Gudmundson and Ostlund (1992a,b) have presented an analytical model for the prediction of laminate stiffness and thermoelastic properties of laminates containing matrix cracks. The model is based on the known solution for stresses around a flaw or crack in an infinite isotropic medium. Comparisons to experimental data for cross-ply laminates showed very good agreement.

McCartney (1989, 1990 a,b, 1991) has developed an improved theory of stress transfer between the fibers and matrix in uniaxial and cross-ply fiber-reinforced composites. Thermal residual stresses are included in this analysis, and all required field equations, boundary conditions, and stress–strain–temperature relations are either satisfied exactly or in an average sense. McCartney also makes an extensive comparison of the shear-lag approach and the more "realistic" models of McCartney (1991) and Hashin (1986 a,b). Tao and Sun (1996) have used a finite-element model to study the reduction of lamina moduli with matrix cracking and highlighted the effect of the shear stiffness in the cracked ply on the global stiffness variation. They found significant differences in the variation of ply-level shear stiffness compared to the stiffness change in the direction of the normal stress perpendicular to the fibers caused by matrix cracking in that ply. Their predicted global stiffness changes matched experimental values well.

5.3 TIME-DEPENDENT STIFFNESS CHANGE

In the discussion above, we have been concerned with stiffness changes (and corresponding internal stress redistributions) that occur as a result of quasi-static loading or repeated (cyclic) loading. These stiffness changes are time dependent indirectly, only in the sense that they respond to time-dependent inputs. However, there is an important class of stiffness changes that are directly time dependent; that is, they change with time in the presence of applied conditions that are constant or unchanging with time.

Many of the general features of this topic are usually discussed under the mechanics genre of viscoelasticity. The genesis of this concept is in the identification of physical behavior that is neither purely elastic (in which energy is stored without loss) or purely (linearly) viscous (in which no energy is stored—all is dissipated). These materials respond to applied excitations by both storing and dissipating energy (Christensen, 1971). The literature on this subject is remarkably well developed and is highly rigorous and replete, albeit, somewhat incomplete in its representation of materials that change their fundamental state with time (perhaps due to the development of damage).

The purpose of the present address of this subject is not to present the fundamental philosophy of the subject but to apply the concepts to the problem at hand. Primarily, we plan to use the representations in the literature to determine how this phenomenon causes the stresses in the critical element to change with time. If the matrix relaxes, for example, the fibers must sustain the load "shed" by the matrix, and if the fibers creep, the stresses in the matrix change. Moreover, at elevated temperatures, any local nonuniformity of stress (e.g., near a joint, cutout, edge, defect, or broken fiber) in a composite will attempt to homogenize itself by viscoelastic response. In continuous-fiber-reinforced composites, this internal stress redistribution is often large and can greatly influence or even control the strength and life of the material.

Viscoelasticity is concerned with the study of the response of materials that "creep" under a constant load, or "relax" their resistance to loading under a constant strain. Moreover, such materials generally have "memory." When an external stimulus is removed, they tend to return to the original shape they had before the stimulus was applied. Finally, viscoelastic materials may display time-delayed fracture or rupture, known as *creep rupture* or *stress rupture*. Many of the features of this response that we are concerned with can be demonstrated by considering *linearly* viscoelastic materials. Linearly viscoelastic materials satisfy the conditions of proportionality and superposition of response (Schapery, 1974). For isotropic materials, the corresponding constitutive equations for a linearly viscoelastic material can be written as (following the notation of Dillard, 1991)

$$P'(\Phi)s_{ij} = Q'(\Phi)e_{ij} \tag{5.12}$$

$$P''(\Phi)s = Q''(\Phi)e \tag{5.13}$$

where P and Q are operator rational expressions (with a single prime for deviatoric quantities and a double prime for dilatational quantities) and Φ is an operator. Stress and strain tensors are s_{ij} and e_{ij}, respectively, with no subscripts for dilatational quantities. Some of the physical features associated with such equations can be extracted by an examination of a few one-dimensional examples of Equations (5.12) and (5.13). One such example is the *standard linear solid*, represented by the general equation

$$\sigma + \tau_1\dot{\sigma} = E_1(\varepsilon + \tau_2\dot{\varepsilon} \qquad 5.14)$$

in which E_1 is the *relaxed modulus* for quasi-static loading, τ_1 is the relaxation modulus (which controls the rate of relaxation of stress under constant strain), and τ_2 is the creep modulus, which controls the creep rate under constant stress. The solution of (5.14) under constant applied stress conditions has the form

$$\varepsilon = \frac{\sigma}{E_1}(1 - e^{-t/\tau_2} \qquad 5.15)$$

if the constant stress σ is instantaneously applied. For most materials, there would also be some instantaneous elastic response, so that the strain might have the form

$$\varepsilon = \frac{\sigma}{E_1}(1 - e^{-t/\tau_2}) + \varepsilon_0 \tag{5.16}$$

The response represented by (5.16) is shown in Figure 5.25.

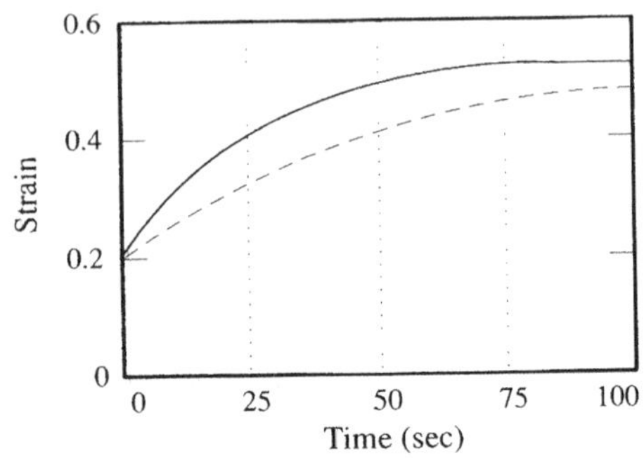

Figure 5.25 Creep response under constant stress for viscoelastic material represented by (5.16).

In Figure 5.25, the initial elastic strain ε_0 has the value of 0.2. The upper curve in this figure has a characteristic time τ_2 of 25 s, while the lower curve has a characteristic time of 50 s. Over the 100 s plotted in the figure, the strain closely approaches an asymptotic value for the short characteristic time but not for the longer one. Also, the characteristic time is not *equal* to the time required for the strain to reach an asymptotic value. However, the characteristic time τ_2 does control the *rate* of the process, including the total time needed to reach any characteristic level of response (such as the asymptotic value shown in Figure 5.25).

This is a good physical example of the characteristic time concept introduced in Chapter 3. The characteristic time of the process being considered represents the rate of the physical events that make up the process and, therefore, controls the rate of the associated mechanical property or performance changes, such as the reduction of stiffness that corresponds to the strain at constant stress shown in Figure 5.25. In Chapter 3 we mentioned that the strength evolution equation [repeated as (5.1) in this chapter] involved a generalized time that must be normalized by a characteristic time, such as τ_2. It was further noted that the choice of the characteristic time was driven by the definition of the controlling process. If viscoelastic response controls the remaining strength and life of the composite, then a characteristic time such as τ_2 (or a corresponding value for the viscoelastic response of the material in question) should be used. Quite often, however, fiber-controlled composites are designed in such a way that the remaining strength and life are *not* controlled by viscoelastic response, but the time-dependent viscoelastic stiffness changes with time enter the problem as an important influence on the (local) stress distributions and redistributions. That is the reason that this subject is introduced in this chapter. *Viscoelastic response is typically an important mechanism for stiffness change and local stress redistribution; it is not a failure mode in the direct sense in most practical cases for fiber-controlled composites unless the fibers undergo significant creep.*

In the strength evolution, (5.1), we assumed that fatigue was the controlling process for remaining strength and life. For this reason, the characteristic time was taken as the instantaneous life of the critical element, N, under the current conditions in the critical element represented by $\mathrm{Fa}[\sigma_{ij}(t)/X_{ij}(t)]$. Reductions in stiffness due to viscoelastic material response change the local stress state in the critical element, $\sigma_{ij}(t)$. They do not affect the characteristic time to failure directly, except through the influence of the changes in $\sigma_{ij}(t)$ on the value of the failure function Fa and, therefore, through the subsequent changes in the instantaneous life N that results from a solution of (5.2).

Now we return to (5.14) and consider the interpretation of some of the material constants (coefficients) in that equation. We can see that if we hold the strain constant, the rate of change of stress is controlled by the coefficient τ_1, called the relaxation modulus. In like fashion, if the stress is held constant, τ_2 controls the rate of strain and is called the creep modulus. If the stress or strain is driven at very high frequencies, then the stress and strain values in (5.14) have negligible values compared to the rates of change of those variables, so that we could approximate the equation as

$$\dot{\sigma} \cong \frac{\tau_2}{\tau_1} E_2 \dot{\varepsilon} = E_1 \dot{\varepsilon} \tag{5.17}$$

where E_1 is the unrelaxed modulus. By the same logic, if the changes in stress and strain are very small compared to the values of those variables, the response is governed by the relaxed modulus E_2. These four limiting cases define the manner in which the material constants τ_1, τ_2, E_1, and E_2 can be measured.

Another physical interpretation of these simple concepts is of considerable importance to the general problem. It has to do with the tendency of a viscoelastic material to dissipate energy (sometimes discussed under the technical banner of damping). This diverse subject deserves comprehensive study by the interested reader. Our limited objective here is to show how the physical and modeling concepts we have introduced in this chapter also lead to dissipative behavior.

For this purpose, we will consider the harmonic excitation and response of a linear viscoelastic material. To demonstrate the principles of such behavior, we will represent the stress (forcing function) as

$$\sigma = \mathrm{Re}\, \sigma_0 e^{i\omega t} = \mathrm{Re}(\sigma' + i\sigma'')[\cos(\omega t) + i \sin(\omega t)] \tag{5.18}$$

where σ' and σ'' are the real and imaginary parts of the driving stress, ω is the driving frequency, and t is time (Re indicates that only the real part of the product is considered in the argument). If (5.18) is substituted into (5.14), we could write

$$(1 + i\omega\tau_1)\sigma = E_2(1 + i\omega\tau_2)\varepsilon \tag{5.19}$$

Now suppose that we also define a *complex modulus* E such that

$$E = E' + iE'' = \frac{\sigma}{\varepsilon} = E_2\left[\frac{1 + \omega^2\tau_1\tau_2}{1 + \omega^2\tau_1^2} + i\frac{\omega(\tau_2 - \tau_1)}{1 + \omega^2\tau_1^2}\right] \tag{5.20}$$

Then, if we define a phase lag δ between the imposed stress and the resulting strain response according to the relationship

$$\frac{\sigma}{\varepsilon} = E_2 e^{i\delta} = E_2[\cos(\delta) + i \ \sin(\delta)] \tag{5.21}$$

and compare (5.20) and (5.21), we can write an expression for the tan(δ) expression commonly used to discuss dissipation and damping as

$$\tan(\delta) = \frac{\omega(\tau_2 - \tau_1)}{1 + \omega^2 \tau_1 \tau_2} \tag{5.22}$$

Since this quantity is a function of ω, we can find the value of ω that corresponds to the maximum value of the phase lag, which is

$$\omega_{\text{max}} = \frac{1}{\sqrt{\tau_1 \tau_2}} \tag{5.23}$$

for which

$$\tan(\delta)_{\text{max}} \doteq \frac{\tau_2 - \tau_1}{2\sqrt{\tau_1 \tau_2}} \tag{5.24}$$

The characteristic relaxation and creep times τ_1 and τ_2, respectively, control the frequency that corresponds to a maximum value of the phase lag and determines the value of that phase lag according to (5.24). To complete this brief introduction of the concept of dissipation, we can also form an expression for the total work done by the forcing function over one cycle of oscillation. Following the development above, we assume that the response strain has the form

$$\varepsilon = \text{Re} \ \varepsilon_0 e^{i(\omega t - \delta)} \tag{5.25}$$

and calculate the work over one cycle as

$$\Delta W = 2 \int_0^{\tau/2} \sigma \frac{d\varepsilon}{dt} \ dt \tag{5.26}$$

Combining (5.18)–(5.26) and remembering that $\omega = 2\pi/\tau$, we obtain

$$\Delta W = \pi \ \delta W_0 \tag{5.27}$$

where $W_0 = \sigma_0 \varepsilon_0$. We could also divide (5.27) by the period of the cycle to obtain the dissipated energy per unit time (per unit volume), which has the form (if δ is small) of

$$\Delta \tilde{W} = \delta \omega \, W_0 \tag{5.28}$$

which has the physical character of the power dissipated by the material. Equation (5.28) shows that the phase lag (for a linear viscoelastic material) controls the fraction of input energy that is dissipated by the material as a result of the time dependence of the elastic stiffness of that material. And in view of (5.23) and (5.24), this dissipated energy is, in turn, controlled by the characteristic time(s) of the viscoelastic process. Hence, physically, the interpretation of the characteristic time introduced in Chapter 3 as normalization for the generalized time concept is clear, at least for this type of material response.

5.3.1 Isothermal Transient Response

Below certain stress levels and up to certain times, most viscoelastic materials exhibit a linearly viscoelastic range of behavior (Findley et al., 1972). Under isothermal conditions, the compliance tensor S of anisotropic materials can be measured by conducting a series p of creep tests with constant applied stress $\hat{\sigma}_{kl}^{p}$ and measuring the resulting strain:

$$\varepsilon_{ij}(t) = S_{ijkl}(t)\hat{\sigma}_{kl} \tag{5.29}$$

Alternatively, the relaxation modulus can be measured by conducting a series of relaxation tests with constant applied strain levels to obtain the resulting stress, as

$$\sigma_{in}(t) = C_{ijkl}(t)\hat{\varepsilon}_{kl} \tag{5.30}$$

If the input values are not constant with time, it is typical to represent these response equations in terms of the Boltzmann integrals:

$$\varepsilon_{ij}(t) = \int_{-\infty}^{t} S_{ijkl}(t-\tau)\frac{d}{d\tau}\sigma_{kl}(\tau)\, d\tau \tag{5.31}$$

$$\sigma_{ij}(t) = \int_{-\infty}^{t} C_{ijkl}(t-\tau)\frac{d}{d\tau}\varepsilon_{kl}(\tau)\, d\tau \tag{5.32}$$

In (5.15) and (5.16), we have seen simple one-dimensional examples of (5.29) for an isotropic material. It is common to use mechanical analog to assist in the interpretation of the forms of $S(t)$ and $C(t)$. Equation (5.15), for example, represents material behavior that can be simulated by the Kelvin model of springs and dashpots, as shown in Figure 5.26 (Christensen, 1971). A series of such models can be used to describe the creep compliance of a given solid over an extended period of time; each

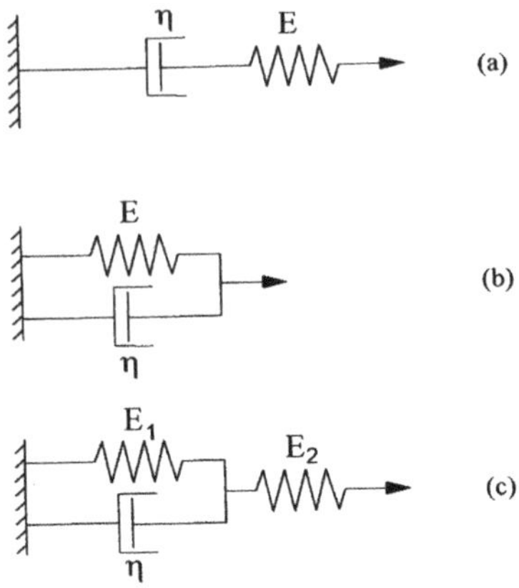

Figure 5.26 Mechanical analog of viscoelastic behavior: (*a*) Kelvin model and (*b*, *c*) two modified Kelvin models [represented by (5.16)].

of the multiple Kelvin elements, for example, can represent the response for different periods of time over the total range. Such a series might have a form similar to the equation

$$S(t) = \sum_{i=1}^{n} \frac{1}{E_i}[1 - e^{-E_i t/\mu_i}] \tag{5.33}$$

where E_i is the stiffness of the *i*th spring and μ_i is the viscosity of the *i*th dashpot. If a free spring is added to that series, the model represents a viscoelastic solid with instantaneous elastic response and no permanent flow, similar to (5.16). If a free dashpot is added, the model describes a viscoelastic fluid. A wide variety of series solutions have been discussed in the literature, with excellent results (cf. Dillard, 1991).

Numerous methods of measurement of time-dependent stiffness and compliance have also been detailed in the literature (e.g., Brinson et al., 1978, 1979; Dillard, 1991; Dowling, 1993; Osiroff, 1990; Richards, 1961). To illustrate the physical essence of the situation, we will briefly outline the method of using instantaneous stress–strain curves at a given moment of time, generally called isochronous stress–strain curves. Of course, since obtaining a stress–strain curve requires a finite period of time, such instantaneous curves cannot be generated directly. Instead, strain data as a function of time at various constant load levels (creep curves) are recorded, as shown schematically in Figure 5.27*a*. Actual compliance data for a unidirectional T650-42/Radel 90° specimen are shown in Figure 5.26*b* (Osiroff, 1990). For each applied level of constant stress σ_i, the strain values ε_i will be different for a given time t_i. These stress and strain values for each t_i can be plotted as an instantaneous stress–strain curve for each such value of t_i. Such an isochronous stress–strain curve

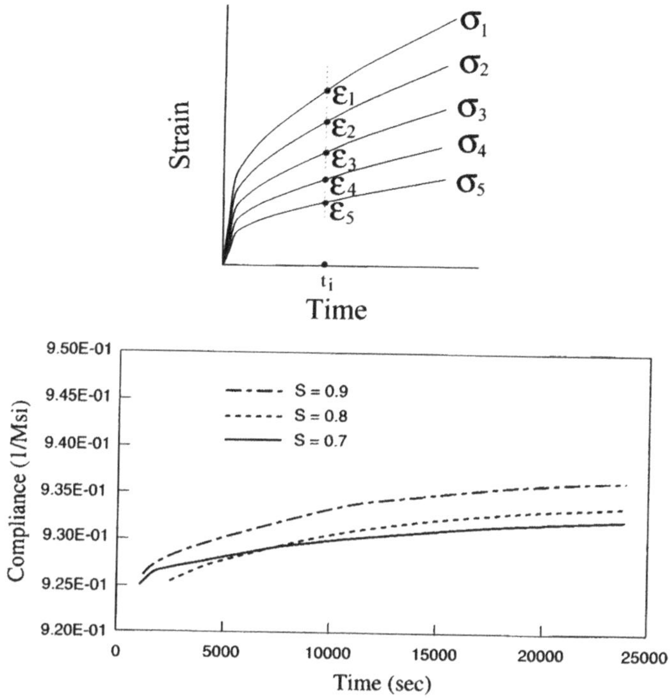

Figure 5.27 (*a*) Schematic diagram and (*b*) data for T650-42/Radel 90° unidirectional specimen (Osiroff, 1990) showing typical creep behavior at constant applied stress.

is shown schematically in Figure 5.28*a*, and the corresponding data are reduced for two times from Figure 5.27*b* to give the result shown in Figure 5.28*b*. As shown in Figure 5.28*a*, for each value of time t_i, an isochronous stress–strain curve can be constructed, and the (secant) slope of that curve can be determined as a function of strain. Finally, for each such strain value, the secant modulus can be plotted as a function of time (usually as a function of the logarithm of time), as shown schematically in Figure 5.29.

Such data reduction methods can be used to determine the material constants in analog models of the creep response, such as the form given in (5.33). A particular case of interest occurs when the isochronous stress–strain curves in Figure 5.28 are linear (as they appear to be for the material data given in the example in Figure 5.28*b*). In that case, the slope does not depend on the strain (or stress) level, and the modulus-versus-time curve in Figure 5.29 is independent of strain or stress level. This is a common assumption for many composite materials in the operating range of behavior of engineering materials; we will examine some important exceptions to that simplification in a later section on nonlinear viscoelasticity. Nevertheless, we will give several examples in a later chapter for which the linear assumption

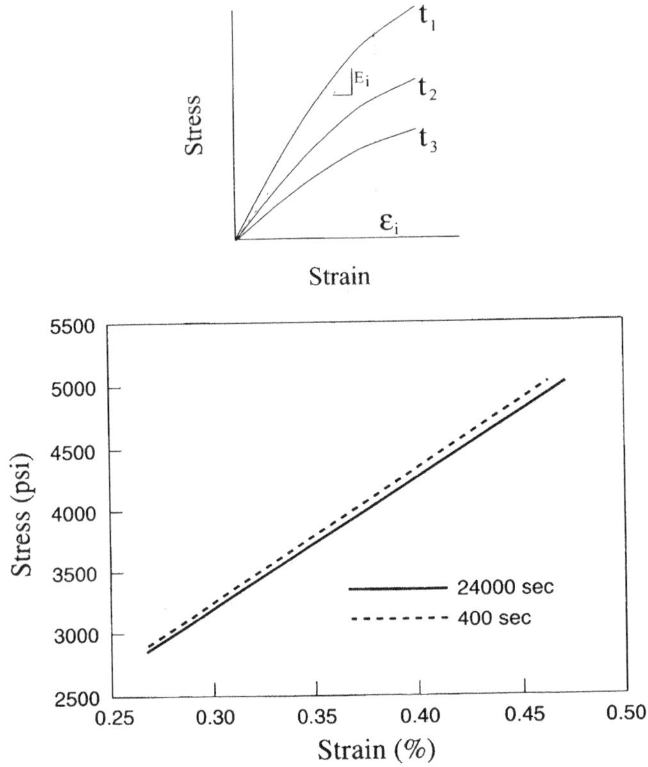

Figure 5.28 (*a*) Schematic diagram and (*b*) data for a T650-42/Radel 90° unidirectional specimen (Osiroff, 1990) showing isochronous stress–strain curves for creep data shown in Figure 5.27.

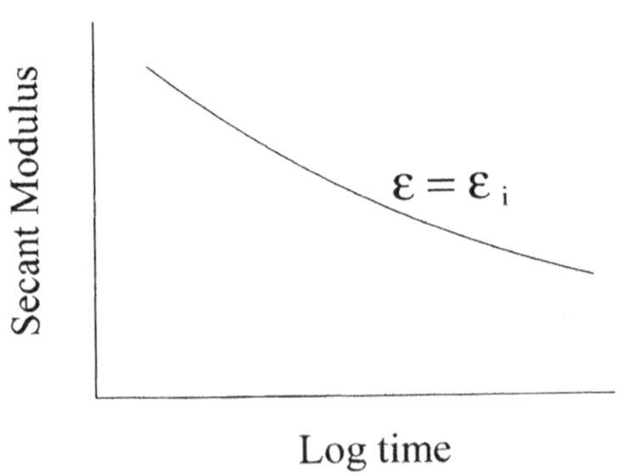

Figure 5.29 Secant modulus of viscoelastic material as function of time for given value of strain determined from creep data and isochronous stress–strain curves.

produces quite satisfactory representations of the viscoelastic behavior of fiber-dominated composites. For those examples, a representation of the form

$$\frac{\text{Modulus(time)}}{\text{Initial modulus}} = 1 + A\hat{t} + B\hat{t}^2 + C\hat{t}^3 + D\hat{t}^4 \tag{5.34}$$

where $\hat{t} = \log(\text{time})$ will be used to represent the curve shown in Figure 5.29.

Of special interest to the present purpose of using these models in the strength evolution (5.1) at the beginning of this chapter is the effect of temperature on creep behavior (and on the representations of that behavior, mentioned above). To understand the physical arguments on which some of those representations are based, we consider the idea of *time–temperature equivalence*.

This concept begins with the argument that creep is an activation process and the rate of that process can be represented by an Arrhenius equation having the form

$$\dot{\varepsilon} = Ae^{-Q/kT} \tag{5.35}$$

where $\dot{\varepsilon}$ is the rate of strain in the present case, k is Boltzmann's constant, T is the temperature (in absolute units), and A and Q are material constants that may depend on the level of applied stress. We examined an expression of this type in Chapter 3. In classical physics, Q is the activation energy, which has the character of the energy barrier that must be overcome for the process to continue. We can rewrite (5.35) in the form

$$\dot{\varepsilon} = A_1\sigma''e^{-Q/kT} \tag{5.36}$$

where we have created an explicit dependence on the applied stress level σ and added another material constant, n. We will examine several features of this type of expression in Chapter 6 in our discussion of stress rupture.

In the present context, we mention that the activation energy Q and the power of n are generally determined by the physical mechanism that controls or dominates the creep process. A great volume of classical literature has been dedicated to the codification of this dependence; Chapter 6 provides a short summary of some of these aspects. In homogeneous metals, for example, typical mechanisms include dislocation creep by vacancy motion (Nabarro–Herring creep for motion through crystalline grains, Coble creep when the motion is along grain boundaries) and diffusion-controlled mechanisms. In composite materials, the mechanisms may also include individual constituent creep, time-dependent differential deformation mechanisms between the constituents, and discrete (time-dependent) damage events such as defect or flaw nucleation or growth. Hence, Q and n (and other material constants for more complete representations) may take on a wide range of values.

Let us return to (5.35) and cast that equation in a different form to represent the time required for the creep process to reach a specific material state, such as a specified value of strain (e.g., which could be the strain to failure). Then we might write

$$t_a = Ae^{Q/kT} \tag{5.37}$$

where t_a is the time in question. Again, following classical methods, if we take the logarithm (base 10) of (5.37) and assume that A is an absolute material constant and Q may depend on stress level, we obtain

$$\log(t_a) = \log(A) + \log(e)\frac{f_1(\sigma)}{kT} \tag{5.38}$$

where Q has been represented as a function of stress, $f_1(\sigma)$. Regrouping terms and redefining constants, (5.38) can be written in the form suggested by Larson and Miller (1952):

$$T[\log(t_a) + C] = f_1(\sigma \qquad 5.39)$$

We will discuss this and other forms in the next chapter. The importance of this type of expression for the present discussion lies in the fact that these simple arguments lead to a useful idea in the interpretation of creep behavior (in the linear viscoelastic range), the idea of time–temperature equivalence. If we read (5.39), it says that, for a given level of applied stress, the time to reach a given strain level depends on the *product* of the temperature and a function of time. The implication is that for a low temperature the time to reach that strain is long but for a high temperature the time to reach the same strain is short (or at least shorter). From a practical standpoint, an immediate consequence of this equivalence is that it may be possible to use short-term, high-temperature tests in the laboratory to determine the long-term behavior of engineering components at lower temperatures associated with service conditions, an "accelerated testing" concept. The idea of accelerating the test results (so that something like a 30-year creep behavior in service can be characterized by a test of only tens or hundreds of hours in the laboratory) is widely used. However, it cannot be assumed that this equivalence concept and accelerated testing as a methodology are generally valid. As noted, changing the temperature is likely to change the controlling creep mechanism and produce data that represent creep behavior that is different than the lower temperature behavior that is of primary interest. For that reason, great care must be taken to ensure that the mechanism of creep does not change if accelerated testing is attempted using temperature or any other accelerating parameter.

Time–temperature equivalence suggests another very widely used scheme for characterizing material behavior and reducing data using shift factors (Cowie, 1991). We consider a series of creep data taken over the time range of about 10^{-4} to 10^4 h (of the order of up to one year for the longest test). That is often a practical limit on such data, although some data are taken over periods of 10 and 20 years in

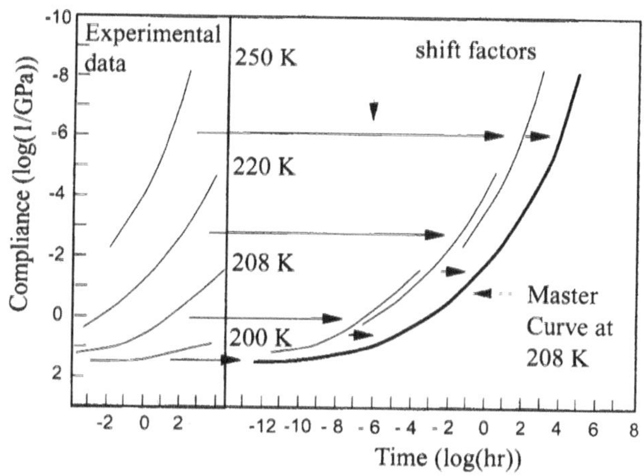

Figure 5.30 Methodology of using time–temperature superposition to establish "shift factors" to create "master curve" of creep response from limited data.

special cases; a common upper limit for laboratory testing is 100 or 1000 h. A hypothetical set of such data is shown on the left of Figure 5.30.

The five creep curves shown in Figure 5.30 are recorded at different temperatures. The concept of time–temperature equivalence suggests that the change in temperature is equivalent to some change in the (log) time scale for the data, that is, that all of the data should fall on a single curve in time at some common (reference) temperature. If true, each of the experimental data curves can be shifted along the time axis until they fall on a common curve. That process is sketched in Figure 5.30. Each of the creep data curves is shifted until the curves overlap and form a common, single compliance curve. Cowie (1991) reminds us that, if the specimens used for the experimental work have different density, the measured values of compliance should be adjusted to account for that by the ratio

$$\text{Reduced compliance}(t) = \frac{T\rho}{T_0\rho_0} \times \text{measured compliance} \tag{5.40}$$

where ρ is the density of the material and T is the temperature in degrees absolute. For each measured data curve, the amount by which the time scale must be shifted is given by

$$\log(t) - \log(t_0) = \log\left(\frac{t}{t_0}\right) = \log\ a_T \tag{5.41}$$

in the notation common to most textbooks. Figure 5.30 shows that the shift factor a_T is a function of temperature. This relationship is conveniently represented by picking a reference temperature (often the T_g of the polymer), and plotting a_T as a function of temperature, relative to that reference temperature, as shown in Figure 5.31. The

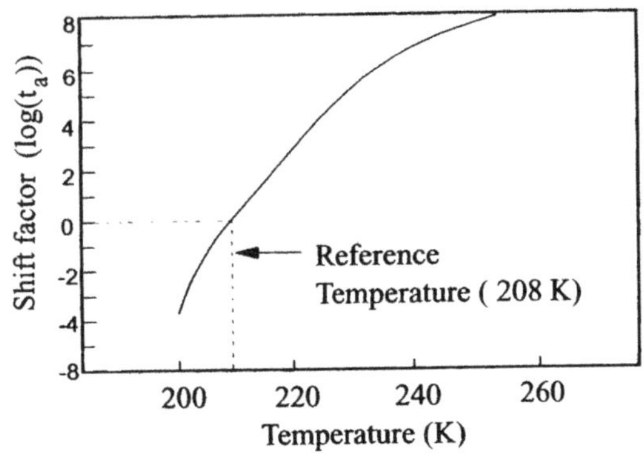

Figure 5.31 Plot of shift factor as function of temperature for data and shifting shown in Figure 5.30.

shift factor is widely discussed in related contexts. Cowie (1991) notes that it can also be defined as the ratio of the retardation times at two temperatures in the form

$$a_T = \frac{\tau}{\tau_0} = \left[\frac{\eta}{\eta_0}\right]\left(\frac{T_0 \rho_0}{T \rho}\right) \tag{5.42}$$

where η is the viscosity of the material. He further notes that if the viscosities obey the Arrhenius equation (as discussed above) and neglecting small correction factors, (5.42) can be cast in the form

$$a_T = \exp\left[b\left(\frac{1}{T} - \frac{1}{T_0}\right)\right] \tag{5.43}$$

where b is a material constant. If the logarithm (base 10) of that expression is taken, it has the form

$$\log(a_T) = -b\left[\frac{T - T_0}{2.303 T_0 T}\right] \tag{5.44}$$

This relationship has a reassuring similarity in form to equations developed by Williams, Landel, and Ferry (cf. Ferry, 1979), which typically have the form

$$\log(a_T) = -b\left[\frac{T - T_0}{c + (T - T_0)}\right] \tag{5.45}$$

where b and c are constants. This equation (and closely related forms) is referred to as the *Williams–Landel–Ferry* (*WLF*) *equation* and is supported by data for various applied conditions. Moreover, the form of these equations can be derived from free-volume theory, which can be used to describe the mechanical behavior of polymers near the glass transition temperature T_g where the behavior cannot be collectively represented by Arrhenius activation concepts. This is of special importance to the present objectives, since we will ultimately apply these methods to polymer composites over temperature ranges that approach T_g.

In general, the dynamic response of materials that demonstrate time-dependent constitutive behavior is strongly dependent on temperature and frequency of excitation, as discussed. For materials such as polymers, such behavior is also strongly dependent on moisture, since moisture can alter the T_g and stiffness of many polymers greatly (Myhre et al., 1982). Shifting procedures for moisture effects have been used (Putter et al., 1982). This extensive subject deserves more attention than the present objectives allow. Readers are referred to excellent discussions by Gao and Weitsman (1998), Smith and Weitsman (1996), Zaffaroni et al. (1998), Dillard (1991), Browning (1972), Wang et al. (1992), and Kibler (1980). Several of these works deal with composites, which brings us to that part of the discussion.

5.3.2 Nonlinear Viscoelastic Response

At least two additional fundamental issues are associated with the consideration of the time-dependent behavior of composites. First, the viscoelastic behavior of the constituents is not likely to be identical, so there is an "adjustment process" that occurs in order for the constituents to come into compliance with the behavior of each other with time. This can create new transient effects that must be accounted for in the analysis; in some cases, these transient effects continue throughout the usable life of the component. Also, many composites are anisotropic, at least at the ply level, so that we must consider the time dependence of the full tensor array of stiffness or compliance properties.

The second fundamental issue is nonlinearity. An easy illustration of a physical condition that can create nonlinear viscoelastic response is provided by the matrix cracking behavior discussed at the beginning of this chapter. The thermally activated molecular motion that is often responsible for the linear-viscoelastic behavior of most polymers can be altered or combined with damage of various types, like matrix cracking (or any of the other damage modes discussed in earlier chapters).

We will consider these two issues in order. We begin by considering how to account for the possibility of anisotropic response of composites and discuss the response of a unidirectional composite lamina. (Laminate behavior will be discussed at the end of the chapter.) We introduce the subject with a simple example. Brinson et al. (1978, 1979) considered the viscoelastic behavior of a graphite–epoxy composite and used independent measurements under various loading conditions to determine the compliance values S_{22}, S_{66} and Poisson's value ν_{21} as a function of time and temperature for a unidirectional laminate. An interesting aspect of their data is shown in Figure 5.32. By inverting the compliance values (to get stiffness units) and normalizing the data to the initial value at room temperature, it can be seen that the behavior of all three matrix-controlled properties (transverse tension, shear, and Poisson contraction) is quite similar. In fact, a master curve constructed by a least-squares fit of all of the data (also shown in Figure 5.32) fits all of the data with a

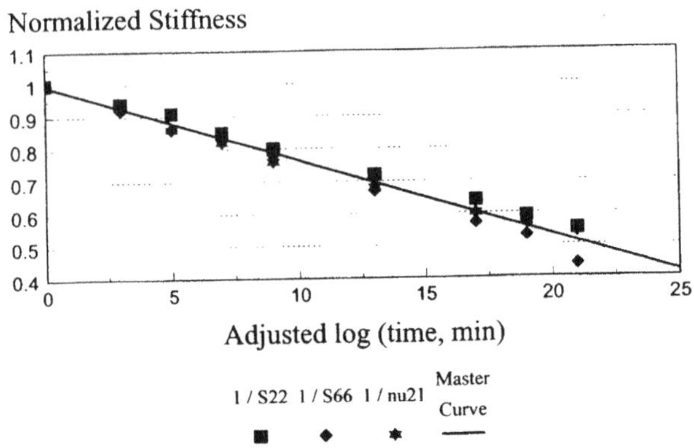

Figure 5.32 Plot of data from Brinson et al. (1979) normalized by value at room temperature for transverse tension, shear, and Poisson contraction compared to best-fit "master curve" for all variations.

maximum error of 14% over 21 decades of time. For that master curve, the shift factor as a function of temperature is shown in Figure 5.33 for a reference temperature of 25°C.

We will see, in a later chapter, that these data can be used to estimate the behavior of epoxy laminates as a function of time using laminate analysis and (5.1). In fact, for later reference, the master curve in Figure 5.32 can be represented by the equation

$$\text{Stiffness}(t) = 1 - 2.1717 \times 10^{-2}t - 7.9692 \times 10^{-4}t^2 + 7.237 \times 10^{-5}t^3 - 1.8081 \times 10^{-6}t^4 \quad (5.46)$$

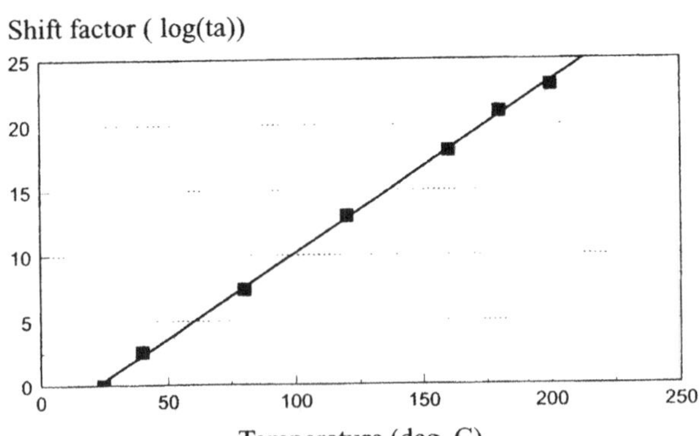

Figure 5.33 Shift factor relative to room temperature for stiffness data shown in Figure 5.32.

and the shift factor in Figure 5.33 can be represented by the equation

$$\log(a_T) = -4.8698 + 0.2345T - 1.7172 \times 10^{-3}T^2 + 1.1229 \times 10^{-5}T^3 - 2.5101 \times 10^{-8}T^4 \tag{5.47}$$

over the temperature range from 23 to 200°C. We will use these results to present an example application in a later chapter.

The possibility of using the shifting methods discussed previously and used above to represent the effects of moisture was already mentioned. An excellent example is provided by the work of Kibler (1980), who measured relaxation curves for $[90]_{15}$, $[0]_6$, and $(+45, -45]_{2S}$ laminae/laminates for wet and dry conditions and reduced the data using shift factors of the type

$$\log(a_{TH}) = \log(a_T) + \log(a_H \qquad 5.48)$$

and found that "although there is no a priori reason why [this equation] should hold, the smoothness of our shifted curves and the overall consistency and reproducibility of results demonstrate its usefulness in the present case." (Kibler, 1980, p. 33). The material systems he characterized were T300/5208 and AS/3502. With an extensive data set, Kibler suggests the possibility of describing all time-dependent results "traceable to matrix effects" in a polymer composite (in which the matrix is the only viscoelastic constituent) with a single set of shift factors, as we have done above, at least for carefully controlled identical conditions. This cannot be viewed as a general rule, but it may be a useful approximation for a number of material systems.

But the contrast Kibler finds between wet and dry conditions for the epoxy results is strong and typical. Some selected results from Kibler's data for the shear modulus are shown in Figure 5.34. Using these results to predict the general response of composite laminae and laminates would appear to be simply a matter of using the compliance values determined from experiment in constitutive relationships of the form

$$\begin{bmatrix} \varepsilon_1(t) \\ \varepsilon_2(t) \\ \varepsilon_3(t) \end{bmatrix} = \begin{bmatrix} S_{11}(t, T, \sigma_1) & S_{12}(t, T, \sigma_1) & 0 \\ S_{12}(t, T, \sigma_1) & S_{22}(t, T, \sigma_2) & 0 \\ 0 & 0 & S_{66}(t, T, \tau_{12}) \end{bmatrix} \begin{bmatrix} \sigma_1 \\ \sigma_2 \\ \tau_{12} \end{bmatrix} \tag{5.49}$$

where S_{ij} is the compliance of an orthotropic lamina and ε_{ij} are the resulting strains from an application of the applied stresses σ_{ij}. The form (5.49) includes the possibility that the compliance values depend not only on time and temperature but also on stress level, a matter we will address at the end of this chapter. An excellent review of this type of approach has been given by Gramoll et al. (1988). Kibler (1980) shows good predictions of laminate behavior for $[0, 90, +45, -45]_S$ and $[0, +45, -45]_S$ stacking sequences using his measured principal compliance values in an elastic lamination code by calculating the response incrementally with small steps in time (for both wet and dry conditions).

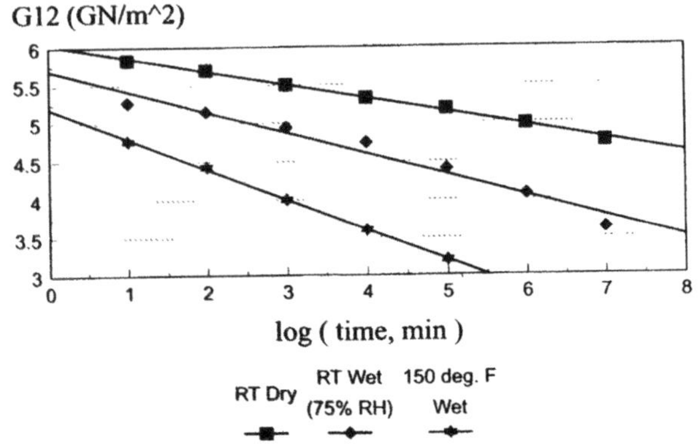

Figure 5.34 Selected data from Kibler (1980) showing effect of moisture and temperature on viscoelastic behavior of graphite–epoxy composites (RT = room temperature; RH = relative humidity).

Dillard (1991) discusses this simplified concept of replacing the modulus in an elastic solution by the relaxation modulus or the inverse of the creep compliance to obtain the time-dependent solution and points out that the approximation using the relaxation modulus is exact if the strains remain constant, and the approximation using the inverse of the creep compliance is exact if the stresses remain constant. The convenience of this quasi-elastic approximation is obvious.

However, as noted earlier, it is quite possible that linear viscoelastic representations do not sufficiently represent the physics of the situation in a real-world application. If damage such as matrix cracking occurs during the time-dependent response, nonlinear effects can be large and may dominate the response. Not so obvious are other nonlinear effects, such as physical aging (changes in free volume, usually in a polymer) or chemical aging (such as corrosion). And other effects are possible.

The literature concerning nonlinear viscoelasticity is replete but not complete. Green and Rivlin (1957) proposed replacing (5.31) and (5.32) with multiple integrals, which necessitates the experimental evaluation of multiple kernel functions making the philosophy very difficult to apply. Schapery (1974) used the concepts of irreversible thermodynamics to modify the single integral representation to obtain the form

$$\varepsilon(t) = g_0 D_0 \sigma + g_1 \int_{\infty}^{t} \Delta D(\Psi - \Psi') \frac{d(g_2 \sigma)}{d\tau}\, d\tau \tag{5.50}$$

(for one-dimensional stress) where D_0 and $\Delta D(\Psi)$ are the initial and transient compliance for the linear viscoelastic response and g_i are material constants used to introduce nonlinear effects. The reduced time parameters Ψ and Ψ' and are given by

$$\Psi(t) = \int_0^t \frac{dt'}{a_\sigma} \qquad \Psi'(\tau) = \int_0^\tau \frac{dt'}{a_\sigma} \tag{5.51}$$

where a_σ is the stress shift factor. Discussions of the experimental determination of these parameters can be found in Tuttle and Brinson (1984) and Osiroff (1990). The transient response $\Delta D(\Psi)$ can be any appropriate time function; the power law t^n is widely used for composites (Lou and Schapery, 1971; Schaffer and Adams, 1981; Tuttle and Brinson, 1984). If the applied stress level is sufficiently high to induce damage events such as matrix cracking and fiber–matrix debonding, the nonlinear visco-
elasticity effects can be large and highly dependent on stress level.

In addition, these representations may be extended to include an important effect called "aging," mentioned earlier. In that case, the momentary creep compliance in (5.48) can be written as

$$D(t, T, t_e) = B(T)D\left(\frac{t}{a}, T_r, t_{er}\right) \tag{5.52}$$

where $B(T)$ is a vertical shift associated with damage or stress level and t_e is the aging time. (The subscript r indicates reference values.) The shift factor in (5.52) has the same interpretation as that shown in (5.42), with the reminder that only time and temperature are being traded off in this case, that is,

$$a(T, t_e) = \left[\frac{\tau(T, t_e)}{\tau(T_r, t_{er})}\right] \tag{5.53}$$

where τ are the retardations discussed in (5.42) (Cowie, 1991; Sullivan et al., 1990). This equation can be arranged to define a shift rate μ by writing it as

$$\log a(T, t_e) = \log(\gamma t_a^\mu) + \log a_T \tag{5.54}$$

where a_T is the shift factor associated with temperature alone and $\log(\gamma)$ is the shift factor at unit aging time. Nominally, all of the parameters may be a function of temperature and aging time. Struik (1978) cast his theory in terms of momentary creep tests, in which the time of the test was short (typically of the order of one-tenth duration) compared to the aging time, and defined an *effective time theory* to relate the compliance in the momentary test to that quantity in a long-term test, that is,

$$\bar{D}(t, T, t_e) = D(\lambda, T, t_e \qquad 5.55)$$

Struik found that the aging shift factor had a power law dependence on the aging time, so that

$$\lambda = t_e \ \ln\left[1 + \frac{t}{t_e}\right] \tag{5.56}$$

for the special case of $\mu = 1$ and

$$\lambda = \frac{t_e}{\alpha}\left[\left(1 + \frac{t}{t_e}\right)^{\alpha} - 1\right] \tag{5.57}$$

for values of $\mu < 1$, where

$$\alpha = 1 - \mu \tag{5.58}$$

The effective time λ replaces the short-term creep time in the prediction of behavior in the presence of aging; that is, the long-term creep compliance can be obtained from the short-term compliance measurements, as noted in (5.55). Experimentally, the conditions of this representation may be difficult to match. The specimen used for the time–temperature superposition characterization must be rejuvenated after each test, and the shift rate μ must be close to unity. In some cases extrapolation of the data may be necessary. Common methods for this purpose include the nonlinear representation suggested by Struik (1978), such that

$$D(t) = D_0 e^{(t/t_0)^m} \tag{5.59}$$

where m has a value of about $\frac{1}{3}$. Alternatively, one can consider the power law representation suggested by Findley and Khosla (1955), Findley et al. (1972), whereby

$$D(t) = D_0 g(\sigma) + m f(\sigma) t^n \tag{5.60}$$

where $g(\sigma)$ and $f(\sigma)$ are stress functions and m, n, and D_0 are fitting parameters. An alternative form of this representation has been used by Findley and Peterson (1958) and discussed by Dillard (1991), for which the power law with nonlinear coefficients is given in terms of strain:

$$\varepsilon(t) = \varepsilon_0 + m t^n \tag{5.61}$$

where $\varepsilon_0 = \varepsilon_0' \sinh(\sigma/\sigma_\varepsilon)$ and $m = m' \sinh(\sigma/\sigma_m)$. The parameters ε_0', m', σ_ε, σ_m, and n are material constants that may be functions of temperature, moisture, and other environments but are assumed to be independent of time or stress level. Findley and Peterson (1958) were able to use this model to represent 10-year data for four different laminated plastic materials with good agreement.

Figures 5.35 and 5.36 show typical results of the shifting process described above for an IM8/ITX system for two laminates, showing the effects of physical aging. Aging reduced the compliance consistently, and the shifting process represented the change. The aging shift rates are also shown in Figure 5.37 as a function of temperature. The values of 0.93–1.3 shown are typical of polymeric materials. The data in Figure 5.37 are particularly interesting in that the physical aging phenomenon persists at temperatures of 220°C, which is 30–40°C above the glass transition

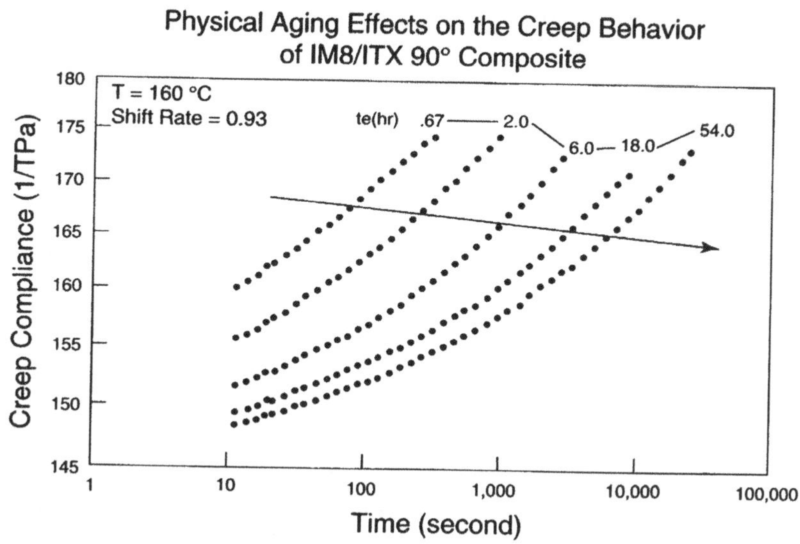

Figure 5.35 Creep compliance of 90° IM8/ITX at 160°C as function of aging time. Last curve is superposition curve after shifting; arrow indicates the shifting direction.

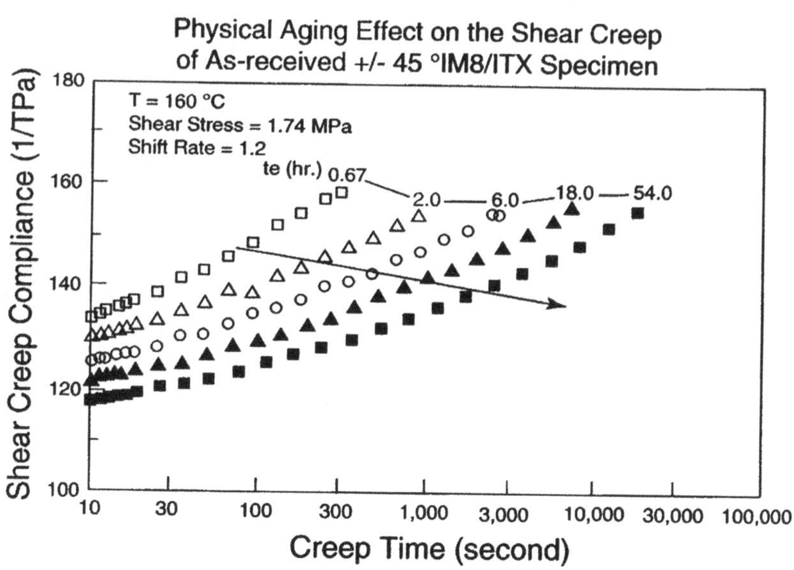

Figure 5.36 Creep compliance of ±45° IM8/ITX at 160°C as function of aging time. Last curve is superposition curve after shifting; arrow indicates shifting direction.

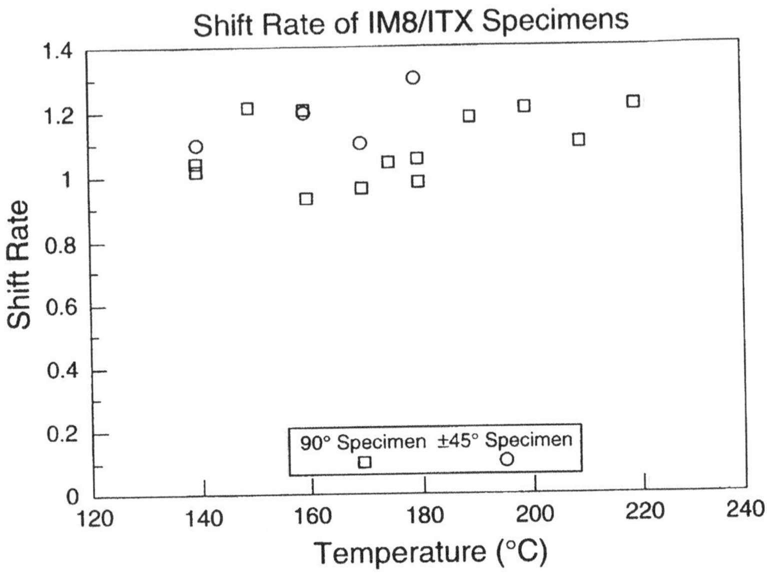

Figure 5.37 Physical aging shift rates versus temperature from two IM8/ITX laminates.

temperature (i.e., the shift rate is still near unity) (Wang et al., 1995). This type of behavior is seen in semicrystalline polymers and filled rubbers. Observations of this type have been reported in polyethyleneterephthalate (PET) (Ania et al., 1989; Struik, 1987), in polyethylene (PE) (Struik, 1987), in polypropylene (PP) (Agarwal and Schultz, 1987; Read et al., 1988; Struik, 1987), Nylon 6 (Struik, 1987), PEEK-based composites, and filled rubber (Struik, 1987). It is postulated that this effect is due to the mobility of the distributed amorphous phase being constrained by the crystalline regions (or the fillers in rubber), at least near the barrier regions. The "average" T_g of the resulting semicrystalline material is higher. Above T_g, regions near the crystalline phases retain their glassy state longer and are therefore subject to aging effects (not seen in amorphous polymers above the T_g). Similar effects are thought to occur in filled rubbers (Wang et al., 1992).

We close our discussion of nonlinear compliance effects by returning to the discussion of laminates, as mentioned in the text associated with (5.49). Dillard (1991) presents a form of the Findley model for which the average matrix stresses are obtained from a modified version of a representation given by Lou and Schapery (1971), according to the lamina relationship

$$\begin{bmatrix} \sigma_1^m \\ \sigma_2^m \\ \tau_{12}^m \end{bmatrix} = \begin{bmatrix} \dfrac{E_m}{E_{11}} & \left(\nu_m - \dfrac{E_m}{E_{11}}\nu_{12}\right) & 0 \\ 0 & 1 & 0 \\ 0 & 0 & 1 \end{bmatrix} \begin{bmatrix} \sigma_1 \\ \sigma_2 \\ \tau_{12} \end{bmatrix} \tag{5.62}$$

where the superscript m indicates matrix stresses, the subscript m indicates matrix properties, and the subscripts 1, 2, 3 represent the lamina level stresses. Dillard notes that the octahedral shear stress is then calculated from the average matrix stress values as

$$\tau_{\text{oct}} \tfrac{1}{3}[(\sigma_1^m - \sigma_2^m)^2 + (\sigma_1^m)^2 + (\sigma_2^m)^2 + 6(\tau_{12}^m)^2]^{1/2} \qquad (5.63)$$

from which both lamina compliance functions S_{22} and S_{66} can be written in the form

$$S(t, \tau_{\text{oct}}^m) = \frac{1}{\tau_{\text{oct}}^m} \quad \varepsilon_0' \sinh\left(\frac{\tau_{\text{oct}}^m}{\sigma_\varepsilon}\right) + m' t^n \sinh\left(\frac{\tau_{\text{oct}}^m}{\sigma_m}\right) \qquad (5.64)$$

where the notation of (5.61) is used.

5.4 TEMPERATURE-DEPENDENT STIFFNESS CHANGE

During our discussion of time-dependent stiffness change in the previous section, we found that the dependence of stiffness on time is inherently related to the relationship of temperature to stiffness. Indeed, we found evidence of time–temperature equivalence in some classes of materials. Several models were described that represent the effect of temperature on stiffness for various ranges and types of material behavior. Arrhenius activation models were found to apply to materials in the glassy state, and the WLF equation is an example of a typical relationship for the polymers in the region of the glass transition temperature. For robust engineering models of strength, stiffness, and life, the focus of this text, two additional modeling capabilities are needed. First, it is necessary to discuss the effect of temperature on the instantaneous modulus of materials, that is, on the stiffness of engineering materials under loading rates that are significantly faster than the internal relaxation rates in a macromolecular material like a polymer, for example. Second, it is necessary to have explicit relationships between stiffness and temperature over the full range of expected engineering service use, not just over a limited range of behavior. We will discuss the rationale behind these two needs and present such a robust model below. This model can, for example, be used directly in (5.1) to represent the effects of temperature on the fiber-controlled strength of polymer matrix composites over ranges of temperature that span glassy to rubbery to flow behavior of the polymer.

In well-designed composite materials, the fiber reinforcement is used to control strength and to limit time-dependent behavior. However, as we will see in the next chapter, even when the fibers are not effected directly by temperature (e.g., carbon fibers in a carbon–epoxy composite), the fiber direction strength and life of a composite can be strongly dependent on temperature. This is a result of the contribution of the matrix to the microfracture process that produces global failure in composites, as discussed in Chapter 4. For example, the fracture of a single fiber in a composite does not generally cause fracture of the composite. It does, however,

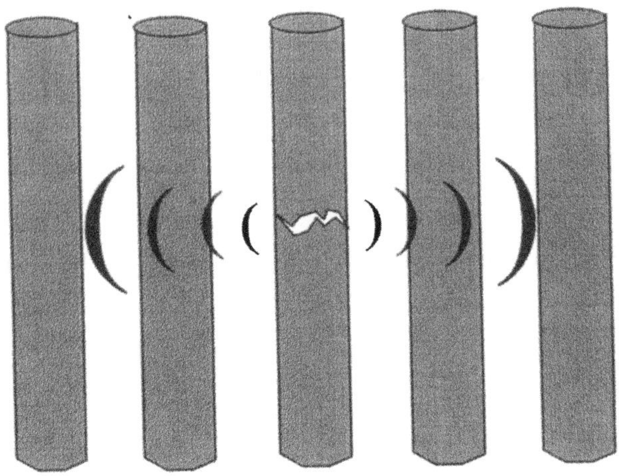

Figure 5.38 Local stress wave in the neighborhood of a fiber fracture in a unidirectional composite.

cause a local redistribution of stress in the neighborhood of the fiber break. That redistribution is a rapid process, and may be a dynamic process, as suggested by Figure 5.38.

The remarkable durability and damage tolerance of fibrous composite materials is, in many ways, due to the complex internal (often local) processes that control degradation and to the strong interactions and interplay between the fibers and matrix. These interactions are most often instantaneous and driven by events that are discrete in time.

Engineering applications of composites are not generally limited to special temperature ranges; polymer composites, for example, are used in components that serve in temperatures that correspond to glassy behavior of the matrix to temperatures well into the rubbery region. The cost and resources necessary to conduct extensive experimental characterization motivate us to construct models of material behavior that represent properties and performance over the full range of expected service conditions. We would like, in the present case, to have a representation of the instantaneous stiffness of a material as an explicit function of temperature over the full range of utility of the material.

The example we will discuss addresses this need for macromolecular materials. Figure 5.39 shows the variation of the instantaneous stiffness for a typical but general polymer as a function of temperature (Young, 1981). For this discussion, the variation has been divided into three regions based on a perception that there are three distinct physical phenomena that control the transitions shown in Figure 5.39. To construct a model that spans the entire range of behavior, we observe that all of the processes that control the transitions are driven by temperature and all of the transitions are, in some sense, the result of changes in mobility of the macromolecules involved at the local level. Increasing the local energy in the macromolecule by

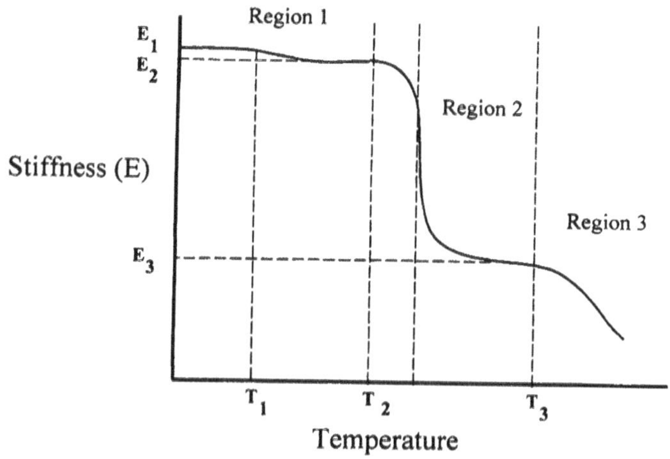

Figure 5.39 Typical variation of instantaneous polymer stiffness with temperature.

increasing the temperature triggers changes in that mobility. Using the concept of relaxation times for each of those processes, as discussed in the previous section, and adopting the notation common to recent discussions of this subject (cf. Nielsen and Landel, 1994), we could write an equation for the stiffness variations shown in Figure 5.39 in the form

$$E(t) = \sum_{i=1}^{N} E_i \exp\left(\frac{-t}{\tau_i}\right) \tag{5.65}$$

where τ_i are the relaxation times for the different transition processes. Alternatively, we can express the relationship in integral form using a distribution function $H(t)$ to obtain

$$E(t) = \int_{-\infty}^{+\infty} H(\tau) \exp\left(\frac{-t}{\tau}\right) d(\ln \tau \qquad \text{5.66)}$$

in which $H(\tau) = \tau E(\tau)$. The distribution functions are determined experimentally from relaxation data according to the relationship $H(\tau) = -d[E_r(t)]/d(\ln t)$, where E_r is the experimental relaxation modulus, or from retardation spectra in a creep experiment using the companion relationship $L(\tau) = d[J(\tau)]/d(\ln t)$, where $J(\tau)$ is the shear-creep compliance (Nielsen and Landel, 1994). Since we wish to represent the modulus across all transitions, we choose a general form for the distribution function that can be used to represent data associated with all processes that may drive the transition behavior. An allowable form is

$$H_k^t(\tau) = \exp\left[-\left(\frac{t}{\tau_k}\right)^{m_k}\right] \tag{5.67}$$

where we have written the distribution function for the kth transition process. We assume that for a given process one can construct some relationship between time and temperature, as discussed in the previous section, and recast (5.67) in the form

$$H_k^t(\tau) = H_k^0 \exp\left[-\left(\frac{T}{\beta_k}\right)^{m_k}\right] \tag{5.68}$$

where the details of the time–temperature relationship have been absorbed into the coefficient H_k^0. Then returning to the form of (5.65) and considering only the three regions (and transitions) shown in Figure 5.39, we use the following equation to represent the stiffness variation across the entire range of service temperature:

$$E = (E_1 - E_2)\exp\left[-\left(\frac{T}{T_1}\right)^{m_1}\right] + (E_2 - E_3)\exp\left[-\left(\frac{T}{T_2}\right)^{m_2}\right] + E_3\exp\left[-\left(\frac{T}{T_3}\right)^{m_3}\right] \tag{5.69}$$

The interpretation of the material parameters in (5.69) is illustrated in Figure 5.40. The T_i correspond to the temperatures at each transition (as given by the maximum of the peaks on the curve of tangent delta versus temperature from a dynamic mechanical analysis (DMA) characterization or by the inflection point of a differential scanning calorimeter (DSC) plot) in degrees Kelvin. The values of E_i are also determined from independent experiments. Here, E_1 is the instantaneous modulus in the glassy region of behavior, E_2 is the instantaneous modulus after the beta transition, and E_3 is the instantaneous stiffness at the beginning of the rubbery region. Van Krevelen (1990) discusses ultrasonic measurement methods for these stiffness values.

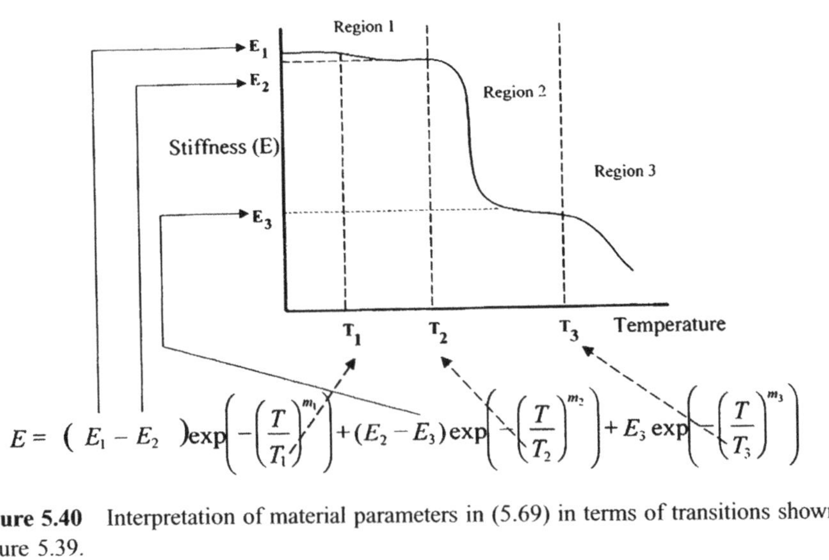

Figure 5.40 Interpretation of material parameters in (5.69) in terms of transitions shown in Figure 5.39.

The dependence of these parameters on the nature of a given polymer is the subject of much literature. Intrinsic characteristics such as cross-link density, crystallinity, molecular weight, chemical structure, and density are primary factors, but extrinsic parameters such as strain, strain rate (or frequency), moisture, time and temperature also influence their values. Mahieux et al. (2001) discuss the influence of these parameters on the present model.

The physical interpretation of the coefficients m_i depends on the process associated with the ith transition. In the rubbery region, entropy controls the internal energy of the polymer, but the physical processes associated with the primary transition (T_g) and secondary transition in region 1 of Figure 5.39 can be discussed in terms of secondary bonding (Ashby and Jones, 1986). Secondary bonds between macromolecules in polymers include "weak" bonding such as hydrogen bonds (dissociation energy of the order of 3–7 kcal/mol), dipole interactions (1.5–3 kcal/mol), van der Waals interactions (0.5–2 kcal/mol), and ionic interactions (10–20 kcal/mol) (Miller, 1966; Platzer, 1969). One can discuss the primary and secondary transitions in terms of mobility of the macromolecules, as the energy provided by increasing temperature overcomes this secondary bonding. Ashby and Jones (1986) speak of "bond melting" in this context. The secondary transition in region 1 of Figure 5.39 is associated with the creation of local degrees of freedom, such as the rotation of subgroups along the macromolecular molecule (Sperling, 1992). The alpha transition is associated with more complete mobility along the length of the macromolecule as many more secondary bonds are overcome by thermal energy. But the mobility of the molecules is still restricted by the presence of entanglements, cross-links, crystallites, fillers, and other local barriers, which constrain the motion of the molecule. This latter situation is called *reptation*, wherein the molecule moves in a snakelike fashion within the constraints mentioned (de Gennes and Leger, 1982).

However, the form of our representation of the stiffness variation has special significance for this secondary bonding case. A close examination of (5.69) reveals that the exponential series has the form first described by Weibull (1951, 1952) for "weak link" behavior. In that form, the power of the argument of the exponent is related to the sharpness of the transition; if m_i is large, the transition is very sharp (i.e., it occurs over a very narrow temperature range), and if m_i is small, the transition occurs over a wider range of temperature. Physically, this corresponds to a process in which all events occur at nearly the same temperature in the first case and over a wide range of temperatures in the second situation. Several physical drivers exist for these variations. If we consider filled polymers or partly crystalline polymers, the m_i values are lower than for homogeneous polymers since the local heterogeneity creates a more nearly random experience for each event in the process of overcoming secondary bonding. A value of m_i equal to 1, for example, corresponds to events that have no relationship to one another, a random process in the Boltzmann sense. However, if the conditions of all secondary bonds are essentially identical or if the breaking of one bond influences (induces) a neighboring bond to release, the events occur at very nearly identical conditions of temperature and time, creating a large value of m_i. These observations have been confirmed by experimental data for a variety of polymers (Mahieux et al., 2001).

Another physical driver for the values of m_i is the number of events that correspond to the transition. We can identify this influence by considering two different transitions in the same polymer for which the factors discussed in the previous paragraph are held constant. For this discussion, we consider the secondary transition in region 1 of Figure 5.39 and the primary transition shown in region 2 of Figure 5.39. We want to compare the distribution of stiffness across those two transitions, which we write, independently, using the first two terms of the right-hand side of (5.69) as follows:

$$E(T)^1 = \hat{E}_1 \exp\left[-\left(\frac{T}{T_1}\right)^{m_1}\right] \qquad E(T)^2 = \hat{E}_2 \exp\left[-\left(\frac{T}{T_2}\right)^{m_2}\right] \tag{5.70}$$

Using standard Weibull statistics, the standard deviation of the distributions can be written as

$$\begin{aligned} S_1 &= \left\{T_1^2\left[\Gamma\left(\frac{m_1+2}{m_1}\right) - \Gamma^2\left(\frac{m_1+1}{m_1}\right)\right]\right\}^{1/2} \\ S_2 &= \left\{T_2^2\left[\Gamma\left(\frac{m_2+2}{m_2}\right) - \Gamma^2\left(\frac{m_2+1}{m_2}\right)\right]\right\}^{1/2} \end{aligned} \tag{5.71}$$

It is possible to relate these two quantities using the central-value theorem. If n chain atoms are involved in the transition in region 1, then we ask how many multiples of n are involved in region 2. According to the central-value theorem, that relationship is

$$S_2 = \frac{S_{\hat{n}}}{\sqrt{\hat{n}}} \tag{5.72}$$

where $\hat{n}$ is the number of multiples of n that are involved in the distribution $E(T)^2$ associated with S_2. Since we are talking about the same secondary bonds in regions 1 and 2, we can assume that we are dealing with the same population in each case. For that situation, we may assume that a reasonable estimate of $S_{\hat{n}}$ is S_1. Then, (5.72) provides a relationship between the distributions shown in (5.64). If we know how many chain atoms or secondary bonds are involved in the transition in region 1 and if we know the values of T_i from measurements, then (5.71) and (5.72) provide us with a relationship between $\hat{n}$ and m_2. To illustrate the utility of these ideas, we will use the data of Sperling (1992). If the number of secondary bonds involved in the secondary transition is of the order of 3–5 (e.g., a typical degree of freedom for rotation of a side group) and if the value of m_1 is about 5 (a typical value for a range of polymers, cf. Mahieux et al., 2001), then Sperling's data suggest that the number of secondary bonds involved in region 2 (across the T_g) is of the order of 10–50. If true, then the values of m_2 from (5.66) are of the order of 30–50. These numbers are in good agreement with the data of Mahieux et al. (2001).

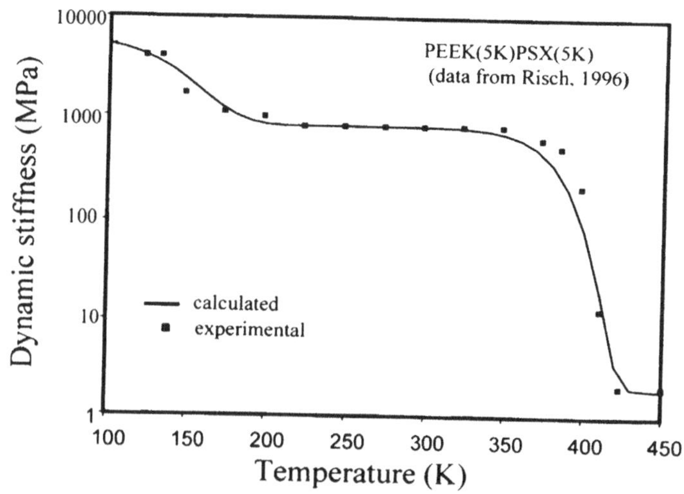

Figure 5.41 Stiffness variation across primary and secondary transition; predicted variation from (5.69) and data from Risch et al. (1996).

An example of the application of (5.69) to data from Risch et al. (1996) is shown in Figure 5.41.

Using the precepts discussed in Chapter 4, we can use the foregoing to estimate the stiffness and strength of composites. A simple rule of mixtures illustrates this effect. Following the work of Mahieux et al. (2001), if the coupling between the fiber and matrix is represented by an efficiency factor (as described by Subramanian, 1994), the fiber direction stiffness of the composite can be written explicitly in terms of temperature in the form

$$E_{11}(T) = \frac{E_m(T)V_m + \sum_{i=1}^{N} \lambda_i \exp[-(T/T_i)^{m_i}]E_f V_f}{V_m + \sum_{i=1}^{N} \lambda_i \exp[-(T/T_i)^{m_i}]V_f} \tag{5.73}$$

where λ_i are the coefficients of (5.69) normalized by the glassy modulus. Mahieux et al. (2001) determined all of the material parameters in (5.69) for PPS in an amorphous and crystalline condition, independently. Then those materials were made into carbon fiber (AS4) reinforced composite materials, and the composite stiffness was measured in the fiber direction as a function of temperature by Walther (1998). The results are compared in Figure 5.42. The sensitivity to the internal structure of the polymer mentioned earlier is very evident in the figure. The predicted results for the two PPS structures bracket the data well, and the sharp drop in modulus of the composite near the secondary transition is reproduced by the model predictions.

Empirical exponential representations of stiffness variations as a function of relaxation times over limited temperature ranges have been proposed by other authors. The "stretched potential" model of Kohlrausch (1847) and Williams and Watts (1970) (KWW) equation can be regarded as a special case of (5.63). Within

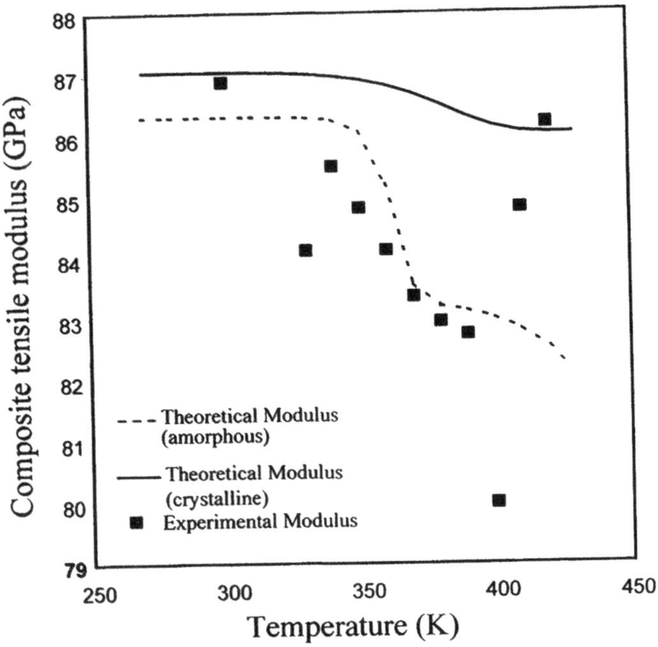

Figure 5.42 Measured and predicted composite stiffness variations with temperature for AS4/PPS unidirectional composite in fiber direction (Walther, 1998).

the constraints of that model, the exponent is discussed in terms of intermolecular coupling, and the characteristic times (in the denominator of the exponent) are identified as characteristic segmental relaxation times. These latter relaxation times have been related to the Hall Helfand relaxation function describing segmental relaxation of an isolated chain (Connolly et al., 1995; Ngai et al., 1986).

Finally, from a thermodynamic standpoint, one can relate the Gibbs energy function G to the absolute temperature T, the pressure P, and the compressibility K using the relationship as stated by McKenna (1989) and others:

$$K = -\frac{1}{V}\left[\frac{\partial^2 G}{\partial P^2}\right]_T \tag{5.74}$$

If the form of G were known in terms of all of the material parameters across the transition temperatures, from glassy behavior to flow behavior, we could use (5.74) directly to find the stiffness relationship. Until such relationships are available, engineering models such as the one above will have to suffice. The need for such relationships is great, not only for the prediction of stiffness and strength as a function of extensive variables like temperature but also for the prediction of other physical properties (e.g., electrical, magnetic, thermal) as explicit functions of extensive variables.

5.5 SUMMARY

We have observed that the stiffness of composite material systems changes with time and with the application of mechanical, thermal, and environmental conditions over time. We have found that good representations of these physical changes can be established for microdamage such as matrix cracking and delamination, viscoelastic behavior, aging, and various combinations of those phenomena. It should be noted that many of the methodologies described for those purposes can be extended to representations of other stiffness change processes such as chemical aging and other microdamage modes with minor additions of philosophy. We also identified a characteristic time concept that further defines that important physical constant introduced in Chapter 3 and repeated in (5.1). Finally, we have introduced an example of the application of the time–temperature superposition representation for a carbon–epoxy system in (5.46) and (5.47) that we will use for calculations in the chapter on applications.

REFERENCES

Agarwal, M. K., and Schultz, J. M. (1981), "The Physical Aging of Isotactic Polypropylene," *Polym. Eng. Sci.*, Vol. 21, No. 12, pp. 776–781.

Allen, D. H., Harris, C. E., and Groves, S. E. (1987a), "A Thermomechanical Constitutive Theory for Elastic Composites with Distributed Damage I: Theoretical Development," *Int. J. Solids Struct.*, Vol. 23, pp. 1301–1318.

Allen, D. H., Harris, C. E., and Groves, S. E. (1987b), "A Thermomechanical Constitutive Theory for Elastic Composites with Distributed Damage II: Theoretical Development," *Int. J. Solids and Struct.*, Vol. 23, pp. 1319–1338.

Ania, F., Martinez-Salazar, J., and Balta Calleja, F. J. (1989), "Physical Aging and Glass Transition in Amorphous Polymers as Revealed by Microhardness," *J. Mater. Sci.*, Vol. 24, No. 8, pp. 2934–2938.

Ashby, M. F., and Jones, D. R. H. (1986), *Engineering Materials 2: An Introduction to Microstructure, Processing and Design*, Pergamon, Oxford (Oxford Shire).

Bailey, J. E., Curtis, P. T., and Parvizi, A. (1979), "On the Transverse Cracking and Longitudinal Splitting Behavior of Glass and Carbon Fibre Reinforced Epoxy Cross-Ply Laminates and the Effect of Poisson and Thermally Generated Strain," *Proc. R. Soc. Lond.*, Vol. A366, pp. 5999–6023.

Brinson, H. F., Morris, D. H., and Yeow, Y. T. (1978), "Experimental Method for the Accelerated Characterization of Composite Materials," Paper presented at the Sixth Conference on Experimental Stress Analysis, Munich Society of Experimental Stress Analysis, Westport, CT.

Brinson, H. F., Morris, D. H., and Yeow, Y. T. (1979), "The Viscoelastic Behavior of the Principal Compliance Matrix of a Unidirectional Graphite/Epoxy Composite," VPI&SU, VPI-E-79-9, Virginia Polytechnic Institute and State University, Blacksburg.

Browning, C. E. (1972), "The Effects of Moisture on the Properties of High Performance Structural Resins and Composites," Technical Report AFML-TR-72-94, Air Force Materials Laboratory, Air Force Systems Command, Wright Patterson Air Force Base, Dayton, OH.

Christensen, R. M. (1971), *Theory of Viscoelasticity*, Academic, New York.

Coleman, B. D., and Gurtin, M. E. (1967), "Thermal Dynamics with Internal State Variables," *J. Chem. Phys.*, Vol. 47, pp. 597–613.

Connolly, M., Karasz, F., and Trimmer, M. (1995), "Viscoelastic and Dielectric Relaxation Behavior of Substituted Poly (*p*-phenylenes)," *Macromolecules*, Vol. 28, pp. 1872–1881.

Cowie, J. M. G. (1991), *Polymers: Chemistry and Physics of Modern Materials*, Chapman and Hall, New York.

de Gennes, P. G., and Leger, L. (1982), *Annual Review of Physical Chemistry*, Annual Review, New York.

Dillard, D. A. (1991), "Viscoelastic Behavior of Laminated Composite Materials," in *Fatigue of Composite Materials*, K. L. Reifsnider, Ed., Elsevier Science, New York, pp. 339–384.

Dowling, N. E. (1993), *Mechanical Behavior of Materials*, Prentice-Hall, Englewood Cliffs, NJ, p. 687.

Ferry, J. D. (1979), *Viscoelastic Properties of Polymers*, 3rd ed., Wiley, New York.

Findley, W. N., and Khosla, G. (1955), "Application of the Superposition Principle and Theories of Mechanical Equation of State, Strain, and Time Hardening to Creep of Plastics under Changing Loads," *J. Appl. Phys.*, Vol. 26, No. 7, pp. 821–831.

Findley, W. N., Lai, J. S., and Onaran, K. (1972), *Creep and Creep Relaxation of Nonlinear Viscoelastic Materials*, North Holland, Amsterdam.

Flaggs, D. L. (1985), "Prediction of Tensile Matrix Failure in Composite Materials," *J. Composite Mater.*, Vol. 19, pp. 29–50.

Fukunaga, H., Chou, T. W., Peters, P. W. M., and Schulte, K. (1984), "Probabilistic Failure Strength Analyses of Graphite/Epoxy Gross-Ply Laminates," *J. Composite Mater.*, Vol. 18, pp. 339–356.

Gao, J., and Weitsman, Y. J. (1998), "Composites in the Sea: Sorption, Strength and Fatigue," Contractors report MAES98-4.0.CM, Office of Naval Research, Arlington, VA, and University of Tennessee, Knoxville, TN.

Garrett, K. W., and Bailey, J. E. (1977), "Multiple Transverse Fracture in 90 Degree Cross-Ply Laminates of a Glass Fibre-Reinforced Polyester," *J. Mater. Sci.*, Vol. 12, pp. 169–174.

Gramoll, K. C., Dillard, D. A., and Brinson, H. F. (1988), "Thermoviscoelastic Characterization and Prediction of Kevlar/Epoxy Composite Laminates," VPI&SU, VPI-E-88-12, and CAS/ESM-88-5, Virginia Polytechnic Institute and State University, Blacksburg.

Green, A. E., and Rivlin, R. S. (1957), "The Mechanics of Nonlinear Materials with Memory," *Arch. Rational Mech. Anal.*, Vol. 1, No. 1, pp. 1–21.

Gudmunson, P., and Ostlund, S. (1992a), "First Order Analysis of Stiffness Reduction Due to Matrix Cracking," *J. Composite Mater.*, Vol. 26, No. 7, pp. 1009–1030.

Gudmunson, P., and Ostlund, S. (1992b), "Prediction of Thermoelastic Properties of Composite Laminates with Matrix Cracks," *Composites Sci. Technol.*, Vol. 44, No. 2, pp. 95–105.

Gudmunson, P., Ostlund, S., and Zang, W. (1991), "Local Stresses and Thermoelastic Properties of Composite Laminates Containing Micro-Cracks," in *Local Mechanics Concepts for Composites Materials Systems*, Springer-Verlag, Berlin, pp. 283–309.

Hashin, Z. (1985a), "Analysis of Cracked Laminates: A Variational Approach," *Mech. Mater.*, Vol. 4, pp. 121–136.

Hashin, Z. (1985b), "Analysis of Orthogonally Cracked Laminates under Tension," *J. Appl. Mech.*, Vol. 54, pp. 872–879.

Highsmith, A. L. (1981), "Stiffness Reduction Resulting from Transverse Cracking in Fibre-Reinforced Composite Laminates," M.S. thesis, College of Engineering, Virginia Polytechnic Institute and State University, Blacksburg.

Highsmith, A. L., and Reifsnider, K. L. (1982), "Stiffness-Reduction Mechanisms in Composite Laminates," in *Damage in Composite Materials*, ASTM STP 775, K. L. Reifsnider, Ed., American Society for Testing and Materials, Philadelphia, pp. 103–117.

Johnson, W. S. (1982), "Mechanisms of Fatigue Damage in Boron/Aluminum Composites, Damage in Composite Materials," in *Damage in Composite Materials*, ASTM STP 775, K. L. Reifsnider, Ed., American Society for Testing and Materials, Philadelphia, pp. 83–102.

Kachanov, L. M. (1986), *Introduction to Continuum Damage Mechanics*, Martinus Nijhoff, Boston.

Kamimura, K. (1984), "Continuum Damage Approach to Mechanical Behavior of Damaged Laminates and a Modeling of Damage Parameters," in *Mechanical Characterization of Load Bearing Fibre Composite Laminates*, A. Cardon and G. Verchery, Eds., Elsevier Science, New York, pp. 36–54.

Kibler, K. G. (1980), "Time-Dependent Environmental Behavior of Graphite/Epoxy Composites," Technical Report AFWAL-TR-80-4052, Materials Laboratory, Wright-Patterson AFB, Dayton, OH.

Kohlrausch, R., 1847, *Pogg. Ann. Phys.*, Vol. 12, p. 393.

Krajcinovic, D. (1984), "Continuum Damage Mechanics," *Appl. Mech. Rev.*, Vol. 37, pp. 1–6.

Kriz, R. D., and Stinchcomb, W. W. (1982), "Effects of Moisture, Residual Thermal Curing Stresses, and Mechanical Load on the Damage Development in Quasi-Isotropic Laminates," in *Damage in Composite Materials*, ASTM STP 775, K. L. Reifsnider, Ed., American Society for Testing and Materials, Philadelphia, pp. 63–80.

Larson, F. R., and Miller, J. (1952), "A Time-Temperature Relationship for Rupture and Creep Stresses," *ASME Trans.*, Vol. 74, pp. 765–771.

Laws, N., and Dvorak, G. J. (1985), "The Loss of Stiffness of Cracked Laminates," in *Fundamentals of Deformation and Fracture* (*Eshelby Memorial Symposium*), B. A. Bilby, K. J. Miller and J. R. Willis, Eds., *Proceedings of the IUTAM, Eshelby Memorial Symposium*, Cambridge University Press, London, pp. 119–127.

Laws, N., Dvorak, G. J., and Hejazi, M. (1983), "Stiffness Changes in Unidirectional Composites Caused by Crack Systems," *Mech. Mater.*, Vol. 2, pp. 123–137.

Lee, J. W., Allen, D. H., and Harris, C. E. (1989), "Internal State Variable Approach for Predicting Stiffness Reductions in Fibrous Laminated Composites with Matrix Cracks," *J. Composite Mater.*, Vol. 23, pp. 273–1291.

Lou, Y. C., and Schapery, R. A. (1971), "Viscoelastic Characterization of a Nonlinear Fibre-Reinforced Plastic," *J. Composite Mater.*, Vol. 5, pp. 208–234.

Masters, J. E. (1982), "An Experimental Investigation of Cumulative Damage Development in Graphite-Epoxy Laminates," Ph.D. Thesis, Materials Engineering Science Department, Virginia Polytechnic Institute and State University, Blacksburg.

Masters, J. E., and Reifsnider, K. L. (1982), "An Investigation of Cumulative Damage Development in Quasi-Isotropic Graphite/Epoxy Laminates," in *Damage in Composite Materials*, ASTM STP 775, K. L. Reifsnider, Ed., American Society for Testing and Materials, Philadelphia, pp. 40–62.

Mahieux, C. A., Reifsnider, K. L., and Case, S. W., (2001), "Property Modeling across Transition Temperatures in PMC's: Part I. Tensile Properties," *Appl. Composite Mater.*, Vol. 8, No. 4, pp. 217–234.

McCartney, L. N. (1989), "New Theoretical Model of Stress Transfer Between Fiber and Matrix in a Uniaxially Fibre-Reinforced Composite," *Proc. R. Soc. Lond.*, Vol. A425, pp. 215–244.

McCartney, L. N. (1990a), "Three-Dimensional Stress Transfer Model for Cross-Ply Laminates of Finite Width Containing Transverse Cracks," in *Proceedings of the Institution of Mechanical Engineers: Fourth International Conference, FRC '90 Fiber Reinforced Composites*, Mechanical Engineering Publications, Edmunds, United Kingdom, pp. 19–26.

McCartney, L. N. (1990b), "Theory of Stress Transfer in a Cross-Ply Laminate Containing a Parallel Arry of Transverse Cracks," NPL Report DMA (A) 189, National Physics Laboratory, Teddington, UK.

McCartney, L. N. (1991), "Analytical Models of Stress Transfer in Unidirectional Composites and Cross-Ply Laminates, and Their Application to the Prediction of Matrix/Transverse Cracking," in *Local Mechanics Concepts for Composites Materials Systems*, J. N. Reddy and K. L. Reifsnider, Eds., Springer-Verlag, Berlin, pp. 251–283.

McKenna, G. B. (1989), "Glass Formation and Glassy Behavior," in *Comprehensive Polymer Science*, Vol. 2: *Polymer Properties*, C. Booth and C. Price, Eds., Pergamon, Oxford, pp. 311–362.

Miller, M. L. (1966), *The Structure of Polymers*, Reinhold, New York.

Moore, R. H. (1988), "Elastic and Time Dependent Matrix Cracking in Cross-Ply Composite Laminates," M.S. Thesis, Engineering Mechanics, Virginia Polytechnic Institute and State University, Blacksburg.

Myhre, S. H., Labor, J. D., and Aker, S. C. (1982), "Moisture Problems in Advanced Composite Structural Repair," *Composites*, Vol. 13, pp. 289–297.

Ngai, K. L., Rendell, R. W., Rajagopal, A. K., Teiltler, S., and Ann, S. (1986), "Three Coupled Relations for Relaxations in Complex Systems," *N.Y. Acad. Sci.*, Vol. 484, pp. 150–184.

Nielsen, L. E., and Landel, R. F. (1994), *Mechanical Properties of Polymers and Composites*, Marcel Dekker, New York.

Nuismer, R. J., and Tan, S. C. (1982), "The Role of Matrix Cracking in the Continuum Constitutive Behavior of a Damaged Composite Ply," in *Proceedings IUTAM Symposium on Mechanics of Composite Materials*, J. N. Reddy and K. L. Reifsnider, Eds., Springer-Verlag, Berlin, pp. 437–448.

Orowan, E. (1948–1949), "Fracture and Strength of Solids," *Rep. Prog. Phys.*, Vol. 12, pp. 185–232.

Osiroff, R. (1990), "Damorheology: Creep-Fatigue Interaction in Composite Materials," Ph.D. Thesis, Materials Engineering Science, College of Engineering, Virginia Polytechnic Institute and State University, Blacksburg.

Parvizi, A., And Bailey, J. E. (1978), "On Multiple Transverse Cracking in Glass Fiber Epoxy Cross-Ply Laminates," *J. Mater. Sci.*, Vol. 13, pp. 2131–2135.

Parvizi, A., Garrett, K. W., and Bailey, J. E. (1978), "Constrained Cracking in Glass Fibre-Reinforced Epoxy Cross-Ply Laminates," *J. Mater. Sci.*, Vol. 13, pp. 195–201.

Pipes, R. B., and Pagano, N. J. (1970), "Interlaminar Stresses in Composite Laminates Under Uniform Axial Extension," *J. Composite Mater.*, Vol. 4, pp. 538–548.

Platzer, N. (1969), "Progress in Polymer Engineering," *Ind. Eng. Chem.*, Vol. 61, No. 5, pp. 10–30.

Putter, S., Buchanan, D. L., and Rehfield, L. W. (1982), "Influence of Frequency and Environmental Conditions on Dynamic Behavior of Graphite/Epoxy Composites," in *Composite Materials Testing and Design*, I. M. Daniels, Ed., ASTM STP 787, American Society for Testing and Materials, Philadelphia.

Read, B. E., Dean, G. D., and Tomlines, P. E. (1988). "Effects of Physical Aging on Creep in Polypropylene," *Polymer*, Vol. 29, pp. 2159–2169.

Reifsnider, K. L. (1977), "Some Fundamental Aspects of the Fatigue and Fracture Response of Composite Materials," in *Proceedings of the Fourteenth Annual Meeting of the Society of Engineering Science*, George C. Sik, Ed., Lehigh University, Bethlehem, PA, pp. 373–384.

Reifsnider, K. L. (1980), "Fatigue Behavior of Composite Materials," *Int. J. Fracture*, Vol. 16, No. 6, pp. 563–583.

Reifsnider, K. L., and Highsmith, A. L. (1981), "Characteristic Damage States: A New Approach to Representing Fatigue Damage in Composite Laminates," in *Materials, Experimentation and Design in Fatigue*, Westbury House, Guildford, Surrey, England, pp. 246–260.

Reifsnider, K. L., Ed. (1982), *Damage in Composite Materials*, ASTM STP 775, American Society for Testing and Materials, Philadelphia, pp. 298–313.

Reifsnider, K. L., and Talug, A. H. (1980), "Analysis of Fatigue Damage in Composite Laminates," *Int. J. Fatigue*, Vol. 1.2, no. 1 pp. 3–11.

Richards, C. W. (1961), *Engineering Materials Science*, Wadsworth, Belmont, San Francisco, CA.

Risch, B. G., Rodrigues, D. E., Lyon, K., McGrath, J. E., and Wilkes, G. L. (1996), *Polymer*, Vol. 37, No. 7, pp. 1229–1242.

Schaffer, B. G., and Adams, D. F. (1981), "Nonlinear Viscoelastic Analysis of a Unidirectional Composite Material," *J. Appl. Mech.*, Vol. 48, pp. 859–865.

Schapery, R. A. (1974), "Viscoelastic Behavior and Analysis of Composite Materials," *Mech. Composite Mater.*, Vol. 2, pp. 85–168.

Schulte, K., and Stinchcomb, W. W. (1984), "Damage Development Near the Edges of a Composite Specimen during Quasi-Static and Fatigue Loading," *Composites Technol. Rev.*, Vol. 6, No. 1, pp. 3–9.

Smith, L. V., and Weitsman, Y. J. (1996), "The Immersed Fatigue Response of Polymer Composites," *Int. J. Fracture*, Vol. 82, pp. 31–42.

Sperling, L. H. (1992), *Introduction to Physical Polymer Science*, 2nd ed., Wiley, New York.

Stinchcomb, W. W., Reifsnider, K. L., Yeung, P., and Masters, J. (1981), "Effect of Ply Constraint on Fatigue Damage Development in Composite Material Laminates," in *Fatigue of Fibrous Composite Materials*, ASTM STP 723, K. N. Lauraitis, Ed., American Society for Testing and Materials, Philadelphia, pp. 64–84.

Struik, L. C. E. (1978), *Physical Aging in Amorphous Polymers and Other Materials*, Elsevier Scientific, New York.

Struik, L. C. E. (1987), "The Mechanical and Physical Aging of Semicrystalline Polymers: 1," *Polymer*, Vol. 28, pp. 1521–1533.

Subramanian, S. (1994), "Effect of Fiber/Matrix Interphase on the Long-Term Behavior of Cross-Ply Laminates," Ph.D. Thesis, Department of Engineering Science and Mechanics, College of Engineering, Virginia Polytechnic Institute and State University, Blacksburg.

Sullivan, J. L., Blais, E. J., and Houston, D., (1990), "Physical Aging in the Creep Behavior of Thermosetting and Thermoplastic Composites," *Composite Sci. Technol.*, Vol. 47, no. 4, pp. 389–405.

Talreja, R. (1985a), *Fatigue of Composite Materials*, Technical University of Denmark, Lyngby.

Talreja, R. (1985b), "Transverse Cracking and Stiffness Reduction in Composite Laminates," *J. Composite Mater.*, Vol. 19, pp. 355–375.

Talreja, R. (1986), "Stiffness Properties of Composite Laminates with Matrix Cracking and Interior Delamination," *Eng. Fracture Mech.*, Vol. 25, pp. 751–762.

Talreja, R. (1990), "Internal Variable Damage Mechanics of Composite Materials," in *Yielding, Damage and Failure of Anisotropic Solids*, J. P. Boehler, Ed., Mechanical Engineering Publications, London, pp. 509–533.

Talreja, R. (1994), Damage Characterization by Internal Variables, in *Damage Mechanics of Composite Materials*, Vol. 9, R. Talreja, Ed., Elsevier Science, Amsterdam, pp. 53–78.

Talug, A. (1978), "Analysis of Stress Fields in Composite Laminates with Interior Cracks," Ph.D. Thesis, College of Engineering, Virginia Polytechnic Institute and State University, Blacksburg.

Tao, J. X., and Sun, C. T. (1996), "Effect of Matrix Cracking on Stiffness of Composite Laminates," *Mech. Composite Mater. Struct.*, Vol. 3, No. 3, pp. 225–239.

Tuttle, M. E., and Brinson, H. F., (1984), "Residual-Foil Strain Gage Technology Applied to Composite Materials," *Exper. Mech.*, Vol. 25, p. 54.

Van Krevelen, V. W. (1990), *Properties of Polymers*, 3rd ed., Elsevier, New York.

Vasil'ev, V. V., Dudchenko, A. A., and Elpat'evskii, A. N. (1970), "Analysis of the Tensile Deformation of Glass-Reinforced Plastics," *Mekhanika*, Kompozitnykh Materialov, Vol. 6, no. 1, pp. 144–147; USSR Polymer Mechanics, Vol. 6, no. 1, 1972, pp. 127–130.

Walther, B. M. (1998), "An Investigation of the Tensile Strength and Stiffness of Unidirectional Polymer-Matrix Carbon-Fiber Composites under the Influence of Elevated Temperatures," M.S. Thesis, submitted to the Department of Engineering Science and Mechanics, College of Engineering, Virginia Polytechnic Institute and State University, Blacksburg.

Wang, J. Z., Dillard, D. A., and Ward, T. C. (1992), "Temperature and Stress Effects in the Creep of Aramid Fibers under Transient Moisture Conditions and Discussions on the Mechanisms," *J. Polym. Sci.: Part B: Polym. Phys.*, Vol. 30, pp. 1391–1400.

Wang, J. Z., Parvatareddy, H., Shang, T., Iyengar, N., Dillard, D. A., and Reifsnider, K. L. (1995), "Physical Aging Behavior of High-Performance Composites," *Composite Sci. Technol.*, Vol. 54, pp. 405–415.

Weibull, W. (1951), "A Statistical Distribution Function of Wide Application," *J. Appl. Mech.*, Vol. 18, pp. 293–297.

Weibull, W. (1952), "Statistical Design of Fatigue Experiments," *J. Appl. Mech.*, Vol. 19, pp. 100–113.

Williams, G., Watts, D. C., (1970), "Non-Symmetrical Dielectric Relaxation Behavior Arising from a Simple Empirical Decay Function," *Trans. Farad. Soc.*, Vol. 66, No. 565, pp. 80–85.

Williams, M. L., Landel, R. F., and Ferry, J. D. (1955), "The Temperature Dependence of Relaxation Mechanisms in Amorphous Polymers and Other Glass-Forming Liquids," *J. Am. Chem. Soc.*, Vol. 77, pp. 3701–3706.

Young, R. J. (1981), *Introduction to Polymers*, Chapman and Hall, New York.

Zaffaroni, G., Cappelletti, C., Mariani, U., and Rigamonti, M. (1998), "Hot Wet Aging of a Glass Reinforced Epoxy Resin," in *Proceedings of the European Conference on Composite Materials, Science, Technologies and Applications*, Vol. 3, C. Visconti, Ed., Woodhead Publishing, Cambridge, pp. 183–190.

EXERCISES

5.1 Consider a carbon–epoxy composite system with the following lamina properties:

$E_1 = 39\,\text{GPa}$	$E_2 = 4.8\,\text{GPa}$	$\nu_{12} = 0.3$	$G_{12} = 6.0\,\text{GPa}$	
$X_t = 1130\,\text{MPa}$	$X_c = 620\,\text{MPa}$	$Y_t = 90\,\text{MPa}$	$Y_c = 600\,\text{MPa}$	$S = 83\,\text{Mpa}$
$\varepsilon_{1t} = 28{,}000$	$\varepsilon_{1c} = 15{,}897$	$\varepsilon_{2t} = 18{,}750$	$\varepsilon_{2c} = 125{,}000$	
$\varepsilon_{12} = 13{,}833$				
$\alpha_{11} = -1.8 \times 10^{-6}/^\circ\text{C}$	$\alpha_{22} = 27 \times 10^{-6}/^\circ\text{C}$			

For a $[0, 45, 90, -45]_{4S}$ laminate, make the following calculations:

(a) What is the percent change in the in-plane tensor laminate stiffness components if the 90° plies are fully saturated with matrix cracks?

(b) What is the percent change in the fiber direction stress in the 0° plies for a given uniaxial applied load (in the 0° direction) if saturation cracking occurs in the 90° plies? How does this percent increase in stress compare with the percent decrease in laminate stiffness in the load direction calculated in part (a)?

(c) What is the percent change in the in-plane tensor laminate stiffness components if the 90° and (both) 45° plies are fully saturated with matrix cracks?

(d) What is the percent change in the fiber direction stress in the 0° plies for a given uniaxial applied load (in the 0° direction) if saturation cracking occurs in the 90° and 45° plies? How does this percent increase in stress compare with the percent decrease in laminate stiffness in the load direction calculated in part (c)?

(e) A discussion of the effects of delamination will be presented in Chapter 7. For the present problem, assume that delamination initiates at the interlaminar interface with the greatest interlaminar normal stress (Pipes and Pagano, 1970). Under uniaxial tensile loading, if delamination occurs in the present laminate, at what interlaminar interface will that delamination occur?

(f) If delamination occurs over 50% of the interface area identified in part (e), estimate the percent change in the laminate stiffness E_x that will result from that separation of the interface? (*Hint*: Approximate the laminate stiffness as the rule-of-mixtures stiffness of the sublaminates formed by the delamination.)

5.2 From experimental measurements it is determined that the stiffness in the direction of the fibers of a SiC-based ceramic composite changes due to creep according to the relationship

$$\frac{E_x}{E_x^0} = 1 - \sigma^{3.5}\Bigg[(3.7 \times 10^{-14}) \log(t)(1 + 0.002T + 0.0004T^2) \exp\left(\frac{-800}{T}\right)^3 + (3.7 \times 10^{-12}) \times (1 + 0.02T + 0.004T^2) \exp\left(\frac{-800}{T}\right)^4 [\log(t)^2 + \log(t)^4]\Bigg]$$

where the stress σ in the equation is in units of megapascals, the temperature T is in units of degrees Centigrade, and time t is in hours. The ultimate strength of the material in the fiber direction is about 210 MPa at room temperature and is relatively independent of temperature up to about 1200°C. For a laminate having ply orientations (or fiber orientations) of $[0, 45, 90, -45]_{4S}$ (as in Exercise 5.1) made from this material, determine the following:

(a) For uniaxial loading of the laminate, determine the laminate stiffness as a function of time (up to 1000 h) for the following three conditions:

Case 1:	$T = 80°C$	$\sigma_x(\text{applied}) = 120$ MPa
Case 2:	$T = 800°C$	$\sigma_x(\text{applied}) = 120$ MPa
Case 3:	$T = 800°C$	$\sigma_x(\text{applied}) = 160$ MPa

Remember that the stiffness change in the fiber direction of each ply depends on the stress in that ply. Plot the three variations on a single chart and compare the results.

(b) How does the stiffness change of the subject laminate caused by matrix cracking (calculated in Exercise 5.1) compare to the laminate stiffness change caused by fiber creep (Exercise 5.2)? What are the fundamental similarities and differences?

5.3 For the laminate defined in Exercise 5.1, using the strength evolution integral introduced in Chapter 3, compare the change in life caused by a 50% reduction in the matrix stiffness of all plies to that caused by a 10% reduction in the fiber direction stiffness of the ply having an orientation that is most closely aligned with the load axis. (You may want to do the calculation for several different types of failure functions in the integral.) Discuss the relative effect of matrix stiffness reduction and fiber stiffness reduction on laminate life.

5.4 Given the results of the calculations conducted in the previous problems, if you were asked to redesign the laminate for which those calculations were made to minimize the effects of stiffness change on life, how would you change the laminate?

CHAPTER 6

STRENGTH EVOLUTION DURING DAMAGE ACCUMULATION

Thesis: Material strength reductions have a major influence on durability and damage tolerance at the global level. Rate equations and data for strength reduction processes are essential elements for modeling damage tolerance and durability.

6.1 PROBLEM DEFINITION

One of the primary inputs to the strength evolution equation in Chapter 3 (that we will use to evaluate the remaining strength and life of composite material systems) is the intrinsic *fast fracture* strength of the composite material(s) involved. We have been calling these principal material strengths X_{ij}. Equation (3.12) had the form

$$\mathrm{Fr}(\tau_1) = 1 - \int_0^1 \left[1 - F\left(\frac{\sigma(\tau)_{ij}}{X(\tau)_{ij}} \right) \right] j\tau^{j-1}\, d\tau \tag{6.1}$$

where the material strengths X_{ij} are understood to be changing during the life of the composite. These changes are the subject of the present chapter.

The scope of this chapter includes the various processes and effects that change the principal material strengths as a function of applied mechanical and environmental conditions, including mechanical loading, thermal variations, moisture or corrosive elements, time, and cyclic or repeated variations of all variables. It does not include a discussion of strength changes under nonuniform stress (such as notched strength, found in Chapter 7), laminate strength, or component strength at the global level (subjects addressed as the application of the present approach in

Chapter 8). We will be concerned with the degradation of the values of strength that we would normally use to characterize a composite material in tension, compression, and shear using *uniform* applied stress usually applied in a single plane to measure the X_{ij} values.

The discussion will begin with a careful examination of the physical behavior that drives this subject, especially the degradation process parameters that must be considered. This discussion will consider the phenomenological representations of the degradation processes as a function of those variables. Then we will examine micromechanical representations of the principal strengths in terms of the constituent material properties, geometry, and arrangement and use those representations to interpret some of the physical data quoted. Finally, we will learn how the information and models are used in the global strength evolution equation (6.1).

In keeping with the general thrust of the book, the discussion will focus on rupture or fracture strength. In contrast to the discussion on stiffness change in Chapter 5, strength is most often dominated by the properties and performance of the fibers. For this reason, Chapter 6 will emphasize the subject of fiber-dominated strength.

In Chapter 3, an alternative form of (6.1) was introduced to make the time dependence of that equation more explicit. The form, which is convenient for the present discussion, is

$$\mathrm{Fr}(\tau_1) = 1 - \int_0^{\tau_1} \left[1 - \mathrm{Fa}\left(\frac{\sigma(t)_{ij}}{X(t)_{ij}} \right) \right] j \left(\frac{vt}{N(t)} \right)^{j-1} d\left(\frac{vt}{N(t)} \right) \tag{6.2}$$

where v is the cyclic frequency of fatigue loading and $N(t)$ is the characteristic "time" to failure. $N(t)$ can be found from the general equation we used to represent fatigue behavior of fiber-controlled composites, written in the form

$$\mathrm{Fa}\left(\frac{\sigma(t)_{ij}}{X(t)_{ij}} \right) = A + B\{\log[N(t)]\}^P \tag{6.3}$$

where A, B, and p are material constants. The inherent assumption is that the characteristic time to failure $N(t)$ is uniquely related to the principal material strength(s) X_{ij}. This reflection of experience in the laboratory accounts for a broad range of materials and the tradition that the fatigue behavior of many material classes is best represented in terms of their rupture or fracture strength.

To implement (6.3) and (6.2), $X(t)_{ij}$ must be specified as a continuous function of time. Figure 6.1 shows a schematic representation of a typical variation (in normalized strength). Several general features of this figure are particularly important. The linear time scale illustrates the emphasis on the nonlinearity of the curves (typically such data are plotted against a logarithmic scale). Strength evolution as a function of hours (of exposure to applied conditions) is quite often highly nonlinear and may even increase or decrease in composites, depending on the conditions (Kerr and Haskins, 1987; Woodford et al., 1992). Thermal aging can produce substantial

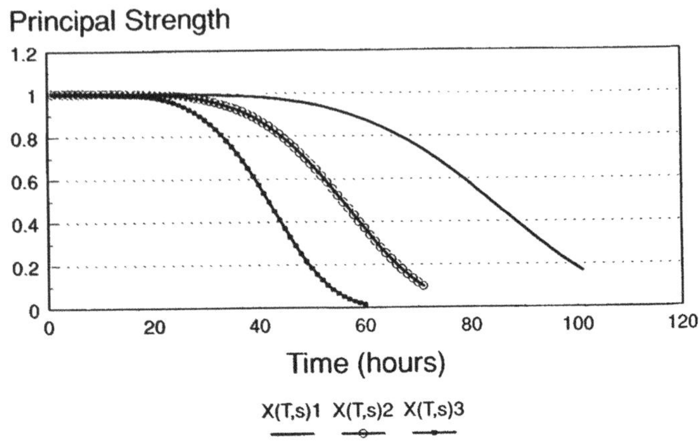

Figure 6.1 Strength variation with time in a fiber-controlled composite.

strength increases in polymer and ceramic composites, even for fiber-controlled failure modes. Of course, a more general result is a reduction of strength, as suggested by Figure 6.1. Moreover, the change in strength with exposure may be sudden; that is, the changes may be very small until quite near the time of fracture, a widely known phenomenon often called "sudden death." The three curves in Figure 6.1 also suggest that the change in strength is a function of several external variables, such as temperature T and stress level s. This sensitivity may be extreme; that is, very small changes in severe conditions may produce very large variations in strength and life.

Figure 6.2 illustrates the manner in which the decrease in principal material strength affects the characteristic life of the specimen, $N(t)$. The decrease of $X(t)$

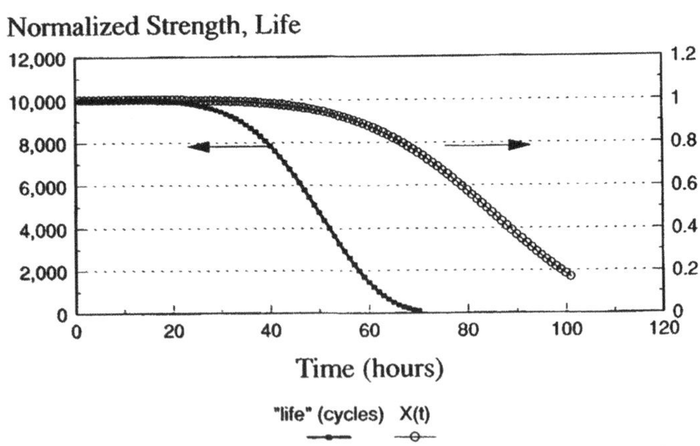

Figure 6.2 Effect of strength reduction (curve and scale on right) on life of specimen (curve and scale on left) using (6.3).

in 100 h produces a sharp reduction of life over a shorter time period, when typical values of $A = 1$, $B = -0.1$, and $P = 1$ are used in (6.3). These changes, in turn, reduce the global strength Fr calculated by (6.2) and discussed in Chapter 3. In the discussion that follows in this chapter, we will address the question of how the principal strengths of composites change as a function of the history of loading and how these changes can be represented in a manner that is compatible with (6.1)–(6.3) and Figures 6.1 and 6.2.

6.2 FACTORS THAT INFLUENCE STRENGTH

6.2.1 Influence of Temperature on Fiber Direction Tensile Strength

The dependence of material properties on the temperature of the material is a fundamental consideration if we are going to construct a methodology for predicting remaining strength and life under combinations of applied mechanical and thermal conditions. For brittle continuous-fiber-reinforced composites, limited data are currently available to serve as the basis for characterizing such dependence. Some experimental data have been collected for a limited range of conditions and systems over the years. A good example is the information and data collected by Kerr and Haskins (1987) on time–temperature–stress capabilities of composite materials during the prior decade. This large data set, and many others (some of which are listed at the end of this chapter) provide many specific answers about behavior but no general philosophy, especially concerning the nature of the constitutive equations that describe the behavior. For example, the fiber direction tensile strength of unidirectional graphite–epoxy coupons typically increases with temperature (even somewhat beyond the use temperature limit) but has a maximum as a function of temperature for a graphite–polyimide composite and is monotonically decreasing for metal matrix composites like boron–aluminum. It is not clear what general constitutive representations or other mechanics models predict such dependence for this very basic situation, but we will see in a subsequent section that micromechanics may provide a feasible method of predicting or at least anticipating such results.

Typical behavior as a function of temperature is shown in Figures 6.3–6.5. Figure 6.3 shows a compilation of data from the literature for two Nicalon fibers, indicating the reduction of stiffness with temperature. Fibers made from other materials will have different behavior (especially as a function of temperature level), but the general nature of the curve can be regarded as generic. The data have been normalized by the initial value of stiffness at room temperature. As such, they can be represented by the equation

$$\frac{E}{E_0} = 1 - 0.4\left(\frac{T - 298}{2350.3 - T}\right) \tag{6.4}$$

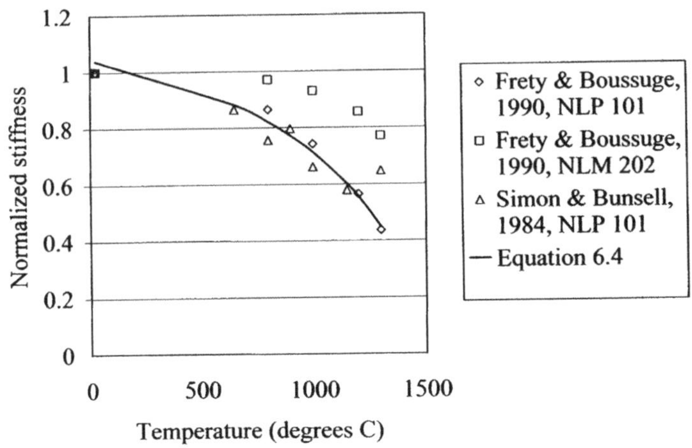

Figure 6.3 Typical data showing dependence of stiffness of SiC fibers on temperature (references listed).

where the temperature T is in degrees Kelvin. The general form of the last term in this equation is quite useful and will appear many times in this book. We can write

$$\text{Scale} = a\left(\frac{T - T_0}{T_{max} - T}\right) \tag{6.5}$$

where "Scale" is a temperature scaling factor and a, T_0, and T_{max} are material constants. The interpretation of the two temperature reference constants is clear;

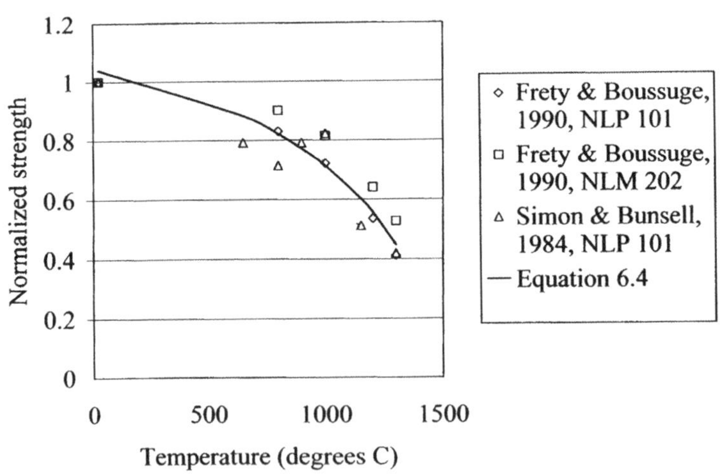

Figure 6.4 Typical data showing dependence of strength of SiC fibers on temperature (references listed).

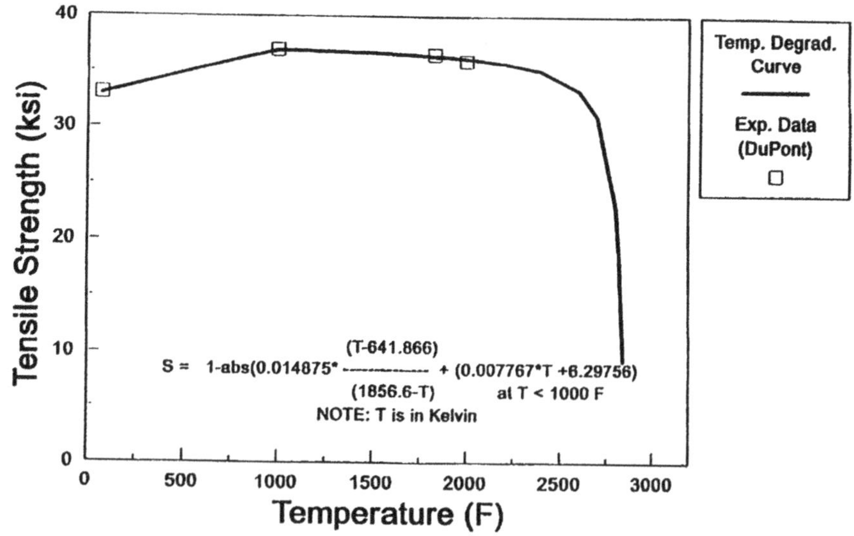

Figure 6.5 Typical data showing dependence of strength of Nicalon–SiC composite on temperature.

T_0 is the value below which essentially no change occurs due to temperature and serves as the initial reference point for the other data and T_{max} is an upper limit on the use temperature of the material since the change becomes unbounded there.

Figure 6.4 shows the same type of plot as Figure 6.3, except that it represents another compilation of data representing the change in *strength* of the Nicalon fiber as a function of temperature. Again, the general form of the variation is generic, but other materials may have quite different behavior. A remarkable similarity exists between the shape of the variation of stiffness shown in Figure 6.3 and the variation of strength shown in Figure 6.4. In fact, Equation (6.4) represents both sets of data very well, as shown by the analytical curve on both plots [which is Equation (6.4)]. Stiffness and strength variations in constituents are not generally identical (in normalized form), but it is not uncommon to find that they are similar, especially for ceramic materials. We will provide further evidence in Chapter 9 that Equation (6.5) is a very useful form for the representation of such temperature dependence and can often be used to represent changes in $X(t)$ as a function of temperature, such as those shown in Figure 6.1.

Figure 6.5 shows the dependence of strength on temperature for a composite made with the same fibers shown in Figures 6.3 and 6.4. The dependence is not monotonic, even though this is a "fiber-dominated" composite. Two aspects of this observation are general. First, composite strength is determined by a complex *interaction* of constituents, not by any simple combination of their properties (as we discussed in Chapter 4). Second, it is not uncommon to find a maximum or minimum in the fiber-dominated strength of a composite as a function of temperature (or other variables). We will examine some of the details of these observations.

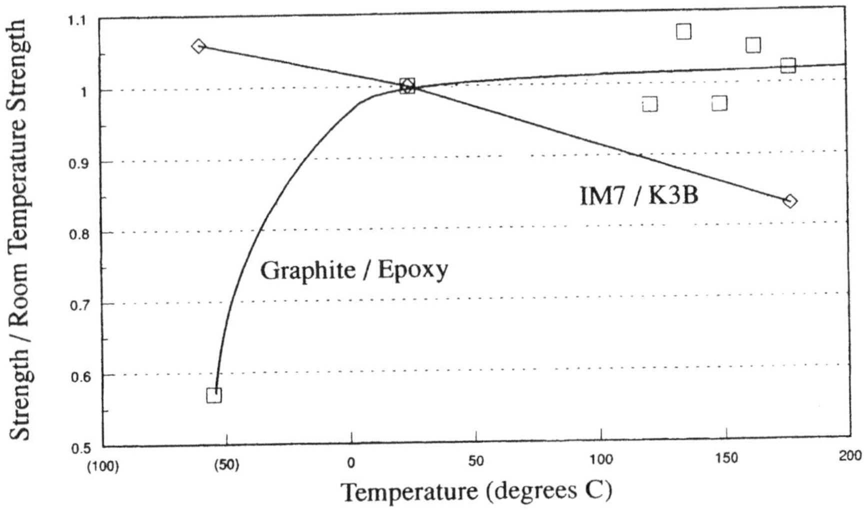

Figure 6.6 Examples of variation of polymer matrix composite strength with temperature (graphite–epoxy data from Kerr and Haskins, 1987).

Figures 6.6 and 6.7 show examples of the influence of elevated temperature on the tensile strength of unidirectional continuous-fiber composites loaded in the fiber direction. Figure 6.6 shows a decrease in tensile strength for an IM7–epoxy system tested in our laboratory. The stiffness of this material is about 166 GPa compared to a rule-of-mixtures value of about 182 GPa. Some evidence shows a rather poor

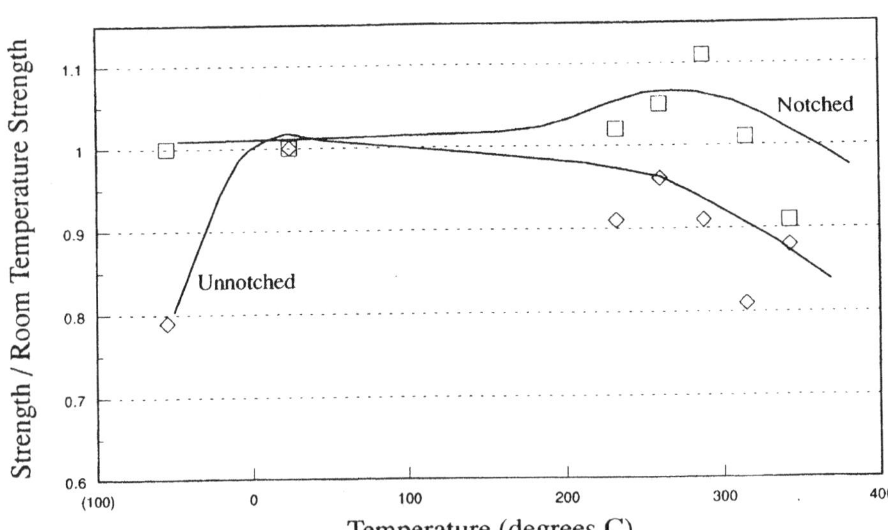

Figure 6.7 Examples of variation of polymer matrix composite strength with temperature (Kerr and Haskins, 1987).

coupling between the elastic response of the fibers and matrix in this polymer matrix system. With increasing temperature, the mechanical gripping of the fibers by the matrix is also reduced since the matrix thermal expansion coefficient is larger than the fibers. Although the fibers are not influenced by the temperature changes, in this case, the change in composite strength is considered to be substantial. We believe this change is caused by a decrease in the fiber–matrix interaction; that is, we are decreasing the "composite effect" on fiber direction strength by increasing the temperature in this system over that temperature range. Figure 6.6 also shows data for a AS/3501-5 system tested by Kerr and Haskins (1987). The rise in strength with temperature for this system is quite large over the temperature range shown. Hence, fiber direction tensile strength may either increase or decrease with temperature, depending on the material system involved and the temperature ranges concerned.

Figure 6.7 shows data for a graphite–polyamide system, also taken from the Kerr and Haskins report (1987). In this case, the system shows a clear maximum in strength with increasing temperature. The figure also shows that the notched (still unidirectional) coupons exhibited an even greater dependence on temperature and a more pronounced maximum in the dependence.

To better understand how composite strength can vary with temperature as a function of the constituents, we use the discussion of micromechanical representations of composite strength introduced in Chapter 4. Representations of this type are now fairly common, albeit philosophically incomplete. The general approach to this enterprise is to identify the failure mode(s) that control fracture and then to set a boundary-value problem that represents the microdetails of the fracture process in terms of the constituents, their geometry, and their arrangement just prior to the fracture event. For our present purposes, we consider such modeling for the tensile strength of a unidirectional continuous-fiber-reinforced composite measured in the fiber direction. The simplest example of such a representation is the rule of mixtures:

$$X_t = v_f X_f + v_m X_m \tag{6.6}$$

where X_f and X_m are the fiber and matrix strengths and the coefficients are the volume fractions of those constituent phases. A linear mixture concept will not produce the effects noted above; that is, a simple combination of constituent behavior will not produce the composite result. In Chapter 4, we introduced models that estimate strength in terms of the geometry, arrangement, properties, and performance of the constituents and the interphase regions between them. For the tensile strength in the fiber direction, we showed that the following representation includes many of the critical physical factors and effects that control strength in fiber-dominated systems:

$$X_t = v_f \; \sigma_0^{m/(m+1)} \left(\frac{2\tau_0 L}{\hat{D}}\right)^{1/(m+1)} \left(\frac{2}{m+2}\right)^{1/(m+1)} \left(\frac{m+1}{m+2}\right) \frac{(1+m)^{1/m}}{[C_n^m + C_n^{m-l} + \cdots + 1]^{1/m}} \tag{6.7}$$

In this representation σ_0, is the characteristic strength of the fibers, m is the Weibull shape parameter of the statistical distribution of fiber strengths, L is the characteristic length of the material, τ_0 is the interfacial shear strength $\hat{D}$, is a fiber diameter, v_f is the volume fraction of the fibers, and C_n are the local stress concentration factors when "n" neighboring fibers are broken (as a precursor to the fracture event).

If we examine (6.6 and 6.7), we see that several parameters are influenced by temperature. The yield strength of the matrix (or interphase region between the fiber and the matrix) can be expected to decrease with increasing temperature. Also, the stiffness of the components will, in general, be a function of temperature. For a polymer matrix material, for example, the shear stiffness will often be strongly temperature dependent.

One way to consider this influence is to talk about the ineffective length that is created in the region of fiber fracture. As discussed earlier, when a fiber breaks, stress is transferred back into the fiber by the material around it. This reaction occurs over a distance that is controlled by the stiffness of the surrounding material (for the elastic case). If the surrounding material is compliant, the ineffective length is large since only a small amount of stress can be transferred, per unit length, by the compliant matrix. But if the ineffective length is large, fiber fracture regions easily interact and connect together to cause fracture. This *bundle strength* concept was discussed in Chapter 4. If the matrix is made very stiff so that the transfer of stress back into a broken fiber occurs over a small distance, the ineffective length is small. Additionally, the stress concentration in the material next to a fiber break is very high, and there is a greatly increased chance of one fiber fracture causing an unstable sequence of neighboring fiber fractures, resulting in global fracture. Hence, to achieve the strongest composite, the material system must be designed so that the ineffective length has a correct (intermediate) value. This is sometimes done by trial and error, often by coating the fibers with a material that controls the properties of the interphase region between the fibers and the matrix. If sufficient constituent data are available, micromechanics representations can be used to estimate the correct value.

However, environmental conditions may alter these local constituent or interphase properties, especially over long service times, and change the ineffective length. For our present example, we consider the effect of temperature on the ineffective length. For the elastic deformation case and the fracture of one fiber, expressions for the ineffective length have been developed. For example, using shear-lag concepts, it is easily shown that

$$\delta = \frac{1}{2}\left[\frac{1}{2}\left(\frac{1 - v_f^{0.5}}{v_f^{0.5}}\right)\left(\frac{E_f}{G_m}\right)\right]^{1/2} \ln\left(\frac{1}{1-\phi}\right) \tag{6.8}$$

where v_f is the fiber volume fraction, E_f is the stiffness of the fibers, G_m is the matrix stiffness, and Φ is an efficiency factor for the stress transfer. (The ineffective length has been normalized by the fiber diameter in this expression.)

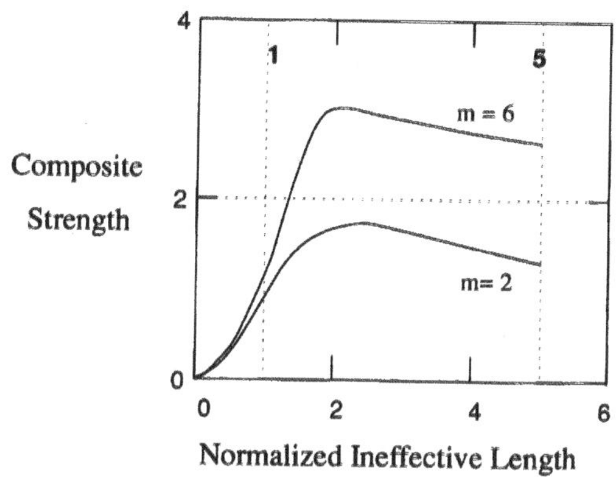

Figure 6.8 Predicted composite strength variations with changes in ineffective length induced by temperature variations using (6.7) and (6.8).

Intuitively, one would expect that if elevated temperature reduces the stiffness of the matrix, then the ineffective length would increase. If this happens, the micromechanical strength equations presented earlier should tell us what happens to the composite strength.

Figure 6.8 shows the strength predicted by (6.7) as a function of variations in the ineffective length for several values of the Weibull shape factor of the fiber strength distribution. The most important feature of Figure 6.8 is the clear maximum that appears in the strength distribution. To the left of this maximum, the small ineffective length creates local stress concentrations leading to brittle fracture. To the right of this maximum, the large ineffective length reduces strength by the coupling of fiber fracture zones.

If increasing temperature causes the ineffective length to increase, the composite strength may either increase or decrease. If the initial strength is to the left of the maximum in strength, the composite strength will increase with increasing temperature since the increase in the ineffective length will reduce the local stress concentrations that limit the strength in that region of the figure. However, if the initial strength is to the right of the maximum, the composite strength should decrease with increasing temperature, since the increase in ineffective length will provide even more opportunity for the fiber fractures to interact and link together in this region where such interactions control the composite strength.

6.2.2 Influence of Environments on Composite Strength

In the previous section, we found that the tensile strength of continuous-fiber-reinforced composites in the fiber direction is generally a complex function of temperature. As an explanation, the physical reasons have to do with the interaction of the composite constituents at the microlevel, a nonlinear composite effect of a substantial nature. As a result, the tensile strength of a composite in the direction of the fibers, dominated by the fibers in many composite systems, can change greatly with temperature, despite the fact that the actual fibers are not affected by the temperature change. Although the analysis methods are not yet complete, micromechanical strength representations have the general capability to describe these interactions and to predict the complex behavior.

In Equation (6.7), for example, if the characteristic strength of the fibers or the interfacial shear strength between the fibers and the matrix is altered by temperature, the tensile strength will change accordingly. In fact, if we are to the right of the maximum in Figure 6.8 (i.e., if the strength is controlled by bundle strength), the strength will monotonically decrease with a decrease in those variables.

Using the micromechancal representations for the compressive strength in the fiber direction as discussed in Chapter 4, we may also make such predictions for the temperature dependence of the principal-strength value. For example, most of the micromechanical compressive strength representations described in that chapter were nearly linearly dependent on the matrix shear modulus G_m. Chapter 5 revealed that a typical variation of G_m (or G_{12} of the composite) is a nonlinear curve, which slopes downward as the upper limit of temperature is reached (e.g., as one approaches or exceeds the T_g of a polymer matrix material). Figure 6.9 shows that type of behavior for the fiber direction compressive strength of an AS4/3501 carbon–epoxy.

Other data collected from the literature for this generic carbon–epoxy system are also shown in Figure 6.9 for the other composite principal strengths. The T_g of this material system is typically about 300°F, depending upon moisture content (a major influence, also to be considered). Essentially, all of the principal strengths degrade as this temperature is approached. Most of the variations are monotonic, except for the fiber direction tension strength (as discussed above) and the in-plane shear strength (S in the figure). Composite properties that are strongly dominated by the matrix properties, such as the tensile and compressive strength in the direction transverse to the fibers and the in-plane shear strength, vary sharply with temperature, losing as much as 80% of their original value. Although the behavior shown in Figure 6.9 is

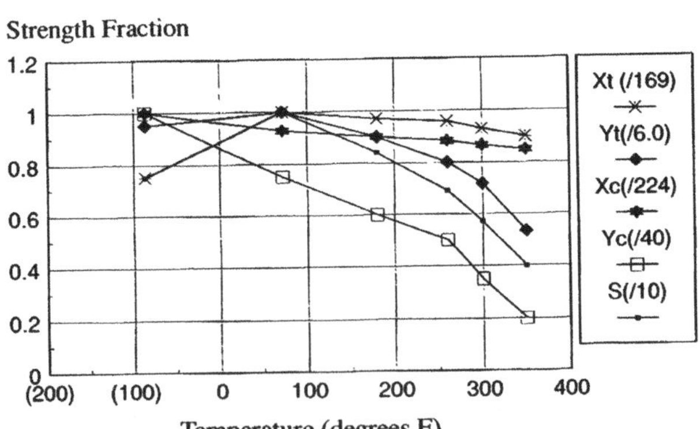

Figure 6.9 Variations of principal-strength values as function of applied temperature for AS4/3501 carbon–epoxy composite.

typical of continuous-fiber composites, variations between material systems is large, and data for a given system should be used to establish the correct dependence on temperature. Strength models such as the ones discussed can be an important help in trying to anticipate behavior under conditions not tested or, in some cases, in guessing how a new system not yet made will respond to temperature changes.

Although temperature is probably the most common environmental variable, other environments may greatly influence the strength of composites, especially the presence of solvents or vapors. The most widely discussed effect of this sort is the influence of moisture (i.e., water vapor) on the properties of polymer composites. In Chapter 5, we learned that moisture can significantly alter the stiffness of polymeric matrix composites and change the T_g of the material by large amounts. However, the combination of elevated temperature and moisture is worse than either of the two environments alone. Figure 6.10 shows one example of the combined temperature–moisture effect for an AS-5/3501 material similar to the one shown in Figure 6.9. The data were developed by Browning (1972) and Browning et al. (1976) and represent reductions in the tensile strength perpendicular to the fiber direction, called Y_t in Figure 6.9. For no moisture, the strength reduction is of the order of 50% in both figures. But if only 1.05% moisture is present, the initial strength at room temperature is reduced to about 60% of its dry value, and as temperature increases to levels near the T_g, the transverse strength drops to only about 30% of its room-temperature value. Increasing the moisture by more than half (to 1.6%) further reduces the values, but the largest effect is created by the initial introduction of moisture; that is, the *presence* of moisture causes a large effect. The addition of more moisture does not make it proportionately worse, at least in this case.

We mentioned in the previous section that the effect of temperature (alone) on quasi-static properties could sometimes be represented by equations such as (6.4) and (6.5). In the chapter on applications, we will use these forms to represent a variety of material systems. However, the scaling factors (and many others typically used in the literature) are monotonic decreasing and will not correctly represent

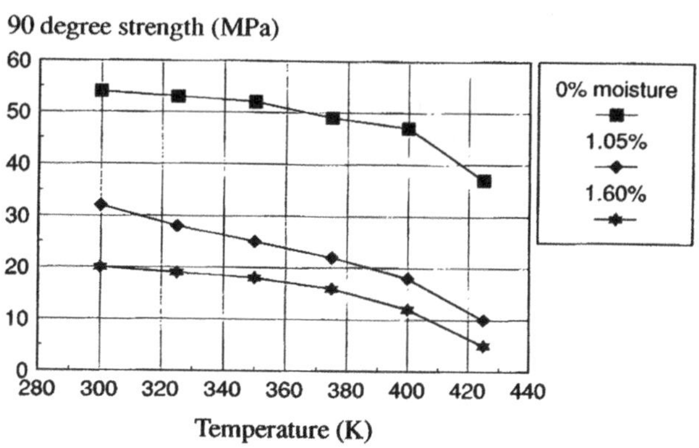

Figure 6.10 Variation of transverse strength of unidirectional composite with temperature and moisture content (Browning, 1972; Browning et al., 1976).

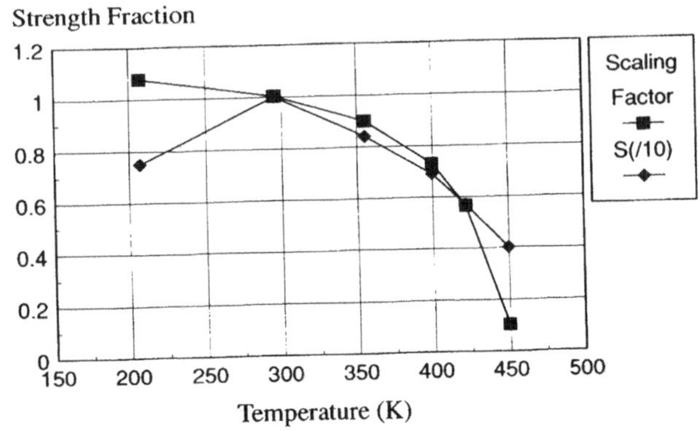

Figure 6.11 Comparisons of predicted variations of composite shear strength using Equation (6.6) with data from Fig. 6.9.

changes that are not uniform in their behavior. To illustrate this point, we use Equation (6.5), which was found to fit the change in stiffness of a Nicalon fiber and a Ni–SiC composite made with that fiber, with only one change in the material constants, to obtain the form

$$\text{Scale} = 1 - 0.23872\left(\frac{T - 298}{490 - T}\right) \tag{6.9}$$

where all temperatures are in degrees Kelvin. In this case, only the maximum temperature has been changed to accommodate the polymer matrix. Figure 6.11 compares this equation with the (normalized) data for the change of in-plane shear strength with temperature for the carbon–epoxy system shown in Figure 6.9.

Above room temperature, the data and the scaling factor are quite similar, but the fact that the shear strength has a maximum is not recovered by the simple form in (6.9). This is a strong motivation for increasing our understanding of the dependence of the strength on the properties, geometry, and arrangement of the constituents so that mechanistic relationships such as the micromechanical models discussed earlier can be used instead of empirical curve fits such as (6.9). It also serves as a warning that empirical relationships can misrepresent data badly, especially at the extremes of the representations.

6.2.3 Time-Dependent Strength

The need for lightweight polymer composite components in high-speed commercial aircraft sparked renewed interest in the use of polymer composites for transportation components that serve at elevated temperature in air or in other aggressive environments. This need also accounts for some new requirements for such service, compared to previous applications that were oriented toward defense hardware.

The most important element of these requirements is the period of service for which components must be designed. For commercial transportation devices, components may be designed for 9000 h between inspections, and total design lives may be as high as 60,000 h.

For polymer composites, these time periods present a special challenge. First, typically, insufficient data exist to define the changes that such materials may undergo during service for such long-term conditions. Second, many of the primary candidates for material systems to be used are almost continuously under development. Third, with limited experience with the specific material systems of interest, it is difficult to establish what models can or should be used to describe the long-term damage tolerance and durability of components made from the material systems. These problems are the focus of our discussion.

When time enters the response equations, the picture is even more complex and uncertain. Physical aging, for example, may compensate for long-term creep effects for amorphous and semicrystalline thermoplastic polymers (below T_g) (Dean et al., 1992; Struik, 1978; Wang, et al., 1975). In composites, nonlinear (nonrecoverable) creep (at elevated temperature) is common and may be a combination of viscoplastic flow, creep, physical aging, and localized plastic strain associated with the stress concentrations around fibers (Nichols et al., 1990). Is it possible (or realistic) to construct constitutive equations that follow this behavior in detail? To construct such constitutive equations (if the answer is "yes"), what experimental data should be recorded, and can we use standard accelerated testing concepts (Wang et al., 1995) to recover the data?

There are more difficult questions. For example, beyond the questions of constitutive equations introduced above, we must also ask, what are the mechanics of long-term environmental effects in a unified sense? Some of the basic elements of this subject are well-established and familiar subjects. At a minimum, they include the following:

- *Viscoelastic–Plastic Behavior.* There has been a virtual rebirth of this subject in recent years (Ha and Springer, 1987; Schapery, 1974; Sun and Chen, 1993). After nearly 50 years of development, the basic philosophy of this field and the associated mechanics are well established and reasonably mature.
- *Aging.* Many good data sets are available that describe this behavior well for many materials and material systems, but in many cases explanations (and mechanics-based models) are not available. Physical aging, generally discussed in terms of free volume, is widely described (Bernatz et al., 1999; Bradshaw and Brinson, 1997; Kerr and Haskins, 1987; Read, et al., 1988; Struik, 1987). Chemical aging and associated processes such as diffusion, capillary action, and corrosion have also been extensively investigated (Emanuel and Buchanchenko, 1987).
- *Stress Rupture.* This physical behavior is typically characterized by the materials community, but the related mechanics are not widely discussed (Kerr and Haskins, 1987).

From the standpoint of engineering applications, these (and other) types of long-term behavior rarely occur independently; they often occur in various combinations. We are concerned here with the question of what mechanism or mechanics approach can we use to discuss them together and to represent their collective effects?

6.3 MODELS OF STRENGTH EVOLUTION

6.3.1 Allowable Deformation Models

In Chapter 5, we discussed time-dependent deformation. Models for the stiffness of various materials as a function of applied stress level and temperature were developed. In general, the focus of the representations associated with creep in that chapter was the redistribution of stress, especially at the local (fiber–matrix or ply) level.

But the deformation associated with creep is also an important engineering variable. We mentioned in Chapter 2 that strain to failure is often a good failure criterion for brittle materials, especially fibrous composites (and especially in directions controlled by the fiber response). The concept of using total strain as a criterion for failure is actually a very broad one. We find evidence of the generality of this concept in several forms.

Figure 6.12 illustrates the breadth of this concept with two data sets. One set represents the final strain to failure during the tension–tension *fatigue* loading of unidirectional glass–epoxy composite coupons (Hahn and Kim, 1976). Hahn and Kim argued that the secant stiffness of the coupons was reduced to roughly the same value during loading in either a quasi-static test to failure or during cyclic loading in a fatigue test, and when that value is reached, failure occurs. However, analysis of their data provides the results shown in Figure 6.12. The *strain to failure* for cyclic loading over a fairly wide range of stress amplitudes is quite similar and not very

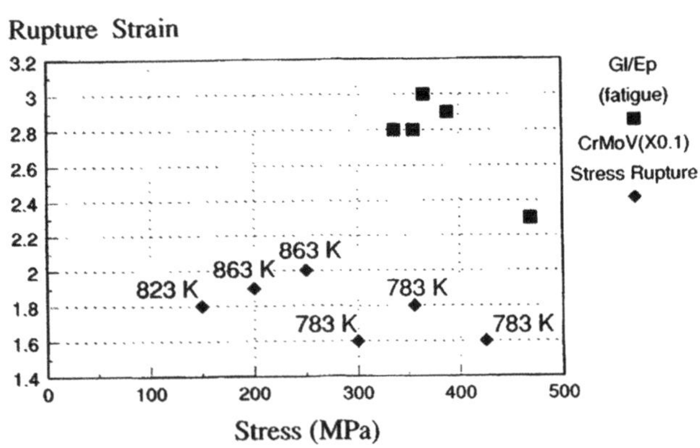

Figure 6.12 Strain at failure for fatigue loading and for constant applied loading for two composite materials (Evans et al., 1993; Hahn and Kim, 1976).

different from the average static failure strain. Highsmith and Reifsnider (1982) recorded similar data for graphite–epoxy materials. The other data set in Figure 6.12 represents the rupture strain in a series of creep rupture tests at constant load levels (the strain is 10 times the ordinate values). Again, for widely different applied stress levels, temperatures ranging from 783 to 863 K, and times to failure ranging over four decades, the strain to failure is remarkably similar (Evans et al., 1993).

A classical representation of creep data (primarily in metals) has a form suggested by this generality. The Monkman–Grant equation that is universally discussed has the general form

$$\dot{\varepsilon} t_f = C \tag{6.10}$$

where $\dot{\varepsilon}$ is the strain rate (in the steady-state region), t_f is the time to failure, and C is a material constant. If C does not depend on applied variables, the implication of Equation (6.10) is that the rupture strain is constant for all strain rates. While the generality of this concept should not be overstated, the strain to failure under conditions in which damage accumulation controls failure (especially long-term conditions of that type) is a good measure of the ability of the material to withstand damage development and can be an important material characteristic for the representation of such behavior.

Of course, an allowable strain may be a design limit apart from the rupture strain of the material. The clearance of a rotating element such as a turbine is fixed by the expected expansion of the diameter of the member during operation, and any creep deformation must be included in that value to avoid contact of the rotor with the housing. Many other scenarios of a similar nature are common. The safety codes that control the design of boilers and pressure vessels around the world typically use criteria such as limiting the applied pressure to two-thirds of the value that would cause rupture (at the operating temperature) in 100,000 h, but an alternative design limit based on deformation and deformation rate is usually allowed. Typically, the design limit pressure might be set at the level that causes a *creep rate* of 0.01% in 1000 h. Hence, creep deformation and deformation rate are common, useful representations of the creep rupture—or stress rupture—phenomenon.

Most of the deformation representations of stress rupture response focus on the deformation rate, especially in the steady-state region. Equations of the general form

$$\dot{\varepsilon} = A\sigma^n \tag{6.11}$$

are often used, where σ is the applied stress level and A and n are material constants. This type of equation is often attributed to Norton (1929). The relationship of this equation to representations of rate process such as the common Arrhenius equation was discussed in Chapter 5.

A variety of associated forms have been offered in the literature for homogeneous and inhomogeneous materials. In the present context, the creep of homogeneous materials is of interest for constituents in a composite. For high-temperature applications, where this subject is of special importance, the creep of ceramics is particularly pertinent. An example of such behavior has been discussed by Corman (1990), who investigated the creep behavior of single-crystal ceramics including yttria-stabilized zirconia, throia, yttrium–aluminum–garnet, beryllia, alumina, and silicon carbide. He was successful in using a form widely quoted for ceramic materials, stated below:

$$\dot{\varepsilon} = \frac{ADGb}{kT}\left(\frac{b}{d}\right)^{p}\left(\frac{\sigma}{G}\right)^{n} \tag{6.12}$$

In this expression, G is the shear modulus of the material, D is the diffusion coefficient for the controlling mechanism, b is the Burgers vector, d is the grain size, A, p, and n are material constants, T is temperature in degrees absolute, and k is Boltzmann's constant (cf. Cannon and Langdon, 1983). In this expression, D and G may be temperature dependent, with forms such as

$$G = G_0 - \partial G\, T \tag{6.13}$$

where ∂G is the rate of change of the shear stiffness with temperature, and

$$D = D_0 \exp\left(\frac{-Q}{RT}\right) \tag{6.14}$$

where Q is the activation energy of the controlling process and R is the universal gas constant.

Some of the data from Corman (1990) are shown in Figure 6.13 to illustrate another feature of this behavior. The data in Figure 6.13 show some of the steady-state creep rate behavior of yttria-stabilized zirconia single crystals for two temperatures and for two active slip systems in the crystals. A clear separation of the data as a function of temperature and as a function of the active slip system occurs. This separation illustrates two general features of importance. First, anisotropic materials such as the crystals in the case of Figure 6.13 or for the general case of composites can be expected to have different creep rates in different directions and, therefore, different stress rupture lives when creep in those directions controls the failure. Second, creep rupture characterization to determine the constants in (6.8)–(6.10) must be done for each independent physical controlling mechanism, and our definition of mechanism must account for anisotropy if it is there. Determining the material constants for a "slip mechanism" in a given material is not enough if there are several different slip planes that may control the behavior (Mura, 1987). In a manner similar to the earlier discussion of damage modes and failure modes, stress rupture and creep behavior in general are very sensitive to the physical controlling microbehavior; no analytical representation will be any better than the precision with which the supporting physical details are determined.

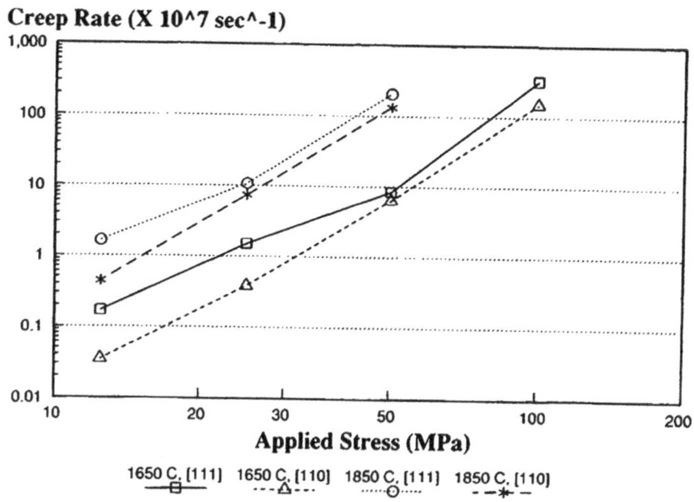

Figure 6.13 Data showing dependence of steady-state creep rate on level of applied stress for yttria-stabilized zirconia (Corman, 1990).

State variable approaches are also common. A general example is the model of Ding et al., (1994), based on a flow rule for inelastic strain rate as a function of two state variables, the applied stress and temperature, in the form

$$\dot{\varepsilon} = \frac{f(\sigma/\sigma_0)e^{-Q/RT}}{\delta} \tag{6.15}$$

where σ_0 is a reference stress, Q is the activation energy of the active process, and δ is a hardening coefficient that is taken to be a state variable. Two evolution rules are assumed for the state variables. The equation for the hardening coefficient has the form

$$\dot{\delta} = \frac{\dot{\delta}_0}{\delta^m} e^{-Q_\delta/RT} = \frac{\beta}{\delta^m} \tag{6.16}$$

where δ_0 is a reference value and Q_δ is the activation energy for hardening. The evolution of the other state variable, which is taken to be a damage variable, ω, is written in a similar form

$$\dot{\omega} = \frac{\dot{\omega}_0(\sigma/\sigma_0)^v e^{-Q_\omega/RT}}{\delta(1-\omega)} = \frac{\gamma}{\delta(1-\omega)} \tag{6.17}$$

where v is a material parameter and Q_ω is the activation energy for damage development. With the use of an explicit damage parameter ω, these equations may be combined and interpreted in terms of the time to failure, which is defined by the time at which $\omega = 1$. With this condition, Ding et al. find the expression

$$t_r = \frac{1}{\beta(1+m)}\left[\left(\frac{m\beta}{2\gamma} + k^m\right)^{(m+1/m)} - k^{m+1}\right] \tag{6.18}$$

where k is the initial value of δ at the inception of creep loading and t_r is the time to rupture. We will return to this expression in a later discussion of damage mechanics approaches and introduce a possible physical interpretation of the damage parameter ω. For the expression above, it has the usual interpretation of having the value zero initially (for no damage present), and the value of unity when the material fails. However, a careful examination of (6.15)–(6.18) shows that they represent a modern version of the Monkman–Grant expression and have the advantage of being explicit in the time to rupture.

Many other forms of the deformation rate equations have been developed. A number of popular forms are based on fracture mechanics. Lange (1976) proposed a form of (6.12) in terms of the crack growth rate to obtain

$$\dot{a} = BY^m \sigma^m a^{m/2} \tag{6.19}$$

where B and m are material constants, Y is the geometric factor in the field stress intensity expression, and a is the crack length. By comparing the crack length obtained from an integration of (6.19) with the critical crack length that corresponds to the fracture toughness, K_{Ic}, of the material, the time to rupture can be estimated.

6.3.2 Time-to-Failure Models

Many stress rupture representations are based on the calculation of the time to rupture t_r. Such data are usually obtained from creep tests conducted at constant load and temperature, most often with unidirectional loading. The general behavior commonly observed is sketched in Figure 6.14. High applied stress levels are generally associated with large-scale deformations (e.g., visoelasticity, plasticity, necking) while lower stress levels may produce material degradation by microdefect (or other) mechanisms for which failure is defined by damage accumulation (Leckie and Hayhurst, 1978).

For multiaxial stress states, the time to failure can be represented as an *isochronus surface*, in terms of the local controlling stress values (principal stresses for isotropic materials). For metals, such surfaces may have the form (Hayhurst, 1972)

$$\alpha\frac{\sigma_1}{\sigma_0} + \beta\frac{\bar{\sigma}}{\sigma_0} + \gamma\frac{J_1}{\sigma_0} = 1 \tag{6.20}$$

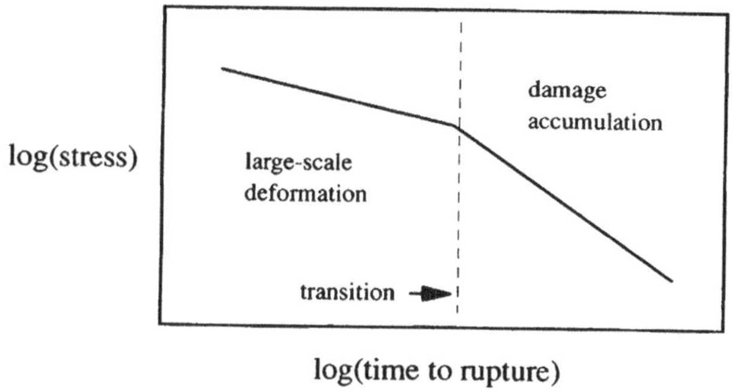

Figure 6.14 Typical relationship of constant applied stress and time to rupture.

where $\bar{\sigma}$ is the effective stress given by

$$\bar{\sigma} = \frac{1}{\sqrt{2}}[(\sigma_1 - \sigma_2)^2 + (\sigma_2 - \sigma_3)^2 + (\sigma_3 - \sigma_1)^2]^{1/2} \tag{6.21}$$

J_1 is the first invariant of stress given by

$$J_1 = \sigma_1 + \sigma_2 + \sigma_3 \tag{6.22}$$

and σ_0 is a material constant. It is required that

$$\alpha + \beta + \gamma = 1 \tag{6.23}$$

As noted by Hayhurst, if maximum stress controls the behavior, then $\alpha = 1$, $\beta = 0$, and $\gamma = 0$. If biaxial states of stress influence the rupture life, then J_1 may enter the equation and α, β, γ will have other values.

Expressions such as the ones above are most often applied to materials to represent their behavior at substantially elevated temperatures (especially for metals). However, similar concepts have been used for the representation of the behavior of materials at room temperature (or ambient operating temperatures, in general). A good example is the work associated with polymer-based materials. The fundamental basis for this work is of general interest to our applications. McClintock and Argon (1966) discusses activation processes in which the rate of change in the observable (e.g., atoms leaving a surface, bonds breaking between atoms) can be expressed as

$$\frac{\partial N}{\partial t} = N_s \nu_a e^{-U_a/kT} \tag{6.24}$$

where N is the observable, N_s is the number of activation sites, T is the temperature (in absolute degrees), k is Boltzmann's constant, ν_a is the frequency of the process being activated, and U_a is the activation energy, usually regarded as the change in

Gibbs free energy necessary to drive the process. The discussion of this equation in various forms for activation processes is extensive in the literature (cf. Glasstone et al., 1941; Zhurkov, 1965, Zhurkov and Korsukova, 1974). Most of these models have their origins in the representations of breaking of bonds or other atomic phenomena such as dislocation motion or diffusion. However, Zhurkov adapted the equation to the prediction of lifetime in the form

$$t_r = t_0 e^{(U_0 - \alpha\sigma)/kT} \tag{6.25}$$

where U_0 is the activation energy, σ is the stress (or effective stress) that is driving the process, t_0 is a characteristic time, and α is a material constant. We will discuss the use of (6.24) and (6.25) for mechanistic modeling in a subsequent section. At this point, let us consider the engineering utility of this approach to stress rupture behavior representation.

6.3.3 Parameter Methods

We address the question of how to estimate the time to rupture (or the time to reach a critical level of creep strain, as discussed earlier) under conditions of constant stress (or effective stress) and constant applied temperature. For such a condition, one might write (6.25) in the form

$$t_r = Ae^{U_0/kT} \tag{6.26}$$

where A is a material constant. Taking the logarithm (to base 10) of this equation, we obtain

$$\log(t_r) = \log A + \log(e)\frac{U_0}{kT} = B + b\frac{U_0}{kT} \tag{6.27}$$

The data we wish to represent might have the form shown in Figure 6.15 (Goldhoff, 1959).

In general, one could consider the possibility that B and U_0 depend on the level of applied stress. If all the data on plots such as Figure 6.15 converge to some common intercept (at a large negative value in the case of Figure 6.15), then B is independent of stress, and we need only consider the possibility of U_0 being stress dependent. Regrouping constants in (6.27) and rearranging, we could then obtain

$$\log(t_r) + C = c\left(\frac{1}{T}\right) \tag{6.28}$$

where c is a function of stress. Then rearranging again, we obtain

$$T[\log(t_r) + C] = P_{\mathrm{LM}}(\sigma \qquad 6.29)$$

where P_{LM} is the Larson–Miller parameter widely used for engineering descriptions of common metals (Larson and Miller, 1952). From data such as those in Figure 6.15, it is possible to create a master curve for the parameter P_{LM} as a function of stress, having the general form shown in Figure 6.16. Although limited use of this approach appears to have been made for composites, the data in Figure 6.16 repre-

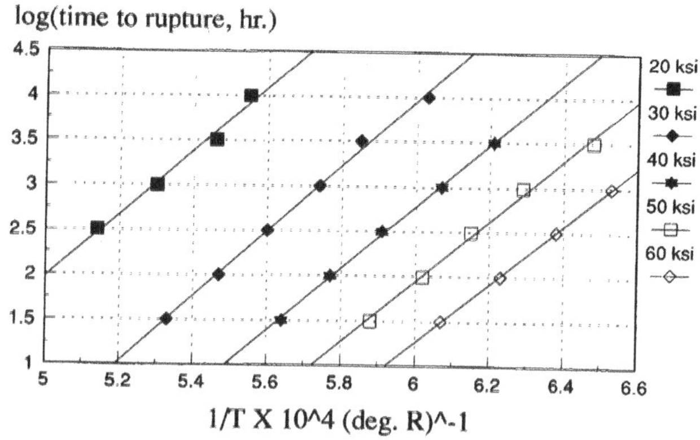

Figure 6.15 Typical stress rupture data as function of stress and temperature (Goldhoff, 1959).

sent the behavior of a high-temperature alloy steel with a plasma sprayed coating, a metal–metal composite of a special type (Woodford, 1992).

The practical implications of (6.29), and others like it, are significant. For a given stress level and a given process (i.e., for a given fixed creep mechanism), (6.29) indicates that time and temperature are equivalent extensive variables. A low temperature will produce a long time to failure that is "equivalent" to a short time to failure produced by a high temperature for a given applied stress. This *time–temperature equivalence* is widely used to *accelerate* testing, since it is difficult to obtain data for creep failures when the design life is long (such as 30 years or more). Hence, the practical mechanism of using elevated temperature to shorten the test time and then using the data to estimate longer lives at lower (design) temperatures is popular.

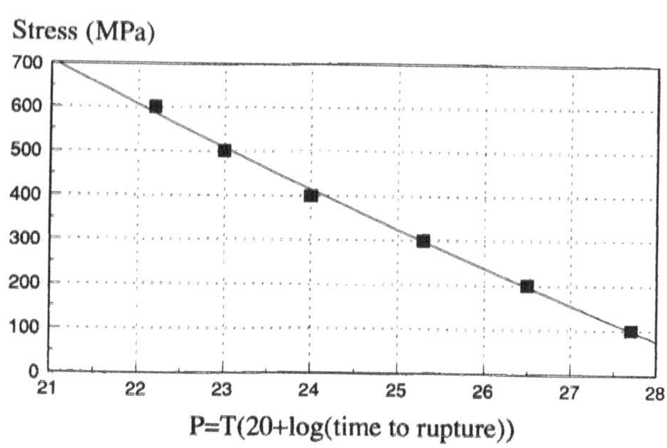

Figure 6.16 Larson–Miller master curve for stress rupture behavior of metal–metal composite (Woodford, 1992).

The foundations of the above equations indicate that this type of accelerated testing must be done with great care and that the constraints on its use are rigid. If the mechanism (i.e., the activation process) is different at different temperatures, the equivalency is lost. And if the process itself cannot be described by an activation equation such as (6.22), then the scheme is invalid. Nevertheless, it is remarkable how often this approach produces quite usable results.

A generalization of this method used for the design of composite components at elevated temperatures is based on an empirical form postulated by Manson and Haferd (1953). They observed that data such as that shown in Figure 6.15 may not have a common intercept but may, instead, have a common point of reference when plotted as a function of temperature (instead of the inverse of temperature). Figure 6.17 shows the concept schematically.

Using the same assumptions about time–temperature equivalence, Manson and Haferd postulated the form

$$P_{\mathrm{MH}} = \frac{T - T_a}{\log(t) - \log(t_a)} \tag{6.30}$$

where T_a is a reference temperature and t_a is a reference time to rupture, as shown in Figure 6.17. Here, P_{MH} is a function of stress, as before, and can be tabulated on a master curve in the same sense as the Larson–Miller scheme noted earlier. Unfortunately, most of the data for the representation of composites in these forms is not in the public domain at this writing. However, application of the Manson–Haferd equation to polymer and ceramic composites has shown good success to date.

6.3.4 Mechanistic Models

A variety of mechanistic models have been developed for various material types for the description of creep and stress rupture behavior. A preponderance of these is associated with homogeneous materials, predominantly metals. However, such models are especially valuable for composite materials since they can be used to suggest relationships between the properties, performance, geometry, and arrange-

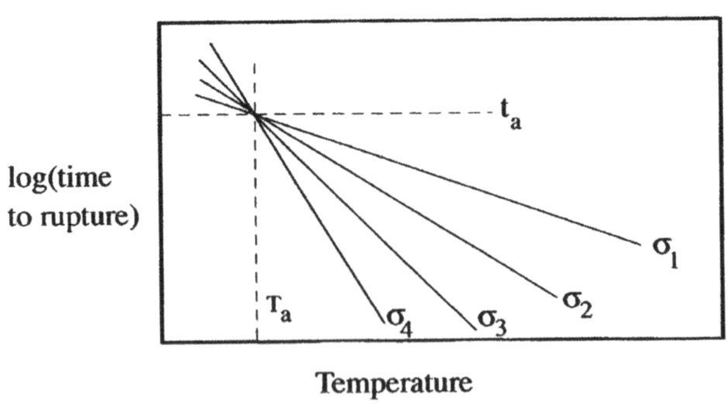

Figure 6.17 Form of stress rupture behavior represented by Manson–Haferd (1953) model.

ment of the constituents and the interphase regions between them and the global composite properties and performance. This field is in its infancy at the moment, but a few salient approaches will be indicated in this section.

Christensen (1981) considers the growth of a flaw (crack) under a *stress induced kinetic state*. He begins by invoking the familiar Griffith argument whereby a crack of length $2a$ in a body of infinite extent releases an energy equal to

$$\frac{\sigma^2 \pi a^2}{E} \tag{6.31}$$

(where σ is the uniform field stress in the direction of unidirectional loading before crack formation and E is the Young's modulus of the material), which is balanced by the energy of formation per unit area, Γ, times the newly formed area of the crack. When the crack begins to grow, these energies match exactly, and the critical stress to cause crack growth can be written as

$$\sigma_c^2 = \frac{2E\Gamma}{\pi a} \tag{6.32}$$

For a viscoelastic material (with a crack in an infinite strip bounded by rigid boundaries), Christensen postulates the corresponding form

$$\sigma^2 = \frac{2\Gamma}{hJ(h/\beta c)} \tag{6.33}$$

where h is the crack length, β is a factor that accounts for the gradient of stress at the crack tip, and c is the time rate of growth of the crack, that is,

$$c = \frac{dh}{dt} \tag{6.34}$$

The parameter $J(t)$ is the uniaxial creep function of the viscoelastic material. Infinitesimal viscoelasticity theory is involved in the derivation of (6.33). To extend this formulation to the calculation of the life of a component, the crack length is assumed to be small compared to the size of the body throughout the life, and the stress σ is not changed by the growth. If the material in question obeys a simple power law equation in creep, then

$$J(t) = J\left(\frac{h}{\beta c}\right) = \tilde{J}t^n + J(0 \qquad 6.35)$$

where the material constants $\tilde{J}$, $J(0)$, and n are adjusted to fit the creep response of the material. Then, combining (6.33)–(6.35) and integrating, one obtains

$$\frac{t}{\alpha} = \int_{h_0}^{h_1} \left(\frac{\gamma}{\sigma^2 y} - 1 \right)^{l/n} \frac{dy}{y} \tag{6.36}$$

where $\alpha = \beta(J(0)/\tilde{J})^{1/n}$, and $\gamma = 2\Gamma/J(0)$, and t is the time required for the crack to grow from the length h_0 to h_1. If the stress is normalized by the fast fracture strength of the material, given by

$$\sigma_s^2 = \frac{2\Gamma}{h_0 J(0)} \tag{6.37}$$

and other groupings are made, then Equation (6.36) becomes

$$\tau = \int_1^{h_1/h_0} \left(\frac{1}{\bar{\sigma}^2 \xi} - 1 \right)^{1/n} \frac{d\xi}{\xi} \tag{6.38}$$

Christensen also notes that the fast fracture residual strength for a crack length of h_1 is

$$\sigma_R^2 = \frac{2\Gamma}{h_1 J(0)} \tag{6.39}$$

In the present context, one can talk about the time to failure given by (6.36) and (6.38) in terms of the applied stress σ. For integer values of n, Christensen was able to find closed-form integrations of those equations. In terms of the normalized variables $\hat{h} = h_1/h_0$ and $\hat{\sigma} = \sigma_R/\sigma_s$ and for a value of $n = 5$, for example, the resulting form is

$$\tau = -\frac{1}{5\hat{\sigma}^{10}} \left(\frac{1}{\hat{h}^5} - 1 \right) \frac{5}{4\hat{\sigma}^8} \left(\frac{1}{\hat{h}^4} - 1 \right) - \frac{10}{3\hat{\sigma}^6} \left(\frac{1}{\hat{h}^3} - 1 \right) + \frac{5}{\hat{\sigma}^4} \left(\frac{1}{\hat{h}^3} - 1 \right)$$
$$- \frac{5}{\hat{\sigma}^2} \left(\frac{1}{\hat{h}} - 1 \right) - \ln(\hat{h} \qquad 6.40)$$

which is the normalized life or characteristic life of the material for a given normalized crack length $\hat{h}$. Christensen goes on to include the statistical aspects of the material behavior in the analysis and to use somewhat more general compliance relationships in place of (6.30). Comparisons were made with creep–lifetime data developed by Chiao and Moore (1986) and Hahn and Chiao (1980) for the stress rupture of composite arimid–epoxy strands acted upon by constant uniaxial loads. For that situation, $n = 0.21$, $\sigma_s = 3450$ MPa, and the normalization factor α, was adjusted to best fit the data. Figure 6.18 shows some of the data and predictions listed by Christensen (1981). Another micromechanical modeling approach is illu-

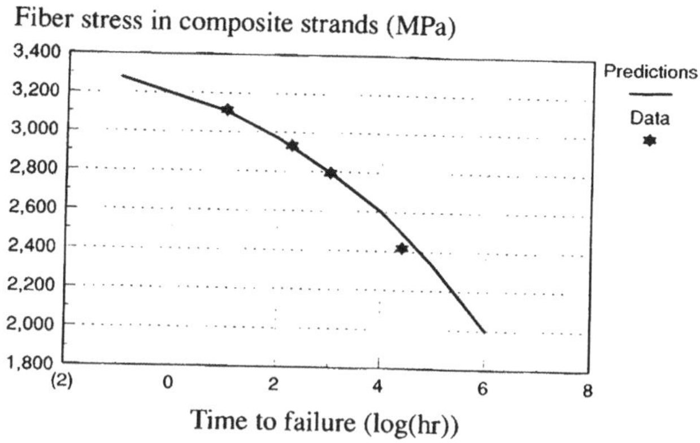

Figure 6.18 Predicted and observed times to failure for composite strands (Christensen, 1981).

strated by the work of Curtin et al. (1996). Their model begins with the assumption that the fiber and the matrix have viscoplastic behavior represented by the equations

$$\dot{\varepsilon}_m = \frac{\dot{\sigma}_m}{E_m} + G\sigma_m^q \qquad \dot{\varepsilon}_f = \frac{\dot{\sigma}_f}{E_f} + H\sigma_f^p \tag{6.41}$$

where G, H, p, and q are material constants. It is assumed that the response being considered results from a uniaxial loading in the direction of the fibers in a unidirectional composite bundle. For this situation, the strain in the fibers and matrix are assumed to be equal, so that

$$\varepsilon_m(t) = \varepsilon_f(t \qquad 6.42)$$

and the corresponding strain rates are equal. Equilibrium requires that

$$\sigma_a(t) = (1 - v_f)\sigma_m(t) + v_f\sigma_{f-\text{total}} \tag{6.43}$$

where σ_a is the global applied stress, v_f is the volume fraction of fibers in the bundle, and $\sigma_{f-\text{total}}$ is the total stress supported by the fibers in the bundle. Curtin (1991) discusses the determination of the quantity $\sigma_{f-\text{total}}$ for a ceramic composite in which the stress transfer between the fibers and the matrix is controlled by frictional sliding forces generated in regions near fiber or matrix cracks. If the recovery length near such breaks is l_f, then fibers within this length carry some fraction of the remote fiber stress σ_f. Then, for an arbitrary cross section of the composite, the total stress supported by the fibers in the group will consist of the stress σ_f supported by the unbroken fibers and the frictional forces of those that are broken within $\pm l_f$. Curtin showed that if the fraction of fibers that have more than one break within $\pm l_f$ of the cross-sectional plane is relatively small, if there are no significant stress concentrations in the fibers near a fiber or matrix crack, and if the finite population of fiber strengths is adequately described by a Weibull distribution, then the total stress supported by the fibers can be written as

$$\sigma_{f-\text{total}}(t) = \left[1 - \frac{1}{2}\left(\frac{\sigma_f^{\max}(s)}{\sigma_c}\right)^{m+1}\right] v_f \sigma_f(t) \tag{6.44}$$

where m is the Weibull shape factor of the fiber strengths and $\sigma_f^{\max}(s)$ takes the maximum value of the fiber stress for all values of time up to the current time t, that is, $-\infty \leq s \leq t$. The characteristic strength in (6.44), according to Curtin, is given by

$$\sigma_c = \left(\frac{\sigma_0^m L_0 \tau}{r_f}\right)^{1/(m+1)} \tag{6.45}$$

where σ_0 is the characteristic strength of the Weibull distribution of fiber strengths, L_0 is the characteristic length, τ is the frictional shear strength, and r_f is the radius of the fibers. Combining (6.41)–(6.45), an expression for the change in the stress in the unbroken fibers (normalized by σ_c) can be obtained in the form

$$\frac{d\hat{\sigma}_f}{d\hat{t}}(\hat{t}) = \frac{\{\hat{\sigma}_a - v_f\hat{\sigma}_f(\hat{t}) + (v_f/2)[\hat{\sigma}_f(\hat{t})]^{m+2}\}^q - [H\sigma_c^{p-q}(1-v_f)/G]][\hat{\sigma}_f(\hat{t})]^p}{(1-v_f)^{q-1}(E_c/E_f - \{v_f(m+2)[\hat{\sigma}_f(\hat{t})]^{m+1}/2\})]} \tag{6.46}$$

where

$$\hat{t} = E_m G \sigma_c^{q-1} t \tag{6.47}$$

and the stresses have been normalized by the characteristic strength. Equation (6.46) can be integrated numerically and used in (6.41) to estimate the rate of strain for the fibers. This rate can be integrated subsequently to get the strain in the fibers as a function of time and compared to the rupture strain for the material in question to estimate the time to fail the bundle. Figure 6.19 shows the general nature of such a calculation compared with data from Weber et al. (1996).

Predictions for 750 MPa were made assuming that the heat-treated fiber strength of 1.29 GPa is appropriate, and the 800-MPa predictions assumed that the processed fiber strength of 1.47 GPa is appropriate (Curtin et al., 1996; Weber et al., 1996).

Du and McMeeking (1995) provide a review of several creep models that brings several important features of the underlying physical behavior to light. A few salient points from their discussion will be summarized here to illustrate the physical features. The reader is referred to the literature for a more complete discussion. For the special case when the fibers are elastic [$\dot{\varepsilon}_f = 0$ in (6.41)] and assuming that the composite relaxation due to matrix creep is unaffected by fiber failure (or

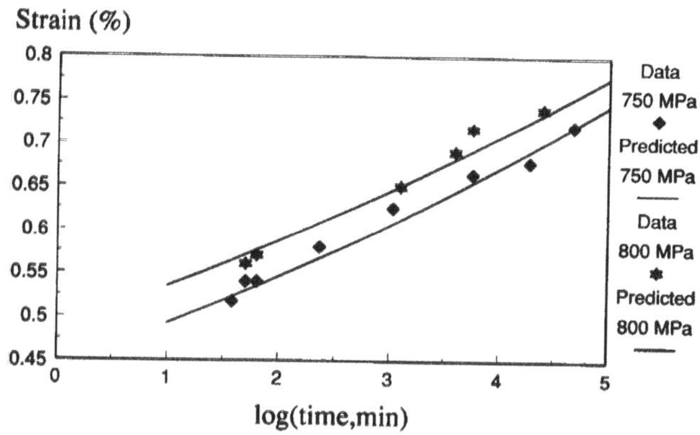

Figure 6.19 Creep strain versus time for titanium matrix composite at applied stress levels of 750 and 800 MPa (Weber et al., 1996).

vice versa), McLean (1985) provides a solution to the remainder of (6.41)–(6.43) in the form

$$\varepsilon(t) = \frac{\sigma}{v_f E_f} - \frac{1 - v_f}{v_f E_f}\left[\left(\frac{E}{E_m \sigma}\right)^{n-1} + \frac{v_f(n-1)E_f E_m G t}{E}\right]^{-1/(n-1)} \tag{6.48}$$

for a constant applied axial stress σ, where E is the rule-of-mixtures composite modulus. Equation (6.48) ignores the effect of fiber failure on the creep or creep rupture process. To illustrate the manner in which this creep law can be used to estimate creep rupture life, we use several additional elements of the models developed by Curtin (1991) and Du and McMeeking (1995). First, consider the length of the fiber from the location of a fiber break that has a lower stress due to the stress-free requirement at the broken end. If the matrix transfers stress back into the fiber by friction or by a (constant) plastic flow stress over that distance, equilibrium requires that

$$\pi D l_f \tau = \pi \frac{D^2}{4} E_f \varepsilon \Rightarrow l_f = \frac{D E_f \varepsilon}{4\tau} \tag{6.49}$$

where D is the diameter of the fiber and τ is the flow stress or friction stress (which is constant over l_f). The average stress in the fiber, including the effect of the broken ends, can be given by

$$\bar{\sigma}_f = (1 - q)E_f \varepsilon + q\tfrac{1}{2}E_f \varepsilon \tag{6.50}$$

in which q is the probability of having a fiber break within l_f. Following Curtin, we use a truncated Weibull representation of q in the form

$$q = \frac{2l_f}{l_0}\left(\frac{E_f \varepsilon}{\sigma_f^0}\right)^m \tag{6.51}$$

where l_0 is the characteristic length of the fiber for which the probability is given, σ_f^0 is the characteristic strength of fibers of that length, and m is the Weibull shape factor. Then combining Equations (6.45) and (6.49)–(6.51), one can write a representation of the average stress in the fibers as

$$\bar{\sigma}_f = E_f \varepsilon\left[1 - \frac{1}{2}\left(\frac{E_f \varepsilon}{\sigma_c}\right)^{m+1}\right] \tag{6.52}$$

If one argues that fracture of the composite is defined by the maximum of Equation (6.52) with respect to strain, then that maximum value is given by

$$\varepsilon^{\max} = \frac{\sigma_c}{E_f}\left(\frac{2}{m+2}\right)^{1/(m+1)} \tag{6.53}$$

This strain value can be used in Equation (6.48) to define a time to failure and a stress rupture life of the composite. In this instance, it is a strain-based stress rupture life, or creep rupture life. Du and McMeeking (1995) refer to this as the strain-based McLean rupture model. If the maximum strain in Equation (6.53) is used in Equation (6.52), one can write the Curtin expression for composite strength in the form

$$\sigma_{\text{composite}} = \sigma_c\left(\frac{2}{m+2}\right)^{1/(m+1)}\left(\frac{m+1}{m+2}\right) \tag{6.54}$$

If one makes the assumption that the composite breaks when the stress in the fibers reaches this value of effective strength, then Equation (6.48) can be used to get a stress-based stress rupture condition, as noted by Du and McMeeking (1995).

An improved representation is achieved by assuming that the creep of the matrix is influenced by fiber fracture. A simple version of such a model is achieved by using the average fiber stress equation (6.52) in Equation (6.43), solving the resulting expression for the matrix stress, and using that matrix stress equation in the first of Equation (6.41). (The response of the fibers is assumed to be elastic, as before.) The result of this exercise is the creep rate expression

$$\dot{\varepsilon} = \frac{\left\{\dfrac{\sigma}{\sigma_c} - f\left(\dfrac{E_f\varepsilon}{\sigma_c}\right)\left[1 - \dfrac{1}{2}\left(\dfrac{E_f\varepsilon}{\sigma_c}\right)^{m+1}\right]\right\}^n}{1 + \dfrac{v_f}{1 - v_f}\dfrac{E_f}{E_m}\left[1 - \left(1 + \dfrac{m}{2}\right)\left(\dfrac{E_f\varepsilon}{\sigma_c}\right)^{m+1}\right]} \frac{G\sigma_c^n}{(1 - v_f)^n} \tag{6.55}$$

It can be seen that the strain rate becomes unbounded (creating a tertiary creep region and a "natural" rupture prediction) when the denominator approaches a null value. Solving for that strain value yields

$$\varepsilon = \frac{\sigma_c}{E_f}\left[\frac{2E}{(2 + m)E_f v_f}\right]^{1/(m+1)} \tag{6.56}$$

Du and McMeeking (1995) note that, at this strain, extensive fiber fracture is taking place and creep of the matrix has essentially ceased.

Du and McMeeking (1995) also use a cell model to represent the complete relaxation of the total fiber due to creep of the matrix, not just the relaxation of the stresses near the end of the fibers (over the ineffective length), as was done to obtain equation (6.55). This analysis is more complex and introduces a time dependence into the value of l_f, that is, $l_f \to l_f(t)$. Du and McMeeking (1995) call this a fiber relaxation model; they present several comparisons of the representations mentioned above. Figure 6.20 illustrates some of the salient features of this comparison. When the matrix creep is not affected by the fibers, the curve marked "Equation (6.48)" results. Rupture can be predicted based on a strain or stress criterion, as discussed in Equation (6.53) and (6.54), and results in predicted rupture points on that curve. When the creep of the matrix is influenced by fiber fracture and stress relaxation occurs only near the fiber ends, the response marked "Equation (6.55)" is

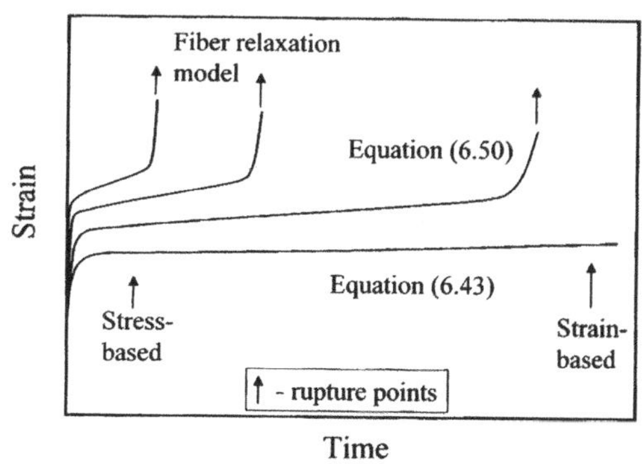

Figure 6.20 Several models of composite creep and stress rupture behavior.

obtained, with a tertiary region. When total fiber relaxation is allowed, the curves marked "fiber relaxation" are obtained. The general character of the comparative results is suggested by Figure 6.20; full details can be found in the Du and McMeeking paper (1995).

6.3.5 Rate Equations for Fatigue Behavior

When we opened this chapter with a problem definition, we mentioned that a general rate equation for fatigue behavior would be used in Equations (6.1) and (6.2), which has the form

$$\frac{S}{S_u} -> \mathrm{Fa}\left(\frac{\sigma(t)_{ij}}{X(t)_{ij}}\right) = A + B\{\log[N(t)]\}^p \tag{6.57}$$

where the local applied stress normalized by the material strength [S/S_u for unidirectional loading, which becomes $\mathrm{Fa}(\sigma(t)_{ij}/X(t)_{ij})$ for complex local stress states] is related to the life under those instantaneous conditions, $N(t)$, and the material constants A, B, and p. If we are going to use this equation as a universal relationship, a more complete examination is warranted.

The most important implication of (6.57) is that the fatigue behavior of materials scales to the ultimate strength in the corresponding direction of loading. This generality has some important exceptions, but it applies to a remarkable number of situations. For example, in metals, especially for long-life (high-cycle) behavior, such an assumption is almost universally made, typically in the one thousand to one million cycles of life regime. In composites, particularly fiber-dominated composites (having short-fiber or long-fiber reinforcement), the generality has also been shown to be quite useful. Figure 6.21 illustrates the generality of such a representation.

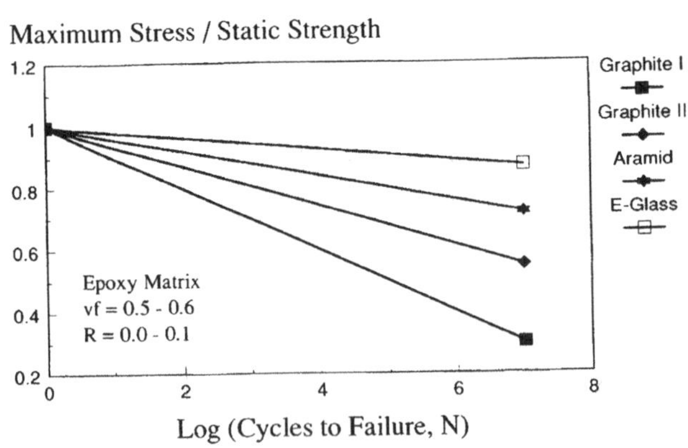

Figure 6.21 Tensile fatigue life versus applied stress level normalized by static strength for several continuous-fiber composite systems (Mandel, 1982).

The first important features of Figure 6.21 are the straight lines on the log-linear scales; that is, the form of (6.57) is appropriate. Second, the slopes [which are represented by the value of B in (6.57)] are quite different, in spite of the fact that the matrix material is the same. Some of the fibers in those composites are quite insensitive to cyclic loading, especially the carbon fiber examples. Hence, the fatigue behavior is being controlled by the interaction between the matrix (which is fatigue sensitive) and the fibers and by the fiber–fiber interaction, which depends on the fibers. Mandel (1982) notes that the ordering of the materials in Figure 6.21 generally follows the difference in the longitudinal modulus of the composites, which controls the amount of strain that the matrix must sustain, an idea also advanced by Dharan (1975). But that explanation is not a complete one. The composites fail only when the fibers fail; the matrix is important as it influences the fiber failure process.

In some cases, unimpregnated strands (of the order of 100 fibers or more) show the same fatigue behavior as the corresponding unidirectional composite (Mandel, 1991). In this case, the fiber–fiber interaction, through mechanical friction and through the statistics of load sharing, obviously dominates the fatigue process. In other cases, evidence shows the matrix controls the fiber fracture process under cyclic loading by controlling the growth rate of matrix cracks that grow from one fiber fracture to the neighboring ones to create a final fracture path (Razvan et al., 1990; Razvan and Reifsnider, 1991). There appears to be little doubt that the fiber–fiber interaction controls fiber-dominated materials. Figure 6.22 (from Mandel,

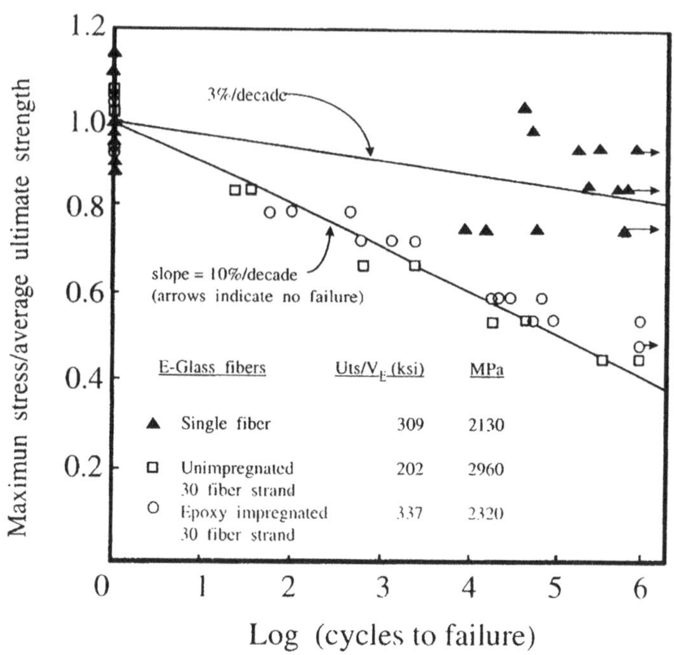

Figure 6.22 Comparison of S–N data for single E-glass 25-μm fibers with that of 330-fiber strands with and without epoxy matrix (Mandel, 1991).

1991) shows a comparison of the fatigue behavior of single fibers compared to strands impregnated with epoxy. The single fibers are much less fatigue sensitive than the composite, in the sense that the slope of their *S–N* curves is much lower.

In some cases, the fibers may be degraded by the cyclic loading or by so-called static fatigue; that is, time-dependent failure (stress rupture) effects. A very common example is the behavior of glass fibers, which are subject to time-dependent creep and chemical effects. The fast fracture strength of E-glass fibers, for example, typically degrades by 3–5% per decade of cycles (at structural frequencies of the order of 10 Hz) in the presence of moisture (Mandel, 1991). Fretting (or contact fatigue) of fibers on fibers (especially in woven materials) can also contribute to fiber degradation.

If we accept the fact that the fatigue behavior we wish to represent scales to the fracture strength of the materials, as suggested by (6.57), then we can question the generality of the form of the equation. Indeed, the form is selected in the present text only for convenience, and other forms may be more suitable for some material systems. The form shown in (6.57) does offer some convenient features. If the initial fracture strength of the material is defined in the first cycle (or quarter cycle) of loading, as we normally would do for an engineering test of strength, then we would expect a value of A in Equation (6.57) that is close to unity. And if the data for the *S–N* curve of a material plot as a straight line on semi-log coordinates, then we would expect the power of the second argument, p, to be close to unity also. In fact, these two values are close to unity for many materials and material systems.

Figure 6.23 shows a typical example of such a representation. The value of A is about 1.04, and the value of p is 1.24. The regression fit to the data results in a nearly linear plot. The data are spread over nearly a decade of scatter, which is also typical. And the slope of the curve, represented by B in (6.57), is about 0.045. More will be

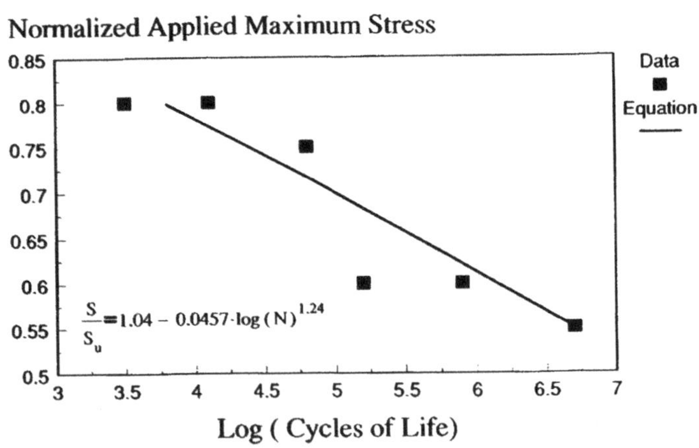

Figure 6.23 Fatigue life data versus maximum applied stress divided by initial fracture strength for fully reversed cyclic loading at 10 Hz of center-notched graphite–epoxy coupons (loading adjusted so that fraction of ultimate strength in tension and compression were the same).

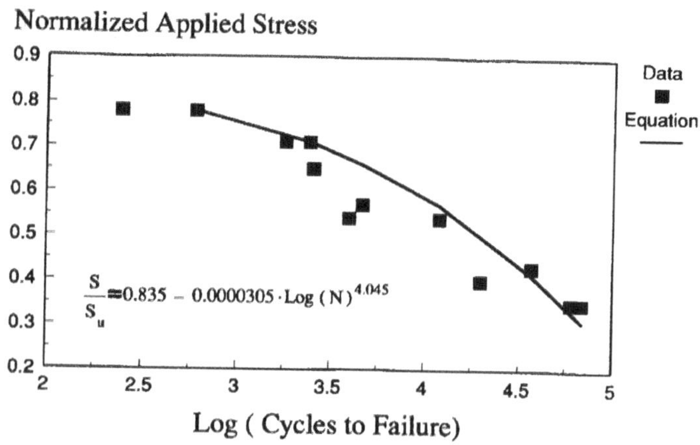

Figure 6.24 Fatigue data for fully reversed loading of Nicalon-reinforced SiC plain-weave composites tested at 1000°C.

said of the relationship of B to material parameters later, but this value is also typical. Interestedly, Figure 6.23 represents data for a notched specimen. This example was chosen to illustrate the utility and generality of the form of (6.57) for notched or unnotched specimens.

Figure 6.24 shows a contrasting representation of fatigue behavior for a ceramic composite at 1000°C. The slope of the fatigue curve is much greater than that of the material shown in Figure 6.23. The S–N curve is also not a straight line on the log-linear plot. The value of A is still fairly close to unity, having a value of 0.835 for the regression fit to the data. But the power of the logarithmic function is 4.045, a very large number and an unusually large value. The coefficient of the second term in the equation is small for this case [primarily because of the large power of $\log(N)$], having a value of 0.0000305. This illustrates several features of (6.43). It can represent distinctly nonlinear S–N behavior (on a set of linear axes, the data in Figure 6.24 drop very sharply for short-life, high-stress conditions). The value of B is greatly affected by the power p. And the value of A can vary noticeably around the value of unity. Nevertheless, with the normalized representation given in (6.43), the values of A, B, and p are generally very well bounded and range over a modest span of values, making it a very convenient representation of fatigue behavior.

Figure 6.25 shows a variation on the representation in (6.43). In this case, the material is carbon-fiber-reinforced PEEK, and the S–N curve demonstrates two distinct regions, a bilinear behavior. Although this type of behavior is somewhat unusual, it demonstrates a very important fact. The slope of the S–N curve reflects the micromechanical process that degrades the material as a function of cycles of loading. If the mechanism changes as a function of load level (or as a function of other extensive variables like temperature), the slope will reflect the change. If this happens as a function of load level, then the S–N curve may have several distinct regions. The material constant B in Equation (6.57) will also be a function of the

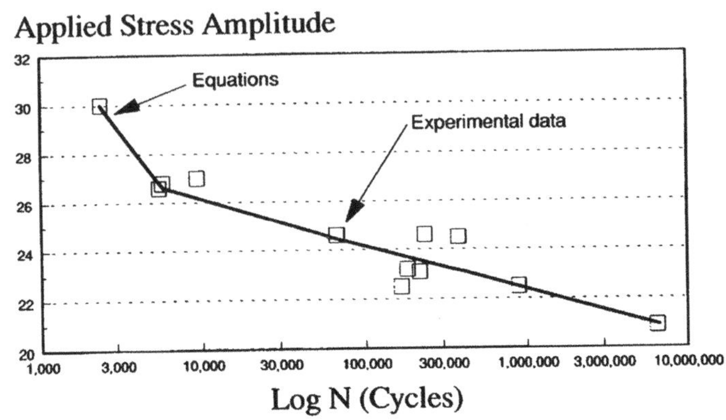

Figure 6.25 Tensile S–N behavior ($R = 0.1$) for carbon-reinforced PEEK composites showing bilinear behavior.

extensive variable, but most often it will be constant within certain ranges and change discontinuously between those ranges, as shown in Figure 6.25.

This brings us to the question of the generality of the slope B. "Universal" values of B have been proposed for various situations. Mandel (1991) noticed that a universal value of B was about -0.1 for composites reinforced with glass fibers. Figure 6.26 shows his plot of supporting data.

The results are quite remarkable. Glass-fiber composites with epoxy, toughened epoxy, polyester matrix materials, and strands without any matrix material show fatigue behavior that fits the form of (6.57) with $A = 1$, $p = 1$, and $B = -0.1$ with good fidelity, even though their static strengths vary by nearly an order of magnitude. Generally, B values between about -0.014 and -0.1 are common. It is the authors' experience that many engineering materials have S–N curves with values of B between these numbers if $p = 1$ in (6.57).

The last remaining major feature of (6.57) that is essential to the representation of fatigue behavior is the ratio of applied (maximum) stress (or stress amplitude for $R = 0$) divided by the ultimate strength of the material in that direction of loading. There are at least two fundamental questions associated with that form.

1. Can the equation be used if the stress is multiaxial; that is, can that ratio be replaced by a suitable failure function (as described in Chapter 2)?
2. As mentioned in the beginning of this chapter and in Chapter 3, this equation (and all such rate equations) are to be applied to the critical element, which may be in a substantially different stress state than the ones used to achieve the characterization represented by (6.57) from laboratory data. Is it valid to use the representation (and that equation) for a critical element that may be a highly constrained small volume of material in the interior of the specimen (or component)?

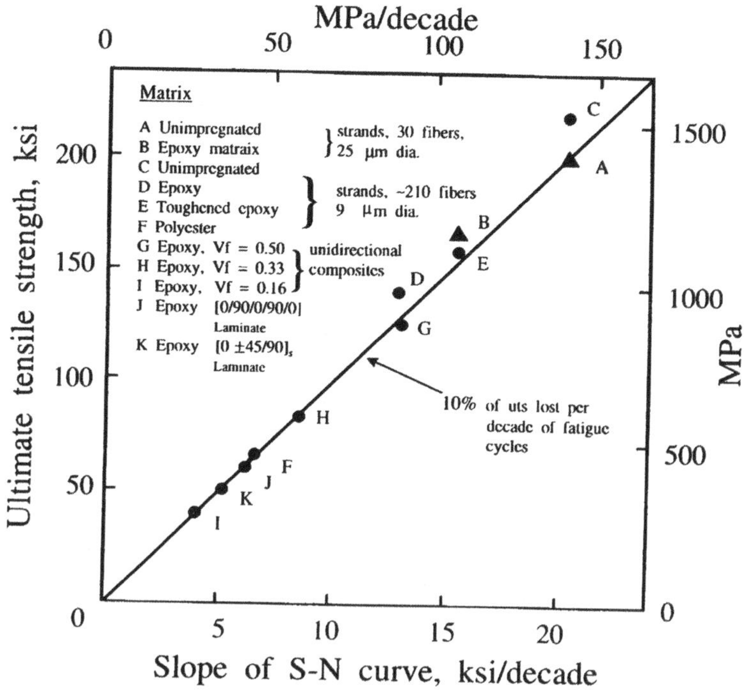

Figure 6.26 Tensile strength versus slope of S–N curve for composites and strands dominated by glass fibers (Mandel, 1991).

To answer the first question, we notice that considerable evidence exists in the literature that supports the idea that it is possible to use a generalized failure function to represent the loading when the resulting fatigue life data are presented by an equation such as (6.57). A substantial amount of early work on this problem has been done (Hashin and Rotem, 1973; Rotem 1979; Rotem and Hashin, 1975). Rotem (1979) discusses applying a fatigue function in the plies of a laminate using the stresses in the coordinate system of the material in the plies. He proposes several such functions, including one for the fiber failure mode and one for matrix-controlled failure. Hashin used a form that is a simple rearrangement of (6.57) of the type

$$X_{ij}^{x} = \left(\frac{\sigma_{ij}}{k_{ij}}\right) f_A(R, N, c, \theta \qquad 6.58)$$

in which σ_{ij} is the local ply stress in the principal axes of that ply, X_{ij}^{x} are the lamina material strength(s) (written in the global system), k_{ij} is the array of ratios that relate the ply stresses in the global system to the local coordinates, R is the stress ratio, N is the cycles to failure, c is a material constant, and θ is the orientation of the ply in question. In some cases, the fatigue function f_A was found to be independent of the

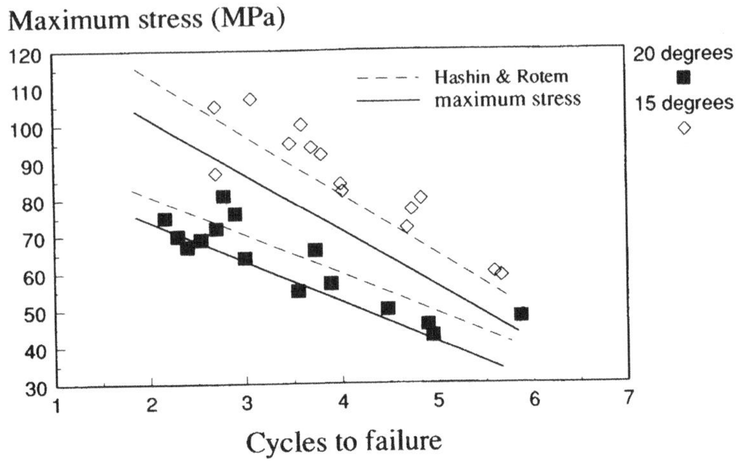

Figure 6.27 Use of failure functions to represent fatigue response of off-axis laminates (Hahn, 1978).

angle θ. Rotem compares such a representation with numerous data sets, with good agreement.

Hahn (1978) makes similar observations using several failure functions for comparison. Figure 6.27 shows two such comparisons using data presented by Hahn for a maximum stress criterion and the Hashin–Rotem (1973) approach. A somewhat different but related approach is suggested by Sims and Brogdon (1977). They define fatigue functions for cyclic loading of several generic types for a given lamina and material system; that is, they use experimental data for (cyclic) longitudinal, transverse, and shear loading of a lamina to define fatigue response functions of the form

$$S = a + \frac{b}{N^x} - \frac{c}{A^y} \tag{6.59}$$

where a, b, c, x, and y are material constants, N is the number of cycles to failure, and A is the ratio of the oscillatory stress amplitude to the mean stress. Then they use these expressions in an appropriate failure function in the form

$$\frac{K_1^2}{\sigma_L^2} - \frac{K_1K_2}{\sigma_L^2} + \frac{K_2^2}{\sigma_t^2} + \frac{K_{12}^2}{\sigma_S^2} = \frac{1}{\sigma_F^2} \tag{6.60}$$

where K_i are the ratios of the local fiber system stresses to the applied stress (assumed to be uniaxial); σ_i are the fatigue functions for longitudinal, transverse, and shear loading [in the form of (6.59)]; and σ_F is the laminate fatigue strength. Implicit in this and the other approaches discussed is the assumption that stresses, strengths, and fatigue behavior are characterized in individual plies of a laminate, even though the basic data for the failure functions (6.59) or their equivalents for the other formulations were obtained for cyclic loading of a lamina with a simpler stress state. The extensive data and comparisons that appear in the literature (including those cited above) suggest that such an approach is viable for the brittle-matrix, continuous-fiber composites tested to date. However, the reader should view this feature of the approach as a working assumption.

In the context of the modeling approach which forms the outline for this text, we must ask if the *S–N* equations, such as (6.57), that are obtained from laboratory experiments for special stress states can be used to represent the fatigue response of individual plies in a laminate, in which they may be subjected to more complex (local) conditions. We have shown evidence that this is possible if the failure function is chosen to represent the correct failure mode (rather than damage mode). Good experimental support exists for the idea of using such a representation when the failure mode is fiber controlled. We will show a variety of examples in Chapter 8 on applications. For this chapter, one important experimental investigation of this question will be examined.

The experiment was specifically concerned with two issues. The first was the one at hand. Can fatigue data from uniaxial loading of unidirectional specimens be used to anticipate the *S–N* behavior of multiaxial laminates under uniaxial loading, for which one ply orientation of the laminate dominates the determination of strength under that loading (a common situation)? Second (to be discussed in Chapter 8 on applications), if we use stiffness change data recovered during the cyclic loading of such a laminate, can these data be used as time-resolved input into the determination of local stresses in Equations (6.1) and (6.57) to continuously improve the prediction of remaining strength and life for a specific specimen under observation? The results of the experiments suggested that the answer to both questions is yes. Let us consider the results of study of the first issue.

Halverson (1996) describes a methodology for determining an intrinsic *S–N* curve for a unidirectional material loaded uniaxially. This is done by using time-resolved measurements to stiffness change during testing of specific specimen to determine the local stress on the unbroken fiber bundles in the material, which act as the critical elements. The logic of this is based on the observation of fiber bundles splitting off as they break, in direct correlation with observed stiffness changes, so that the changes are directly related to a reduction in cross-sectional area. A static distribution of strength is also measured for the coupons. Then, for each specimen tested, (6.57) is used in (6.1) to predict the life of the specific specimen, with only A and B undetermined in (6.57). [The parameter p in Equation (6.57) was taken to have the value of unity.] After testing many such specimens, a least-squares fit was used to select values of A and B that resulted in a distribution of strengths [in the denominator of Equation (6.57)] that matched the observed static distribution. In this way, values of A and B for the critical element were determined. The results of the tests are shown in Figure 6.28. As one might expect, the intrinsic *S–N* curve has higher fatigue strengths than the coupons. However, the intrinsic *S–N* curve also has a larger slope; that is, the critical elements show more fatigue sensitivity than the coupons.

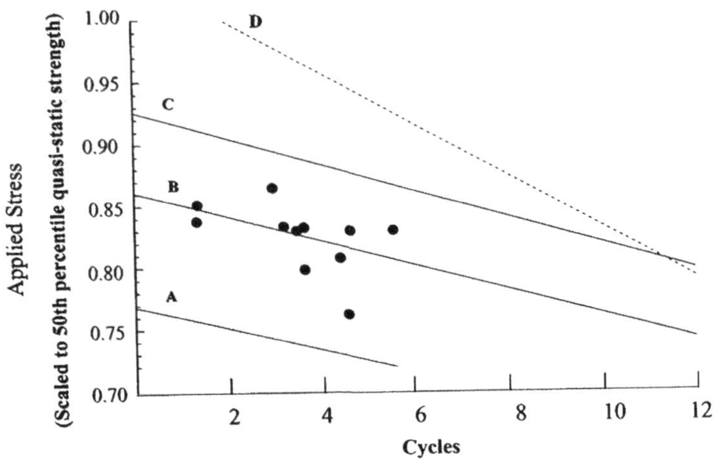

Figure 6.28 Results of unidirectional fatigue tests in standard *S–N* format. Line *B* corresponds to measured, best-fit *S–N* curve of coupons. Lines *A* and *C* are the 10th and 90th percentile initial-strength predictions, respectively. Line *D* is derived intrinsic *S–N* curve of critical elements, unbroken fiber bundles in this case (Halverson (1996)). (Applied stress is scaled to 50th percentile quasi-static strength.)

Halverson (1996) validates the intrinsic *S–N* curve concept by applying the determined relationship to the prediction of the fatigue lives of cross-ply specimens of the same material using the 0° plies as the critical elements. Again, measurements of stiffness change for individual specimens in real time were used to be sure the local stresses were correct in (6.57) and (6.1). The resulting predictions for each such specimen tested are shown in Figure 6.29.

The results are remarkable in the sense that although the applied loading to several of the specimens was the same (although the stresses differed slightly because of different cross-sectional areas), the fatigue lives of these specimens sometimes differed by as much as two orders of magnitude. (Actually, this is not unusual for fatigue life distributions.) However, using the intrinsic *S–N* curve and specific measurements of stiffness change for each specimen, the predicted 10–90 band of life includes the observed life data in all but one case. Halverson also compares predicted results if the simple *S–N* curve of the coupons is used. Using the intrinsic *S–N* curve gives superior results; using the coupon *S–N* curve produced 10–90 bands that did not include several measured values.

The point to be taken from this discussion is that *S–N* curves determined from uniaxial loading of unidirectional specimens can be used to predict the fatigue response of laminates. We may also conclude that significantly improved results can be obtained if an intrinsic *S–N* curve of the critical elements can be determined. Finally, if real-time measurements can be made to continuously update the inputs to (6.1) during the test of a specific specimen or component, predictions of remaining strength for a specific case can be made; these predictions then reflect statistical

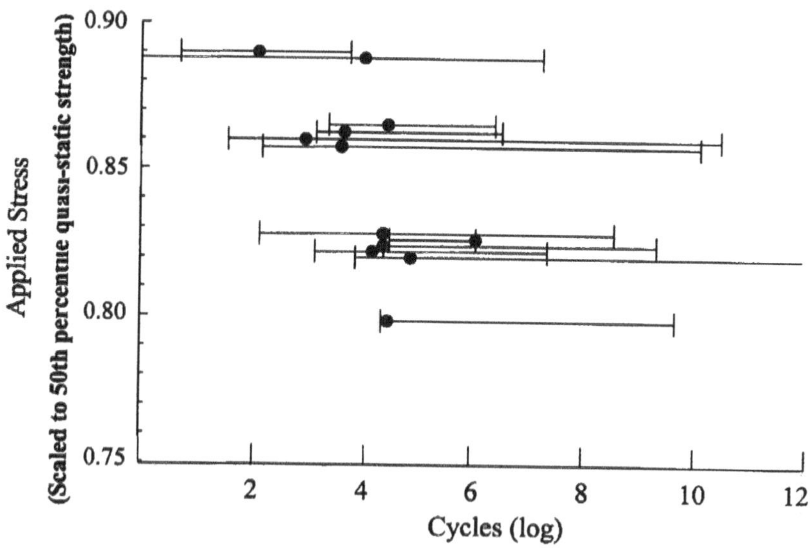

Figure 6.29 Predicted 10–90 probability bounds for fatigue life of cross-ply laminates using intrinsic S–N curve for 0° plies compared to measured fatigue lives. (Applied stress is scaled to 50th percentile quasi-static strength.)

differences in the initial strength and the damage process (or even loading history) of a given specimen and are significantly more accurate than those obtained from average values of behavior. This last concept is an analog to measuring crack length in a homogeneous material as continuous input to a crack growth law.

6.3.6 Other Approaches

Numerous other modeling approaches to the description of creep and stress rupture are available. Virtually all of these focus on strain and strain rate as a function of time. However, the degradation of the material strength is the primary concern of our present approach to the general subject of durability. It is difficult to find models of that behavior. Nevertheless, the time and temperature-dependent material strength variations of the type discussed above can be used directly to specify $X_{ij}(t)$ in the failure function $\mathrm{Fa}(\sigma_{ij}(t)/X_{ij}(t))$ that appears in the evolution equation we wish to evaluate to determine the remaining strength of a component. Moreover, the allowable deformation models and time-to-failure models can serve the same purpose indirectly by specifying the ending point of a decrease in material strength, although they do not provide specific information about the intermediate values. In addition, the micromechanical methods relate the details of how the composites are made to the resulting creep and stress rupture behavior. To illustrate how these results may be used in the present approach, an example follows.

6.4 APPLICATION EXAMPLE

6.4.1 Problem

We know that the strength of a material is a function of temperature and that the design application of the material for a specific component involves variations of temperature that correspond to as much as a 20% change in the material strength. How much would this strength change influence the life of the component?

6.4.2 Solution

We recall from Chapter 3 that we can calculate the remaining strength of our component using the equation

$$\mathrm{Fr} = 1 - \int \left[1 - \mathrm{Fa}\left(\frac{\sigma_{ij}(t)}{X_{ij}(t)}\right)\right] j \left(\frac{vt}{N(t)}\right)^{j-1} d\left(\frac{vt}{N(t)}\right) \tag{6.61}$$

For the situation in question, we assume that the frequency of the load application is 10 cycles per second; that is, that $v = 10$. Our fatigue data provide the relationship

$$\mathrm{Fa} = A + B[\log(N)]^{p} = 1 - 0.1[\log(N)]^{1.0} = f(t) \tag{6.62}$$

The initial value of Fa for the applied stress level is found to be 0.65; that is, the applied loading results in a stress in the component that is initially 65% of the level of the material strength. From (6.62), the life of the component under such conditions is about 3162 cycles, or about 312 s at 10 Hz. Equation (6.61) can be used to calculate the change in strength for this hypothetical situation in which Fa is constant, with a value of 0.65. Figure 6.30 shows the first 10 s of life for this case, which results in a drop of component strength of less than 1%.

However, we know that the temperature changes during the test (or during service) such that a maximum of 20% change in the material strength results. Let us assume

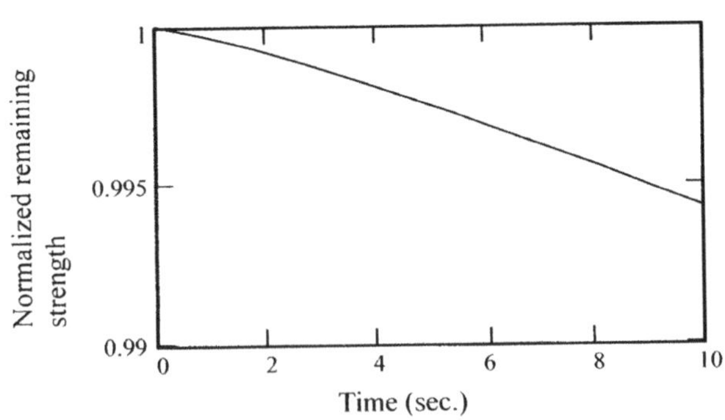

Figure 6.30 Remaining strength for first 10 s of life if initial conditions are constant [from (6.61)].

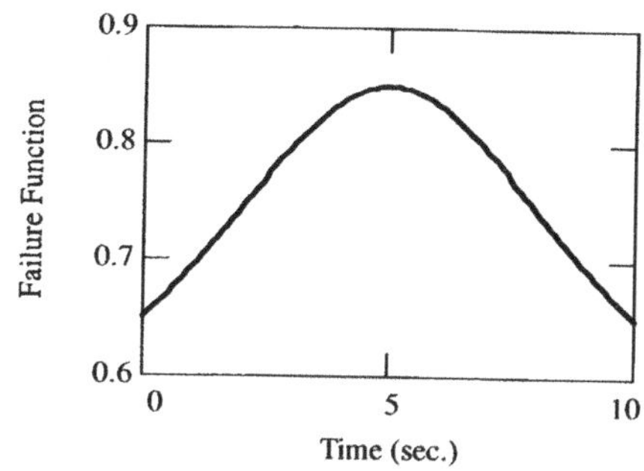

Figure 6.31 Change of local failure function Fa [represented as $f(t)$] caused by change in temperature.

that the local stress remains constant during this change, so that Fa [or actually $f(t)$] changes by the same 20%. Hence, we take $f(t)$ in (6.62) to have the form

$$f(t) = 0.65 + 0.8\left(\frac{0.1t - 0.01t^2}{2 - 4(0.1t - 0.01t^2)}\right) \qquad (6.63)$$

which has the shape shown in Figure 6.31.

As specified in the problem, the strength decreases by 20% during this period, causing an increase in Fa of the same amount if the stresses are constant. Then using (6.63) in (6.62), we see that the instantaneous life of the material element being considered is also a function of time, that is, $N \rightarrow N(t)$. Putting this form into (6.61) (neglecting derivatives of $N(t)$) and evaluating the integral for the first 10 s of application give the result shown in Figure 6.32. The result is quite different than that obtained for the case of constant temperature. From Figure 6.32 we see that the temperature variation that occurred over just 10 s (of the expected life of over 300 s) causes the component to degrade significantly in that period, even though there was only a 20% increase in the local failure function (from 0.65 to 0.85). (We remember that failure is predicted when the remaining strength falls below the level of applied

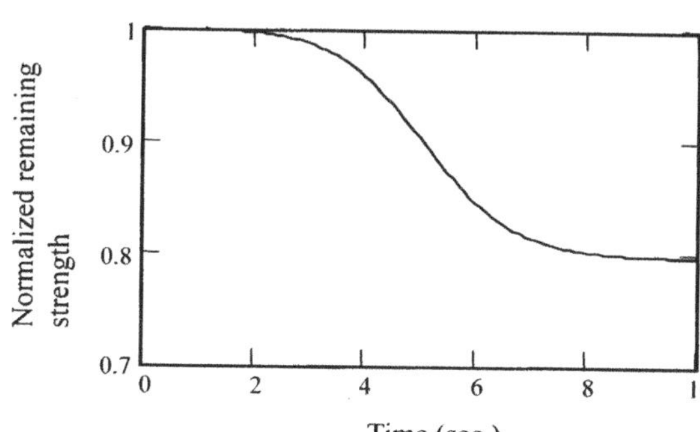

Figure 6.32 Change of component remaining strength caused by change in temperature and induced change in local failure function shown in Fig. 6.31.

loading, 0.65 in this case.) While the data in this example are hypothetical, they are all typical of high-temperature materials and applications, and the result is also typical. Small variations of temperature can, and do, have a large effect on the life of components controlled by creep or stress rupture. Chapter 8, which discusses various example applications, will address these ideas.

REFERENCES

Bernatz, K. M., Giri, L., Simon, S. L., and Plazek, D. J. (1999), "Evolution of Viscosity and Steady State Compliance During Physical Aging by Creep and Recovery Experiments," *J. Chem. Phys.* Vol. 111, No. 5, pp. 2235–2241.

Bradshaw, R. D., and Brinson, L.C. (1997), "Physical Aging in Polymers and Polymer Composites: An Analysis and Method for Time-Aging Time Superposition," *Polym. Eng. Sc.* Vol. 37, pp. 31–44.

Browning, C. E. (1972), "The Effects of Moisture on the Properties of High Performance Structural Resins and Composites," Technical Report AFML-TR-72-94, Air Force Materilas Laboratory, Air Force Systems Command, Wright-Patterson AFB, Dayton, OH, September 1972.

Browning, C. E., Husman, G. E., and Whitney, J.M. (1976), "Moisture Effects in Epoxy Matrix Composites," in *Composite Materials: Testing and Design*, ASTM STP 617, American Society for Testing and Materials, Philadelphia.

Cannon, W. R. and Langdon, T. G. (1983), "Creep of Ceramics; Part 1: Mechanical Characteristics; Part 2: An Examination of Flow Mechanisms," *J. Mater. Sci.* Vol.18, pp. 1–50.

Chen, J. L., and Sun, C. T. (1993), "Plastic Potential Function Suitable for Anisotropic Fiber Composites," *J. Composite Materials*, Vol. 27, No. 14, pp. 1379–1390.

Chiao, T. T., and Moore, R.L. (1986), "Fiber/Epoxy Filament Wound Vessels; Final report," UCID-20845, Lawrence Livermore National Laboratory, Livermore, CA, August 1986.

Christensen, R. M. (1981), "Lifetime Predictions for Polymers and Composites under Constant Load," *J. Rheol.* Vol. 25, No. 5, pp. 517–528.

Corman, G. S. (1990), "Creep of Oxide Single Crystals," WRDC-TR-90-4059, Air Force Materials Laboratory, AFWL, Wright Patterson AFB, OH.

Curtin, W. A. (1991), "Theory of the Mechanical Priorities of Ceramic Matrix Composites," *J. Am. Ceramic Soc.*, Vol. 74, p. 2837.

Curtin, W. A., Fabeny, B., Ibnabkeljalil, M., Iyengar, N. and Reifsnider, K.L. (1996), "Long-Term Performance of Ceramic Matrix Composites at Elevated Temperatures: Modeling of Creep and Creep Rupture," Subcontractors report, Subcontract 19X-SA946C, to Oak Ridge National Laboratory, Oak Ridge, TN.

Dean, D., Miyase, A., Geil, P. H. (1992), "Interaction of Creep and Physical Aging in a Semicrystalline Thermoplastic Matrix Composite above Tg," *Journal of Thermoplastic Composite Materials*, Vol. 5, No.2, pp. 136–151.

Dharan, C. H. (1975). "Fatigue Failure in Graphite Fiber and Glass Fiber Polymer Composites," *J. Mater. Sci.*, Vol. 10, No.10, pp. 1665–1670.

Ding, J. L., Liu, K. C., and Brinkman, C. R. (1994), "A Phenomenological High-Temperature Deformation and Life Prediction Model for Advanced Silicon Nitride Ceramics," in *Durability and Damage Tolerance*, A. K. Noor and K. L. Reifsnider, Eds., AD Vol. 43, American Society of Mechanical Engineers, New York.

Du, Z. Z., and McMeeking, R. M. (1995) "Creep Models for Metal Matrix Composites with Long Brittle Fibers," *J. Mech. Phys. Solids*, Vol. 43, No. 5, pp. 701–726.

Emanuel, N. M., and Buchachenko, A. L. (1987), *Chemical Physics of Polymer Degradation and Stabilization*, VNU Science Press, Utrecht, The Netherlands.

Evans, R. W., Willis, M. R., Wilshire, B., Holdsworth, S., Senior, B., Fleming, A., Spindler, M., and Williams, J. A. (1993), "Prediction of Long-Term Creep and Creep Fracture Data for A 1CrMoV Rotor Steel," in *Creep and Fracture of Engineering Materials and Structures*, B. Wilshire and R. W. Evans, Eds., Bourne Press, Bournemouth, United Kingdom, pp. 633–640.

Frety, N. and Boussuge, M. (1990), "Relationship Between High-Temperature Development of Fibre-Matrix Interfaces and the Mechanical Behaviour of SiC–SiC Composites," *Composites Science and Technology*, Vol. 37, pp. 177–189.

Glasstone, S., Laidler, K. J., and Eyring, H. (1941), *Theory of Rate Processes*, McGraw-Hill, New York.

Goldhoff, R. (1959), *Mater. Design Eng.* Vol. 49, No. 4, pp. 93–97.

Ha, S. K., and Springer, G. S. (1987), "Mechanical Properties of Graphite Epoxy Composites at Elevated Temperatures," in *Proc ICCM & ECCM, 2nd European Conference on Composite Materials*, Vol. 4, F. L. Mathews, N. C. R. Buskell, J. M. Hodgkinson, and J. Morton, Eds., Elsevier Applied Science, London.

Hahn, H. T. (1978), "Fatigue Behavior and Life Prediction of Composite Laminates," AFML-TR-78-43, Air Force Materials Laboratory, Wright-Patterson AFB, Dayton, OH.

Hahn, H. T., and Chiao, T. T. (1980), "Long-Term Behavior of Composite Laminates," in *Advances in Composite Materials*, A. R. Bunsell et al., Eds., Pergamon, New York, pp. 584–596.

Hahn, H. T. and Kim, R. Y. (1976), "Fatigue Behavior of Composite Laminates," *J. Comp. Mater.* Vol. 10, pp. 156–180.

Halverson, H. (1996), "Improving Fatigue Life Predictions: Theory and Experiment on Unidirectional and Crossply Polymer Matrix Composites," M.S. Thesis, Department of Engineering Science and Mechanics, College of Engineering, Virginia Polytechnic Institute and State University.

Hashin, Z., and Rotem, A. (1973), "A Fatigue Failure Criterion for Fibre-Reinforced Materials," *J. Composite Mater.* Vol. 7, pp. 448–464.

Hayhurst, D. R. (1972), "Creep Rupture under Multiaxial States of Stress," *J. Mech. Phys. Sol.*, Vol. 20, p. 381.

Highsmith, A. L., and Reifsnider, K. L. (1982), "Stiffness-Reduction Mechanisms in Composite Laminates," in *Damage in Composite Materials*, ASTM STP 775, K. L. Reifsnider, Ed., American Society for Testing and Materials, Philadelphia, pp. 103–117.

Kerr, J. R., and Haskins, J. F. (1987), "Time-Temperature-Stress Capabilities of Composite Materials for Advanced Supersonic Technology Application," NASA Contractor Report 178272, Langley Research Center, Hampton, VA.

Lange, F. F. (1976), "Interrelations between Creep and Slow Crack Growth for Tensile Loading Conditions," *Int. J. Fracture*, Vol. 12, No. 5, pp. 739–744.

Larson, F .R., and Miller, F. (1952), "A Time-Temperature Relationship for Rupture and Creep Stresses," *ASME Trans.* Vol. 74, pp. 765–771.

Leckie, F. A., and Hayhurst, D. R. (1978), "Creep Rupture," in *Creep of Engineering Materials*, C. D. Pomeroy, Ed., Mechanical Engineering Publications, London, pp. 111–126.

Mandel, J. F. (1982), in *Developments in Reinforced Plastics-2*, G. Pritchard, Ed., Applied Science, New York, p. 67.

Mandel, J. F. (1991), "Fatigue Behavior of Short Fiber Composite Materials," in *Fatigue of Composite Materials*, Elsevier Science, New York.

Manson, S. S., and Haferd, A. M. (1953), "A Linear Time-Temperature Relation for Extrapolation of Creep and Stress-Rupture Data," Nat. Advisory Comm. for Aeronautics Technical Note 2890, Washington, DC.

McClintock, F. A., and Argon, A.S. (1966), *Mechanical Behavior of Materials*, Addison-Wesley, Ontario, Canada.

McLean, M. (1985), "Creep Deformation of Metal Matrix Composites," *Composites Sci. Technol.* Vol. 23, p. 37.

McLean, M. (1989), "Creep Deformation of Metal Matrix Composites," in *Materials and Engineering Design: The Next Decade*, B. F. Dayson and D. R. Hayhurst, Eds., Institute of Metals, London, p. 287.

Mura, T. (1987), *Micromechanics of Defects in Solids*, Martinus Nijhoff, Boston.

Nichols, M. E., Wang, S. S., and Geil, P. H. (1990), "Creep and Physical Aging in a Polyamideimide Carbon-Fiber Composite," *Journal of Macromolecular Science-Physics*, Vol. B29, No.4, pp. 303–336.

Norton, F. H. (1929), *The Creep of Steel at High Temperatures*, McGraw-Hill, London.

Razvan, A., Bakis, C. E., and Reifsnider, K. L. (1990), "S.E.M. Investigation of Fiber Fracture in Composite Laminates," *Mater. Characterization*, Vol. 24. No. 2, pp. 79–190.

Razvan, A., and Reifsnider, K. L. (1991), "Fiber Fracture and Strength Degradation in Unidirectional Graphite/Epoxy Composite Materials," *Theory Appl. Fracture Mech.* Vol. 16, No. 1, pp. 81–89.

Read, B. E., Dean, G. D., and Tomlins, P. E. (1988), "Effects of Physical Aging on Creep in Polypropylene," *Polymer*, Vol. 29, pp. 2159–2169.

Reifsnider, K. L. (1991), *Fatigue of Composite Materials*, Elsevier Science, New York.

Richards, C. W. (1961), *Engineering Materials Science*, Wadsworth, San Francisco, California.

Rotem, A. (1979), "Fatigue Failure of Multidirectional Laminates," *AIAA J.*, Vol. 17, No. 3, pp. 271–277.

Rotem, A., and Hashin, Z. (1975), "Fatigue Failure of Laminates," TAU-SOE/159-75, Department of Solid Mechanics, Materials, and Structures, Tel-Aviv University, Israel.

Schapery, R. A. (1974), "Viscolestic Behavior and Analysis of Composite Materials," *Mech. Composite Mater*, Vol. 2, pp. 85–168.

Simon, G. and Bunsell, A. R. (1984), "Mechanical and Structural Characterization of the Nicalon Silicon Carbide Fibre," *Journal of Materials Science*, Vol. 19, pp. 3649–3657.

Sims, D. F., and Brogdon, V. H. (1977), "Fatigue Behavior of Composites under Different Loading Modes," in *Fatigue of Filamentary Composite Materials*, ASTM STP 636, K. L. Reifsnider and K. N. Lauraitis, Eds., American Society for Testing and Materials, Philadelphia, pp. 185–205.

Springer, G. S. and Shen, C. H. (1977), "Effects of Moisture and Temperature on the Tensile Strength of Composite Materials," AFML-TR-77-82, Air Force Materials Laboratory, AFWL, Wright Patterson AFB, Dayton, OH.

Struik, L. C. E., (1978), *Physical Aging in Amorphous Polymers and Other Materials*, Elsevier Scientific, Amsterdam-Oxford-New York.

Struik, L. C. E. (1987), "The Mechanical and Physical Aging of Semicrystalline Polymers, 1–2," *Polymer*, Vol. 28, pp. 1521–1533, 1534–1542.

Sun, C. T., and Chen, J. L. (1991), "A Micromechanical Model for Plastic Behavior of Fibrous Composites," Composites and Technology, pp. 115–129.

Townley, C. H. A. (1992), "High Temperature Design Procedures—Developments in Europe Since 1986," in *Creep Characterization, Damage, and Life Assessments: Proceedings of the Fifth International Conference on Creep of Materials*, Buena Vista, 18–21 May, D. A. Woodford, C. H. A. Townley and M. Ohnami, Eds., ASM International, Materials Park ,OH, pp. 1–11.

Wang, J. Z., Parvatareddy, H., Chang, T., Iyengar, N., Dillard, D. A., Reifsnider, K. L. (1995), "Physical Aging Behavior of High-Performance Composites," *Composites Science and Technology*, Vol. 54, pp. 405–415.

Weber, C. H., Du, Z. Z., and Zok, F. W. (1996), "High Temperature Deformation and Fracture of a Fibre-Reinforced Titanium Matrix Composite," *Acta Metal.*, Vol. 44, No. 2, pp. 683–695.

Whitney, J. M. (1979), "Fatigue Characterization of Composite Materials," AFML-TR-79-4111, Mechanics and Surface Interactions Branch, Air Force Wright Aeronautical Laboratories, Wright-Patterson AFB, Dayton, OH.

Woodford, D. A. (1992), "Creep Strength and Creep Damage Considerations for Life Cycle Engineering," in *Proceedings of the Fifth International Conference on Creep of Materials*, ASM International, OH, pp. 23–32.

Woodford, D. A., Townley, C. H. A., and Ohnami, M., Eds., (1992), *Proceedings of the Fifth International Conference on Creep of Materials*, ASM International, OH.

Yang, J. N. and Liu, M. D., (1976), "Residual Strength Degradation Model and Theory of Periodic Proof Tests for Graphite Epoxy Laminates," AFML-TR-76-225, Contractors Report, Air Force Materials Laboratory, Air Force Wright Aeronautical Laboratories, Wright Patterson, AFB, Ohio.

Zhurkov, S. N. (1965), *Int. J. Fracture Mech.*, Vol. 1, p. 311.

Zhurkov, S. N., and Korsukov, V. E. (1974), Polymer Physics Edition, "Atomic Mechanism of Fracture of Solid Polymers," *J. Polym. Sci.*, Pt. A-2, Vol. 12, No. 2, pp. 385–398.

EXERCISES

6.1 The time to rupture data, t_r, for the stress rupture of Ni/calcium–alumino–silicate (CAS) for three temperatures (in absolute) are given below as a function of the applied stress normalized by the initial room-temperature ultimate strength, Fa^0, where these times to rupture are in hours

log (t_r, 1255)	log (t_r, 1000)	log (t_r, 800)	Fa^0
2.577	4.184	6.587	0.4
2.362	3.835	6.038	0.45
2.147	3.487	5.489	0.5
1.933	3.138	4.94	0.55
1.718	2.789	4.391	0.6
1.503	2.441	3.842	0.65
1.288	2.092	3.293	0.7
1.074	1.743	2.745	0.75
0.859	1.395	2.196	0.8
0.644	1.046	1.647	0.85
0.429	0.697	1.098	0.9
0.215	0.349	0.549	0.95
0	0	0	1

(a) Determine the material constants if a Zhurkov equation is used to fit those data, that is, to determine A, B, and C in the equation having the generalized form

$$\log(t_r) = A + \left[\frac{B - C\,\mathrm{Fa}^0}{T}\right]$$

in which Boltzmann's constant has been absorbed into B and C and the temperature is in degrees absolute. Determine the constants for all three temperatures:

1255: $A =$ ________ $B =$ ________ $C =$ ________
1000: $A =$ ________ $B =$ ________ $C =$ ________
800: $A =$ ________ $B =$ ________ $C =$ ________

(b) Determine the material constants for the Manson–Haferd representation in the form

$$\frac{T - T_0}{\log(t_r) - \log(t_{r_0})} = 1 - A\,\mathrm{Fa}^0$$

1255: $T_0 =$ ________ $\log(t_{r_0}) =$ ________ $A =$ ________
1000: $T_0 =$ ________ $\log(t_{r_0}) =$ ________ $A =$ ________
800: $T_0 =$ ________ $\log(t_{r_0}) =$ ________ $A =$ ________

(c) We do not know how the material strength is decreasing with time for any given time to rupture, t_r. In the absence of such data, let us assume that $X(t) = X^0\,(1 - t/t_r(\mathrm{Fa}, T)$ that is, that the material strength varies linearly over the calculated life. Let us also assume that the room-temperature fatigue behavior of the material follows the relationship $\mathrm{Fa} = 1 - 0.05\log(N)$ for a frequency of loading of one cycle per second. Then in the strength evolution integral (copied below), $\mathrm{Fa}(t)$ would have the form

$$Fa(t, T) = \frac{\mathrm{Fa}^0}{1 - t/t_r(\mathrm{Fa}, T)}$$

where $t_r(\mathrm{Fa}, T)$ is determined from one of the rate equations mentioned above. Using such a relationship;

(i) Calculate the global remaining strength and life for the combined effects of stress rupture and fatigue at 1000 degrees absolute for $\mathrm{Fa}^0 = 0.6$ (one cycle per second loading), and compare it with the stress rupture life, alone, using the Zhurkov equation.

(ii) Repeat (i) using the Manson–Haferd equation.

(iii) Discuss similarities and differences:

$$\mathrm{Fr}(\tau) = 1 - \int [1 - \mathrm{Fa}(t, T)] j \left(\frac{n}{N[\mathrm{Fa}(t, T)]} \right)^{j-1} d\left(\frac{n}{N[\mathrm{Fa}(t, T)]} \right) \qquad j = 1, 2$$

6.2 Examine and compare the sensitivity of several of the micromechanical models of creep and stress rupture to the physical parameters that represent the volume fraction of fibers, v_f; ratio of fiber stiffness to matrix stiffness, E_f/E_m; and applied stress level σ. Models to be considered include McLean's representation [Equation (6.43)], and the Curtin–McLean expression [Equation (6.50)].

6.3 The following set of data were recorded by Yang and Liu (1976) from room-temperature fatigue ($R = 0.1$) of quasi-isotropic graphite–epoxy laminates.

Specimen Number	Maximum Stress Level (Percent of Average Static Strength)	Number of Cycles to Fracture	Residual Strength (Percent of Average Static Strength)
1	85	1650	Fatigue failure
2	85	1950	Fatigue failure
3	85	1320	Fatigue failure
4	80	2050	Fatigue failure
5	75	50,980	Fatigue failure
6	75	6480	Fatigue failure
7	70	155,000	Fatigue failure
8	70	228,500	Fatigue failure
9	70	88,000	Fatigue failure
10	70	117,580	Fatigue failure
11	70	228,700	Fatigue failure
12	70	221,200	Fatigue failure
13	70	310,000	Fatigue failure
14	70	18,790	Fatigue failure
15	70	3840	Fatigue failure
16	67	161,000	Fatigue failure
17	67	110,000	Fatigue failure
18	65	523,500	Fatigue failure
19	65	863,200	Fatigue failure
20	62	1,346,300	Fatigue failure
21	65	1,007,000	80
22	72.5	30,000	81
23	72.5	30,000	86
24	72.5	30,000	83
25	72.5	30,000	82

(a) For this data set, determine the material parameters B and p in Equation (6.43). How do they compare to the example values given in the text?

(b) Plot Equation (6.43) using your values of A, B, and p and put the experimental values on the same graph for comparison. How well does Equation (6.43) fit the data?

(c) For the sake of illustration, assume that the values of A, B, and p determined in (a) can be used to represent the fatigue response of the critical elements in this laminate. Also assume that no other degradation process is active and enter your representation [Equation (6.43)], into Equation (6.1) to obtain estimates of the remaining strength variation over the life for various applied load levels. Check the predictions against the residual strength data given for tests 21–25. Discuss reasons for similarities and differences between predicted and observed results. Can you suggest methods of improving the agreement with only the data at hand? For example, can you think of a way to estimate the S–N behavior of the 0° plies in this laminate (which are likely to be the critical elements in this case)?

(d) Considering that this is a polymer composite, what other data are likely to be needed to achieve a complete representation of the degradation processes as inputs to Equation (6.1)?

CHAPTER 7

NONUNIFORM STRESS STATES

Thesis: Nonuniform stress states associated with edge effects in laminates and geometric stress concentrations induce multiaxial local stress gradients and consequent damage and failure processes that require the introduction of a characteristic dimension in models of strength. These models enter our methodology as failure functions that represent those failure modes.

7.1 PROBLEM DEFINITION

We have discussed failure modes in terms of failure functions, $\mathrm{Fa}(\sigma_{ij}/X_{ij})$, which have a form that reflects the nature of the failure process. For example, the ratio of fiber direction stress to fiber direction strength in the critical element and a Tsai–Hill function for matrix-controlled failure (as discussed in Chapter 2) are typical failure functions. The failure function is a major feature of the kernel of the strength evolution integral, as discussed in Chapter 3, having the form

$$\mathrm{Fr}(\tau_1) = 1 - \int_0^{\tau_1} \left[1 - \mathrm{Fa}\left(\frac{\sigma_{ij}}{X_{ij}} \right) \right] j\tau^{j-1}\, d\tau \tag{7.1}$$

The focus of this chapter is the form of Fa in this equation when nonuniform stress states introduce multidimensional local stress states (and stress gradients) that cause failure processes and modes that are peculiar to that situation.

The scope of this chapter includes stress states near edges and notches, failure processes and models for these stress states, fracture mechanics and energy methods of representation, and delamination-controlled strength and life. There is a heavy emphasis in the chapter on stress analysis, since obtaining the local stresses in the vicinity of an edge or notch is not a simple matter for a multiaxial laminate, and having the right stress state is the first (and essential) step in calculating correct local values of $\text{Fa}(\sigma_{ij}/X_{ij})$. However, these discussions will only be outlined, since the literature is rich and reasonably mature in this area. The second major emphasis is on the concept of strength in the presence of stress gradients for locally brittle materials that fail by a damage accumulation process. The statistics of this process and the relationship of local damage to global failure require the introduction of a material characteristic length (a material property) for robust representations of the strength of laminae and laminates.

The discussion of fracture mechanics methods will be highly focused since the singular stress field at the tip of a single crack in a highly inhomogeneous material generally controls local flaw growth but rarely controls global failure, per se. However, this analytical approach does have important applications to self-similar flaw growth and is especially useful for the discussion of delamination growth and certain special problems in which crack growth is a rate-controlling feature. In this context, energy methods (especially energy release methods) are also of great utility and will be included in our discussion.

7.2 LAMINATE EDGE-RELATED STRESSES

7.2.1 Straight Edges

We assume that the reader has a working knowledge of classical laminate theory (CLT). Classical laminate theory rests on the same conceptual foundations as plate theory and does not account for interlaminar stresses. For the present discussion, we must understand the origin and nature of such stresses in order to identify the characteristics of the damage and failure process(es) that control strength and the form of $\text{Fa}(\sigma_{ij}/X_{ij})$. Several physical arguments are useful for the development of this understanding. The first is illustrated in Figure 7.1.

In Figure 7.1, the deformation of an angle ply laminae (in which the fibers are oriented at some angle to a unidirectional load axis that defines the x direction) is considered under the influence of a single axial force acting along the x axis. If one considers the relative stiffness of material in the lamina along opposing diagonal directions (through the corners of the coupon, as shown by dotted lines AA and BB), and if the fibers are assumed to be stiffer than the matrix material, then it is clear that the coupon will tend to deform in shear under the simple uniaxial tensile loading, because the change in shape in the matrix-dominated directions, AA, is larger than the change in shape in the fiber-dominated directions, BB, under the same resolved stress in those directions for a lamina. (This is the source of the so-called A_{i6} shear extension coupling in the $ABBD$ matrix discussed in CLT.)

If the two laminae are bonded together to form a laminate, then the shear deformations cannot occur, and the bonded pair (we will assume that there are two

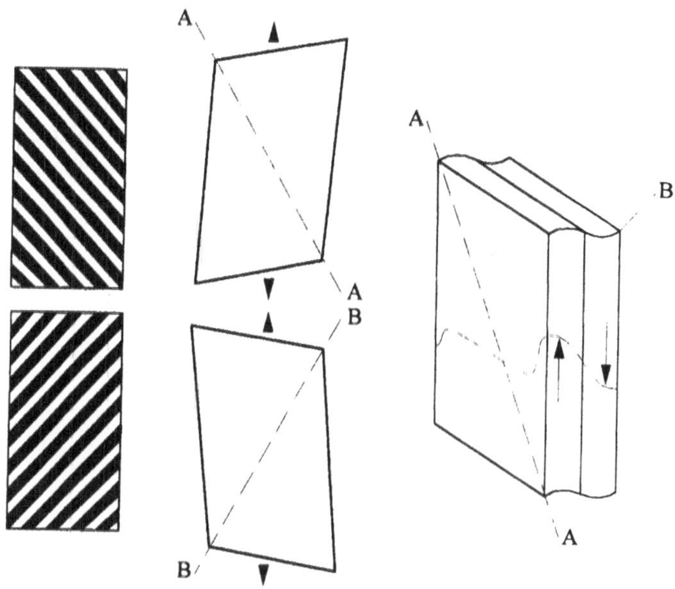

(b) (c)

Figure 7.1 (*a*) Deformation of unbonded angle plies under axial loading; (*b*) bonded laminate having such plies; (*c*) edge deformations when plies are laminated together, caused by relaxation of constraint of each ply on deformation of other at free edge.

symmetric pairs to form a laminate with symmetric stacking sequence) will deform with a simple extensional deformation, as shown in Figure 7.1*b*. However, to conform to the simple extensional deformation, an in-plane shear stress is created in the off axis plies—τ_{xy} in the diagram. But this shear stress cannot exist at the edge of the laminate, since the free edge has no tractions. Hence, the material in those regions tries to deform in the manner shown in Figure 7.1*a* constrained by the material in the interior, which is deforming in simple extension, as in Figure 7.1*b*. Looking at the edge, relative displacements occur in the directions and manner shown in Figure 7.1*c* under that situation. Two aspects of this phenomenon are especially important to the present discussion. First, this relative deformation of the two plies at the edge creates an interlaminar stress, τ_{xz}. Second, this is an "edge effect"; that is, this interlaminar stress disappears in the interior of the specimen where the deformation of the two plies is uniform through the thickness.

Another physical demonstration of these two points is provided by the free-body arguments first advanced by Pagano (1970 a,b) and Pagano and Pipes (1971).

As shown in Figure 7.2*a*, for a $[0, \pm\theta]_S$ laminate, an in-plane loading of the laminate produces in-plane shear stresses τ_{xy} in the angle plies. These stresses, acting on the faces of an element of those plies in the interior, create resultant tractions that are in equilibrium. However, if the free body is moved to the edge, as in Figure 7.2*b*, one component of the stresses on that element must vanish at the free edge, so that the element is no longer in equilibrium. If the interlaminar stresses from Figure 7.1*c* are added to the diagram, as in Figure 7.2*c*, equilibrium can be established by means of those interlaminar stresses acting on the top and bottom surface of the element, near the free edge. Figures 7.1 and 7.2 show that the interlaminar stress τ_{xz} occurs near the edge of a multiaxial laminate and that it is an edge effect caused by the change of local constraint near the edge and does not occur in the interior of the laminate.

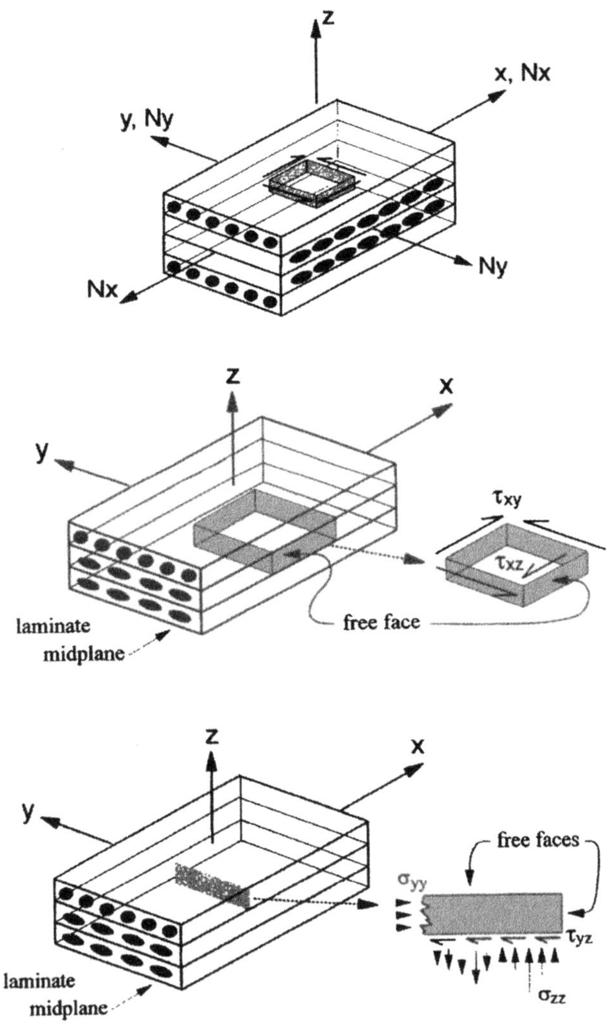

Figure 7.2 (*a*) In-plane loading of $[0, \pm\theta]_S$ laminate; (*b*) free-body diagram of element of one θ lamina near edge of laminate; (*c*) free-body diagram (in *yz* plane) of element of 0° ply near edge of laminate.

Similar arguments can be made for the occurrence of other interlaminar stress components. A free-body diagram of an element from the top (0°) ply, near the laminate edge, drawn in the *yz* plane, is shown in Figure 7.2*c*. The transverse normal stress σ_{yy}, which is nonzero in the interior of the 0° ply (and generally compressive for the laminate we have chosen), must vanish at the free edge of the laminate (another edge effect). For our free-body diagram to be in equilibrium in the *y* direction, there must be shear-stress resultants on the bottom of the element (since the top is also free—the top of our laminate). These shear resultants are created (as the edge is approached and the σ_{yy} stresses vanish) by interlaminar τ_{yz} stresses, as shown. But these shear stresses must act only on the bottom of our equilibrium element so that the element would not be in rotational equilibrium unless normal stresses were acting on the bottom of the element. There must be a nonlinear distribution of interlaminar normal stresses σ_{zz} on that surface. Finally, since the top of the element is a free surface, the interlaminar normal stress must form a self-equilibrating couple in the *z* direction.

We will use these physical effects to discuss damage and failure modes. For example, if the Poisson contraction of the center plies of our $[0, \pm\theta]_S$ laminate is greater (under axial tensile loading) than that of the outer plies, then the σ_{yy} stress in the interior of the surface plies is compressive. Figure 7.2*c* shows that the interlaminar normal stress at the edge of the laminate, σ_{zz}, will also be compressive. However, if the interior plies have an orientation of 90°, then their Poisson contraction would be less than the 0° plies and they would resist contraction in the *y* direction. In that case, the transverse normal stress σ_{yy} would be tensile in the surface plies, and the interlaminar normal stress at the edge would be tensile, a driving force for the initiation and propagation of delamination. This logic, and the consequent occurrence of delamination, is well established. We will address the specifics of delamination in greater detail later in this chapter.

Analyses of the stress distributions near the free edge of a laminate are now abundant in the literature, but estimates of the fully three-dimensional stress fields are still the domain of approximate methods, most often finite-element formulations. The discussions and contributions of Pagano to this problem are particularly enlightening (cf. Pagano, 1970a; Pagano and Pipes, 1971; Pipes and Pagano, 1970). Early work of interest also includes Puppo and Evensen (1970), Srinivas and Rao (1970), Whitney and Sun (1973), and Wang and Choi (1982a,b). The utility of finite-element methods for this problem has also been widely discussed (cf. Whitcomb, 1998).

To understand the basic nature of the stresses in the vicinity of the edge of a laminate, we use the results of an extension of Reissner's variational principle to laminated bodies (Pagano, 1977). Pagano constructs a self-consistent field theory that satisfies the following requirements:

1. All six stress components are nonzero, in general.
2. Traction and displacement continuity conditions at interfaces between adjacent layers are satisfied.

3. The computed stress field acting on the surfaces of an arbitrary region bounded by any two of the parallel interfaces of the laminate, in conjunction with the prescribed pointwise traction boundary conditions on those portions of the external laminate boundary that lie in that region, satisfy the conditions of vanishing resultant force and moment identically (layer equilibrium).

Pagano constructs such a theory in which the force variables are force and moment resultants (per unit length, which are taken to be linear functions of z) acting on the cross sections of a lamina layer, and interlaminar stresses acting on the interfacial surfaces. Discontinuous interfaces (such as interface cracks) can be modeled. Also, the number of layers or regions (which satisfy layer equilibrium) need not coincide with the number of plies in a laminate. It is possible to introduce additional conceptual interfaces and resulting regions to improve the accuracy of the solution. We will examine some results of this analysis as a function of the number of regions, N (data from Pagano, 1977). Comparison data from finite-element analysis are taken from Wang and Crossman (1977).

Figure 7.3 shows the variation of the in-plane shear stress in an angle-ply (symmetric angle-ply) laminate as a function of distance from the centerline of the laminate to the free edge. We reasoned, in Figure 7.2*b*, that this stress component must seek a value of zero near the free edge and that this variation was an edge effect. Figure 7.3 shows this result. Even for this laminate, which was chosen to demonstrate a large edge effect, the variation is confined to a distance equal to only

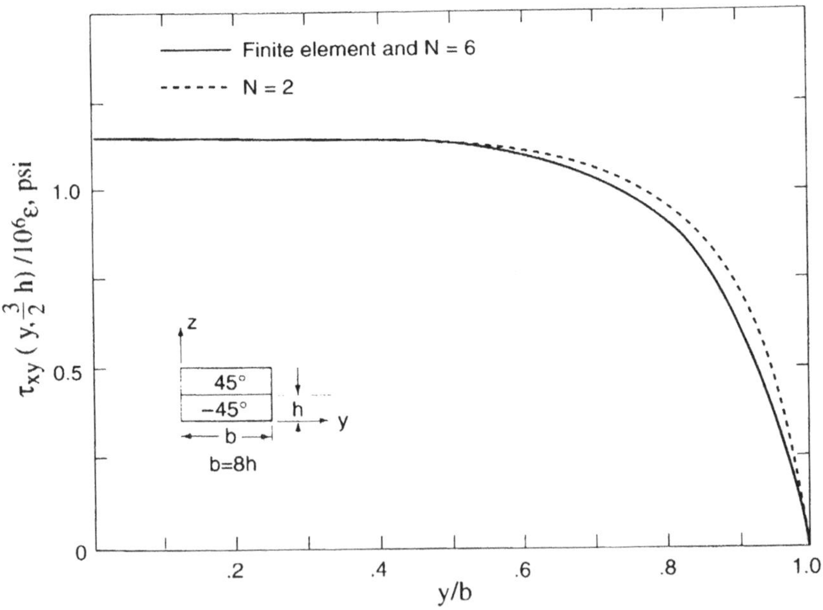

Figure 7.3 In-plane ply-level shear-stress distribution in symmetric angle-ply laminate as function of distance from centerline of specimen ($y/b = 0$) to free edge of specimen ($y/b = 1$) (Pagano, 1977).

two or three ply thickness from the edge. The in-plane shear stress still has more than half of its value within one ply thickness from the edge. Given the arguments we made in Figures 7.1 and 7.2, we might expect the interlaminar shear stresses to appear over exactly those distances, to ensure equilibrium of any representative free body in that region. Figure 7.4 shows that the interlaminar shear stress in the *xz* plane looks like a reflection of the results in Figure 7.3, as expected. The interlaminar stress is theoretically singular at the laminate edge, but neither of the approximate analyses shown predicts this (physically unrealistic) value.

In Figure 7.2*c* we used a physical argument to anticipate the occurrence of an interlaminar normal stress along the edge of the laminate and noted that it should occur as a self-equilibrating couple. The analysis (this time for a cross-ply laminate) produces just such a result, as shown in Figure 7.5. We also reasoned in our discussion of Figure 7.2*c* that if the interior lamina of an angle-ply laminate had an orientation of 90° and the outer ply (or plies) had lesser angles of orientation, then the interlaminar normal stress at the laminate edge would be tensile for a tensile global loading (assuming that the fibers are stiffer than the matrix). This result is also shown in Figure 7.5. And if one were to compare the area under the distribution curve on the negative side of the zero-stress line (in the interior) to the positive area near the edge, they would be found to be equal; that is, the interlaminar normal stress forms a self-equilibrating couple.

Finally, the distribution of the interlaminar shear stress in the *yz* plane near the laminate edge for a cross-ply laminate is shown in Figure 7.6. Since this stress has a

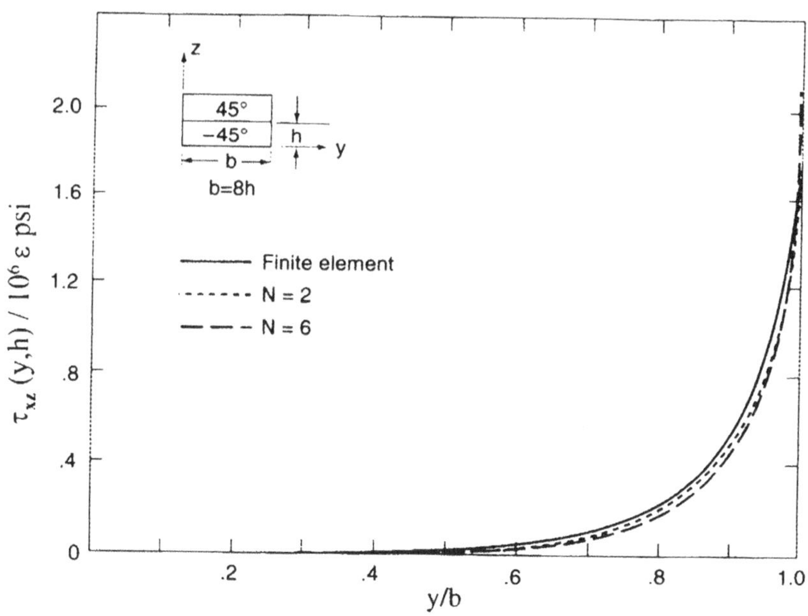

Figure 7.4 Variation of interlaminar shear stress in *xz* plane of symmetric angle-ply laminate as function of distance from centerline of specimen ($y/b = 0$) to free edge of specimen ($y/b = 1$) (Pagano, 1977).

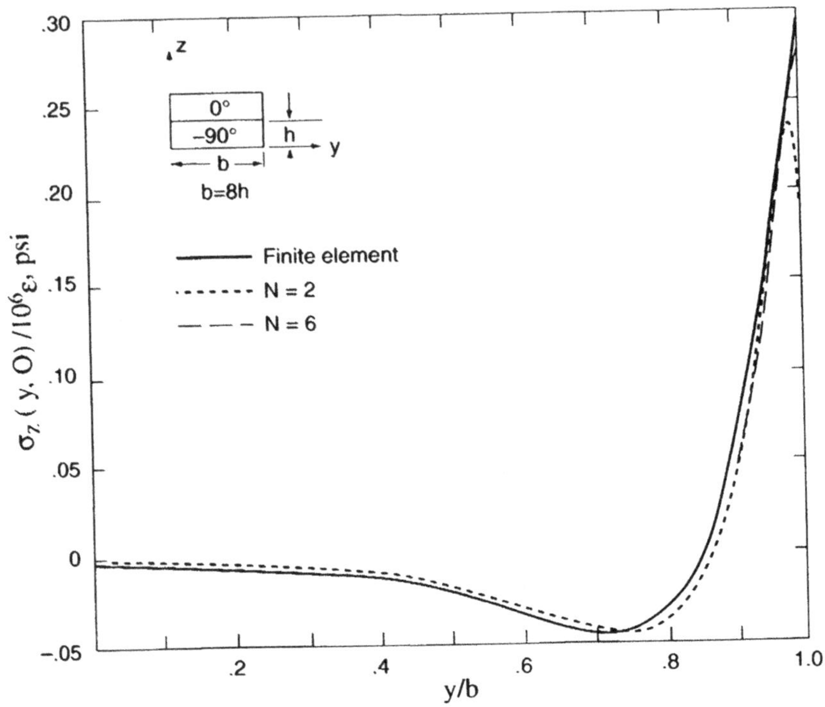

Figure 7.5 Interlaminar normal stress distribution in symmetric cross-ply laminate as function of distance from centerline of specimen ($y/b = 0$) to free edge of specimen ($y/b = 1$) (Pagano, 1977).

component that acts in the y direction, it, too, must vanish near the free edge with that normal vector. This effect is seen in Figure 7.6 as well. The effect of adding extra layers to the approximate analysis of Pagano (1977) is significant in some cases, and the agreement between the different analysis methods shown is generally good. As we will see below, getting the right value for any of these stresses quite near (or at) the edge is not generally a physically meaningful or necessary enterprise, at least in the sense that defect initiation and growth are found to depend on the stresses in a *region* near the edge, not on the absolute value at the actual boundary of the edge. This idea, very different from the stress concentration at the surface typically discussed for homogeneous materials, is a critical feature of this chapter and of the philosophy of the subject and will be discussed in detail in a later section.

7.2.2 Notches, Holes, and Cutouts

As mentioned earlier, notches (such as holes) are edges, and they are also stress concentrators. We have looked at the effect of the edge above. But the effect of the stress concentration is also not intuitive. An illustration is shown in Figure 7.7. The

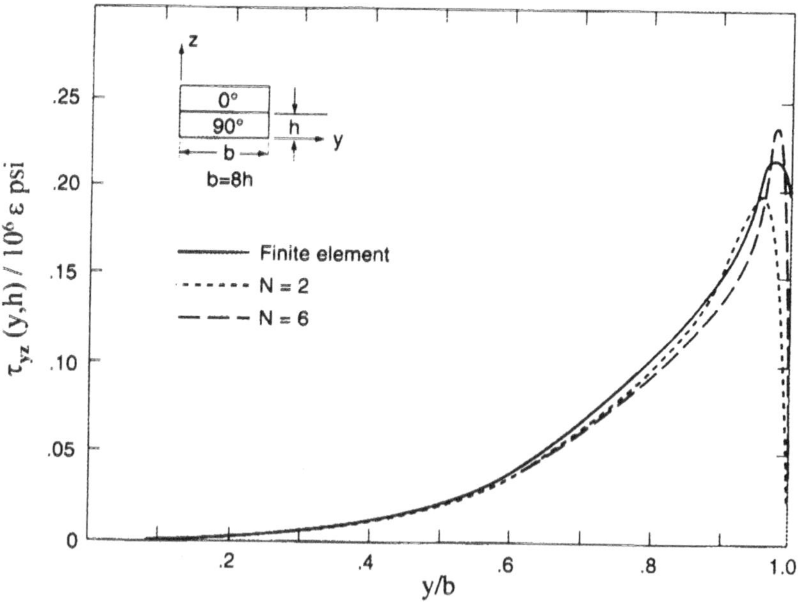

Figure 7.6 Distribution of interlaminar shear stress in *yz* plane of symmetric cross-ply laminate as function of distance from centerline of specimen ($y/b = 0$) to free edge of specimen ($y/b = 1$) (Pagano, 1977).

stress concentration (here quoted at the edge) of a center hole in a finite coupon lamina loaded in axial tension is strongly influenced by the orientation of the fibers in that lamina relative to the loading axis. (Here we assume that the fibers are much stiffer than the matrix they reinforce.) The data shown are from the classical work of Greszczuk (1972).

The figure contains three kinds of information. The stress concentration numbers are obtained by dividing the maximum azimuthal normal stress (having a direction tangent to the center hole) by the value of applied remote stress. The azimuthal location of the maximum stress concentration is given by the abscissa and indicated on the figure as the angle θ. The orientation of the fibers in the lamina coupon (relative to the direction of loading) is given by the angle φ. For example, if we consider the case of a unidirectional coupon with fibers aligned with the loading direction (the x axis in Fig. 7.7), then (for this boron–epoxy material system) there is a stress concentration of about 6.8 located at 90° to the loading axis along the y axis. The location of this maximum is intuitive, in the sense that it occurs in a direction transverse to the loading axis as it would for an isotropic material. However, the value of the stress concentration, 6.8, is remarkably higher than the classical value of 3 that we would have obtained for the isotropic case. For material systems with even higher degrees of anisotropy, the stress concentration can be even higher; values of 8 or 9 are not uncommon. For the design engineer, this result means a different stress

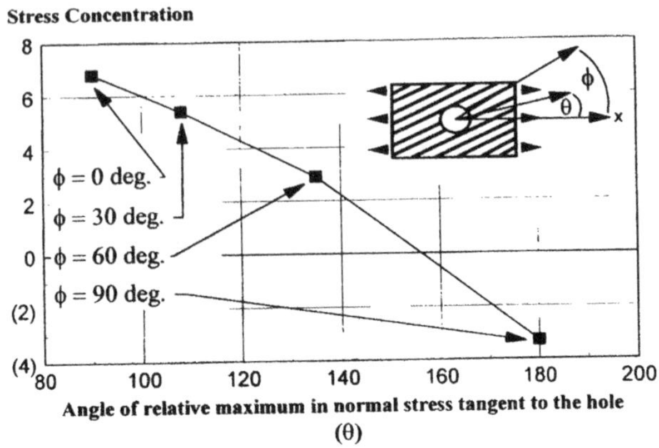

Figure 7.7 Location and value of stress concentration at edge of hole in coupon having center hole, made from lamina with fiber orientation φ under uniaxial tensile loading.

concentration value exists for each material system, and a comprehensive stress analysis must be used to determine this value if it is to be used for design purposes.

Returning to Figure 7.7, we now consider the case when the reinforcing fibers have an orientation of 30° or 60° to the loading direction. The stress concentration drops rather quickly as the angle of the fibers to the loading direction increases. (We are reminded that the example is for a material with fibers much stiffer than the matrix.) Moreover, the azimuthal location of the maximum tangent normal stress is no longer at 90°, as for the first example. Instead, the maximum moves around the edge of the hole, to 108° for the 30° fiber orientation and to about 135° for a fiber orientation of 60°. These are not intuitive angles; if they are needed for modeling of strength, they must be determined from a proper analysis. Finally, the case of fibers oriented at 90° to the loading direction produces a maximum value along the loading axis of about −3.3 for this material system. Again, this is substantially different from the isotropic case.

However, having the value of the stress concentration at the edge of the hole is not sufficient for an estimate of strength in these materials. What do the stresses away from the hole look like? For present purposes, we will examine a few azimuthal distributions and look at the edge effects to compare to the simpler case of a straight laminate edge. The classical analysis of an elliptical hole in an anisotropic plate with arbitrary in-plane loading has been in the literature for many years and provides a useful means for obtaining pointwise global stress resultants and strains (Lekhnitskii, 1981). However, these stress distributions, like the stress concentrations in Figure 7.7, are not intuitive and challenge our logic in the construction of models of strength.

Mollenhauer (1997) conducted an extensive experimental investigation of strain distributions near cutouts in laminates. Tensile coupon specimens with center holes were formed from carbon–epoxy laminates with two stacking sequences. Subply variations in displacement, through the thickness of the laminate, were measured

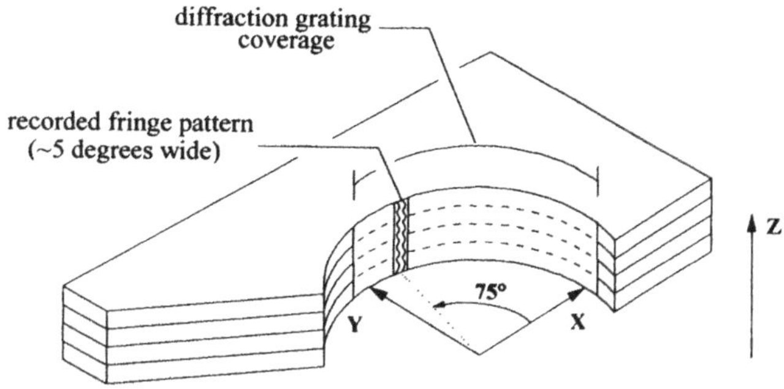

Figure 7.8 Geometry examined by experimental Moiré methods by Mollenhauer (1997). (Diagram shows a cut-away of one-quarter of the actual experimental specimen.)

from high-resolution Moiré patterns as a function of azimuthal angle around the inside of the center hole. The experimental strains were compared with a three-dimensional modeling method, known as the spline variational elastic laminate technology (SVELT) model, as described by Iarve (1996). A diagram of the region examined is shown in Figure 7.8.

Experimental displacement (and calculated strain) measurements were made from 45° to 84° to the axis of uniaxial loading (the x axis in Fig. 7.8). Average values through the thickness were checked against known solutions for uniform, isotropic materials. Figure 7.9 shows an example of the measured data for the azimuthal normal strain for an aluminum specimen compared to the results for laminate 1 (with a stacking sequence of $[+30_2, -30_2, 90_4]_{3S}$) and laminate 2 (with a stacking

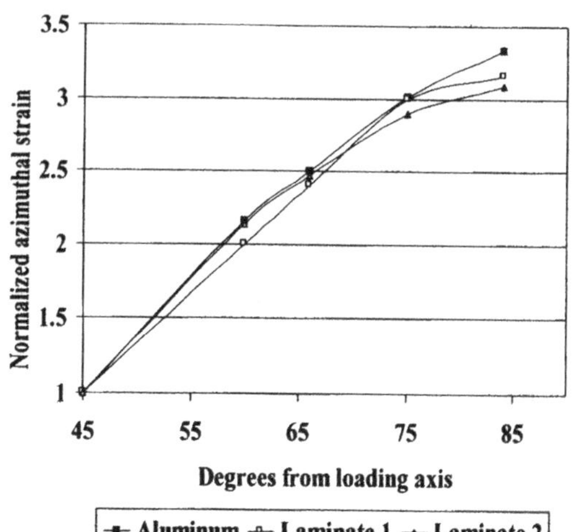

Figure 7.9 Experimental data for average azimuthal strain (averaged through thickness) as function of angle from loading axis for aluminum and two carbon–epoxy laminate types.

sequence of $[0_4, 90_4]_{3S}$. (Remember that the data are normalized by the value at 45°, so that the stress concentration of 3 at 90° is not recovered in the plot.)

For the range of values shown in Figure 7.9, the average azimuthal normal strain values are similar, with important differences as one approaches the transverse axis. However, the variation of the normal strains is large as a function of angle, and this variation (in all of the stresses and strains) must be considered in the construction and evaluation of a failure function when the stress state is nonuniform. The variation of the field quantities is also large through the thickness, from ply to ply. Figure 7.10 shows the variation of the theta-z shear strain (through the thickness shear in the plane of the surface of the center hole) for laminate type 1. (Experimental data and analytical predictions are shown and compared.) Values range from more than 15,000 microstrain to −15,000 microstrain at various positions throughout the thickness. Moreover, changes in the variation are discontinuous, with sharp peaks and valleys occurring at the interfaces between the plies. One can imagine (and we have shown in earlier figures) that damage (such as delaminations) might tend to initiate at the interlaminar positions. The azimuthal normal strain distribution (in microstrain units) for that laminate (at the same observation angle of 75°) is shown in Figure 7.11.

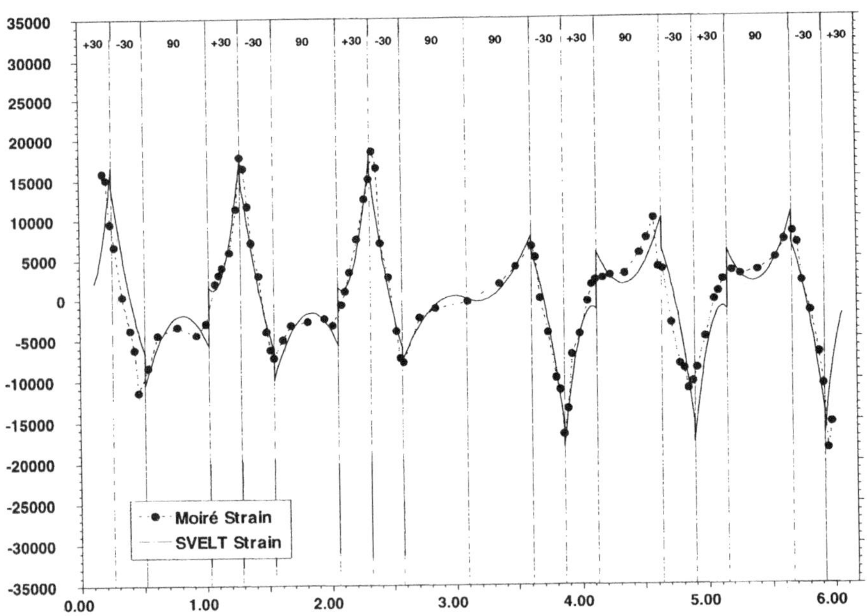

Figure 7.10 Theta-z shear-strain distributions (observed and predicted) for $[+30_2, -30_2, 90_4]_{3S}$ carbon–epoxy laminate under uniaxial loading along x axis (Mollenhauer, 1997).

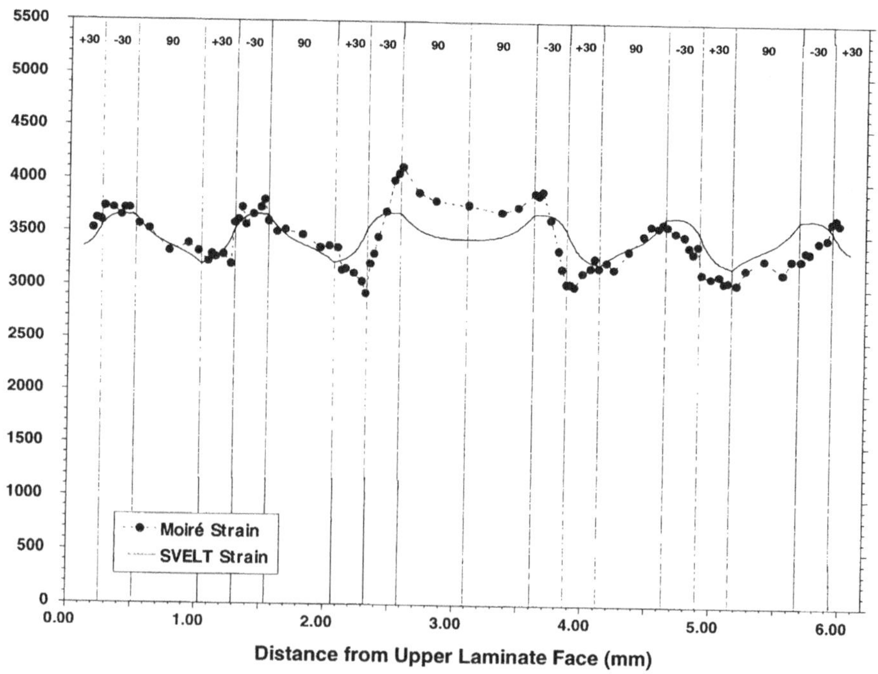

Figure 7.11 Azimuthal normal strain distributions (observed and predicted) for $[+30_2, -30_2, 90_4]_{3S}$ carbon–epoxy laminate under uniaxial loading along x axis (Mollenhauer, 1997).

These variations are quite sensitive to the nature of the laminate. If the same measurements and predictions for a $[0_4, 90_4]_{3S}$ laminate are made, one obtains the results shown in Figures 7.12 and 7.13.

The through-the-thickness shear strains, shown in Figure 7.12, are remarkably variable, with sharp discontinuities in slope at the ply boundaries. The azimuthal normal strain, shown in Figure 7.13, varies by more than 30% for this specific case, about the same as for the prior laminate type.

These variations are important to our efforts to identify damage and failure modes and to establish their appropriate failure function. If the local variations give rise to damage and failure modes, then we must construct analysis methods that provide accurate representations of the variations as inputs to an appropriate failure function. To make an accurate prediction of strength, the local stresses and strains that drive the active failure mode must be determined. This can be a very demanding requirement, especially when the stress-and-strain field is highly nonuniform and the damage and failure modes are complex. A further experimental demonstration of the nature and effect of stresses around a center hole in a composite coupon is provided by Figures 7.14*a*,*b*. Using the basic thermoelastic philosophy of Biot (1956) and the anisotropic elasticity solution for global strains in an infinite, homo-

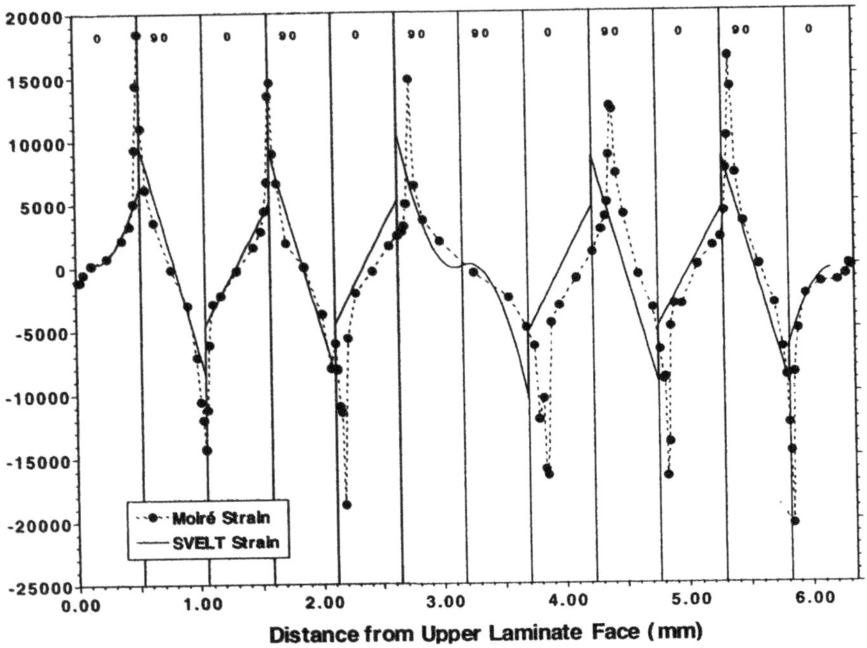

Figure 7.12 Theta-z shear-strain distributions (observed and predicted) for $[0_4, 90_4]_{3S}$ carbon–epoxy laminate under uniaxial loading along x axis (Mollenhauer, 1997).

geneous plat with an elliptical opening of Lekhnitskii (1981), Bakis (1988) was able to construct a micromechanical model of the volumetric strains in the constituents at the ply level of a laminate under uniaxial loading. Figure 7.14*a* shows the distribution of adiabatic heat emission predicted by a micromechanical analysis of deformations in a quasi-isotropic laminate coupon (38.1 mm wide, 254 mm long, 1 mm thick) with a 6.35-mm-diameter center hole subjected to a unidirectional global loading. Under these conditions, the lamina (pointwise, averaged through the thickness) strains in this quasi-isotropic laminate are the same as those in an isotropic plate; that is, they are symmetric about the centerline axis of loading. However, the predicted (Fig. 7.14*a*) and observed (Fig. 7.14*b*) patterns of heat emission are clearly not symmetric. This unsymmetry is a direct result of the anisotropic nature of the surface ply, which has an orientation of 45° (in the clockwise direction from the loading axis) in this figure. Although the ply's global strain is symmetric with respect to the load axis, the microstrains in the ply's constituent materials are not symmetric with respect to its axis. The matrix material contributes most of the thermoelastic response, so that the unsymmetry seen in the patterns corresponds, for the most part, to the unsymmetric local strains of the matrix material around the notch. In particular, heat emission is enhanced in the directions in which most of the deformation is concentrated in the matrix material (from bottom right to top left in the figures)

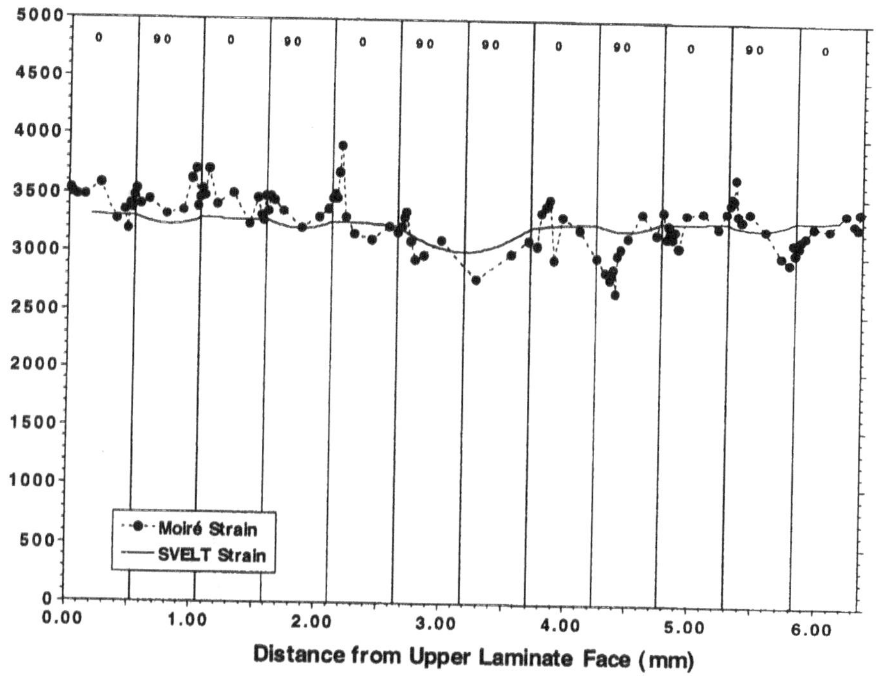

Figure 7.13 Azimuthal normal strain distributions (observed and predicted) for $[0_4, 90_4]_{3S}$ carbon–epoxy laminate under uniaxial loading along x axis (Mollenhauer, 1997).

and suppressed in the directions in which the deformation of the matrix and fibers is forced to be the same (as in bottom left to top right in the figures).

This illustrates a physical reality in inhomogeneous materials that is critical to our understanding of strength and damage development. Although ply-level deformations may be uniform or at least equal in various directions, the microconstituents may (and generally will) be under a nonuniform state of stress and strain. Moreover, this nonuniformity will generally be proportional to the anisotropy of the composite, which in turn is proportional to the difference between the stiffness of the microconstituents, their geometry, and their arrangement. In the present context, when the global strains are nonuniform, the microstresses and strains at the constituent level have greater differences, which have the effect of making damage modes associated with separation of the constituents (such as fiber–matrix debonding and delamination) more likely and more common. Any failure criterion, especially one applied at the ply level, must be supported by an exact and precise analysis of the local stresses and strains. Finally, the variation of the stress pattern along any radius direction is very different for a given azimuthal angle. If a failure criterion that requires stress and strain inputs is used along one of those directions, it is necessary to evaluate the criterion for a range of radial directions to find the condition that is most likely to cause local failure.

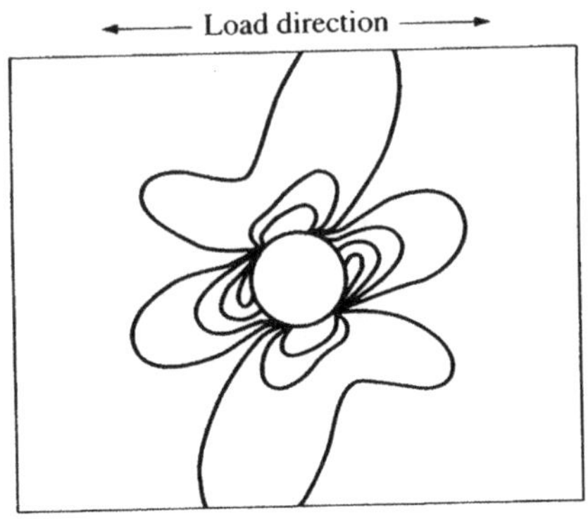

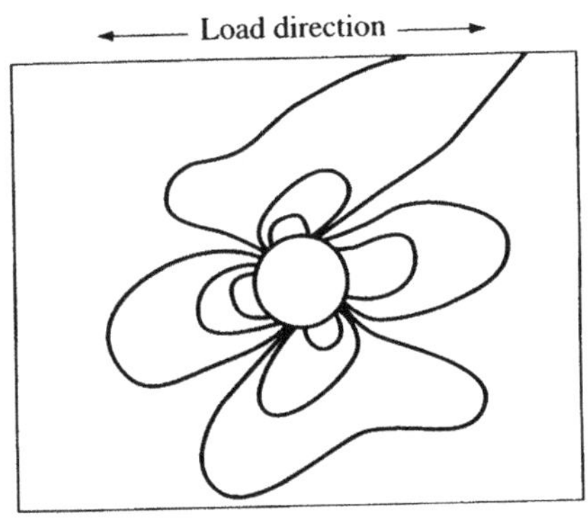

Figure 7.14 (*a*) Thermoelastic adiabatic heat emission pattern predicted by micromechanical theory for uniaxial cyclic loading of quasi-isotropic laminate with center hole; (*b*) observed temperature contours.

7.3 UNDAMAGED NOTCHED STRENGTH

7.3.1 Tensile Strength

In the preceding sections, the basic nature of the nonuniform stress state at the edge of a multiaxial laminate and near the edge of a notched lamina and laminate was described. In general, we found the ply-level stress state is fully three dimensional and varies sharply over a small region, sometimes called a boundary layer, near the edge. In the presence of a notch, in-plane values of the stress field are also elevated to create a stress concentration. However, the distribution of the individual stress components as a function of azimuthal angle around the edge of the notch is complex and not necessarily intuitive.

The details of this distribution are critical to the determination of the strength of continuous-fiber-reinforced composites in the presence of notches or other stress concentrators. Indeed, as stated in the thesis of this chapter, the fundamental nature of the stress distribution controls the strength of notched fibrous composites and motivates the introduction of a new material parameter into representations of notched strength. With homogeneous materials, surface stress concentration factors or crack-tip field stress intensity values in the singular region are typically the basis for notched (or cracked) strength. With fibrous composites, strength is usually determined by the details of the field stress distribution at the fiber–matrix or ply level over a characteristic distance that typically has the same dimensional order as the ply thickness (sometimes as large as the characteristic dimension of the notch or hole itself). Figure 7.15 illustrates the nature of the physical behavior that gives rise to this fact. The figure is a schematic diagram of the distribution of the fiber direction normal stress in a lamina near a notch, a circular hole with radius R in this case. This distribution is superimposed over images of the fibers themselves (not to scale in this drawing). Also shown (along the y axis, which is assumed to be the axis of uniaxial loading in this example) is the assumed distribution of strength of the fibers, on which is marked the specific values of the strength of the first 10 fibers along the x axis, away from the hole. A value of remote loading has been assumed that will begin to cause some local fiber fractures near the hole. The strength of the individual fibers, in general, will be statistically distributed within the strength distribution (as suggested by the arrows located along the strength distribution curve drawn next to the y axis). In the case sketched, the first fiber (nearest to the edge of the hole) happens to have a strength that is very high in the distribution, while fiber 4 has a strength near the bottom of the distribution. (It should be noted that, in reality,

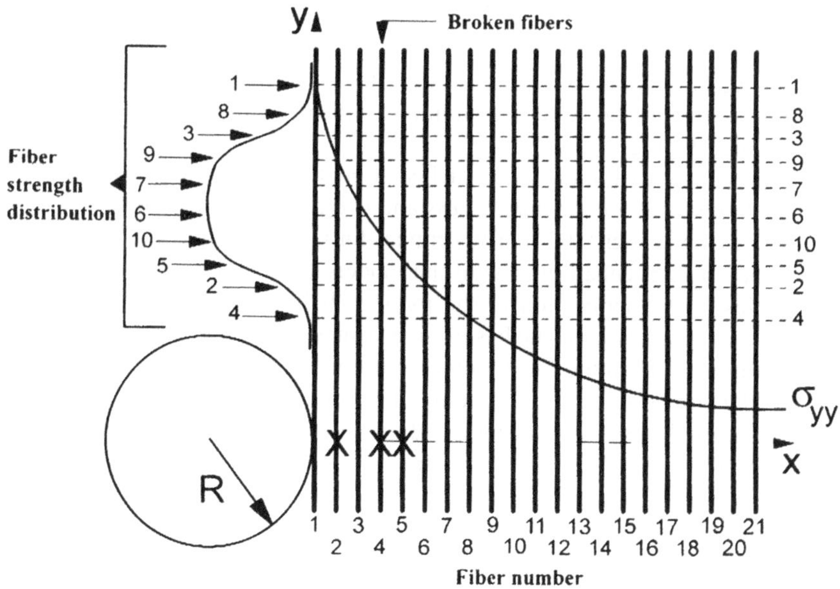

Figure 7.15 Statistical nature of fiber failure process that determines notched strength of continuous-fiber reinforced composite laminae and laminates.

more of the fibers would have strengths near the middle of the distribution than near the ends, but it was not possible to show the arrows that way in Fig. 7.15.)

The failure process and the resulting lamina or laminate strength are determined by the interaction of the stress distribution and the strength distribution, an argument outlined by Wu (1978). The basis of the process is illustrated by Figure 7.15. The first fundamental question is "will the fibers break near the hole?" For the case shown in Figure 7.15, fiber 1 shows a strength that is just slightly larger than the stress in that fiber (the value of σ_{yy} in the first fiber for the distributions shown). So the first fiber will not break. However, the strength of the second fiber is very low in the fiber strength distribution, and the stress in that fiber greatly exceeds its strength, so it fractures, as suggested by the "X" on that fiber in the figure. The strength of the third fiber is more near the average, but the σ_{yy} value has dropped below that value at that position, so fiber 3 does not break. However, fiber 4 does break, and even though the value of σ_{yy} has dropped quite a bit by the time fiber 5 is reached, its strength is just below that value, so that both fibers 4 and 5 break. Hence, the local pattern of fiber breaks is determined by the statistical distribution of strength of the fibers, and by the local stress field which is a function of the lamina and laminate properties. This local region over which fiber damage develops is widely discussed in the literature and is given various names, including *process zone*, *damage zone*, and numerous other derivatives. The important fact is that this region and its characteristic dimension are the essential feature of any model that is to correctly represent the behavior of this class of materials in the notched condition.

We will examine the most commonly used (and generally successful) models of notched strength and observe how this characteristic dimension enters the problem. We will focus on tensile strength for the moment. It should be mentioned that the present approach includes several important additions to the classical form of the model that we have found to be especially effective in the many applications that form the basis for this book. These additions will be clearly identified and can be used or ignored by the reader.

The most widely used models of notched strength in continuous-fiber-reinforced laminae and laminates are the *point-stress* and *average-stress* models first described by Whitney and Nuismer (1974). The original form of the models will be summarized, and then several enhancements of the models will be described, followed by a few examples of the application of the models. It is assumed that the unnotched strength of the lamina or laminate has been measured; we will use the notation X_L for that strength value. Then the point-stress criterion states that the notched composite fails when the normal stress in the loading direction along the axis transverse to that direction exceeds X_L over a characteristic dimension (a in Fig. 7.16), which is taken to be a material parameter. To demonstrate this concept, Whitney and Nuismer approximated the stress along the x axis by the elasticity solution for an isotropic material normalized by the remote applied stress σ_0:

$$\frac{\sigma_{yy}}{\sigma_0} = 1 + \frac{1}{2}\left[\frac{R}{x}\right]^2 + \frac{3}{2}\left[\frac{R}{x}\right]^4 \tag{7.2}$$

Then the argument becomes that failure occurs for the notched laminate when

$$\sigma_0 = X_N \qquad \sigma_{yy} \geq X_L \quad \text{for all } R \leq x \leq R + a$$

Then using these conditions in (7.2) and solving for the notched strength, X_N yields

$$\frac{X_N}{X_L} = \frac{2}{2 + \xi^2 + 3\xi^4} \qquad \text{where} \quad \xi = \frac{R}{R + a} \tag{7.3}$$

The average-strength criterion is based on a similar concept, except that the basic premise is that the notched laminate fails when the average stress over a characteristic distance (b in Fig. 7.16) is equal to (or greater than) the unnotched laminate strength:

$$X_L = \frac{1}{b} \int_R^{R+b} \sigma_{yy}(x,\ 0)\, dx \tag{7.4}$$

Then introducing the variable $\zeta = R/(R + b)$, changing the variable in (7.4), integrating by parts, and rearranging yield the form

$$\frac{X_N}{X_L} = \frac{2(1 - \zeta)}{2 - \zeta^2 - \zeta^4} \tag{7.5}$$

The two criteria, (7.3) and (7.5), have several interesting features, illustrated in Figure 7.17. First, the effect of the notch decreases with notch radius (or hole size). This is consistent with observed data for inhomogeneous brittle materials (including such things as cast iron). When the notch becomes small enough, its characteristic dimension (and region of disturbance of the local stress field) becomes comparable with the dimensions of the constituents and their local stress field

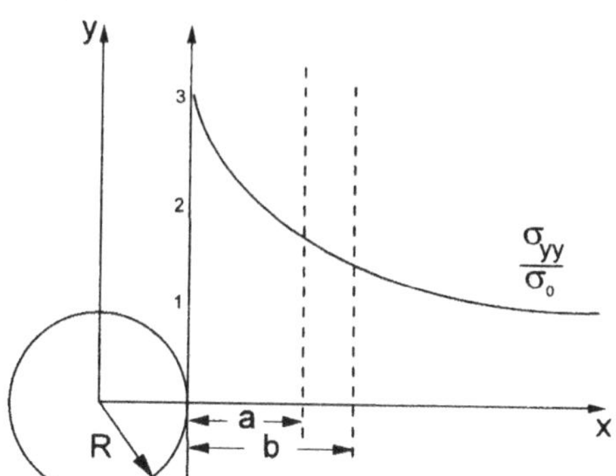

Figure 7.16 Load direction normal stress distribution near hole with radius R in composite laminate showing two characteristic distances a and b along axis transverse to loading direction.

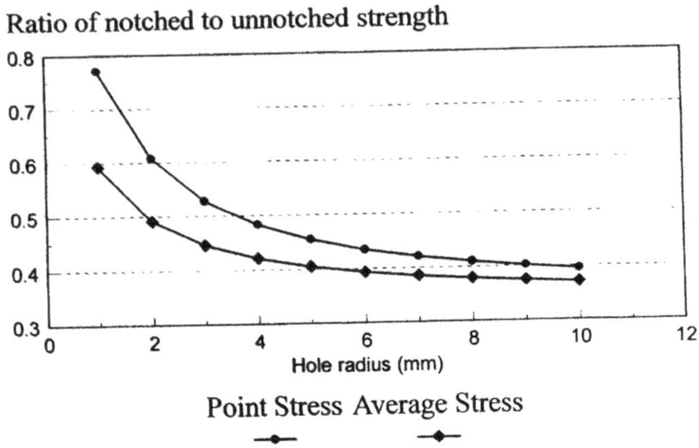

Figure 7.17 Ratio of notched to unnotched strength as function of hole size for average- and point-stress models.

disturbances. In a sense, the notch then becomes no worse than the local microstructure, and no global reduction in strength results. This is very different from the standard engineering practice of using stress concentrations that depend only on geometry and have no sensitivity to material length scales. Every engineer knows that the stress concentration at the edge of a hole in a (wide) isotropic plate under uniaxial loading has the value of 3 and under normal circumstances would expect the global strength to be reduced to about one-third of the unnotched value. Figure 7.17 shows that such a reduction, even for the analysis described above, which uses the isotropic stress distribution, occurs only for large hole radii. A second important feature is the dependence of the notched strength on the actual details of the stress distribution near the hole, not just on the value at the edge of the hole. Both expressions (7.3) and (7.5) were derived from the equation that describes the stress field in the vicinity of the hole (albeit, a rather rough approximation in the original presentation of the criterion). For a given material system and laminate and even for a different loading, that distribution and the predicted notched strength will change.

A third essential feature is the material characteristic length, called a and b in the equations above. A value of 0.8 mm and 1 mm were used for a and b, respectively, in Figure 7.17. These values are determined by comparing the predictions with a small set of pilot experiments, typically for one or two hole radii. Then all other results are predicted. These parameters are material constants, in some sense. Figure 7.18 shows predictions of this model compared to experimental results taken from the original paper by Whitney and Nuismer (1974). With values of 1 and 3.75 mm for the material characteristic length in the point-stress and average-stress criteria, respectively, the agreement with data is quite satisfactory, and the physical trends are recovered well.

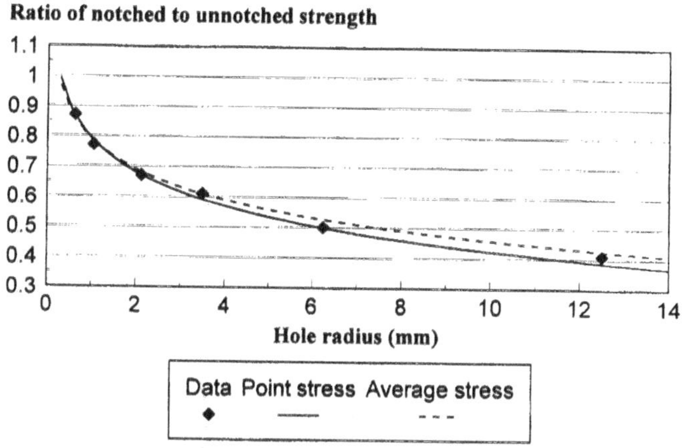

Figure 7.18 Comparison of point-stress ($a = 1$ mm) and average-stress ($b = 3.75$ mm) predictions of notched tensile strength for quasi-isotropic glass–epoxy laminate to observed values (Whitney and Nuismer, 1974).

Since the original introduction of this model, there have been several enhancements of the approach that add to its accuracy and utility. One of the most obvious changes was recognized by Konish and Whitney (1975), who made an attempt to correct the (isotropic) stress state used in the original formulation to reflect the actual stress state in a laminate that had some finite anisotropy. We will summarize their addition below because it is instructive and is sometimes still used because of its convenience. However, using the exact values of the field stress for a given material system and laminate and for a given stress concentration (e.g., notch, hole, cross-sectional change) is now rather simple to do because of the many stress analysis methods available for this purpose, as mentioned in an earlier section. Hence, approximate forms are to be viewed as a convenience with some attendant cost in accuracy.

The approach of Konish and Whitney is, in fact, very convenient. They start with the general solution for an elliptical hole in an anisotropic plate for arbitrary in-plane loading, provided by Lekhnitskii (1981), mentioned earlier. This analysis provides an exact description of the stress field near the hole and can be used directly in place of (7.2), with conditions (7.3) and (7.5) to construct a criterion for each situation. Alternatively, they provide a convenient engineering approach by noticing that the essence of the stress distribution for a given laminate could be captured in part by the correct stress concentration factor K_T for a generally anisotropic plate. Along the transverse axis, for unidirectional loading of a symmetric laminate with orthotropic in-plane stiffness properties, that factor has the convenient form

$$K_T = 1 + \left[\frac{2}{A_{22}}\left(\sqrt{A_{11}A_{22}} - A_{12} + \frac{A_{11}A_{22} - A_{12}^2}{2A_{66}}\right)\right]^{1/2} \tag{7.6}$$

where A_{ij} are the components of the usual A matrix in CLT. Then, they postulate a correction to (7.2) in the form

$$\frac{\sigma_{yy}}{\sigma_0} = 1 + \frac{1}{2}\left[\frac{R}{x}\right]^2 + \frac{3}{2}\left[\frac{R}{x}\right]^4 + C_1\left[\frac{R}{x}\right]^6 + C_2\left[\frac{R}{x}\right]^8 \tag{7.7}$$

where C_1 and C_2 are determined by setting the integral of (7.2) and (7.7) over the x axis in Figure 7.16 equal to each other, in satisfaction of equilibrium requirements, and by setting the value of (7.7) equal to the stress concentration K_T at the hole boundary. The resulting expression is

$$\frac{\sigma_{yy}}{\sigma_0} = 1 + \frac{1}{2}\left[\frac{R}{x}\right]^2 + \frac{3}{2}\left[\frac{R}{x}\right]^4 - \frac{K_T - 3}{2}\left[5\left(\frac{R}{x}\right)^6 - 7\left(\frac{R}{x}\right)^8\right] \tag{7.8}$$

Then (7.8) can be applied to the point-stress and average-stress conditions to get equivalent expressions to (7.3) and (7.5). For the point-stress case, the result is simply

$$\frac{X_N}{X_L} = \frac{2}{2 + \xi^2 + 3\xi^4 - (K_T - 3)(5\xi^6 - 7\xi^8)} \tag{7.9}$$

and for the average-stress case, the result is

$$\frac{X_N}{X_L} = \frac{2(1 - \zeta)}{2 - \zeta^2 - \zeta^4 - (K_T - 3)(\zeta^6 - \zeta^8)} \tag{7.10}$$

A few example calculations will illustrate typical behavior of these criteria.

7.3.2 Example Calculations

Consider a graphite–epoxy system that has the following lamina properties: $E_1 = 275$ GPa, $E_2 = 8.5$ GPa, $G_{12} = 25$ GPa, and $\nu_{12} = 0.25$, for which the A matrix has the values

$$A_{ij} = \begin{bmatrix} 68 & 0.53 & 0 \\ 0.53 & 2.13 & 0 \\ 0 & 0 & 6.25 \end{bmatrix} \times 10^6$$

Then for this material system for a unidirectional lamina under uniaxial loading, (7.6) gives $K_T = 5.654$ for the stress concentration factor at the edge of the hole. Let us predict the lamina strength for a hole with radius $R = 6.25$ mm and assume that for this material system our characteristic dimensions are $a = 1.25$ mm and $b = 3.75$ mm. Then (7.9) and (7.10) yield the predicted ratios of notched to unnotched strength as 0.499 and 0.551, respectively. Hence, this lamina, made from this material system, with a 6.25-mm hole will be about half as strong as the unnotched lamina under uniaxial tensile loading.

Suppose we now make a $[90,\ 0_2]_S$ laminate out of the same material and keep the radius of the center hole the same. The A matrix becomes

$$A_{ij} = \begin{bmatrix} 140 & 1.6 & 0 \\ 1.6 & 73 & 0 \\ 0 & 0 & 18.75 \end{bmatrix} \times 10^6$$

and (7.6) gives $K_T = 4.192$ for the stress concentration factor at the edge of the hole, a smaller value than the one for the unidirectional material. For this situation, the ratio of notched to unnotched strength is 0.49 for the point-stress criterion and 0.531 for the average-stress criterion. The stress concentration factor itself does not have a major effect on the strength of the laminate. Equations (7.9) and (7.10) show that, clearly, since only the quantity $K_T - 3$ enters the equations as an effect of anisotropy (since $K_T = 3$ for the isotropic case), and that factor multiplies fractions raised to a large power.

However, for this example, suppose we double the radius of the hole from 6.25 to 12.5 mm. The stress concentration factor is unchanged, but the predicted values of notched to unnotched strength ratios change significantly:

$X_N/X_L = 0.49 \Rightarrow 0.37$ for the point-stress criterion and
$X_N/X_L = 0.531 \Rightarrow 0.482$ for the average-stress criterion.

The *distribution* of stress near the hole, not just its value at the edge of the hole, is controlling the predicted and observed notched strength of these brittle, anisotropic composite materials.

7.3.3 Generalized Approach to Quasi-Static Notched Tensile Strength

Two more features must be added to the model described above to make it a more general and robust model of notched strength. First, all of the preceding discussion was based on the laminate-level properties unless the laminate was a single lamina by default. If the notched strength of a laminate is to be predicted by the preceding analysis, then the stress state for a notched laminate is determined and used in the point-stress and average-stress criteria, and the characteristic material parameters are determined for the laminate, not really for the lamina material system. If the ply orientations in the laminate change, even though they are made from the same material, one can expect the values of a and b to change as well. This is not true if the entire concept is applied at the ply level. For this approach, the following application sequence is used. (We will discuss the concept for a fiber-controlled failure mode.)

1. For a given specimen and notch geometry and global loading, solve for the laminate pointwise field strain values using analytical or approximate methods as necessary.
2. Resolve these strains into ply-level stresses.
3. Determine the ply in the laminate for this geometry and loading that has the highest fiber stress levels, the critical ply.
4. Determine the azimuthal radius direction ϕ that lies perpendicular to the fiber direction in the critical ply.
5. Define a failure function (as discussed in previous chapters) that is appropriate for the description of the failure mode in that ply. (This may be as simple as the fiber stress divided by the ply strength in the fiber direction.)
6. Average the failure function over the distance b along the radius vector with angle ϕ to the x axis, as shown in Figure 7.19.
7. Compare with pilot data to determine the material constant b; that is, determine the value of b for which the averaged value of the failure function is exactly equal to unity when the applied loading just causes failure of the notched laminate.
8. Use that value of b and the procedure outlined in 1–7 to predict the notched strength of all other laminates, geometries, and loadings for that material system.

This process is easily automated in a computer code, as is done in the MRLife code series, to be discussed in Chapter 8. After finding the stresses in the material system in all of the plies, a scan around the hole or notch defines the angle ϕ. Then the material constant is adjusted to make the average value of the selected failure function equal to unity over that distance, b, when the applied quasi-static loading just causes failure. Several data points may be used to account for the statistical scatter in this step; in fact, confidence intervals can be defined at that point. Then, all other cases can be predicted.

There are some special features of this approach that should be noted. The loading of the laminate can be multiaxial; it does not need to be uniaxial as in the

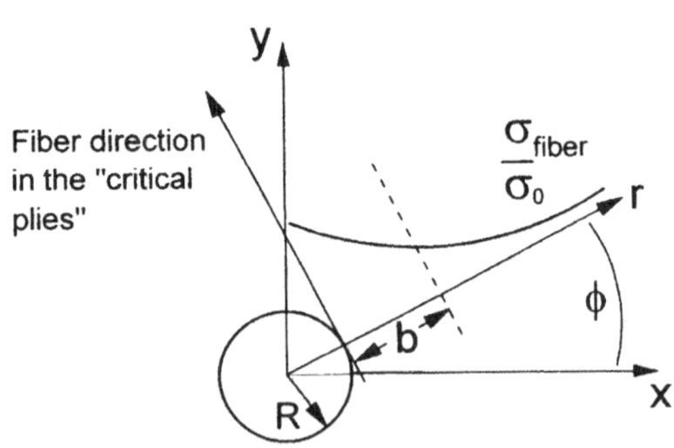

Figure 7.19 Fiber stress distribution along radial direction r, forming angle ϕ with transverse x axis.

original Whitney–Nuismer formulation. In general, the plies of a laminate are under multiaxial stress, even when a uniaxial load is applied at the global level. The use of the appropriate failure function at the ply level accounts for this feature, and the approach does not rely on the values or distributions of only one stress component. Care does have to be taken that the failure mode of the critical ply is correctly modeled and the correct failure function is selected, since the failure mode of a ply in a laminate may not be the same as the failure mode of that ply loaded alone, without the constraint of neighboring plies. However, for fiber-controlled failure and when failure of the notched laminate is defined by fracture, the simple ratio of fiber stress to fiber strength in the critical ply has remarkable utility for the prediction of notched laminate fracture strength. Note that when the characteristic material length parameters (a and b above) are determined for a given material system, it is implied that they can be used (in the strength-controlling plies) for all laminates made from that material system, eliminating the need to evaluate that parameter for every laminate type.

7.3.4 Example Calculations

Several examples of the manner in which this analysis can be applied are given below. In all cases, the global stresses were found for uniaxial loading P of a coupon specimen with a center hole. For this situation, the laminate stresses (for the anisotropic plate with circular hole) are given by the Lekhnitskii relationships:

$$\begin{aligned} \sigma_x &= P + 2\,\mathrm{Re}[\mu_1^2 \phi_1' Z_1 + \mu_2^2 \phi_{21}' Z_2] \\ \sigma_y &= 2\,\mathrm{Re}[\phi_1' Z_1 + \phi_2' Z_2 \\ \sigma_{xy} &= -2\,\mathrm{Re}[\mu_1 \phi_1' Z_1 + \mu_2 \phi_2' Z_2] \end{aligned} \qquad 7.11)$$

where the prime denotes differentiation and μ_1 and μ_2 are the complex roots of the characteristic equation

$$A_{11}\mu^4 - 2A_{16}\mu^3 + 2(A_{12} + A_{66})\mu^2 - 2A_{26}\mu + A_{22} = 0 \qquad (7.12)$$

in which A_{ij} are components of the compliance matrix. The other terms in (7.11) have the following definitions:

$$\begin{aligned} &\mu_1 = \alpha + \beta i && \mu_2 = \gamma + \delta i \\ &\phi_1(Z_1) = \frac{\beta_1}{\mu_1 - \mu_2}\frac{1}{\xi_1} && \phi_2(Z_2) = \frac{\beta_1}{\mu_1 - \mu_2}\frac{1}{\xi_2} \\ &\beta_1 = -\frac{1}{2} irP && i = \sqrt{-1} \\ &\xi_1 = Z_1 + \frac{\sqrt{Z_1^2 - r^2 - r^2\mu_1^2}}{r - ir\mu_1} && \xi_2 = Z_2 + \frac{\sqrt{Z_2^2 - r^2 - r^2\mu_2^2}}{r - ir\mu_2} \\ &Z_1 = x + \mu_1 y \qquad Z_2 = x + \mu_2 y && r = \text{hole radius} \end{aligned} \qquad (7.13)$$

For the unidirectional lamina case, these global stresses are also the ply-level stresses (in the global coordinate system). In this case, the stresses can be integrated directly following the eight steps discussed above and in Figure 7.19 to evaluate the average-stress criterion in the critical element. The failure function can be chosen to match the failure mode. Let us consider the strength of such a unidirectional lamina for orientations equal to or greater than 5° off of the loading axis. Laboratory experiments suggest that the failure mode of such a notched lamina is by shear, at least for angles of alignment up to 40° or so. Hence, we chose the failure function to be

$$\text{Fa} = \frac{\sigma_{12}(0,\ r)}{X_{12}} \tag{7.14}$$

where X_{12} is the in-plane shear strength of the lamina, as discussed in Chapter 2, and σ_{12} is the in-plane shear stress in the fiber system (not the global system). Data for such a situation were recorded using an AS4/3502 carbon–epoxy system with a 65% volume fraction of fibers. The center hole had a diameter of 3.1925 mm. The specimen dimensions were 19.15 mm wide, 652.3 mm long, and 0.14 mm thick. Figure 7.20 shows the results of the predicted strengths compared to the measured values. The material parameter (the characteristic length over which the stress in the fiber direction is averaged) was obtained by fitting the data for the 5° orientation test; therefore, the agreement for that case is guaranteed. The other predictions are acceptably close to the data but begin to diverge above 40°, where the failure mode becomes more closely associated with transverse strength than with shear. However, the methodology described above works reasonably well for this case. A more robust failure function than (7.14) would improve the agreement.

Now let us consider laminate examples. First, let us consider laminates of the type $[0,\ \pm\theta]_S$. In particular, for this example, we will consider six configurations in

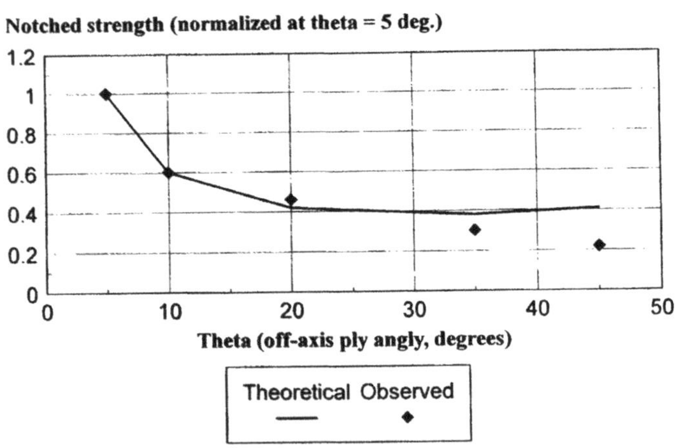

Figure 7.20 Comparison of predicted and observed notched tensile strength (normalized at $\theta = 5°$) as function of fiber orientation angle for unidirectional specimens.

which $\theta = 10°$, $25°$, $35°$, $45°$, $65°$, $90°$. This is a fairly common engineering situation, in which the 0° plies have a major influence on the strength of the laminate (generally, laminates would be designed to ensure fiber domination of properties). Now, the stresses obtained from (7.11)–(7.13) must be translated to the ply level, point by point, using CLT or an equivalent method. For the single-axis loading considered in this example (for which the load P is applied along the global "0° axis," which coincides with the orientation of the 0° plies), the stresses along the transverse axis (the x axis in Fig. 7.19) will be averaged to evaluate the average-stress criterion in the 0° plies.

The characteristic material length parameter b in Figures 7.16 and 7.19 can be determined by matching a single data point, say the 10° case for this example. We can then use this result for the rest of the laminate predictions. However, El-Zein and Reifsnider (1990) have found a simple scheme to improve the agreement for the level of stress concentration near the hole, in the same spirit discussed above for the Konish and Whitney (1975) model. In the notation of this example, the stress along the transverse axis at the edge of the hole for a given laminate is given by

$$\sigma_{y'y'} = P[1 + (\beta + \delta)] \tag{7.15}$$

that is, the sum of the absolute values of the imaginary roots of the characteristic equation determines the stress concentration at the edge of the hole. Consequently, El-Zein and Reifsnider (1990) postulated that

$$(\beta + \delta)_1 b_1 = (\beta + \delta)_2 b_2 \tag{7.16}$$

where the subscripts 1 and 2 correspond to two choices of the off-axis ply orientations (θ_1 and θ_2 for this example) and b_i are the material characteristic length parameters discussed earlier.

Table 7.1 shows the results of applying this methodology to the $[0, \pm\theta]_S$ laminates compared to experimental results. The results are also plotted in Figure 7.21. The experimental data represent single specimen tests; more will be said of the statistical variation in experimental data in a subsequent example. The value of the material characteristic length b is determined from matching the data for the 10° case. The other results are close to the observations, with the worst difference

TABLE 7.1 Comparison of Predicted and Observed Notched Tensile Strengths

Laminate	Observed (MPa)	Predicted (MPa)	B (mm)	$\beta + \delta$	R (mm)
$[0, \pm10]_S$	815.1	815.1	1.92	4.76	3.19
$[0, \pm25]_S$	536.18	515.5	3.06	3.0	3.19
$[0, \pm35]_S$	433.8	419.7	3.83	2.3	3.19
$[0, \pm45]_S$	440.4	378.9	4.60	2.0	3.19
$[0, \pm65]_S$	448.06	447.3	4.34	2.108	3.19
$[0, 90]_S$	463.0	372.06	2.81	3.29	3.19

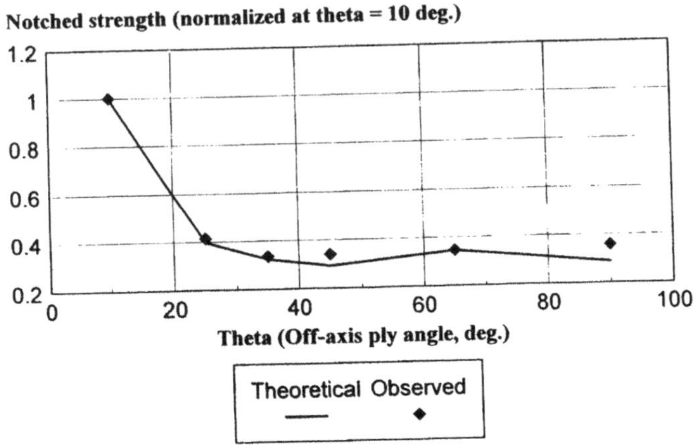

Figure 7.21 Comparison of predicted and observed notched tensile strength (normalized at $\theta = 10^\circ$) of $[0, \pm\theta]_S$ laminates as function of orientation of off-axis plies.

being for the 90° laminate (a difference of about 20%) for which the failure mode is likely to be somewhat different. The failure function used in this example was the fiber direction stress divided by X_t, as suggested earlier.

Figure 7.21 shows some very interesting effects. For example, there is a relative minimum (predicted and observed) at angles in the 45° range, an interesting physical feature correctly detected by the methodology. The plot also illustrates how quickly the contribution of the off-axis plies drops off, a matter of interest to designers. Finally, Table 7.1 shows that the variation of the characteristic material length parameter for this case is about 30%.

A third example involves the effect of hole size, a feature mentioned in our earlier discussion of the average- and point-stress criteria. For this comparison, data from experimental tests conducted by Whitney and Kim (1977) were used. They investigated the strength of carbon–epoxy quasi-isotropic laminates with two different stacking sequences and three different center-hole radii, as shown in Table 7.2. As it happens, the roots of the characteristic equation are the same for these conditions, so

TABLE 7.2 Comparison of Predicted and Observed Notched Tensile Strengths

Laminate	Observed (MPa)	Predicted (MPa)	B (mm)	$\beta = \delta$	R (mm)
$[\pm 45, 0, 90]_S$	335.88	373.44	2.68	2.0	1.277
$[\pm 45, 0, 90]_S$	277.11	277.11	2.68	2.0	3.831
$[\pm 45, 0, 90]_S$	260.1	228.88	2.68	2.0	7.662
$[90, 0, \pm 45]_S$	367.37	365.17	2.68	2.0	1.277
$[90, 0, \pm 45]_S$	296.27	277.11	2.68	2.0	3.831
$[90, 0, \pm 45]_S$	254.93	228.88	2.68	2.0	7.662

the material length parameter b was constant. The predicted results are within about 10% of the observed values.

The results shown in the examples above illustrate the versatility and robustness of this approach. Statistical variability of the experimental data for the tensile strength of coupons with center holes is typically of the order of 25% (cf. Lagace, 1986). The results obtained using the present approach are usually within the experimental scatter.

For an historical review of other quasi-static tensile strength models, we refer the reader to an extensive literature on the subject (cf. Awerbuch and Madhuker, 1985).

7.3.5 Notched Compressive Strength

In Chapter 4, micromechanical models of the principal material strengths were reviewed. Several models for compressive strength that reflect different failure modes, including crushing, localized shearing, microbuckling, and postbuckling/"crippling" behavior, were noted. These models can be applied to the notched problem, with the addition of a global model such as the Whitney–Nuismer formulation discussed earlier. The literature is replete with examples of such applications. Bergmann and Stavrinidis (1988) find a particularly comprehensive data set in an ESTEC report. (See also Mikulas, 1980; Rhodes et al., 1982; Starnes and Williams, 1982.) The approach of Mikulas (1980) is particularly illustrative. Figure 7.22 is extracted from data that appear in a report by Starnes and Williams (1982) and is presented in the form introduced by Mikulas (1980). Reading from the top of the figure to the bottom, the "notch-insensitive" line is simply the strength of a notched coupon in compression based on the net cross section left when the hole is removed from the cross section. Hence, it has the value of unity when there is no hole (when the hole diameter is zero) and goes to zero when the hole diameter is the same as the

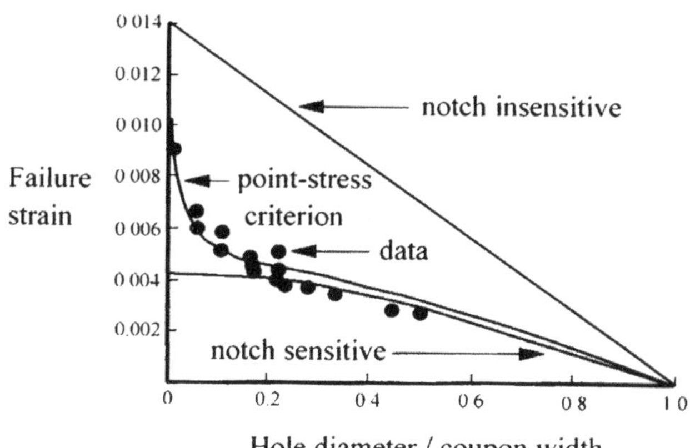

Figure 7.22 Comparison of predicted and observed failure strain in compression for notched coupons as function of center-hole diameter normalized by coupon width (Starnes and Williams, 1982).

specimen width. Then, if a point-stress criterion is used at the laminate level, the second curve from the top is obtained. The data seem to follow that functional form well. Finally, the lowermost curve is the "notch-sensitive" curve, which reduces the predicted strength in proportion to the stress concentration factor at the edge of the hole for the anisotropic plate (as discussed above). The results shown in Figure 7.22 are very typical. For small hole sizes, the effect of the notch is minimal, while for the large holes, the effect begins to approach standard results for homogeneous materials.

As in the tensile case, the model is further improved if the analysis is done at the ply level rather than the laminate level. A good demonstration is provided by the work of Gurdal and Haftka (1987). They were successful in identifying a shift in the critical ply (or critical element) in the failure of multiaxial laminates with a center hole under combined compressive and shear loading (50% of compression) as a function of angle of rotation of the material axes relative to the principal loading axes. They also emphasized the importance of using a failure function that matches the failure mode. In particular, they were successful in predicting the variation of the compressive strength of a quasi-isotropic laminate as a function of loading angle relative to the material axes using a failure function that includes the effect of shear failure as well as fiber direction stress.

7.4 NOTCHED STRENGTH AFTER DAMAGE

Even under quasi-static loading, some damage usually develops in the neighborhood of a notch in brittle-fiber-reinforced composite laminae and laminates. A typical example is shown by the radiograph in Figure 7.23. Matrix cracks appear in all of the plies of the laminate and can be seen running parallel to the fibers in the 90°, 45°, and 0° ply orientations. Quasi-static loading may also produce local delamination initiation (which will be discussed in some detail in a subsequent section). Minor amounts of this damage, such as that shown in Figure 7.23, cause only minor changes in the local stress distribution, and these changes are absorbed into the determination of the material length parameter when notched strength models are calibrated by comparison to pilot data. This situation is analogous to the laminate strength discussion in Chapter 2; that is, ply cracking (first-ply cracking, second-ply cracking, etc.) is part of the process that leads to the last ply fracture that defines the strength. We get a more robust representation if we can follow that progression of damage, even for quasi-static loading, and many such progressive failure models have been developed (cf. Daniel and Ishai, 1994; Eggers, 1985; Kanninen et al., 1975; Whitcomb, 1998). In general, the initiation of damage can be predicted well by such methods, but they are much less reliable when the amount of damage becomes substantial. The challenge with such an approach is to correctly represent the local boundary-value problem when internal surfaces are intricately interlaced. Figure 7.24 illustrates the complexity of the damage, even for a small number of loading cycles. The radiographs in that figure show matrix cracks (Fig. 7.24*a*) and delaminations (Fig. 7.24*b*) produced by cyclic compressive loading in a

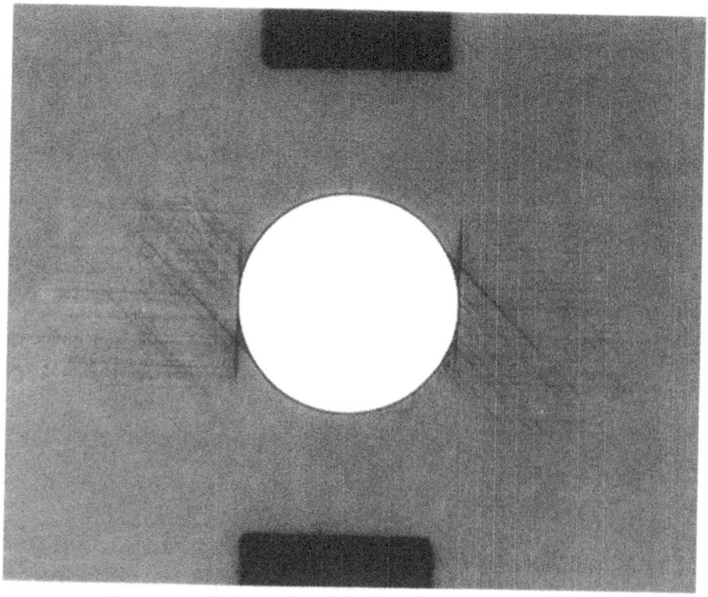

Figure 7.23 Radiograph of matrix cracking caused by uniaxial quasi-static tensile loading (in vertical direction) of quasi-isotropic carbon–epoxy laminate showing matrix cracks and two tangent regions of matrix shear parallel to loading direction.

carbon–epoxy laminate with the ply stacking sequence of [±45, 90, −45, 22.5, −67.5, −22.5, 67.5, ±45, 67.5, 22.5, −67.5, −22.5, 67.5, ±22.5, 22.5]$_S$ after about 10% of its life. Even at this point, the microdamage is complex and interactive. The delamination initiation occurs preferentially near the surface because of reduced constraint in the region, even for interfaces that have the same strain energy release rate as an identical interface in the interior of the thickness; that is, the distribution of the damage is highly three-dimensional in character.

Figures 7.25 and 7.26 show more highly developed damage patterns for fully reversed fatigue loading ($R = -1$) of edge-notched quasi-isotropic carbon–epoxy laminates made from an AS4/1808 (an American Cyanimid toughened epoxy) system. (The uniaxial loading direction is horizontal in these figures.)

When about half of the life of the specimen has been reached (Fig. 7.25), the individual features of the damage can still be identified, that is, matrix cracks and regions of matrix shear, matrix cracking in the 90° plies, shear cracks in the 45° plies, and delamination initiation in near-surface interfaces. However, late in the life (Fig. 7.26), the interaction of damage is severe, and it is a very great challenge to represent all of the details in any progressive failure analysis of all of the microdamage. This remains a frontier at this writing.

The general distribution of damage in the neighborhood of the notch is illustrated in Figure 7.26*b*. The stacking sequence of that laminate is [0, 45, 90, −45]$_{S4}$.

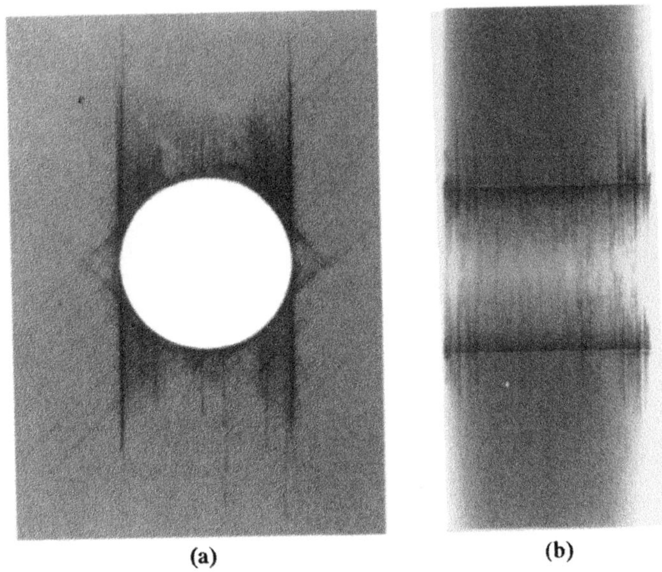

Figure 7.24 (*a*) Radiograph of matrix cracking caused by uniaxial compressive fatigue loading (in vertical direction) of quasi-isotropic carbon–epoxy laminate showing matrix cracks and regions of matrix shear parallel to loading direction; (*b*) thickness direction radiograph of hole region (magnified) showing distribution of matrix cracking and delamination through thickness of 48-ply laminate.

Figure 7.26*b* clearly shows the transverse matrix cracks in the 90° plies and the diagonal shear cracks in the double −45° plies. Delamination is again more prominent near the surfaces of the laminate than in the interior of the thickness. Most of the delaminations are connected in some way to transverse matrix cracking in either the 90° or 45° plies. This coupling is evident in Figure 7.26*a* in another way. The shadowy regions are delaminations that have grown from the notch. Close examination reveals that the boundary of the delaminated region corresponds to the boundary of dense matrix cracking that is associated with the notch; that is, the matrix cracks appear to be drivers for the growth of the delamination. This is a typical result. Matrix cracks do not necessarily degrade strength per se. (They usually occur when the laminate's initial strength is measured to obtain the reference value, for example.) But the greatest danger associated with matrix cracks may be their capability to nucleate or drive delamination. However, delamination is typically a damage mode and atypically a failure mode. More will be said of this situation in the next section.

Figures 7.27 and 7.28 illustrate another important feature of damage development that we must consider in our effort to construct a model of remaining strength after damage development and characterize the long-term behavior of notched laminates. The feature of interest is the effect of load level. Laboratory characterization is most often accelerated, since it is often not possible to run a test for the entire anticipated length of the service life of the component (which may be 60,000 h or more). The

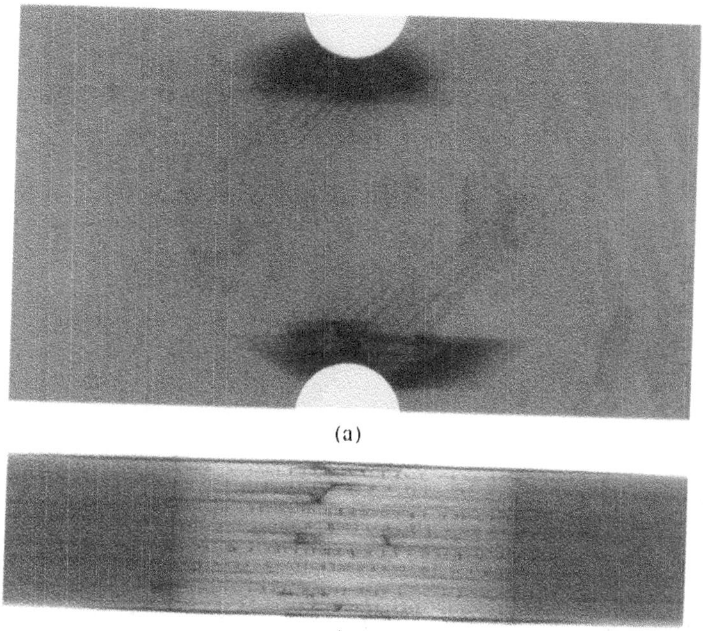

(a)

(b)

Figure 7.25 (*a*) Radiograph of matrix cracking caused by uniaxial fully reversed fatigue loading (in horizontal direction) of quasi-isotropic carbon–epoxy laminate showing matrix cracks and regions of matrix shear parallel to loading direction; (*b*) edge radiograph of notch region (enlarged) showing matrix cracking in 90° plies, shear cracks in 45° plies, and delamination initiation in near-surface interfaces early in life of specimen.

most common method of getting data more quickly is to use higher applied load levels to characterize the material and then extend the data by extrapolation, heuristic models, or mechanistic methods.

In Figures 7.27 and 7.28, the high-load condition produces lives between 1000 and 30,000 cycles, while the low-load condition yields lives in the range of 300,000–1,000,000 cycles. The radiographs were taken at about 10% ("early"), 50% ("middle"), and 90% ("late") of the average specimen life at those levels of loading. The most striking and important difference in the results is that the damage for the low-load case is much more extensive; the lowest cyclic load level produces the most damage for the same percent of life. This is certainly a logical result, since the number of cycles sustained by the specimen at the low-load levels is greater but not intuitive since the level of applied load is lower. This example emphasizes a critical fact that must drive the philosophy used to predict the life of composite components in real engineering applications.

For a specimen to serve the full life at low levels of stress specified in the design of a component, it must sustain and tolerate *more* damage than specimens tested in the laboratory at higher stress levels during characterization of the materials used. For composite systems made from brittle constituents, more damage often means

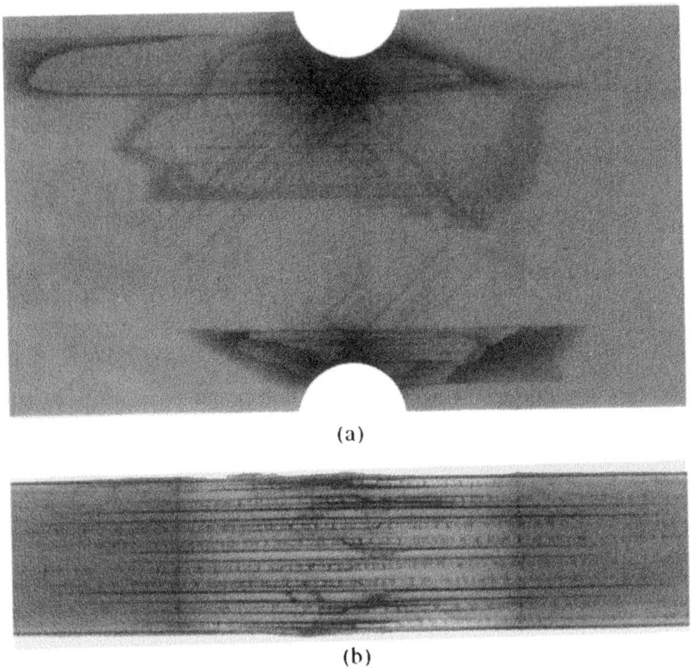

Figure 7.26 (*a*) Radiograph of matrix cracking caused by uniaxial fully reversed fatigue loading (in horizontal direction) of quasi-isotropic carbon–epoxy laminate showing matrix cracks and regions of matrix shear parallel to loading direction; (*b*) edge radiograph of notch region (enlarged) showing matrix cracking in 90° plies, shear cracks in 45° plies, and delamination initiation in near-surface interfaces late in life of specimen.

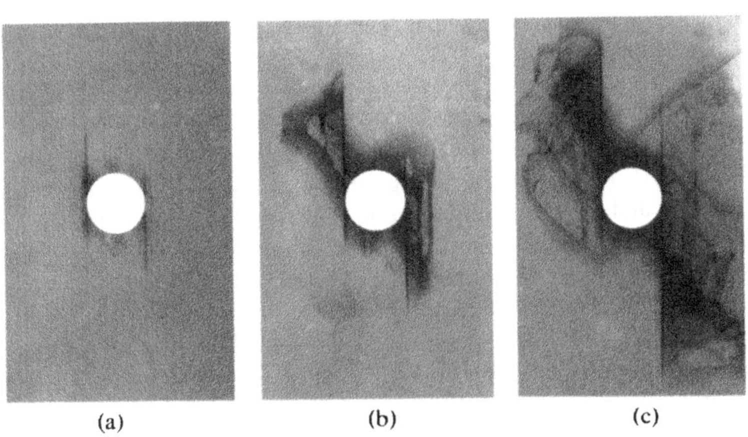

Figure 7.27 Radiographs of center-notched, AS4/1808, $[0, 45, 0, 45]_{S4}$ coupons during high-load fatigue for (*a*) early, 907 cycles; (*b*) middle, 3930 cycles; and (*c*) late, 8410 cycles; conditions in life of specimens.

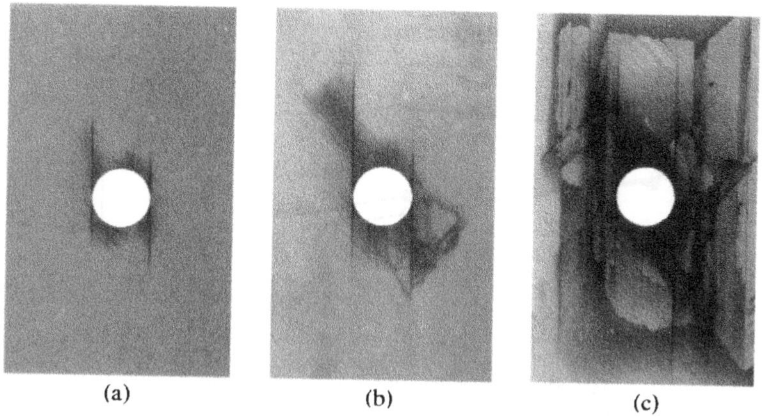

Figure 7.28 Radiographs of center-notched, AS4/1808, [0, 45, 0, 45]$_{S4}$ coupons during low-load fatigue for (*a*) early, 25,860 cycles; (*b*) middle, 100,000 cycles; and (*c*) late, 291,000 cycles; conditions in life of specimens.

larger numbers of microdefects that are more widely and evenly distributed and more likely interactive and coupled than for the high-load (laboratory) case. High-load conditions usually produce highly concentrated, localized, and more directional (if the laminate is anisotropic) damage. Since the object is to model the real case, models of remaining strength must account for large amounts of damage and the changes in material states and (especially) changes in stress states caused by that damage.

Perhaps the most important result of those changes for notched laminates is the relaxation of the stress concentration caused by the notch as a function of cycles of loading and time. Time-dependent behavior was discussed in the chapter on stiffness change; viscoelastic deformation of the matrix will contribute to the relaxation of stress near the notch in the manner discussed earlier and can be handled by entering time-dependent stiffness into the analysis used to solve for the global and ply stresses around the notch.

But cycle-dependent stress relaxation around a notch is less well understood. Figures 7.24–7.28 illustrate that microcracking is extensive and progressive as cyclic loading occurs. Furthermore, the details of the cracking are characteristic of the orientations of the plies in the laminate involved, with cracks typically forming along fiber directions in matrix material that is sustaining high normal tensile or shear stress at the local level. However, the damage patterns have some general global similarities for essentially all laminates. One is the tendency for the high-shear regions (tangent to the hole in the loading direction, as discussed in Figures 7.7–7.14) to develop severe damage (especially obvious in Figs. 7.24*a* and 7.27*a*). This shear damage has an unusual effect for notched composites, especially during cyclic loading in tension and compression. Figure 7.29 illustrates this phenomenon (Bakis and Stinchcomb, 1986, 1988; Reifsnider et al., 1988). The data show that for this laminate, under fully reversed cyclic loading at a level that results in a life of

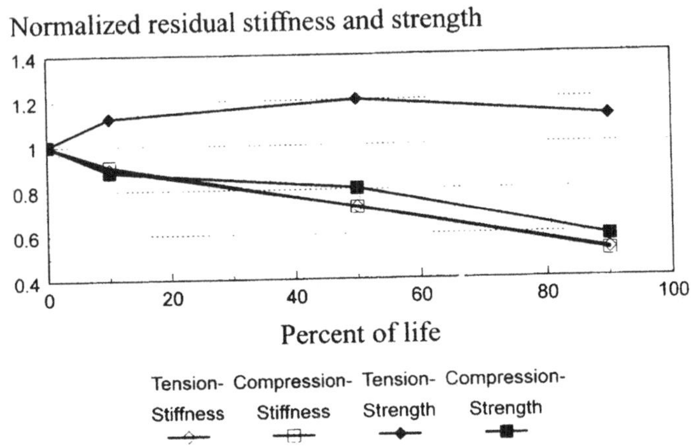

Figure 7.29 Remaining stiffness and strength for AS4/1808, center-notched $[0, 45, 90, -45]_{S4}$ carbon–epoxy laminates under fully reversed cyclic loading as function of percent of life.

about 100,000 cycles, the stiffness of the specimens (measured across the hole) is monotonic decreasing (as expected) with a nearly identical change in the tensile and compressive stiffness at 10, 50, and 90% of the life of the specimens. (Each data point in the figure is an average of data from six specimens.) The change in remaining strength is similar, with a loss of nearly 40% over 90% of the life. However, the change in strength in tension is positive over the first half of life, with an increase of about 20% for this case.

This result is typical. The stress relaxation around a notch in composite laminates is typically large, especially for low stress levels commonly used in engineering applications (for which damage accumulation can be quite large, as discussed in Figs. 7.23–7.28). This relaxation can result in a substantial decrease in the stress concentration near the notch and an increase in the notched strength, especially in tension. This phenomenon is seen in nearly all continuous-fiber-reinforced composite systems, unless the modulus of the fibers is essentially identical to that of the matrix. It is most distinct for systems that are highly anisotropic.

Modeling of this increase in strength can be thought of as having two parts. Modeling of viscoelastic, viscoelastic–plastic, or elastic–plastic relaxation under quasi-static loading has become commonplace, even with commercial finite-element codes such as ANSYS and ABAQUS. Progressive damage methods have been discussed by countless authors (cf. Chamis, 1984; Dunyak, 1991; Reddy and Miravete, 1995; Reifsnider et al., 1988). Nonlinear approximate analysis of the notched problem in composites for quasi-static loading is capable of estimating the ply-level stress and strain using local stiffness reductions and accumulated strains (in the case of plasticity) to account for matrix cracking and microdefect formation. The local stresses and strains can then be used in the notched-strength models (such as the Whitney–Nuismer scheme discussed earlier) to estimate the quasi-static strength in the presence of damage.

The second part of the problem, however, is not so well understood. Although it is possible to estimate the rate of self-similar single-crack growth (based on established characterization methods) in homogeneous and some inhomogeneous materials, the estimation of the rate of accumulation of multiple cracks and defects in composite systems is difficult, especially when those defects interact with each other. As clearly seen from Figures 7.23–7.28, the accumulation of interaction results in the redistribution that causes the stress relaxation and residual strength increase shown in Figure 7.29.

If all of the ply stresses are known as a function of damage accumulation, then it is clear from the literature, as recounted in the present discussion, that the remaining strength can be predicted well by applying a model such as the average-stress or point-stress method at the ply level in the last ply to fail, or the critical element as we have called it, bearing in mind that the azimuthal location around the hole at which that criterion is evaluated must be determined by searching for the maximum in the failure function being used for the criterion. If a rate of damage initiation and accumulation can be established by experiment and represented in a progressive damage model, then this approach is operative. However, direct observation and interpretation of all of the damage events in a practical situation are not possible.

Alternatively, if a measurable external variable, such as some feature of the collective damage that is easily measured, or stiffness change at the global level can be related to the fundamental nature of the local stress relaxation, then the local stresses can be estimated as a function of service history of the notched component. Two engineering methods have proven to be particularly useful and will be outlined here. Both methods depend on the concept of effective ellipticity, as sketched in Figure 7.30. As damage accumulates near a notch, the stress concentration relaxes in proportion. The limit of that relaxation is the ligament strength of the coupon or component, that is, the net section condition. Conceptually, one could simulate the condition by defining an effective ellipticity of the hole (e.g., the ratio b/a in Fig. 7.30) that would seek the limit of infinity when only the net section was holding the load. The initial value of such an effective ellipticity is unity. If a correct ellipticity is chosen for a given damage condition, then the local stress distribution should be changed accordingly. If we change ellipse 1 in Figure 7.30 (the initial circular condition) to ellipse 2, then the distribution of the local normal stress in the principal loading direction might change from curve 1 to curve 2, as suggested in that figure. If that stress distribution is correct (or nearly correct), then we can use one of the criteria discussed to estimate the strength for that state of damage.

The first fundamental question is: Do the stress distributions in laminates actually change in the manner estimated by imposing an effective ellipticity? For a series of laminates tested under uniaxial loading, the estimated stress distributions appear to follow observed ones well (Reifsnider et al., 1995). However, establishing a method for estimating what ellipticity to use is a challenge. One approach is to use the length of the shear-damage zones along lines tangent to the hole in the principal loading direction to estimate, directly, the length of the effective semimajor axis of the ellipse, b in Figure 7.30. Wagnecz (1987) and Bakis (1988) have investigated this method. Using radiography, for example, Bakis measured an average predicted b/a

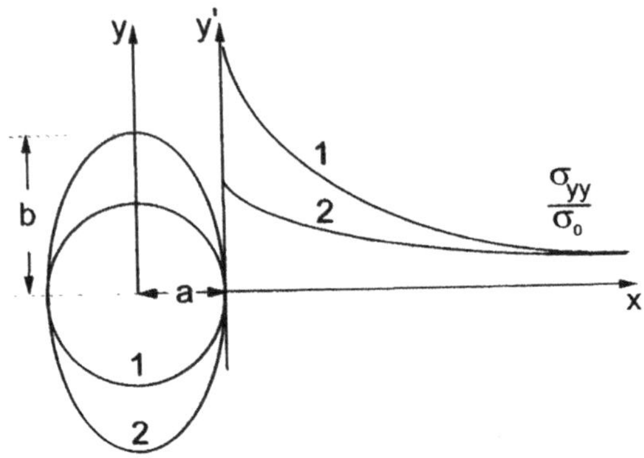

Figure 7.30 Dependence of local stress distribution on ellipticity of hole in infinite plate.

of 2.01 with a standard deviation of 0.736 for a series of laminates, which is to be compared to the value of 1.348 with a standard deviation of 0.12 implied by the comparison of residual strengths to initial strengths of the same specimens, a 30% difference. Obviously a scaling method for such an approach (or a better definition of the correct physical metric) could be sought.

A second method is based on stiffness. For example, if the stiffness of a coupon specimen is measured across the hole as a function of loading history, then the effective ellipticity of the hole can be adjusted in any analysis (such as the Lekhnitskii solution) to match the plate's measured stiffness at those points. The resulting stress distribution is then approximated throughout the plate. Variations of this method have been used with considerable success by the present authors.

7.4.1 Example Calculation

Consider the IM7/K3B material system mentioned above with the properties given in Table 7.3 (Case et al., 1997). Coupon specimens 25.4 mm wide with a center hole

TABLE 7.3 Lamina Properties Used in Example Calculations

Property	Room temperature	350°F (177°C)
E_1	24.3 ± 0.3 msi (168 ± 2 GPa)	25.7 ± 1.6 msi (177 ± 11 GPa)
X_t	392 ± 18 ksi (2700 ± 124 MPa)	325 ± 27 ksi (2240 ± 186 MPa)
E_2	1.38 ± 0.03 msi (9.51 ± 0.21 GPa)	1.27 ± 0.02 msi (8.76 ± 0.14 GPa)
Y_t	8.63 ± 1.17 ksi (59.5 ± 8.1 MPa)	5.15 ± 0.42 ksi (35.5 ± 2.9 MPa)
G_{12}	0.721 ± 0.043 msi (4.97 ± 0.3 GPa)	0.610 ± 0.04 msi (4.21 ± 0.28 GPa)
S	26.4 ± 0.6 ksi (182 ± 4 MPa)	17.1 ± 0.5 ksi (118 ± 3 MPa)
$[0, \pm 45]_{2S}$ notched tensile strength	77.9 ± 3.4 ksi (537 ± 23 MPa)	76.6 ± 0.6 ksi (528 ± 4 MPa)

Note: Number of replicates: 5.

6.35 mm in diameter were made from $[0, \pm 45]_{2S}$ laminates. EP-08-045PG-120 strain gates were fastened to the specimen along the x axis in Figure 7.30 to determine the strain–stress distribution as a function of distance from the hole initially and after cyclic (fatigue) loading at 10 Hz ($R = 0.1$) at 70% and 80% of their tensile ultimate strength. Residual strength was determined for specimens subjected to loading at room temperature and at 177°C. Using the measured strain distributions (for pilot specimens), an effective b/a ratio was determined as a function of stress level and temperature using the equation

$$\frac{b}{a} = A + B\ \exp[-D(\text{Fa})^p n \qquad 7.17)$$

where n is the current cycle count, Fa is the initial ratio of the applied stress level to the ultimate strength of the laminate, and A, B, D, and p are fitting parameters. In addition to modifying the effective axes of the center hole in the Lekhnitskii solution, the values of the matrix-controlled stiffness, E_2 and G_{12}, were modified according to the equation

$$\frac{G_{12}(n)}{G_{12}(0)} = \frac{E_2(n)}{E_2(0)} = C - (C - 1)(1 + \alpha n)\ \exp(-\beta n) \qquad (7.18)$$

where C, α, and β are fitting parameters. The parameters in (7.17) and (7.18) were determined by comparing (at room temperature) the strain–stress distributions for pilot tests and used for all other conditions. An example of the changes in ellipticity as a function of various loading conditions as a function of cycles of loading is shown in Figure 7.31. For high load levels at elevated temperature, the ellipticity of the hole is extreme, with b/a values up to 17. Room-temperature results are more modest, with typical values of the order of 4 or 5 for this material.

An example of the comparison of actual strain distributions with predicted values appears in Figure 7.32. The predicted and observed strain distributions are in good agreement for the undamaged as well as the damaged condition (for which the values of strain are very different). The physical damage in these specimens was quite severe and included matrix cracking and substantial local delamination, not unlike the specimens shown in Figures 7.2*c* and 7.28*c*.

In Chapter 8 we will discuss this example more completely, since there are several features of the influence of temperature that should be discussed more fully. However, in the present context, using room-temperature fatigue behavior, the average stress criterion, and (7.17)–(7.18) in (7.1), the predicted and observed remaining strength result shown in Figure 7.33 is produced. For this situation, although the strength of the critical element (the 0° ply in this case) is decreasing monotonically, resulting in an initial drop in global strength for loading at the 80% level, the global strength is generally increasing for the range of these tests. This occurs because the stress relaxation around the hole is increasing the global strength (by lowering the local stress in the critical element) more rapidly than the degradation of the strength of the critical element itself. The increase is highest for the lowest loading level, and

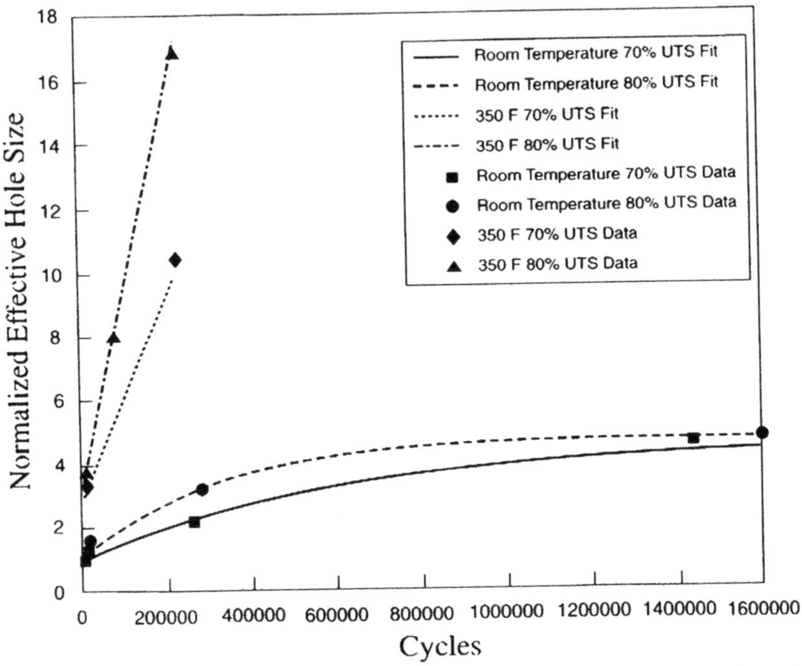

Figure 7.31 Normalized effective major axis as function of cycles as determined from measurements of strain distributions after various loading histories.

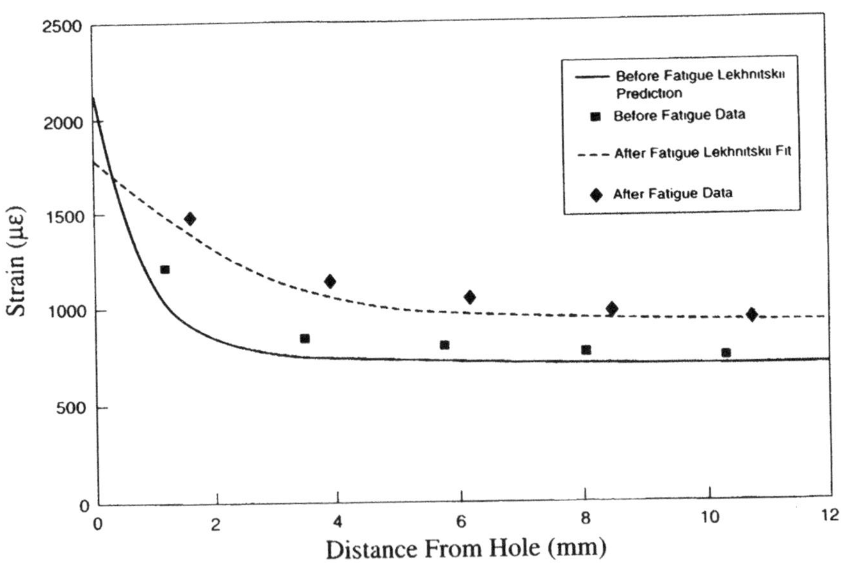

Figure 7.32 Predicted and observed normal strain distributions in loading direction as function of distance from edge of center hole in $[0, +45, -45]_S$ IM7/K3B laminates before and after fatigue damage development.

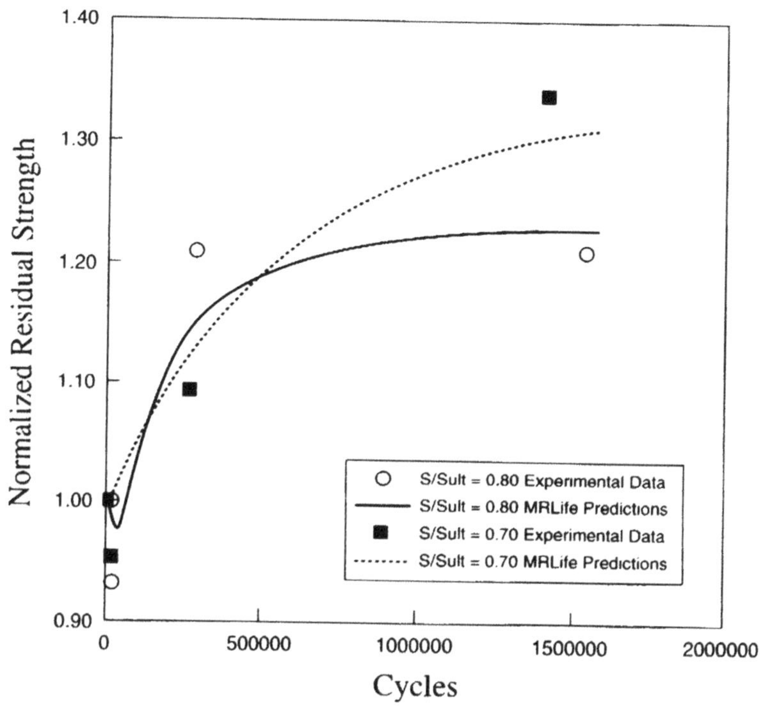

Figure 7.33 Predicted and observed remaining strength of notched IM7/K3B coupons fatigued at room temperature as function of cycles.

the predicted values follow the data well. Equation (7.1) makes it possible to correctly predict the net effect of the two competing physical processes, the relaxation of the global stresses and the degradation of the critical element. In other situations, the higher loading will produce the greatest increase in remaining strength. This is a predictive capability in the direct sense, since none of the input data relate to this issue.

7.5 FRACTURE MECHANICS AND ENERGY METHODS

7.5.1 Fundamental Concepts

The general field of fracture mechanics is firmly founded and described widely and well in the literature. General introductory works include Anderson (1995) and Broutman (1974), with signature literature in a series of Special Technical Publications of the American Society for Testing and Materials (cf. Smith, 1979). Starting points for more advanced study include Kanninen and Poplar (1985), Freund (1990), and Rice (1968b). Only the concepts of the basic philosophy will be outlined here to support the applications of interest to the present discussion.

Fracture mechanics, in the modern interpretation, concentrates on the nature of the stress concentration associated with geometrically sharp notches or defects, especially stress–strain distribution in the regions near the tips of cracks. Inglis (1913) is generally credited with the identification of this problem. Inglis was able to show that the normal stress in the direction of the applied loading at the tip of an elliptical notch in an isotropic material has the value given by

$$\sigma = \sigma^{\infty}\left(1 + \frac{2a}{b}\right) \tag{7.19}$$

where σ^{∞} is the remote applied stress and a and b are the semimajor and semiminor axes of the elliptical notch, as shown in Figure 7.34. When the notch is circular ($a = b$), the classical stress concentration of value 3 is recovered, but as the ellipticity of the notch increases, the local stress concentration becomes much higher. Using a geometric identity, (7.19) can be stated in terms of the notch root radius ρ (also shown in Fig. 7.34) to obtain

$$\sigma = \sigma^{\infty}\left(1 + 2\sqrt{\frac{a}{\rho}}\right) \tag{7.20}$$

which becomes

$$\sigma \cong \sigma^{\infty} 2\sqrt{\frac{a}{\rho}} \tag{7.21}$$

for $a \gg b$. Inglis showed that (7.21) is a good approximation for essentially any notch that is elliptical at the notch tip. The stress in (7.19)–(7.21) is a point function; that is, it is the analytical estimate of the normal stress at a point on the surface of the notch at the notch tip.

These equations predict that if the local notch root radius approaches the value zero, that is, if the notch becomes infinitely sharp, then the local point stress becomes infinite in magnitude. But we must ask what physical behavior corresponds to this situation? If the material has any ductility, local yielding may prevent the radius of curvature from becoming infinitely sharp. If the material is absolutely rigid, then one might expect that the local limit on the radius of curvature is the atomic spacing and the resulting predicted stress might match the stress needed to separate two atoms of the material. A simple one-dimensional analysis (found in most elementary materials science and physics texts) provides an estimate of the separation (cohesive) stress in the form

$$\sigma_c = \sqrt{\frac{E\gamma_S}{\chi_0}} \tag{7.22}$$

where E is Young's modulus of the material, γ_S is the surface energy per unit area of the separation surface, and χ_0 is the initial atomic spacing of the atoms. Then for this physical interpretation, equating (7.21) and (7.22) suggests that the remote stress needed to cause a crack to form at the tip of a notch is equal to

$$\sigma_I = \sqrt{\frac{E\gamma_S}{4a}} \tag{7.23}$$

where σ_I is the crack initiation stress. There are many approximations embodied in (7.23), including the fact that the approach of Inglis can be applied to the atomic level of material analysis. However, similar forms have been found by simulations (Gehlen and Kanninen, 1970).

One might expect the initiation stress given in (7.23) to be a limiting value, since the material behavior has been idealized as rigid–elastic with no operating ductility. Only under those conditions can we expect a local point-stress value at an infinitesimal point on the material to control physical behavior, that is, the formation of a crack. For other types of physical behavior, such as local plasticity, a larger region of material is involved and energy arguments are needed.

Griffith (1920) provided the classical foundation for energy approaches to this philosophy. Griffith argued on thermodynamic grounds that the energy of crack formation or growth must be balanced by the energy needed to form the new crack surfaces, that is, the total change in stored strain energy and the work of external applied forces must match the total surface energy of the newly formed crack.

Using simple one-dimensional arguments, Griffith postulated that the total stored strain energy released when a crack having total length $2a$ is formed in a plate (as in Fig. 7.34) could be approximated by the expression $-\pi a^2\sigma^2/E$, which implies that, that all of the stored energy in a circle with radius equal to the half-crack length was released. He also argued that this energy must be exactly balanced by the surface energy required to form the crack, which he set equal to $4a\gamma_S$ (in a plate with unit thickness). If the crack is in equilibrium upon formation, then

$$\frac{\delta}{\delta a}\left(4a\gamma_S - \frac{\pi a^2\sigma^2}{E}\right) = 0 \tag{7.24}$$

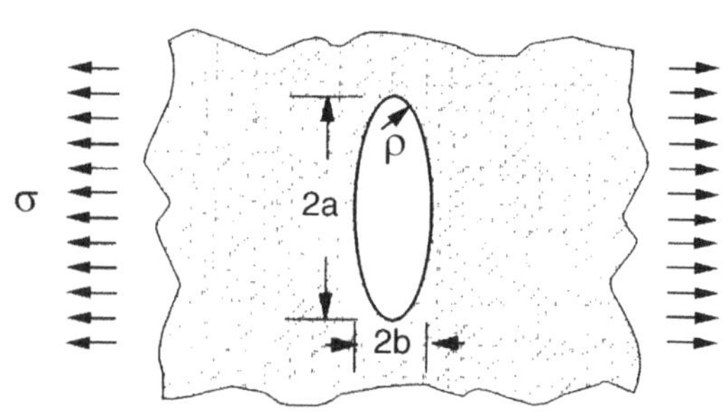

Figure 7.34 Elliptical hole in flat plate of infinite extent under uniform remote loading.

from which Griffith obtained the equation

$$\sigma_f = \sqrt{\frac{2E\gamma_S}{\pi a}} \tag{7.25}$$

where σ_f is the *fracture stress*, in the notation of Griffith, and a is the half-crack length. This equation was further modified by Irwin (1948) and Orowan (1948) to explicitly represent the work per unit area of plastic deformation, γ_P, by writing the form

$$\sigma_f = \sqrt{\frac{2E(\gamma_S + \gamma_P)}{\pi a}} \tag{7.26}$$

The contrast between the brittle initiation behavior estimated by (7.23) and the ductile growth by (7.26) is clear. One can expect brittle glasslike fracture or cleavage to have strengths closer to (7.23) and materials for which a region of material behavior (like a local plastic zone) controls fracture to have strengths closer to (7.26). Such is the case. And it is common for these two equations to be discussed in terms of crack initiation [(7.23)] and crack propagation [(7.26)] (e.g., Anderson, 1995).

These equations can be further combined to estimate the remote stress needed to initiate a crack with half length equal to just one atomic spacing at the tip of an elliptical crack (Fig. 7.34) by setting $\rho = a$ in (7.26) and equating (7.26) and (7.21) to obtain

$$\sigma^{\infty} = \sqrt{\frac{\gamma_S E}{2\pi a}} \tag{7.27}$$

the form discussed by Orowan (1948).

An important physical interpretation of fracture mechanics results from a rearrangement of (7.27). Recognizing that $\sqrt{\gamma_S E}$ is a material constant, a material property may be defined, called the *fracture toughness* for crack growth, by the equation

$$K_C = \beta \sigma_C^{\infty} \sqrt{a} \tag{7.28}$$

where σ_C^{∞} is the critical remote stress necessary to cause global fracture for a crack of length $2a$ and β is a geometric factor that we will define below. Fracture toughness, as the physical resistance to crack growth, is determined from laboratory experiments and is an important design parameter for brittle-material behavior in which the local stresses that drive crack growth controls global fracture strength.

The concept of strain energy release rate is also an energy argument that is useful in our discussion of crack growth in the present context. Consider the thought experiment represented in Figure 7.35, wherein a crack with initial length $2a$ extends

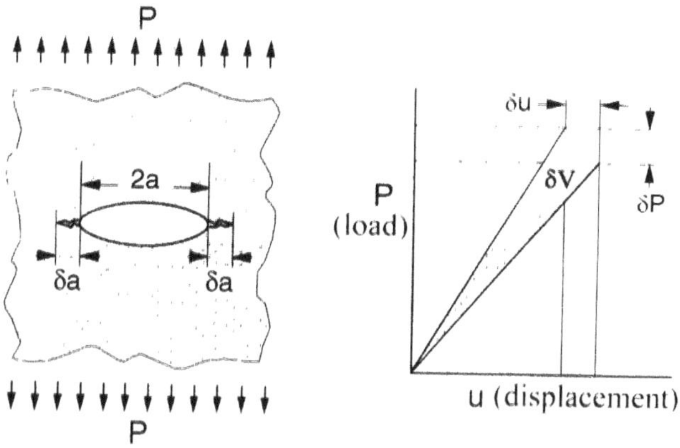

Figure 7.35 Incremental extension of crack in infinite plate and crack and corresponding changes in load–displacement diagram.

by an amount δa due to the application of the load P. The process of crack extension changes the stored energy by an amount δV. The applied force does an amount of work $\delta W = P\,\delta u$, where u is the global displacement of the point of action of the force P. For a plate with thickness B, if the rate of strain energy release per unit area of crack extension is G, then one can write an energy balance in the form

$$G(B\,\delta a) = \delta W - \delta V \tag{7.29}$$

which represents the work done in separating two planes of material to form the new crack increment of area, $B\,\delta a$. If we introduce the compliance of the plate, C, so that $u = CP$, then

$$\delta V = \delta(\tfrac{1}{2}CP^2) \qquad \delta W = P\,\delta u = P\,\delta(cP) \tag{7.30}$$

Combining (7.29) and (7.30) and solving for G, one obtains the classical equation

$$G = \frac{1}{2}P^2\frac{\delta C}{\delta A} \tag{7.31}$$

where $A = B\,\delta a$, the increment of the new crack area.

We can think of the strain energy release rate as the driving force for crack extension. The remarkable physical generality represented by (7.31) is that the driving force for crack extension is proportional to the global (structural) compliance change of the element. This is a robust result that can be applied to multiple-defect development and for more complex loading situations (Rice, 1968 a,b; Rice, 1988). It can also be used for the analysis of crack growth with limited local plasticity (Dowling, 1993).

The applicability of (7.23), (7.26)–(7.28), and (7.31) depends on several other assumptions that are important to the present discussion. The assumption is that self-similar crack initiation and growth occurs, that is, that the local geometry of the crack is always the same. It is also assumed that the materials are homogeneous (single phase or single material), uniform (no spatial dependence of properties from point to point), and isotropic. Many of these assumptions can be relaxed with more rigorous formulations, as we will see in subsequent discussions. But one major feature always remains. Fracture mechanics is always concerned with local behavior at a small scale. Physical behavior such as local crack growth is likely to be described well by such an approach. But the global strength of composite materials often depends on the accumulation and interaction of multiple microfracture events rather than on their initiation per se. Therefore, care is needed in the application of fracture mechanics to the estimation of the global strength of composite materials. If the formation and self-similar propagation of a single crack controls global strength, then fracture toughness can be defined uniquely and fracture mechanics representations are likely to be consistent with physical observations of strength. However, if many cracks interact, slow stable crack growth occurs, or substantial material constitutional nonlinearity occurs at the local or global level, linear elastic fracture mechanics may have limited use in the description of the behavior.

We will examine two examples of the use of fracture mechanics for the description of the behavior of laminates. The first employs fracture mechanics for the estimation of the strength of notched composite laminates and the second examines the use of fracture mechanics for the description of the growth of delamination.

7.5.2 Fracture Mechanics Representation of Laminate Notched Strength

We mentioned that it is possible to formulate more complete mechanics representations of the general problem of a crack in an infinite plate, and in some cases, closed-form solutions to those formulations can be obtained. The most typical equations are solutions for plane-strain deformations of cracked bodies of infinite extent. These solutions have the general forms

$$\lim_{r \to 0} \sigma_{ij}^{k} = \frac{K_k}{\sqrt{2\pi r}} f_{ij}^{k}(\theta \qquad 7.32)$$

where i, j refer to the (x, y, z) coordinates in Figure 7.36, k refers to the mode of crack deformation ($k = 1$ for mode I crack opening, $k = 2$ for mode II in-plane shear, $k = 3$ for mode III out-of-plane shear), and K is the field stress intensity (the amplitude of the local stress field). The early solutions of this type were provided by Westergaard (1939), Sneddon (1946), Irwin (1957), and Williams (1957). A general discussion can be found in Anderson (1995) and Kanninen and Poplar (1985).

The field stress intensity has the form discussed in (7.28); that is, it can generally be written as

$$K = \beta\sigma\sqrt{\pi a} \qquad (7.33)$$

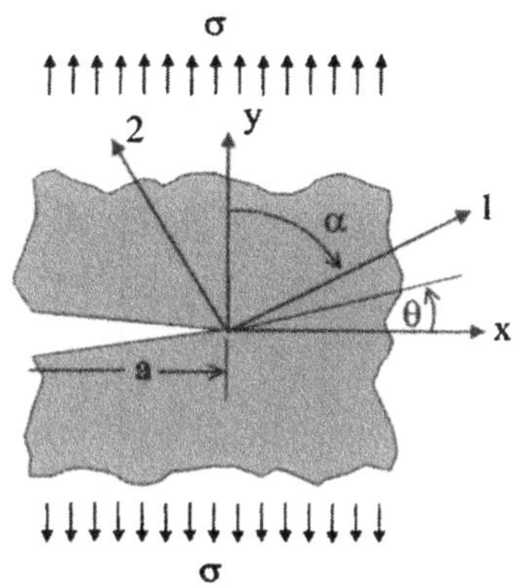

Figure 7.36 Coordinate systems used for description of crack in anisotropic solid.

where it is understood that K may include contributions from combined loading modes of I, II, and III, in general. For mode I loading of an isotropic finite plate with width $2W$ and crack length $2a$, for example,

$$K = \sigma\sqrt{\pi a}\left[\frac{2W}{\pi a}\tan\left(\frac{\pi a}{2W}\right)\right]^{1/2} \tag{7.34}$$

For a given crack length, if the applied stress σ is increased monotonically, fracture of the finite plate will occur when the value of K reaches the critical value K_C, the measured fracture toughness of the material. If fracture mechanics analysis correctly describes the physics and mechanics of the fracture process, that critical value will be a material property. It will not depend on the specimen geometry, the mode of loading, or the geometry of the crack. The form of (7.34) will, however, change as these parameters change.

For present purposes, consider mode I loading of a unidirectional (orthotropic) plate having a crack with length $2a$, as shown in Figure 7.36. We will illustrate the subject concept by outlining the approach proposed by Poe and Sova (1980). The singular stress field for a crack with length $2a$ in such a material under mode I loading (shown in Fig. 7.36) is given by

$$\begin{bmatrix} \sigma_x \\ \sigma_y \\ \tau_{xy} \end{bmatrix} = \frac{K}{E_y\sqrt{2\pi r}}\begin{bmatrix} \sqrt{E_x E_y} \\ E_y \\ 0 \end{bmatrix} \tag{7.35}$$

where Young's modulus values refer to the global directions shown in Figure 7.36. The corresponding strains can be given by

$$\begin{bmatrix} \varepsilon_x \\ \varepsilon_y \\ \gamma_{xy} \end{bmatrix} = \frac{K}{E_y\sqrt{2\pi r}}[\kappa]\begin{bmatrix} \sqrt{E_x E_y} \\ E_y \\ 0 \end{bmatrix} \qquad [\kappa] = \begin{bmatrix} \frac{1}{E_x} & \frac{-\nu_{yx}}{E_y} & 0 \\ \frac{-\nu_{yx}}{E_y} & \frac{1}{E_y} & 0 \\ 0 & 0 & \frac{1}{G_{xy}} \end{bmatrix} \tag{7.36}$$

However, many engineering composite systems are fiber dominated; that is, a small group of fibers are critical elements, as we discussed previously. When that group of fibers breaks in a local region, global fracture occurs. This concept and the concept of a characteristic material length (or volume) over which the local fracture process is defined (associated with Wu and Whitney–Nuismer in our earlier discussion) can be used in the present approach. First, we can restate the strains in the fiber coordinate system, the (1, 2) system in Figure 7.36, by using the usual transformation matrix

$$\begin{bmatrix} \varepsilon_1 \\ \varepsilon_2 \\ \gamma_{12} \end{bmatrix}_i = [T]_i\begin{bmatrix} \varepsilon_x \\ \varepsilon_y \\ \gamma_{xy} \end{bmatrix} \qquad [T] = \begin{bmatrix} \sin^2(\alpha) & \cos^2(\alpha) & \sin(2\alpha) \\ \cos^2(\alpha) & \sin^2(\alpha) & -\sin(2\alpha) \\ -\frac{1}{2}\sin(2\alpha) & \frac{1}{2}\sin(2\alpha) & \cos(2\alpha) \end{bmatrix} \tag{7.37}$$

where the subscript i refers to the ith ply in the laminate, having orientation α, as shown in Figure 7.36. Then combining (7.36) and (7.37), one can write

$$\begin{bmatrix} \varepsilon_1 \\ \varepsilon_2 \\ \gamma_{12} \end{bmatrix}_i = \frac{K}{E\sqrt{2\pi r}}\begin{bmatrix} \xi_1 \\ \xi_2 \\ \xi_{12} \end{bmatrix}_i \tag{7.38}$$

where

$$\begin{bmatrix} \xi_1 \\ \xi_2 \\ \xi_{12} \end{bmatrix}_i = [T]_i[\kappa]\begin{bmatrix} \sqrt{E_x E_y} \\ E_y \\ 0 \end{bmatrix} \tag{7.39}$$

in ply i. Poe and Sova (1980) postulate that laminate failure occurs when the fiber direction strains in the critical load-carrying ply reach a limiting value. This assumption is consistent with three elements of the general philosophy described in this book. First, the critical ply in Poe and Sova is the critical element in the failure criterion in the present case. Second, failure criteria for a laminate seem to work best when applied at the ply level. Third, the maximum-strain criterion often matches physical behavior well for laminates, which display basically brittle failure.

Considering (7.38) and using the same physical reasoning that we used to write (7.28), it can be argued (as did Poe and Sova) that $\varepsilon_1\sqrt{2\pi r}$ is a constant at failure for all laminae independent of orientation. With this idea, it is possible to define a toughness parameter with the form

$$Q_C = \frac{K_Q(\xi_1)_i}{E_y} \tag{7.40}$$

where K_Q is a candidate value material constant and $(\xi_1)_i$ is the fiber direction value in the critical ply. For the 0° plies of a laminate, for example, $(\xi_1)_0 = 1 - \nu_{xy}\sqrt{E_x/E_y}$. If Q_C is a material property, it should have constant values for all laminate types made from a given composite material system. Table 7.4 shows some values for Q_C and for ξ_1 for the boron–aluminum system examined by Poe and Sova.

For general laminates (such as the middle three in Table 7.4), the toughness of the material is essentially constant when defined in this way. For the special cases of unidirectional and entirely off-axis laminates, the toughness is not constant. In the latter cases, the assumption is that the failure mode was not consistent with the assumptions discussed above and not the same as the other laminates. Table 7.4 also shows there are at least two other fundamental issues in this formulation. First, how is K_Q obtained?

Poe and Sova argued along the following general lines for the form of K_Q. Using the general tenets of linear elastic fracture mechanics, one could write a standard field stress intensity factor for a slit in a finite plate in the form

$$K = S\sqrt{\pi a \sec\left(\frac{\pi a}{w}\right)} \tag{7.41}$$

where S is the remote stress, a is the half-slit (or crack) length, and w is the width of the plate. Then, it is possible to introduce an initial flaw concept, with dimension a_O,

TABLE 7.4 Values of Parameter in Poe–Sova Model

Laminate orientation	ξ_1	$\xi_1 K_Q/E_y$	$\xi_1 K_Q/E_{uy}$
$[0]_6$	0.8409	0.012	0.01347
$[0_2, +45, -45]_S$	0.7841	0.00764	0.01223
$[+45, -45, 0_2]_S$	0.7806	0.00771	0.01233
$[0, +45, -45]_S$	0.7375	0.0077	0.01477
$[+45, -45]_{2S}$	0.6771	0.0035	0.1001

Source: Poe and Sova (1980).

such that the critical field stress intensity K_Q is defined by the ultimate strength S_C^{un} when the half-crack length a goes to zero. Using (7.41), one can define a_O as

$$a_O = \frac{(K_Q/S_C^{un})^2}{\pi} \tag{7.42}$$

which has the physical interpretation of the damage zone size associated with fracture in the composite system being modeled. Then, for the notched condition,

$$K_Q = S_C^n \sqrt{\pi(a + a_O)\ \sec\left(\frac{\pi a}{w}\right)} \tag{7.43}$$

and if we define the standard elastic isotropic critical stress intensity factor as

$$K_{QE} = S_C^n \sqrt{\pi(a)\ \sec\left(\frac{\pi a}{w}\right)} \tag{7.44}$$

and use (7.41)–(7.44), we can solve to find

$$K_Q = \frac{K_{QE}}{\sqrt{1 - K_{QE}^2/\pi a(S_C^{un})^2}} \tag{7.45}$$

which can be used in the toughness equation (7.40).

The other fundamental question is what to use for the laminate stiffness E_y in the toughness equation. If damage development is significant in the presence of a notch (or without one), then the stiffness of the laminate just before laminate failure (defining strength) may be quite different from the initial value, as we have seen in the chapter on stiffness change. In that case, the stiffness at fracture (in some sense) should be used. Poe and Sova made such a correction, calling the reduced stiffness E_{uy}, and obtained the results shown in the last column in Table 7.4. With this correction, only one laminate (which is likely to have failed primarily in shear rather than by fiber failure) was clearly not described by the modeling method. A variety of other laminates and material systems were examined by Poe and Sova with similar results. With a model such as this one, the strength of any notched laminate can be estimated if the toughness parameter is determined from pilot tests. Figure 7.37, reproduced from the work of Poe and Sova, shows typical results for such a prediction for two other material systems.

This example illustrates the general nature of a rational application of linear elastic fracture mechanics to the prediction of the notched strength of composite laminates. Its success is a result of several general rules discussed widely in this text. First, the failure mode of the laminates is well defined and specifically modeled. Second, the idea of strength is evaluated at the local level, the ply level in this case. Hence, the 0° ply becomes the critical element for this model. Third, for this non-uniform stress state, a characteristic material length is introduced to account for the

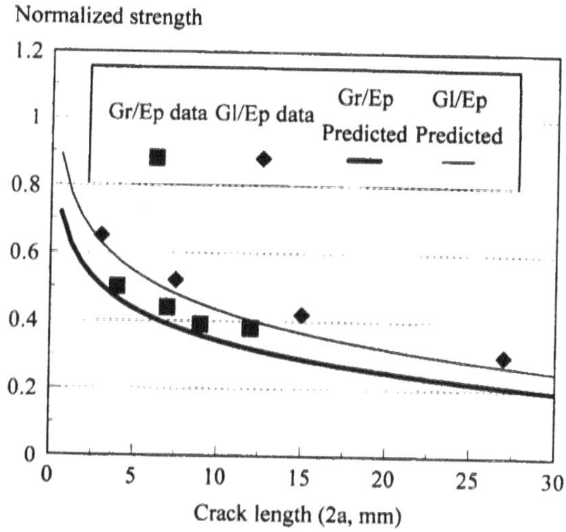

Figure 7.37 Predicted and observed strength of graphite–epoxy (Gr/Ep) and E-glass–epoxy (Gl/Ep) cross-ply coupon laminates as function of slit length (Poe and Sova, 1980).

finite volume of material that is involved in the process of damage and failure of the composite material. With these corrections, linear elastic fracture mechanics seems to be a useful representation of strength, at least for the quasi-static case. However, we should notice that such a model is based on the fiction of an infinite singular stress field at the notch or crack tip, which is even more at variance with physical reality in inhomogeneous materials. The model is a highly local one and is less likely to be useful for the prediction of remaining strength after long periods of damage development than the damage accumulation approaches discussed earlier.

7.5.3 Delamination

Perhaps the most successful application of fracture mechanics to composites has been for the purpose of analyzing delamination formation and growth. In fact, the literature on this subject is remarkably prolific. Reviews have been given by O'Brien (1991) and in an ASTM Special Technical Publication on "Delamination and Debonding of Materials" (1985). Since the literature is so mature in this area, only some conceptual arguments will be presented in this section. The presentation will feature a discussion of the effects of lamination, the initiation of delamination, the growth of delamination, and some comments on applications in the present context.

To appreciate the effects and consequences of delamination, we begin with a reminder of the consequences of lamination. We will limit the discussion to lamination of planar solid bodies for which the in-plane dimensions are generally large compared to the laminate thickness, a typical situation for applications of composites. The most important effect of lamination for thin layers of such materials is the result that the in-plane strains of all of the plies are the same under in-plane loading and follow classical plate theory for out-of-plane deformations. The mechanics of such deformations are well developed for infinitesimal and finite deformations (Hyer, 1998). For our purposes, however, the consequences of lamination include the effects on defect initiation and growth.

Composite materials, especially those that involve aligned fiber reinforcement, usually develop widely distributed damage events that are arrested by the inhomogeneous nature of the material. For a unidirectional lamina, however, the defect growth is only arrested in directions that intersect the fibers. Cracks or debonds that travel in the direction of the fibers are free to propagate in an unstable manner and cause failure in a highly brittle fashion. Lamination inhibits such growth by providing strength in weak directions when fibers in different laminae have different orientations and by arresting crack growth in those directions. As it happens, the distributed nature of damage in fiber-reinforced composite materials is one of the keys to the success of such systems in engineering components. This is especially true for the long-term behavior (durability and damage tolerance) of these materials. Distributed damage avoids damage localization, a primary cause of time-dependent and cycle-dependent failure in homogeneous materials.

The most important exception to this generality is delamination. The physical process of delamination is the formation and self-similar propagation of a cracklike defect, usually along material principal directions at the local or ply level. When delamination occurs, there are at least two important aspects of the degradation. First, there is a loss of stiffness of the laminate, since the strains of all the plies are no longer necessarily the same in the plane; that is, the composite system is less constrained at the local level. Second, there is generally a loss of strength because the effect of ply-to-ply reinforcement is lost and defects can grow in weak directions uninhibited by the constraint of adjacent plies. The change in stiffness caused by delamination may be quite large, as we will see, so the attendant stress redistribution (flow of stress into the remaining unbroken material) may be large as well. However, while delamination is a common damage mode in composite laminates, it is a less common failure mode if failure is defined by fracture. Delamination-induced failure is more commonly caused by stiffness loss, especially since many designs are, in fact, stiffness designs. In this context, compression failure by local or global instability can be induced by delamination and is a serious failure mode associated with this damage mode.

Initiation Delamination is initiated by the formation of a local crack, in the mechanics sense. Physically, this process is driven by stresses and energy that tend to create internal surfaces by deformations characterized as mode I, mode II, or mode III cracking. Figure 7.38 illustrates those deformation modes. Mode I deformation is driven by crack-opening normal stresses, such as the interlaminar tensile stresses in Figure 7.5. Mode II and III deformations can be driven by interlaminar shear stresses such as those shown in Figure 7.2. We will discuss the manner in which delamination initiates as a result of such deformations using two concepts, a stress-based argument and an energy-based method.

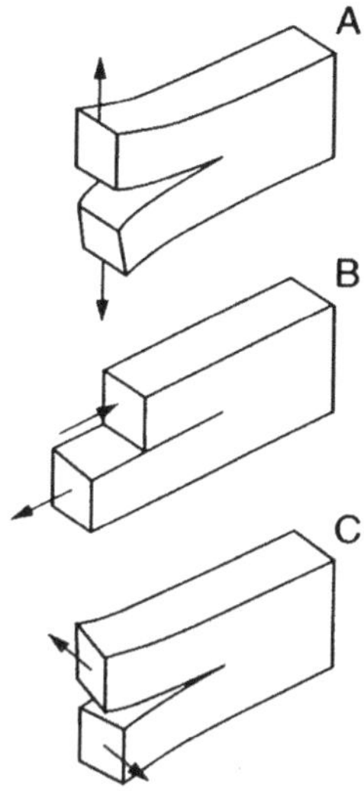

Figure 7.38 (*a*) Mode I, (*b*) mode II, and (*c*) mode III crack deformation types.

It is well known that the order of the singularity associated with finite edge effects in composite laminates is affected by the properties of the plies adjacent to the delamination and the properties of the laminate and that the order can be different from the $1/\sqrt{r}$ dependence noted in (7.32). If we wish to use a stress-based argument to estimate the initiation of delamination, the correct local stresses can only be found if that order is determined. Numerous investigators have addressed this problem. An outline of the approach used by Shalev and Reifsnider (1990) will be presented here, and the results will be compared to experimental data for the initiation of delamination around the periphery of a center hole in a coupon.

The details of the geometry and coordinates used to discuss this approach are shown in Figure 7.39. Near the edge of the laminate (either the free edge or the boundary of the center hole), the stress field is dominated by the singular stresses. Within the boundary layer near the edge, changes with respect to x_1 and x_2 are considered to be large compared to changes with respect to x_3, so that within that region, changes with respect to x_3 are neglected. Beyond the boundary region, the field stresses must match the remote solution, which is taken to be the form given by Lekhnitskii (1981). End effects are neglected. Inserting the small-strain displacement equations into the resulting equilibrium equations and integrating result in the forms

$$
\begin{aligned}
u_1 &= -\tfrac{1}{2}A_1S_{33}x_3^2 - A_4x_2x_3 + U_1(x_1, x_2) + \omega_2x_3 - \omega_3x_2 + u_{10} \\
u_2 &= -\tfrac{1}{2}A_2S_{33}x_3^2 - A_4x_1x_3 + U_2(x_1, x_2) + \omega_3x_1 - \omega_1x_3 + u_{20} \\
u_3 &= (A_1x_1 + A_2x_2 + A_3)S_{33}x_3 + U_3(x_1, x_2) + \omega_1x_2 - \omega_2x_1 + u_{30}
\end{aligned}
\tag{7.46}
$$

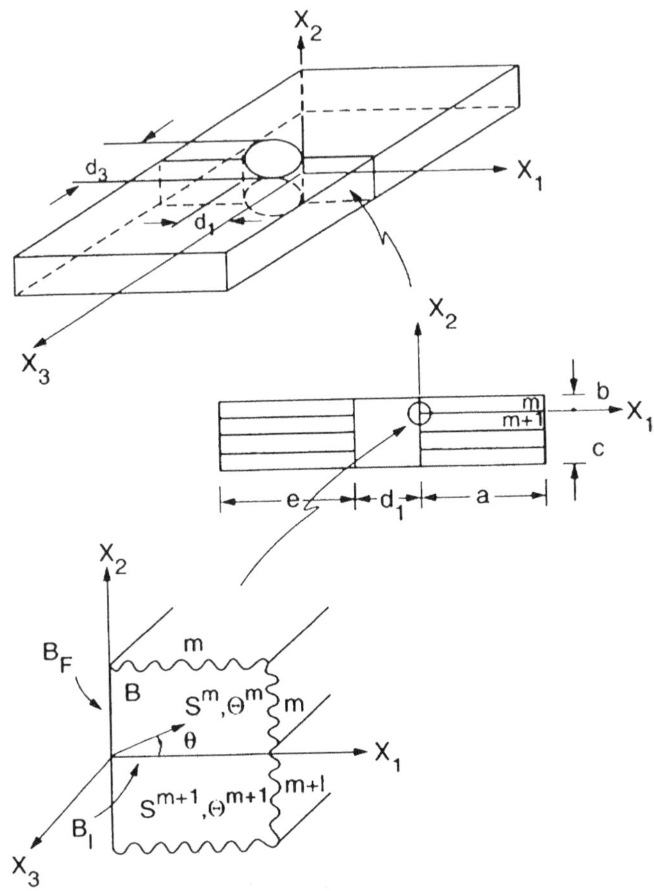

Figure 7.39 Geometry and coordinates of composite laminate with embedded elliptical hole.

where u_{i0} and ω_i are rigid-body translations and rotations, respectively, S_{ij} are the components of the compliance tensor, and A_i are constants of integration. Then, following the usual notation of Wang and Choi (1982a,b) and Lekhnitskii (1981), the resulting governing equations have the form

$$\begin{aligned} L_3F + L_2\Psi &= -2A_4 + A_1S_{34} - A_2S_{35} \\ L_4F + L_3\Psi &= 0 \end{aligned} \tag{7.47}$$

where L_i are operators having the usual definitions and F and Ψ are the stress potentials defined by Lekhnitskii. Considering a laminate of finite extent, traction-free boundary conditions are imposed at the edge of the laminate and at the boundary of the elliptical hole. Static equilibrium with the remote loading is imposed by integrals of the form suggested by Lekhnitskii (1981),

$$\iint_B \sigma_{33} x_2 \, dx_1 \, dx_2 = M_{11}$$
$$\iint_B \sigma_{33} \, dx_1 \, dx_2 = P_{33} \tag{7.48}$$

over the cross-sectional area B. In the vicinity of the hole, the laminate is converted into an anisotropic plate by means of its effective elastic constants, $(E_x, E_y, G_{xy}, v_{xy})$ calculated by CLT. These properties are used in the Lekhnitskii solution to obtain global strains that are, in turn, used in CLT to obtain the ply-level stresses that apply beyond the boundary region where edge effects are negligible.

Continuity of traction and displacements at the interfaces between the plies is then required. Homogeneous and particular solutions to the resulting system of equations are sought. For the homogeneous solution, the stress potentials are assumed to have the forms

$$F(x_1, x_2) = \sum_{k=1}^{6} F_k(Z_k) \qquad \Psi(x_1, x_2) = \sum_{k=1}^{6} \eta_k F_k'(Z_k) \tag{7.49}$$

where $Z_k = x_1 + \mu_k x_2$, μ_k are the roots of the characteristic equation, η_k are ratios of components of the characteristic equation, and $F_k'(Z_k)$ are the derivatives with respect to the argument Z_k. Then the stresses (for the homogeneous solution) have the form

$$\sigma_{11} = \sum_{k=1}^{6} \mu_k^2 F''(Z_k) \qquad \sigma_{12} = -\sum_{k=1}^{6} \mu_k F''(Z_k) \qquad \sigma_{13} = \sum_{k=1}^{6} \mu_k \eta_k F''(Z_k)$$
$$\sigma_{22} = \sum_{k=1}^{6} F''(Z_k) \qquad \sigma_{23} = -\sum_{k=1}^{6} \eta_k F''(Z_k) \tag{7.50}$$

Then following Wang and Choi (1982a,b), the functions $F_k(Z_k)$ are expressed in the form

$$F_k(Z_k) + c_k \frac{Z_k^{\delta+2}}{(\delta+1)(\delta+2)} \tag{7.51}$$

where c_k are undetermined constants. Then (7.50) and (7.51) yield stresses of the form $\sigma = Kr^{\delta}$. The exact order of the local singularity is determined by solving for δ as r goes to zero. The value of δ is determined from the specific geometry in the vicinity of the edge as well as by the elastic constants of the two adjacent plies. The interfacial and free-edge conditions yield three equations each for each ply in a ply pair. The powers δ_i become the eigenvalues of the system of 12 equations, determined from the characteristic equation. The algebraically smallest eigenvalue is a real number in [−1, 0] and is the order of the singularity, as discussed by Wang and Choi (1982a,b). The properly truncated set of eigenvalues is used in the particular solution to ensure convergence. The full expressions for the stresses are then used in satisfaction of the end conditions at the boundary on which the applied stresses are applied and to match the Lekhnitskii solution as one moves away from the hole. The solution of this overdetermined set of equations requires some care, but convergence can be obtained. The method has been validated by comparison with closed-form and finite-element solutions for specific cases. Figure 7.40 shows stress distributions near a hole in a $[+45, -45]_S$ laminate made from a graphite–epoxy system. Convergence was achieved using 25 eigenvalues and about 200 integration points. Figure

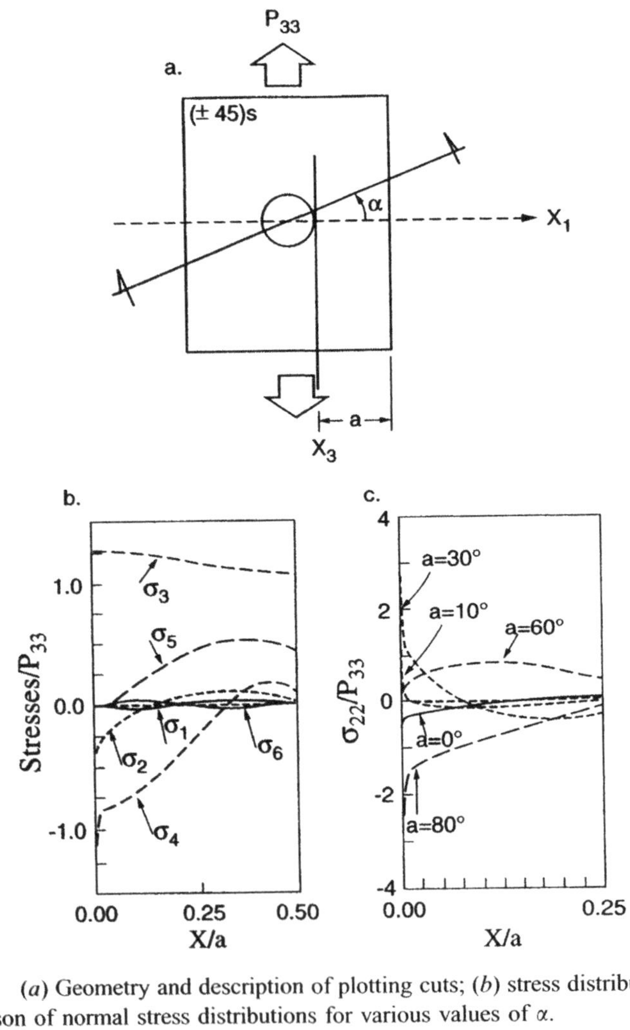

Figure 7.40 (*a*) Geometry and description of plotting cuts; (*b*) stress distribution at $\alpha = 0°$; (*c*) comparison of normal stress distributions for various values of α.

7.40*c* shows distributions of the normal stress σ_{22} for several values of the angle α. The values of stress are highly dependent upon the azimuthal direction.

With the local stresses in hand (using the above method or other means), it is now possible to examine their relevance to the initiation of delamination. Following Shalev and Reifsnider (1990), we start with the analysis for the stress singularities at all of the interfaces at different azimuthal angles. Then we calculate the near-field parameters by the completion of the particular solution. A comparison of stresses at two interfaces yields the form

$$\frac{\sigma_i^{(1)}}{\sigma_i^{(2)}} = \frac{K_i^{(1)}}{K_i^{(2)}} r^{(\delta_1^{(1)} - \delta_1^{(2)})} \tag{7.52}$$

which is to be evaluated as r approaches zero. For distinct values of the order of the singularity at the two locations, (7.52) approaches either zero or infinity (since the stresses are infinite at $r = 0$). But if $\delta^{(1)} = \delta_1^{(2)}$, then as r approaches zero,

$$\frac{\sigma_i^{(1)}}{\sigma_i^{(2)}} = \frac{K_i^{(1)}}{K_i^{(2)}} \tag{7.53}$$

For the case of similar interfaces (with similar values of the singularity), Shalev and Reifsnider used the ratio of the field stress intensities [as in (7.53)] or a strain energy release rate calculation to identify the location of most likely delamination initiation.

A brief examination of this stress-based approach is instructive. Data from Shalev (1988) and Shalev and Reifsnider (1990) will be used; details of the experimental method that produced the data should be found in those references. Material system I was AS4/3501-6 and material system II was an AS4/1808 combination. A quasi-isotropic laminate with stacking sequence $[0, 45, 90, -45]_{S4}$ (laminate A) and an orthotropic laminate with stacking sequence $[0, 45, 0, -45]_{S4}$ (laminate B) with center holes were extended in tension with progressive loading. The tests were interrupted and enhanced x-ray radiographs were taken to determine the location of the onset of delaminations as a function of loading. Figure 7.41 shows how the data are compared to predictions in Figure 7.42. Only a few examples of the comparison will be shown here. Solid bars in the region of the hole show the ply boundary and azimuthal angle for which delamination was observed. Large X's show the location of predicted delamination initiation.

The predicted locations match the observed behavior very well for all cases examined. Also, the sequence of appearance of the initiations could be estimated. As mentioned, the order of the singularity was used to determine which interface was most likely to delaminate and to approximate the azimuthal position. Then the field stress intensity values were used to make a final determination of the location. The singular stresses cannot actually occur in a physical specimen. But the order of the singularity does control the local region of influence, that is, the size of the local volume of material affected by the singular stress field. As we discussed for Figure 7.15, that region is known to be important to the process of failure of composite

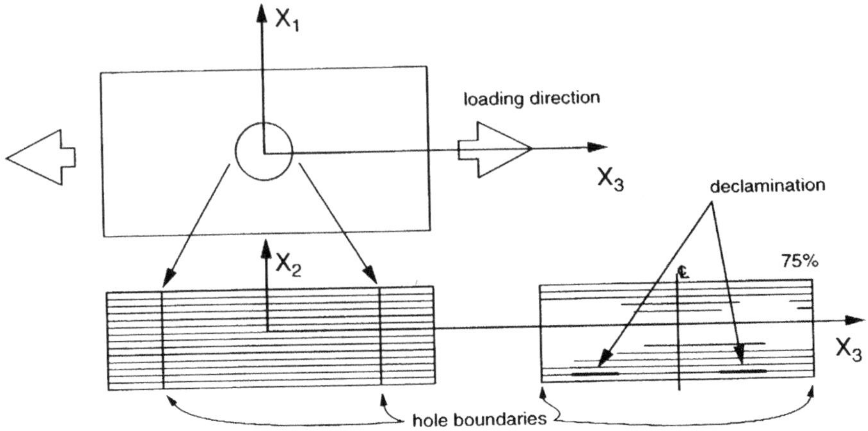

Figure 7.41 Presentation method for delamination initiation data.

materials. The success of this approach suggests that the region of influence and the local intensity parameters control the initiation of delamination.

Several questions are left unanswered by the stress-based approach mentioned above. One question is at what applied load level will delamination initiate. Brewer and Lagace (1988) postulated one answer to that question. They defined average

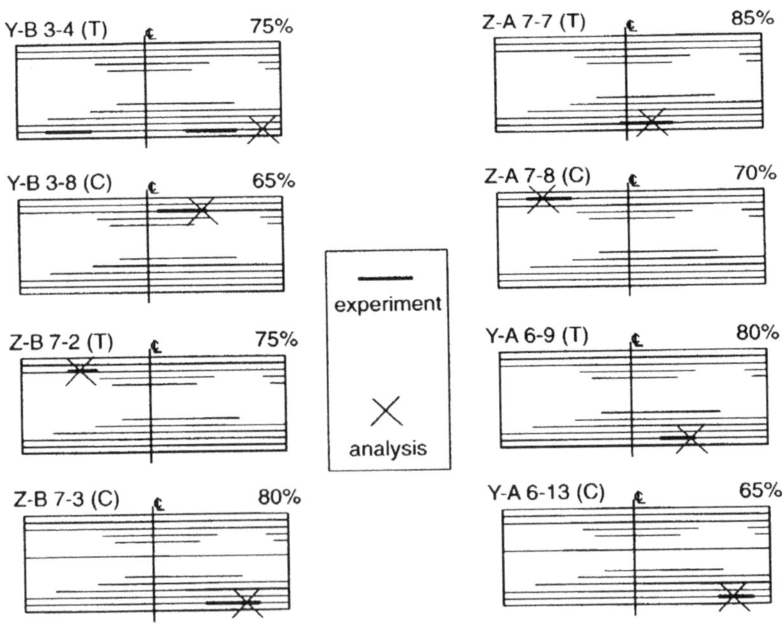

Figure 7.42 Predicted and observed locations of delamination initiation for two material systems and two stacking sequences (Shalev and Reifsnider, 1990).

stresses over the edge-affected region, defined by a characteristic length in the sense of the Whitney and Nuismer model discussed earlier. Using an averaging dimension χ, the average stresses have the form

$$\bar{\sigma}_{ij} = \frac{1}{\chi}\int_0^{\chi} \sigma_{ij}\, dx \tag{7.54}$$

Then they proposed a failure criterion for the initiation of delamination of the type

$$\left(\frac{\bar{\sigma}_{1z}}{Z_{S1}}\right)^2 + \left(\frac{\bar{\sigma}_{2z}}{Z_{S2}}\right)^2 + \left(\frac{\bar{\sigma}_{zz}^t}{Z_t}\right)^2 + \left(\frac{\bar{\sigma}_{zz}^c}{Z_c}\right)^2 = 1 \tag{7.55}$$

where Z_{si} are the interlaminar shear strengths, Z_t is the interlaminar tensile strength, and Z_c is the interlaminar compressive strength. Only the tensile or compressive normal stress elements of (7.55) are used for a given situation, depending upon the sign of the interlaminar normal stress for the interface considered. Brewer and Lagace (1988) collected a large data set for graphite–epoxy laminates with a variety of stacking sequences and ply groupings. They found that values of the material interlaminar strengths could be chosen such that the predicted delamination onset stress as a function of, for example, the number of grouped plies, n, in a $[0_n, (+15, -15)_n]_S$ laminate was nearly always within the scatter band of the physical observations. Typical values of the material constants for their system were $Z_t = 53.9\,\text{MPa}$, $Z_{S1} = 105\,\text{MPa}$, and $\chi = 0.178\,\text{mm}$.

Stress-based criteria for the estimation of the initiation of delamination appear to be consistent with physical behavior, provided considerable care is taken to determine the correct values of local stress in the plies and between the plies in the edge-affected regions of a laminate and initiation criteria are carefully chosen. However, the requirement of determining the exact local interlaminar stresses is an obstacle to the engineering use of such an approach.

Energy-based methods are, in some cases, more amenable to convenient approximate solutions and are widely used in the applied community. A good example of this approach is the method suggested by O'Brien (1991). The geometry considered by O'Brien is shown in Figure 7.43. It is supposed that under in-plane uniaxial loading along its length L the laminate in Figure 7.43 delaminates along the left- and right-hand side edges, forming sublaminates (three in the case illustrated). If an energy approach to the description of this process is adopted, then the strain energy release rate expression (7.31) can be applied to this problem. If we define a compliance C, the elongation of the laminate in the direction of the in-plane load P is given by $\Delta L = CP$. Then if the modulus of the laminate in the direction of loading is E and the cross-sectional area of the laminate (perpendicular to the direction of

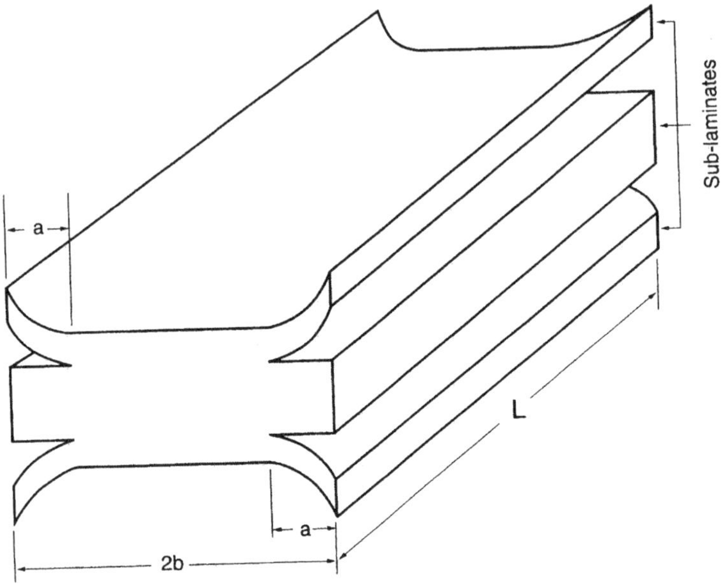

Figure 7.43 Schematic diagram of symmetric edge delamination of laminate into sub-laminate layers.

loading) is A, the compliance is given by $C = L/AE$, and we can write the strain energy release rate in terms of the total volume of the laminate, V, as

$$G = \frac{P^2}{2} \frac{\delta C}{\delta A_d} = -V \frac{\varepsilon^2}{2} \frac{\delta E}{\delta A_d} \tag{7.56}$$

where A_d is the total area of the delamination surfaces. O'Brien uses a simplified form to estimate the effect of the delamination on the laminate stiffness. He approximates the stiffness of the fully delaminated laminate as

$$E_{\text{delam}} = \frac{\sum_{i=1}^{n} E_i t_i}{t} \tag{7.57}$$

where t is the total laminate thickness, t_i is the thickness of the ith sublaminate, and E_i is the stiffness of the ith sublaminate when the laminate is fully delaminated. For example, for a laminate having the stacking sequence $[\pm 30, \pm 30, 90, \overline{90}]_S$, which will most likely delaminate along the $-30/90$ interface to form two sublaminates having $[\pm 30, \pm 30]$ plies and a third sublaminate having $[90]_3$ plies, the fully delaminated stiffness can be estimated by the expression

$$E_{\text{delam}} = \frac{8E(\pm 30)_2 + 3E(90)_3}{11} \tag{7.58}$$

Then, based on comparisons with experimental data, O'Brien postulates an expression for the stiffness of the laminate as a function of the length of the delaminated regions, a, in the form

$$E = (E_{\text{delam}} - E_0)\frac{a}{b} + E_0 \tag{7.59}$$

where E_0 is the initial modulus of the laminate before any delamination has occurred. In the instance when the delamination does not propagate with a straight edge, an equivalent expression can be written in terms of the area of delamination A_d and the total interfacial area $A_i = 2Lb$ in our case. Then (7.59) takes the form

$$E = (E_{\text{delam}} - E_0)\frac{A_d}{A_i} + E_0 \tag{7.60}$$

Combining Equations (7.56), (7.57), and (7.59), O'Brien obtains an expression for the total strain energy release rate for all (combined) modes of deformation for the edge delamination initiation problem in the form

$$G = \tfrac{1}{2}\varepsilon^2 t(E_0 - E_{\text{delam}} \qquad 7.61)$$

where ε is the global laminate strain in the direction of the applied loading. This remarkable result implies that the strain energy release *rate* does not depend on the length or the shape of the delamination itself. It also suggests experimental methods for the determination of a critical strain for the initiation of delamination. If (7.61) is used to determine the critical total strain energy release rate G_c for a given material system and laminate, then for that material system (if the tenets of fracture mechanics apply in this case), it should be possible to predict the critical strain value for delamination initiation in any other laminate by solving (7.61) to obtain

$$\varepsilon_c = \sqrt{\frac{2G_c}{t(E_0 - E_{\text{delam}})}} \tag{7.62}$$

An interesting feature of this prediction is the dependence on the total laminate thickness t. Equation (7.62) predicts that the critical strain for the initiation of delamination is inversely proportional to the square root of the thickness of the laminate; that is, thicker laminates initiate at a lower level of applied strain than thin ones. This is in sharp contrast to the stress-based approach wherein the dependence on laminate thickness is not as explicit. For the local stress discussion outlined above, for example, interlaminar stresses in the boundary layer near the edge of the laminate are more highly dependent upon local details than on global dimensions. The results of this approach are illustrated in Figure 7.44. The predicted results were obtained from (7.62). This approach matches data well for this situation and for many other cases discussed by O'Brien (1991). When matrix cracking occurs, O'Brien shows that (7.62) can be modified to describe the growth of a delamination

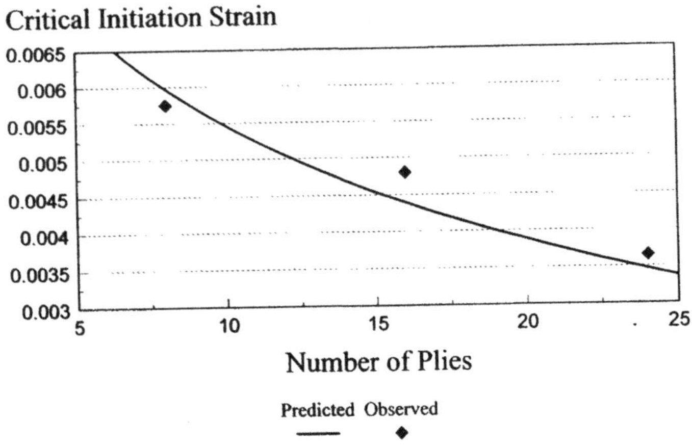

Figure 7.44 Predicted and observed delamination initiation strain for $[+45_n, -45_n, 0_n, 90_n]_S$ graphite–epoxy laminates (O'Brien, 1991).

from a transverse matrix crack by introducing the thickness t_{DL} and modulus E_{DL} of the uncracked, locally delaminated regions to obtain

$$\varepsilon_c = \frac{1}{E_0 t}\sqrt{\frac{4G_c}{1/t_{DL}E_{DL} - 1/tE_0}} \tag{7.63}$$

Figure 7.45 shows the results of this correction for a $[25, -25, 90_n]_S$ laminate as a function of the number of repeated plies, n. For the graphite–epoxy laminate considered in this figure (O'Brien, 1991), there is a transition from edge delamination to

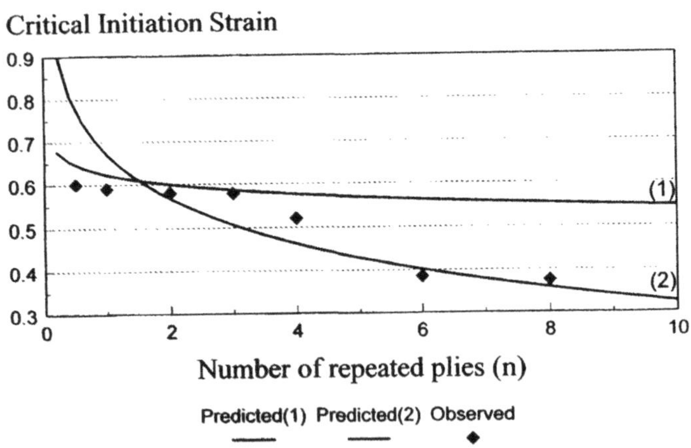

Figure 7.45 Delamination initiation strain levels for $[25, -25, 90_n]_S$ laminates predicted from (7.62) (1) and (7.63) (2) compared to experimental observations (O'Brien, 1991).

local delamination growing from matrix cracks as the number of repeated plies, n, goes from 3 to 4.

The ability to predict the initiation of delamination in thick laminates from measurements on thinner or smaller ones is a traditional advantage of the fracture mechanics approach. Physically, it is not clear why a strain energy release rate approach, based on the discussion of a crack with a specific geometry and size, should describe the initiation of cracking (delamination) in a laminate. It can be argued that inherent interlaminar defects of microscopic dimensions are present in the initial condition and form detectable cracks only when a critical strain energy release rate is reached at the local level (O'Brien, 1982; O'Brien et al., 1982; Wang et al., 1983).

The role of matrix cracking in the initiation should be emphasized. An illustration of a general situation at the local level is shown in Figure 7.46. The figure shows the edge of a cross-ply laminate in which a matrix crack has formed in the 90° plies (which lie along the centerline of the thickness of the laminate in this example). If the laminate is under extensional forces, the axial normal stress in the 90° plies is relaxed near the crack face. Along the centerline of the 90° plies, the crack is free to open, but as one approaches the ply boundaries, the 0° plies exert a shear stress on the local material elements in the 90° plies. A sketch of the major features of the local stress state in the 90° plies quite close to the ply boundary is provided on the right-hand side of Figure 7.46. If the local material element is also near the point of bonding between the 90° and 0° plies (the first point in the x direction, as a function of distance from the crack face, where the shear stress is applied by the 0° plies), then the local stress state is dominated by the local interlaminar shear stress, as sketched in the figure. The resulting normal stresses in the (x', y') system will create

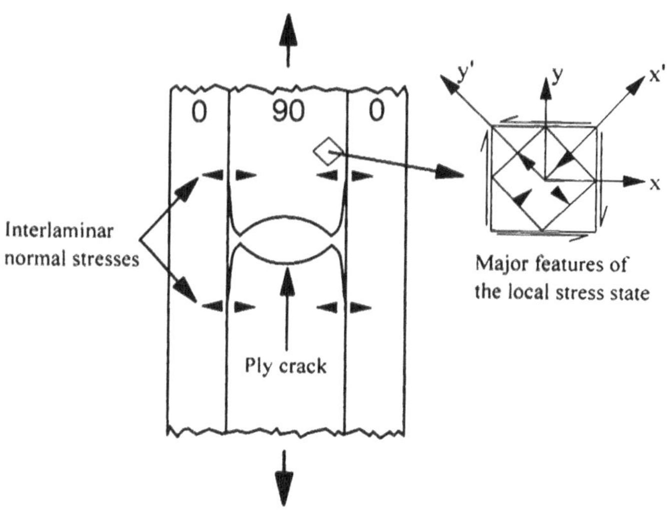

Figure 7.46 Local deformations and interlaminar normal stresses near transverse matrix crack in cross-ply laminate.

a deformation pattern that is similar to the form shown for the 90° ply, resulting in a tendency to cause delaminations emanating from the tips of the transverse crack, as shown in Figure 7.46. A more thorough discussion of this behavior follows.

For laminates and material systems of common interest, delamination is often initiated by local stresses near the tips of intralaminar matrix cracks. This behavior is pervasive and can be one of the most serious possible consequences of matrix cracking. There are two typical embodiments of this damage mode, internal delamination initiation near the crossover points of cracks in plies having different orientations (along the interface of those plies) and at the edge of a laminate. The general features of the internal stress states associated with these situations are illustrated by Figure 7.47.

Perhaps the most widely discussed influence of such a crack is the concentration of normal stress in the crack opening direction immediately ahead of the crack tip, σ_x in Figure 7.47. Although this particular aspect of the problem has been widely discussed for homogeneous materials, it is much less common to find descriptions of that situation in composite materials. Figure 7.48 shows an example of the increase in σ_x in the neighboring plies as a result of a crack in the 90° plies of a quasi-isotropic graphite–epoxy laminate calculated using a quasi-three-dimensional finite-difference scheme. The direct proof of the existence of these concentrated stresses was provided by the work of Highsmith (1981, 1984) wherein a very-high-resolution Moiré diffraction device was constructed and used to resolve strain distributions in the 0° ply of several different laminates in regions near cracks in adjacent off-axis plies during quasi-static loading. An example of those results is shown in Figures 7.49*a*,*b*. Figure 7.49*a* was produced by the interference between a reference beam and a beam that was incident on a diffraction grating bonded to the specimen surface which had about 800 lines per millimeter. The cracks in the off-axis 90° plies of the $[0, 90_3]_s$ glass–epoxy specimen can be seen as white horizontal bars with a spacing of about 4 mm in the original photograph. The constant-displacement diffraction lines are denser in the region of the off-axis cracks, which indicates a strain concentration in the 0° ply surface that is being observed. The strain distribution calculated from that pattern between two of those cracks is shown in Figure 7.49*b* along with strain plots calculated from a simple

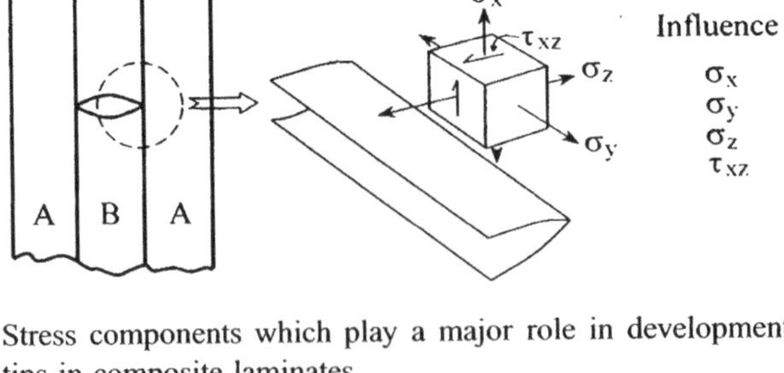

Figure 7.47 Stress components which play a major role in development of damage near primary crack tips in composite laminates.

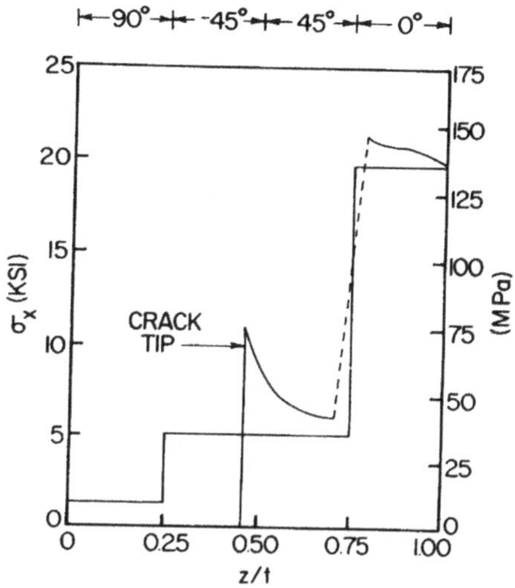

Figure 7.48 Through thickness variation of σ_x for quasi-isotropic laminate of graphite–epoxy with 90° and −45° ply cracking (Talug and Reifsnider, 1979).

shear-lag model for three choices of the value of the parameter G/b, the only parameter in that model.

The stress concentration in the crack opening normal stress component is first in the order of influence because of its importance to the fiber failure damage mode. The (chronologically) first and possibly most significant effect of the elevation of this stress component is the nucleation of fiber fractures in adjacent plies in regions next to the crack tips. As fibers ultimately define the strength of these laminates, this is, in a sense, the beginning of the end of the damage process. Surprisingly, as we noted earlier, this fiber fracture initiation begins early in the sequence of damage development, usually in the first one-third of the life of a laminate.

The second stress component in our "order of influence" is the normal stress parallel to the crack front, σ_y. To appreciate the physical situation which gives rise to this stress, recall that the primary matrix cracks generally extend through the thickness of a ply and run parallel to the fibers over large distances in the y direction, commonly across the entire width of a test specimen. Hence, from the standpoint of the local crack opening behavior, we can assume that the deformation near the tip of the crack occurs under plane-strain conditions in the y direction. The local increase in the normal stress σ_x causes a corresponding region of increased local strain in that direction, which causes the material element in that state of stress to contract in the y direction (in an attempt to maintain its volume) by an amount greater than the surrounding material further away from the crack tip. This contraction is, of course, resisted by the surrounding region, which results in the creation of a tensile normal stress σ_y in the material along the edge of the crack. This behavior is not peculiar to the stress fields around primary cracks in composite laminates but is well

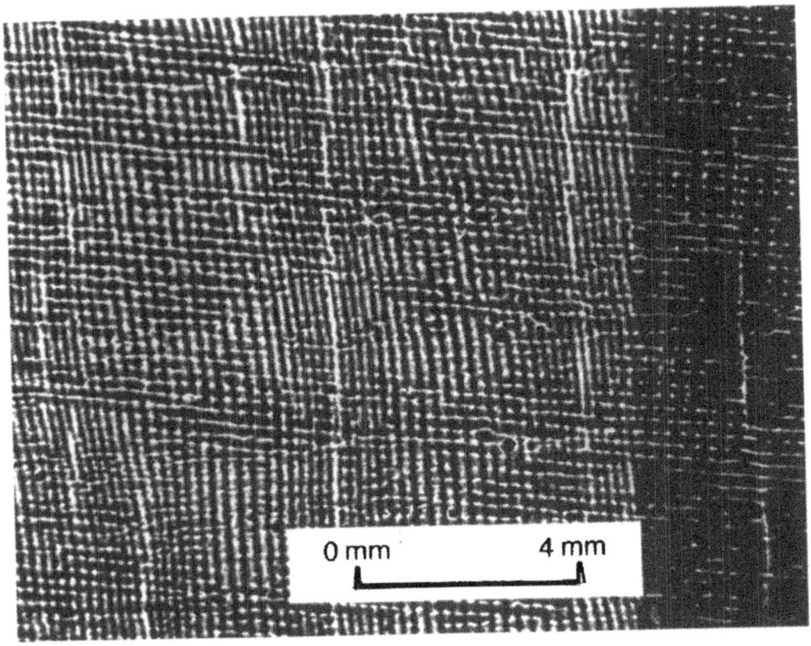

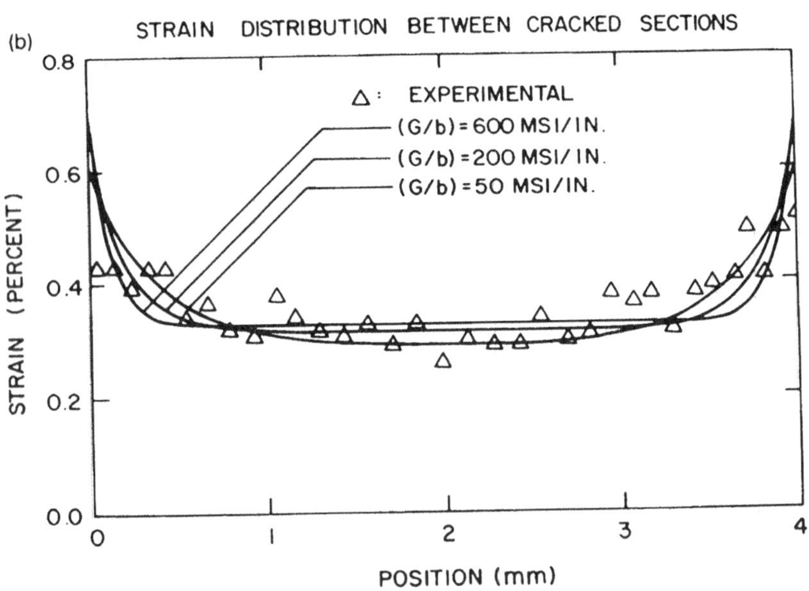

Figure 7.49 (*a*) Moiré diffraction pattern showing increased strain in 0° plies in regions near cracks in underlying 90° plies of $[0, 90_3]_S$ laminate; (*b*) comparison of measured values of those strains between two such cracks with analytical predictions of those strains using shear-lag model with three different values of parameter G/b in that model.

established in homogeneous materials where fracture mechanics solutions can be used to obtain estimates of the magnitude of this stress component. This stress component is a primary contributor to the occurrence of secondary cracking, as demonstrated in Chapter 1. It is listed as the second most important stress component in our order of influence, as secondary cracking behavior is ubiquitous in our observations of damage development in composite laminates that have fairly brittle matrix materials such as epoxy (Jamison et al., 1981; Reifsnider et al., 1988). This is true even in situations where the global ply stress in the y direction (calculated from laminate analysis) is compressive as a result of Poisson mismatches between the individual plies. Hence, although the strains associated with this stress component have not been accurately measured, its influence on the development of damage is thought to be large and significant based on the high density of secondary cracks that is observed in so many situations.

The next two components of stress to be discussed are the interlaminar stresses σ_z and τ_{xz}. These two components will be discussed together because they act to produce similar damage events. Examples of these stress components, as noted by Talug and Reifsnider (1978, 1979), are shown in Figures 7.50 and 7.51. The physical situation associated with these stress components can be appreciated by considering the deformations around the crack shown in Figure 7.47. These are two physical tendencies of particular interest. First, the material directly above and below the free faces of the crack in material B will tend to translate in the load direction (opening the crack) and is prevented from doing so (in part) by shear forces that develop along the interfaces of the A and B materials and keep the fractured portions of the B ply from "slipping out." That physical situation is associated with the τ_{xz} components

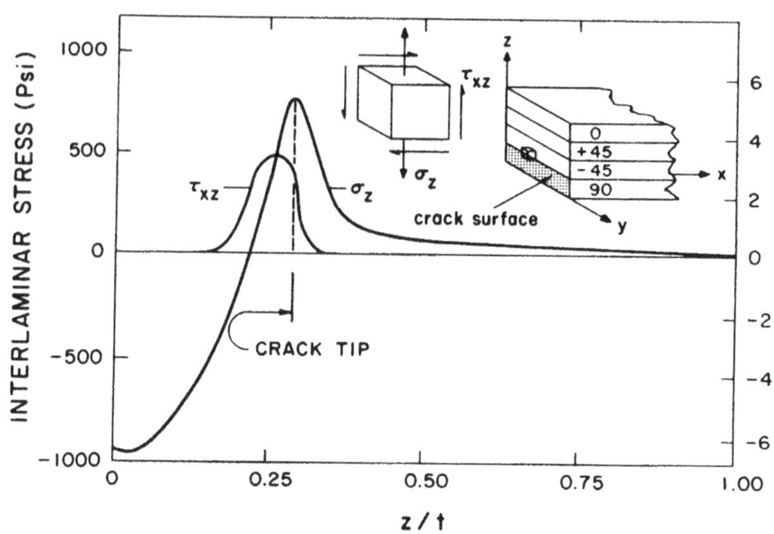

Figure 7.50 Calculated distribution of interlaminar normal and shear stress through thickness of eight-ply quasi-isotropic laminate with crack in 90° ply next to centerline.

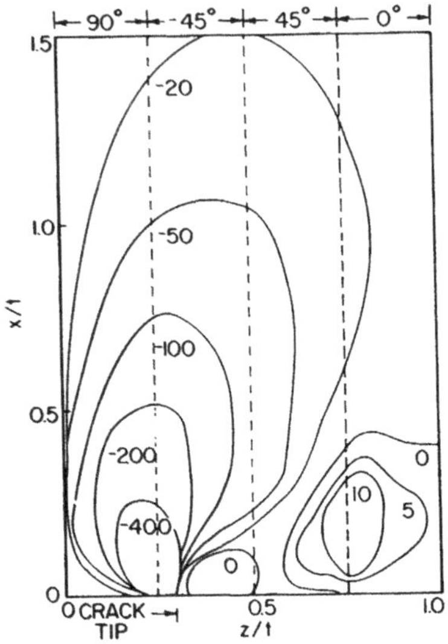

Figure 7.51 Calculated contours of interlaminar normal and shear stress (in psi) for graphite–epoxy quasi-isotropic laminate with 90° crack as shown in Fig. 7.50. The plot is in a plane, which contains the thickness dimension and the length dimension of the laminate.

shown on the material element on the right-hand side of Figures 7.47 and 7.50. Second, as the material along the crack face translates in the load direction, it tends to pull the crack tips inward toward the center of the ply in the sense that the tension in a string is increased by pulling a point in the center of the string in the direction transverse to its axis. This physical situation is reflected in the stress component σ_z, an interlaminar normal stress that acts to pull the interface between plies A and B apart near the crack tips.

The primary result of these stresses is the separation of the plies along the ply boundaries to form a local edge delamination, as sketched in Figure 7.46. However, such local delaminations also form in the interior of laminates and may link up to form global delaminations that propagate quickly. Highsmith (1984) has examined the mechanics of this situation.

In a dissertation on the topic, Highsmith (1984) considered the problem of determining the local stress around the crossing point of a transverse crack in the 90° plies with a longitudinal split in the 0° plies within a $[0, 90_2]_s$ graphite–epoxy laminate. This so-called crossed-crack problem in that particular laminate was picked for consideration because of physical observations, which indicated a high tendency for internal delamination formation in that situation. Figure 7.52*a* shows the idealized damage state that was analyzed.

The damage state is assumed to consist of uniformly spaced transverse cracks in the 90° plies and uniformly spaced longitudinal splits in the 0° plies. In the early stages of damage development, the longitudinal splits may be limited to short distances in the *x* direction and may not extend through the total thickness of the

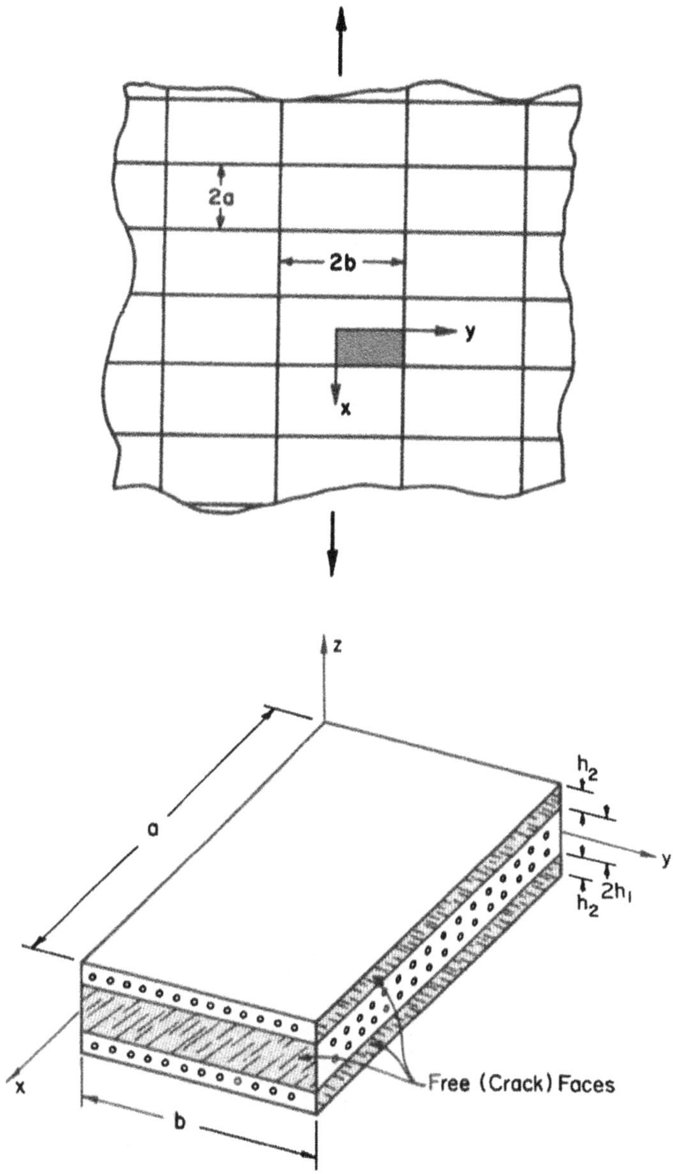

Figure 7.52 (*a*) Crossed-crack pattern in cross-ply laminate showing quadrant of representative volume that was analyzed by Highsmith (1984); (*b*) equilibrium element used for analysis with corresponding coordinate systems and dimensions.

0° plies. However, when internal delaminations begin to form, the idealization shown in Figure 7.52 is a reasonable approximation of a typical state of damage. Figure 7.52*b* shows a schematic drawing of the representative quadrant from the idealized damage state. Cracks in the 90° and 0° plies are treated as free faces, a situation that contributes greatly to the eventual consequences of the internal delamination. Highsmith used a structural theory for the mechanical response of a laminate developed by Pagano (1977) that invokes Reissner's (1997) variational principle, a variational form that reduces to the governing equations of elasticity. Pagano's structural theory is based upon the assumption of linear variations of in-plane stresses through the thickness of a layer of material. The theory produces a system of 23 equations per layer that describes the equilibrium conditions, constitutive behavior, and interlayer continuity conditions imposed on each layer of the laminate. The reader is referred to Highsmith (1984) for the details of this treatment. The method was validated using experimental measurements of actual deformation fields from the surface of a specimen with crossed-crack patterns similar to those analyzed.

A sample of the results of particular interest to us here is shown in Figures 7.53–7.55. Figure 7.53 shows the predicted distribution of axial stress σ_x in the 0° ply at the 0°–90° interface where the cracks cross. The crossing point of the cracks is in the rear corner of the three-dimensional plot. The two cracks in the 90° and 0° plies run along the back faces of the diagram. The diagram clearly shows a stress concentration in the axial normal stress caused by the 90° crack, as we might expect. There is a small but perceptible increase in this stress component near the position where the

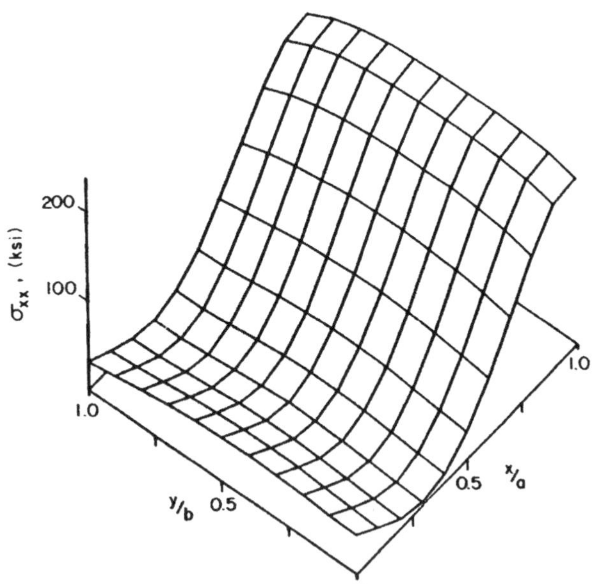

Figure 7.53 Contour plot of axial normal stress component in unbroken ply of $[0, 90_2]_S$ graphite–epoxy laminate near intersection of cracks in 0° and 90° plies.

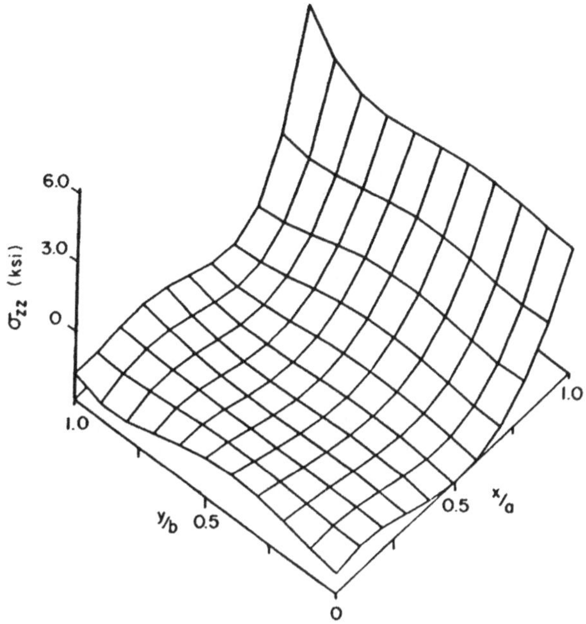

Figure 7.54 Contour plot of interlaminar normal stress at interface between 0° and 90° plies near intersection of cracks in those plies for laminate shown in Fig. 7.52*b*.

cracks cross, but there is no concentration caused by the longitudinal crack in general. Figure 7.54 shows the predicted distribution of interlaminar normal stress σ_z at the 0°–90° interface over the same region. Again, the stress component is elevated near the transverse crack in the 90° plies, as we might expect from our observations of Figure 7.50 for the two-dimensional analysis. However, there is also a distinct synergism between the crossed cracks, shown by the elevated value of σ_z at the crossing point. These elevated values of tensile interlaminar normal stress are a major driving force for the formation of the local delamination in the region that we observed earlier. Figure 7.55 shows a plot of the other interlaminar stress of primary concern to us. This shear stress is also greatly elevated at the position of the transverse crack and is relatively uninfluenced by the presence of the longitudinal crack in this case.

Although these results have been developed for a specific laminate as a matter of expediency, they are by no means peculiar to this situation. In a limited study of the parameters that influence the results of the analysis, Highsmith (1984) has shown that whereas the amplitudes of these stresses are influenced by geometry effects such as ply thickness and material properties, the general nature of the results is unchanged for other crossed-crack situations. The physics of the situations, in particular the boundary-value problem of mechanics, is fixed by the definition of the crossed-crack problem, whereas parameters such as the material properties and ply dimensions influence only the magnitudes of the local effects.

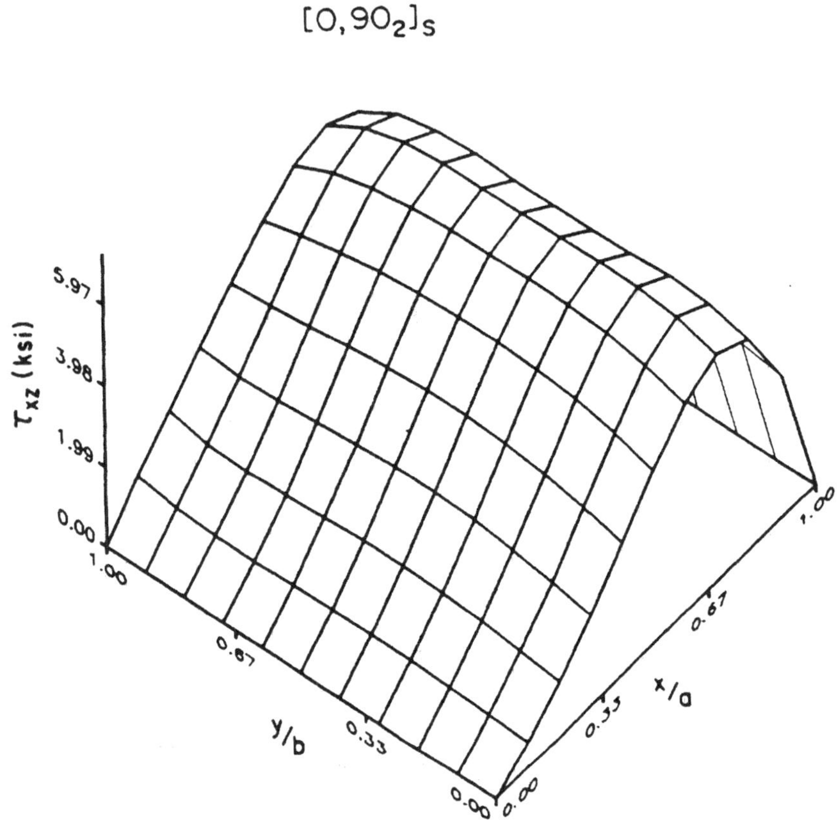

Figure 7.55 Contour plot of interlaminar shear stress at interface between 0° and 90° plies at intersection of matrix cracks in those plies for laminate shown in Fig. 7.52*b*.

These results clearly indicate that the internal stresses near a matrix crack, which stops at a ply interface, tend to cause separation of the interface. This tendency is significantly elevated when two such cracks cross at the interface, as is common in composite laminates. From the standpoint of the mechanics of this situation, we are left with the question of how that local stress state changes if such a delamination forms. If a delamination formed in the interior of a specimen between two plies in the absence of any matrix cracks, continuity conditions at the boundary of such a delamination would force the deformations in that region to be the same as the surrounding undelaminated material. However, the presence of the matrix cracks (which cause the delaminations to form) relaxes the constraint of the boundary conditions around the delamination and contributes greatly to the nature of the local stress state in the region of the delamination, especially at the delamination front.

This discovery was also noted by Highsmith (1984). Figure 7.56, taken from that reference, is a portion of a penetrant-enhanced X-ray radiograph taken of a $[0,\ 90_2]_s$

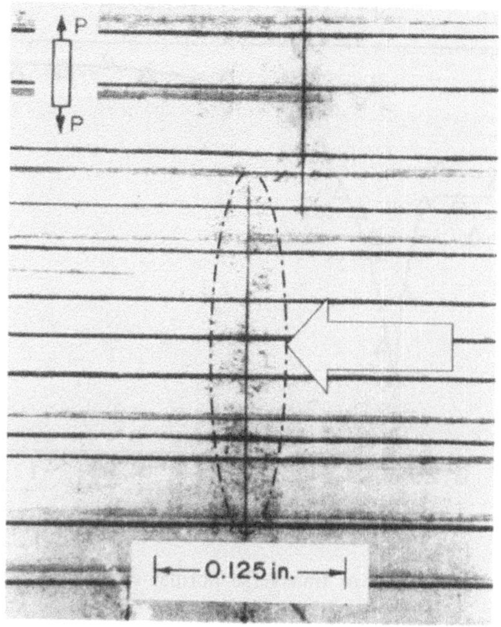

Figure 7.56 Radiograph of crossed-crack pattern showing region of internal delamination near vertical secondary cracks, indicated by arrow.

specimen that had been subject to tensile fatigue loading. The dark blotches, which look like ink spots, are actually dye penetrant that seeped into the Moiré grid that was bonded to the specimen surface. We can see evidence of several transverse cracks and two longitudinal cracks in the radiograph. To the right of the centrally located longitudinal split, there is a shadowy area that is semi-elliptical in shape, indicated by an arrow. This area is a small internal delamination, which was found to be at the 0°–90° interface nearest the specimen grating. Using the high-resolution Moiré diffraction scheme mentioned earlier, the deformations in that region were analyzed. Figure 7.57 shows a typical result of this analysis (further details are included in Highsmith, 1984). The longitudinal split is located at the origin of the plot in Figure 7.57. The delamination front occurs at a distance of 0.875 mm from the split. Within the delamination, there is a plateau of nearly constant transverse strain of about −0.28%. For linear elastic response of that laminate a longitudinal strain of 0.728% is expected. The corresponding Poisson's ratio is quite large, 0.385. At the delamination tip, the transverse strain is concentrated, with a value of −0.65% indicated. Beyond that position, the transverse strain returns to the global value that we would expect for the laminate. Highsmith has reasoned that, as the average axial stress σ_z along a specimen section passing through the delamination must be the same as the applied stress to satisfy equilibrium and as the delaminated material is less stiff than its undelaminated surroundings, there must be a region of stress concentration to ensure equilibrium. This concentration of stress is observed to

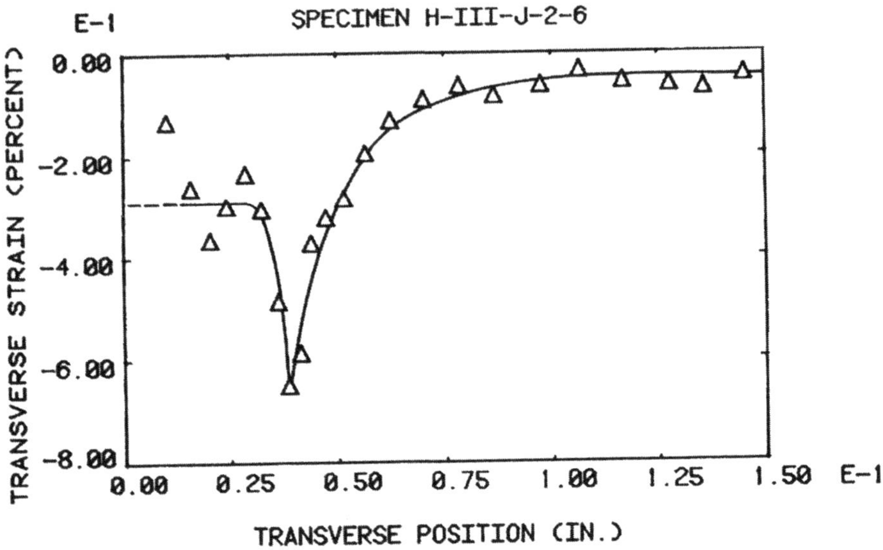

Figure 7.57 Measured values of transverse normal strain across delamination boundary shown in Fig. 7.56.

occur at the delamination front, not only in the direction transverse to the load axis but also in other directions, including the load axis direction.

Although this problem still requires a great deal of further study, we know that internal delaminations that form around matrix cracks can cause significant elevations in the in-plane values of normal stress in the plies surrounding the delamination. We might expect these elevated stress values to contribute to further damage and possibly to the failure of the laminate. Furthermore, fiber fracture occurs preferentially in regions of internal delamination (Razvan and Reifsnider, 1991). Whether this fracture occurs at the front of the delamination as it grows over the region as part of the growth process or the fibers break in the delaminated region after delamination has occurred remains unclear.

We mentioned that the damage development process discussed above could be greatly accelerated in certain circumstances by combined tension–compression loading. This is certainly the case with either internal delamination or edge delamination. The nature of this acceleration is not entirely intuitive. Although such things as local instability (buckling) may cause local delaminations to grow rapidly, it is less obvious that other types of damage such as matrix cracking may also be accelerated by combinations of tension and compression. Schultz and Reifsnider have found, for example, that secondary cracks in 45° plies that form along the tip of primary cracks in the 90° plies of a quasi-isotropic graphite–epoxy laminate can initiate and grow during compressive load excursions under cyclic tension–compression loading (Reifsnider and Schultz, 1984; Schultz, 1984). The mechanics of that situation is not different from the crossed-crack problem described above.

Application of Concepts The primary question of interest here is how can we use understandings and representations of the events and effects associated with delamination to predict remaining stiffness, strength, and life?

In Chapter 5 we discussed the importance of stiffness change. Deformations and strains are measurable extensive variables. But we can only estimate stresses if we know the field values of stiffness as a function of time and history. Any stress analysis that we use to estimate strength or life can be no better than our knowledge of stiffness.

In composite laminates, matrix cracking and delamination are most often combined effects, and their collective influence on ply-level stresses (which are frequently the basis for strength estimates) can be quite large. A typical example of the combined effect of cracking and delamination on the axial stiffness of a quasi-isotropic coupon during cyclic loading ($R = 0.1$) to a maximum stress of about 60% of the ultimate strength of the laminate is shown in Figure 7.58. The stacking sequence of the laminate was $[0, 90, +45, -45]_S$. Matrix cracking appears early and begins to saturate in the midlife region. Near the end of life, a sharp drop in laminate stiffness is observed, corresponding to the onset of significant delamination growth (between the 90° and +45° plies) from the edge of the laminate. For the material system used in this test, the following reference stiffness values apply:

$$E(\text{undamaged}) = 54.4 \text{ GPa}$$
$$E(\text{fully matrix cracked}) = 47.8 \text{ GPa}$$
$$E(50\% \text{ cracked and delaminated}) = 40.7 \text{ GPa}$$
$$E(\text{fully cracked and delaminated}) = 35.8 \text{ GPa}$$

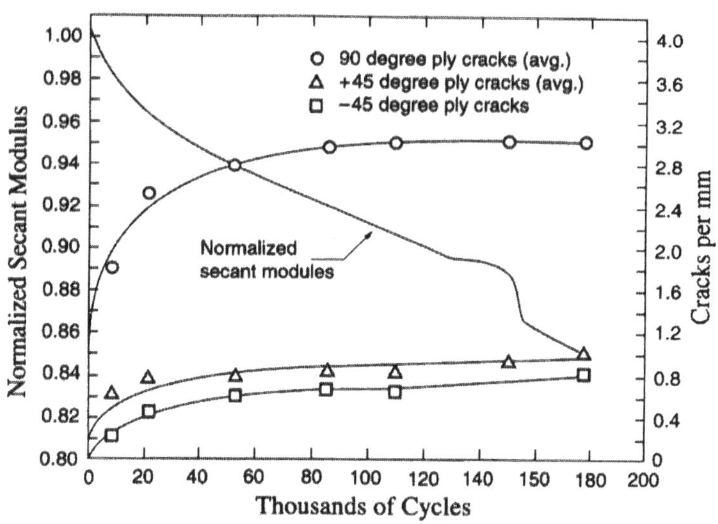

Figure 7.58 Measured crack densities and stiffness change in quasi-isotropic graphite–epoxy laminate.

In this case, delamination and matrix cracking together can produce a 34% change in laminate stiffness. In the context of structural engineering, this is a massive change in stiffness that would greatly influence the design and function of a component. Since many structures are stiffness designs, such a change could be "failure" in that sense.

Table 7.5 shows some estimates of the stiffness changes versus measured values. Measurements of crack densities and delamination growth were made during the test to assist in the estimates. Even with the large increments of change used in the discount method in this case, the estimates are quite close to the measured values of laminate stiffness change. It is clear that the expected changes in stiffness calculations of ply-level stresses, based on observed damage development, are occurring. Such estimates are quite useful for engineering-level ply stress calculations. Table 7.6 shows the stresses in individual plies of this example laminate before and after damage develops. As cracks develop in the off-axis plies, the axial stress in the 0° plies increases by about 14%, which is also almost exactly the same percent change as the (decrease in) laminate modulus. When the laminate is fully delaminated and all off-axis plies are cracked, the only plies remaining that are "holding the load" are 0° plies, so (for the applied stress of 1000 MPa) the ply stress is 4000 MPa, an increase of 52%! The change in laminate stiffness for that fully damaged condition is 34%.

From this simple example, we see that one of the most important effects of delamination is the stiffness change it can cause and the resulting local and global redistribution of stresses caused by those stiffness changes. These stress redistributions can significantly affect strength and life (in a manner to be discussed below). For this reason, although delamination is infrequently a failure mode per se, it can be a serious damage mode that is an indirect cause of failure, either by loss of stiffness or by redistribution of stresses that overload the plies that are most highly stressed in a laminate.

We have also seen from the discussions in this section and in Chapter 5 that it is possible to estimate the amount of stiffness change from a knowledge of the type and extent of damage that develops and that these stiffness changes at the global level (measured or implied) can be used to calculate local stress states and their changes.

This brings us to the question of strength under these conditions. We will discuss several definitions of strength. The first definition might be the initiation of delamination. We have discussed the use of local stress, strain, and strain energy release

TABLE 7.5 Stiffness Change Due to Delamination

Condition	A_{delam}/A	E (GPa)	ΔE_{delam} (%)	ΔE_{total} (%)	$\Delta E_{measured}$ (%)
Stage I	0.083	50.5[a]	1.7	7.1	5.2
Stage II	0.186	45.8[b]	4.2	15.8	14.7
Stage III	0.189	45.5[b]	4.8	16.4	16.7
Near failure	0.303	44.2[b]	7.6	18.8	19.7

[a] Off-axis E_2 and G_{12} discounted 50%; [b] Off-axis E_2 and G_{12} discounted 100%.

TABLE 7.6 Laminar Stresses Before and After Damage

Laminate Condition	Engineering Moduli (GPa)				0° Ply Stresses (MPa)			90° Ply Stresses (MPa)			45° Ply Stresses (MPa)		
	E_L	E_T	G_{LT}	ν_{LT}	σ_x	σ_y	τ_{xy}	σ_x	σ_y	τ_{xy}	σ_x	σ_y	τ_{xy}
Undamaged	54.4	54.4	20.5	0.32	2632	−2.3	0	167	−797	0	600	400	418
90's cracked	52.4	54.6	19.5	0.31	2735	0.1	0	0	−852	0	632	426	442
90's and 45's cracked	47.8	50.0	19.0	0.33	2993	−4.7	0	0	−1000	0	503	503	503
Delaminated													
[0, 90]	77.2	77.2	4.1	0.04	3131	59	0	217	−59	0	—	—	—
$[+45, -45]_S$	15	15	37.2	0.81	—	—	—	—	—	—	325	0	136
Fully delaminated and cracked	35.8	—	—	—	4000	0	0	0	0	0	0	0	0

rate as a criterion for delamination initiation and shown correlations with each. But in each case it must be assumed that there is a unique material property (interlaminar strength, strain to failure, and critical strain energy release rate—possibly for modes I, II, and III) that can be used to define the critical level of applied load. Defining these physical quantities may be difficult. An example for the strain energy criterion will illustrate this difficulty.

Figure 7.59 shows an example of the (total) critical strain energy release rate as a function of edge delamination length for quasi-static tensile loading of a graphite–epoxy laminate. Although the magnitude of the increase is large for this special choice of laminae orientations, the rising resistance curve and resulting stable delamination growth with increasing load is a typical result. Generally speaking, the interaction of delaminations with matrix cracks, bridging of fibers across delamination interfaces, and interaction of delaminations with the various other modes of distributed damage are responsible for this stability. It is important to notice that elastic analysis [such as Equation (7.62)] predicts that the critical strain energy release rate is independent of delamination length. However, there is some evidence that the critical values measured in the manner illustrated in Figure 7.59 are material characteristics, in the sense that the delamination of other laminates with different plies and stacking sequences, made from the same material system, can be predicted using such data from one laminate measurement (O'Brien, 1991). It should also be noted that if other modes of damage are suppressed by some method such as the use of a toughened matrix material, the slope of the resistance curve can be expected to decrease, and although the initiation of delamination may occur at higher strain

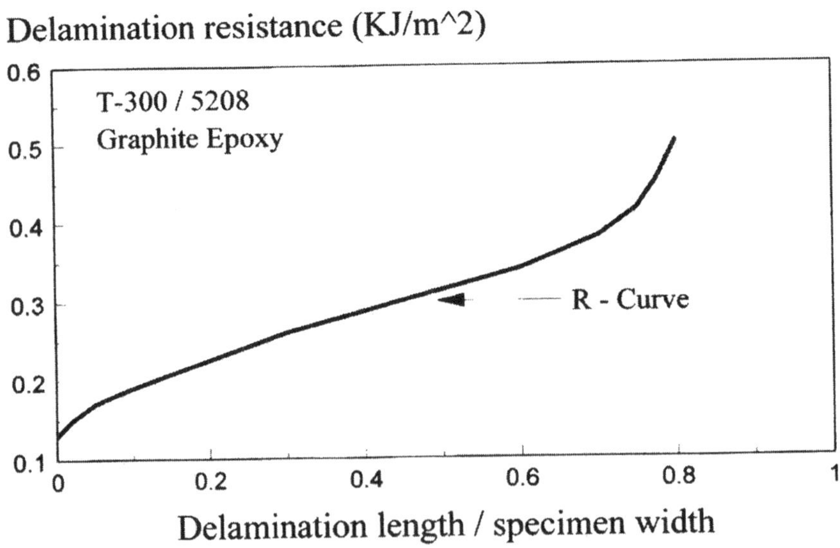

Figure 7.59 Edge delamination resistance curve for $[\pm 30, \pm 30, 90, \overline{90}]_S$ laminate under quasi-static tensile loading (O'Brien, 1981).

energy release values, the delamination may immediately extend through the laminate in an unstable manner (O'Brien et al., 1982). Although the discussion above is in terms of total strain energy release rate, it may be essential to consider the contribution of the individual modes of crack opening, in-plane shear, and out-of-plane shear to the process. We showed in an earlier example (cf. Fig. 7.42) that the strain energy release associated with the crack opening mode (mode I) is generally a good indicator of delamination initiation. Total strain energy release does not match those data in some cases (O'Brien, 1991).

Various test methods are used to separate the effect of different modes of loading. Figure 7.60 shows the results of a *crack-lap-shear (CLS)* test conducted under constant end displacement (Ramkumar, 1983). For this specimen and loading condition, the energy release rate associated with mode II deformation is nearly four times the value for mode I. Typical critical values of strain energy release rate for polymer-based continuous-fiber composites are of the order of 100–700 for G_I and 150–800 for G_{II} in units of joules per meters squared (Ramkumar, 1983; Whitney et al., 1982; Wilkins, 1981). Figure 7.60 also shows data for the rate of crack growth in that test as a function of crack length. Since the test was conducted under constant end displacement, the load applied to the specimen was decreasing as the crack grew and the compliance of the specimen increased. Hence, the crack growth was also decreasing, in proportion to the decrease in the strain energy release rates. This

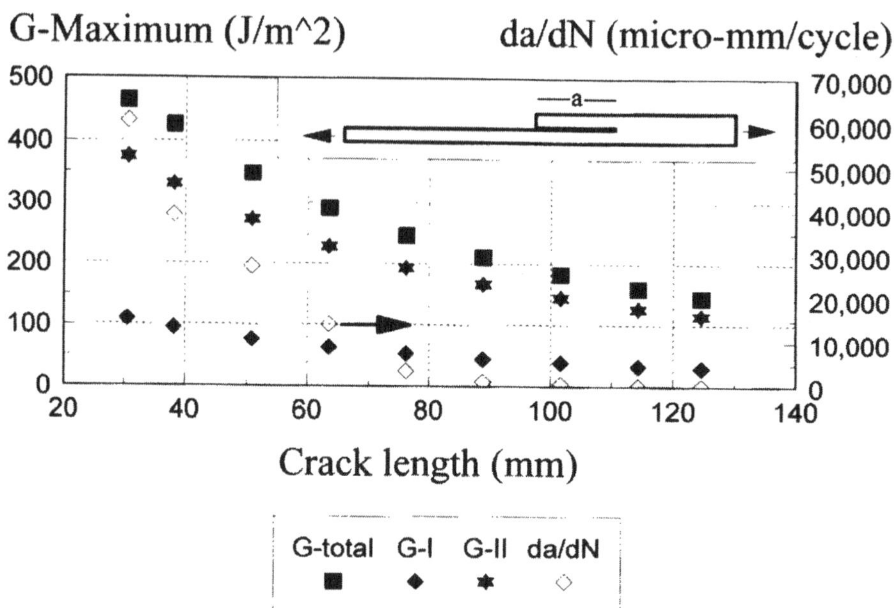

Figure 7.60 Contribution of modes I and II to maximum strain energy release rate and delamination growth rate as function of crack length in crack-lap-shear specimen during displacement-controlled cyclic loading (Ramkumar, 1983).

raises the next strength-related question. We have seen that it is common for the critical strain energy release rate of a material system to increase as a function of increasing crack length for quasi-static loading. Is that typical of cyclic loading?

There is strong evidence that the reverse is true. For example, O'Brien (1991) has shown data for several laminates for which the critical strain energy release rate was measured (for edge delamination) after various cycles of loading. The critical values all decreased with cycles, sometimes by as much as 80%. If we cannot be sure that the *critical strain energy release rates* of laminates are (constant) material properties (as we suggested at the beginning of our discussion of the application of fracture mechanics to composite materials), then what strength criteria can we use for a strength-based life prediction?

One answer to this question is widely accepted. In Chapter 2 we discussed various strength criteria and noted that strain-based methods are often successful for fiber-dominated composites that fail at fairly low strain levels. In the present context, there is good evidence that laminates made from the same material system fail at essentially the same strain value, whether they delaminate or not (cf. O'Brien, 1985). So global, nominal strain normalized by the strain to failure is a possible failure function for use in strength evolution equations. Alternatively, it may be appropriate to use any of the other failure criteria discussed in Chapter 2 if they represent the failure mode that applies to a given laminate used in a stress state dictated by a specific application. Delamination affects the failure function, primarily, by significantly altering the stress state (or strain state) that appears in the numerator of any such function having the form $\mathrm{Fa} = \mathrm{Fa}(\sigma_{ij}/X_{ij})$ or $\mathrm{Fa} = \mathrm{Fa}(\varepsilon_{ij}/\Sigma_{ij})$ where Σ_{ij} are the tensor strain to failure values for the material at the ply level or, more generally, in the critical element. As we have seen in this section, these alterations in stress in the critical element (like the 0° plies in the example in Table 7.6) can be quite large.

Finally, how does this behavior affect the life of a fibrous composite? In the context of the philosophy discussed in this book, the answer to this question has two parts. First, it is necessary to discuss the growth of delaminations, since the stiffness change, resulting stress redistribution, and change in the value of the failure function in the critical element depend on the delamination length or area. Second, it is necessary to incorporate the effect of the growing delamination into the calculation of remaining strength through the use of Equation (7.1).

Although the question of how to use strain energy release rate to predict the initiation of delamination is still incompletely answered, there seems to be general agreement that the growth of delaminations can be described by some variation of the form

$$\frac{da}{dN} = \alpha G^{\beta} \tag{7.64}$$

where a is the delamination length, N is the number of cycles of loading, and α and β are material constants, at least within certain confines defined by the laminate and experiment used to measure them. Figure 7.61 shows an example of such a representation for the specimen discussed in Figure 7.59. The data [taken from a NASA Technical Memorandum (O'Brien, 1981)] represent the average of four specimen

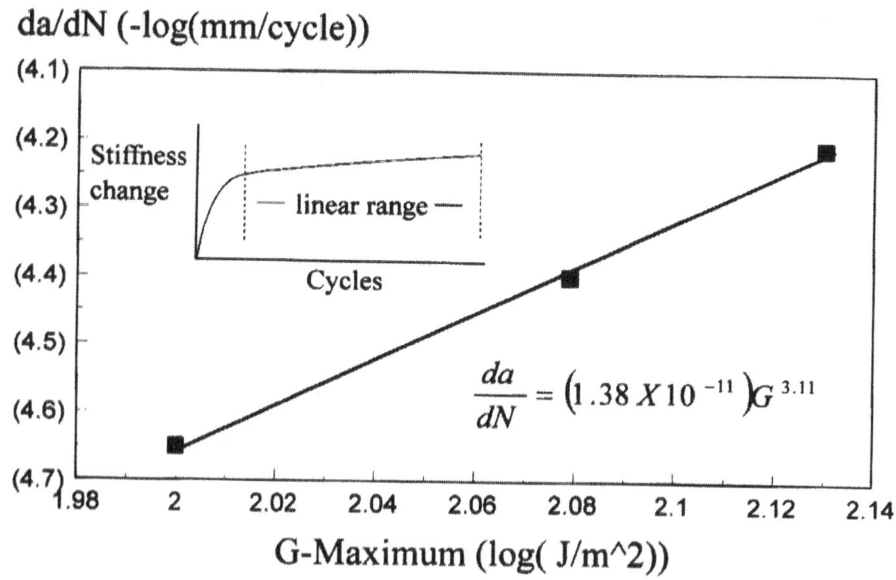

Figure 7.61 Edge delamination growth rate as function of maximum level of strain energy release rate for "linear region" of stiffness change (O'Brien, 1981).

tests each. As noted in the figure, the power law representation is generally applied to the region of delamination growth that has a linear change in stiffness with cycles. It should also be noted that the values of α and β are highly dependent on the physical units used as well as on the nature of the delamination growth.

The literature is replete with representations that have the form of Equation (7.64), with variations for the mode of deformation, material system, delamination interface, and various applied conditions (cf. Ramkumar, 1985; Wilkins, 1983; Wilkins et al., 1982). To use such a representation to calculate the growth of a delamination, several essential steps must be followed:

1. The geometry and loading of the specimen or component and the delamination interface must be precisely determined and represented.
2. A stress analysis must be conducted that correctly represents the global and (especially) the local stress fields as the delamination grows, including nonlinear effects (e.g., damage initiation and accumulation, viscoelasticity, and microgeometry changes).
3. Those stress fields must be used to calculate the correct strain energy release rate for the delamination type and mode observed.
4. Careful experiments must be done to determine the constants α and β for such delamination growth.
5. The generality of the results must be assessed if results for other geometries, loadings, laminates, or material systems are to be estimated.

This is not a simple process, but it is a tractable one, and there is much literature for guidance. As mentioned earlier, it may be that the stiffness change that results from the calculated delamination growth defines "failure," especially if the component is a stiffness design. Or failure may be caused by the growth of the delamination to a specified length, such as across the total section of a component. In that case, the crack length is the answer.

However, a more common situation is that the local stress redistributions act (with other processes and events described in this book) to change the value of the failure function in the critical element and alter the arguments in the strength evolution integral:

$$\mathrm{Fr}(\tau_1) = 1 - \int_0^{\tau_1} \left[1 - \mathrm{Fa}\left(\frac{\sigma_{ij}}{X_{ij}} \right) \right] j\tau^{j-1}\, d\tau \tag{7.65}$$

The failure function will be defined by the failure mode, and for delamination that mode may be a compression failure like microbuckling or global buckling of a remaining section after several surface plies have delaminated. We will examine some examples of this type in Chapter 8. But other failure modes may also be controlling, and delamination may be a contributor to the local stress redistribution that causes failure in some other way. In any case, the methodology discussed above can be used to represent the process and to predict the remaining strength and life.

REFERENCES

Anderson, T. L. (1995), *Fracture Mechanics: Fundamentals and Applications*, CRC Press, Ann Arbor, MI.

Awerbuch, J., and Madhuker, M. S. (1985), "Notched Strength of Composite Laminates: Predictions and Experiments—A Review," *J. Reinforced Plastics Composites*, Vol. 4, No. 1, pp. 3–159.

Bakis, C. E. (1988), "Fatigue Behavior of Notched Carbon Epoxy Laminates During Reversed Cyclic Loads," Ph.D. Thesis, Department of Engineering Science and Mechanics, College of Engineering, Virginia Polytechnic Institute and State University, Blacksburg.

Bakis, C. E., and Stinchcomb, W. W. (1986), "Response of Thick, Notched Laminates Subjected to Tension-Compression Cyclic Loads," in *Composite Materials: Fatigue and Fracture*, ASTM-STP 907, H. T. Hahn, Ed., American Society for Testing and Materials, Philadelphia, pp. 314–339.

Bergmann, H. W., and Stavrinidis, C. (1988), *Development of Fracture Weakness Maps for Composite Materials*, Vols. 1 and 2, German Aerospace Research Establishment (DFVRL), Braunschweig, Germany.

Bergmann, H. W., and Stavrinidis, C. (1988), "Development of Fracture Mechanics Maps for Composite Materials under Compressive Loading," Final Report, ESTEC Contract No. 6400/85/NL/PBSC, German Aerospace Research Establishment (DFVLR), Braunschweig.

Biot, M. A. (1956), "Thermoelasticity and Irreversible Thermodynamics," *J. Appl. Phys.*, Vol. 27, pp. 240–253.

Brewer, J. C., and Lagace, P. A. (1988), "Quadratic Stress Criterion for Initiation of Delamination," *J. Composite Mater.*, Vol. 22, pp. 1141–1155.

Broutman, L. J., Ed. (1974), *Fracture and Fatigue*, Academic, New York.

Case, S. W., Plunkett, R. B., and Reifsnider, K. L. (1997), "Simulation Methods for Life and Remaining Strength Prediction of High-Temperature Polymeric Composites Subjected to Cyclic Loads," in *Second Symposium on High Temperature and Environmental Effects on Polymeric Composites*, STP 1302, T. S. Gates and A. H. Zureick, Eds., American Society for Testing and Materials, pp. 35–49.

Chamis, C. C. (1984), "Simplified Composite Micromechanics Equations for Strength, Fracture Toughness and Environmental Effects," in *Proceedings of the Thirty-Ninth Annual Conference, Reinforced Plastics/Composites Institute, Composites go to Market*, Society of the Plastics Industry, N.Y., pp. 11-D-1–11-D-16.

Daniel, I. M., Anastassopoulous, G., and Lee, J. W. (1989), "Experimental Micromechanics of Brittle-Matrix Composites," in *Micromechanics: Experimental Techniques*, W. N. Sharpe, Ed., AMD, Vol. 10, American Society of Mechanical Engineers, pp. 133–147.

Daniel, I. M., and Ishai, O. (1994), *Engineering Mechanics of Composite Materials*, Oxford University Press, New York.

Dowling, N. E. (1993), *Mechanical Behavior of Materials*, Prentice-Hall, Englewood Cliffs, NJ.

Dunyak, T. J. (1991), "Properties and Performance of a Ceramic Composite Component," Ph.D. Thesis, Engineering Mechanics, Virginia Polytechnic Institute and State University, Blacksburg.

Eggers, H. (1985), "Crack Conditions for Statically Loaded Notched UD-Laminates of Carbon Fibre-Reinforced Plastic," Report IB 131-85, Institut fur Strukturmechanik, Braunschweig.

El-Zein, M., and Reifsnider, K. L. (1990), "The Strength Prediction of Composite Laminates Containing a Circular Hole," *J. Composites Technol. Res.*, Vol. 12, No. 3, pp. 147–154.

Freund, L. B. (1990), *Dynamic Fracture Mechanics*, Cambridge University Press, Cambridge.

Gehlen, P. C., and Kanninen, M. F. (1970), "An Atomic Model for Cleavage Crack Propagation in Iron," in *Inelastic Behavior of Solids*, M. F. Kanninen, W. F. Adler, A. R. Rosenfeld, R. I. Iaffee, Eds., McGraw-Hill, New York, pp. 587–603.

Greszczuk, L. B. (1972), "Stress Concentrations and Failure Criteria for Orthotropic and Anisotropic Plates with Circular Openings," in *Composite Materials: Testing and Design*, ASTM STP 497, H. T. Corten, Ed., American Society for Testing and Materials, Philadelphia, pp. 363–381.

Griffith, A. A. (1920), "The Phenomena of Rupture and Flow in Solids," *Philos. Trans. Ser. A*, Vol. 221, pp. 163–198.

Gurdal, Z., and Haftka, R. T. (1987), "Compressive Failure Model for Anisotropic Plates with a Cutout," *AIAA J.*, Vol. 25, No. 11, pp. 1476–1481.

Hyer, M. W. (1998), *Stress Analysis of Fiber-Reinforced Composite Materials*, McGraw-Hill, New York.

Highsmith, A. L. (1981), "Stiffness Reduction Resulting from Transverse Cracking in Fiber-Reinforced Composite Laminates," Master's Thesis, College of Engineering, Virginia Polytechnic Institute and State University, Blacksburg.

Highsmith, A. L. (1984), "Damage Induced Stress Redistribution in Composite Laminates," Ph.D. Dissertation, College of Engineering, Virginia Polytechnic Institute and State University, Blacksburg.

Iarve, E. (1996), "Spline Variational Three Dimensional Stress Analysis of Laminated Composite Plates with Open Holes," *Int. J. Solids Structures*, Vol. 33, No. 14, pp. 2095–2118.

Inglis, C. E. (1913), "Stresses in a Plate Due to the Presence of Cracks and Sharp Corners," *Trans. Inst. Naval Architects*, Vol. 55, pp. 219–241.

Irwin, G. R. (1948), "Fracture Dynamics," in *Fracturing of Metals*, American Society for Metals, G. Sachs, Ed., Cleveland, pp. 147–166.

Irwin, G. R. (1957), "Analysis of Stresses and Strains Near the End of a Crack Traversing a Plate," *J. Appl. Mech.*, Vol. 24, pp. 361–364.

Jamison, R. D., Highsmith, A. L., and Reifsnider, K. L. (1981), *Composites Technol. Rev.*, Vol. 3, No. 4, p. 158.

Kanninen, M. F., and Poplar, C. H. (1985), *Advanced Fracture Mechanics*, Oxford University Press, New York.

Kanninen, M. F., Rybicki, E. F., Griffith, W. I., and Broek, D. (1975), "Fundamental Analysis of the Failure of Polymer-Based Fiber Reinforced Composites," Final Report, NASA-AMES Grant NSG-2038, Battelle Columbus Laboratories, Columbus.

Konish, H. J., and Whitney, J. M. (1975), "Approximate Stresses in an Orthotropic Plate Containing a Circular Hole," *Journal of Composite Materials*, Vol. 9, pp. 157–166.

Lagace, P. A. (1986), "Notch Sensitivity and Stacking Sequence of Laminated Composites," in *Composite Materials: Testing and Design*, J. M. Whitney, Ed., ASTM STP 893, American Society for Testing and Materials, Philadelphia, pp. 161–176.

Lekhnitskii, S. G. (1981), *Theory of Elasticity of an Anisotropic Body*, Mir Publishers, Moscow.

Mikulas, M. M. (1980), "Failure Prediction Techniques for Compression Loaded Composite Laminates with Holes," NASA CP 2142, Langley Research Center, Hampton, VA.

Mollenhauer, D. H. (1997), "Interlaminar Deformation at a Hole in Laminated Composites: A Detailed Experimental Investigation Using Moiré Interferometry," Ph.D. Thesis, Engineering Science and Mechanics, College of Engineering, Virginia Polytechnic Institute and State University, Blacksburg.

O'Brien, T. K. (1981), "Characterization of Delamination Onset and Growth in a Composites Laminate," NASA Technical Memorandum 81940, Langley Research Center, Hampton, VA.

O'Brien, T. K. (1985), "Analysis of Local Delaminations and their Influence on Composite Laminate Behavior," in *Delamination and Debonding of Materials*, ASTM STP 876, W. S. Johnson, Ed., American Society of Testing and Materials, Philadelphia, pp. 282–297.

O'Brien, T. K. (1991), "Delamination of Composite Materials," in *Fatigue of Composite Materials*, K. L. Reifsnider, Ed., Elsevier, New York, pp. 181–198.

O'Brien, T. K., Johnson, N. J., Morris, D. H., and Simonds, R. A. (1982), "A Sample Test for the Interlaminar Fracture Toughness of Composites," *SAMPE J.*, Vol. 18, No. 4, pp. 8–15.

Orowan, E. (1948), "Fracture and Strength of Solids," *Rep. Progr. Phys.*, Vol. 12, p. 18–232.

Pagano, N. J. (1970a), "Exact Solutions for Rectangular Bidirectional Composites and Sandwich Plates," *J. Composite Mater.*, Vol. 4, pp. 20–35.

Pagano, N. J. (1970b), "Influence of Shear Coupling in Cylindrical Bending of Anisotropic Laminates," *J. Composite Mater.*, Vol. 4, pp. 330–343.

Pagano, N. J. (1977), "Stress Fields in Composite Laminates," AFML-TR-77-114, Air Force Wright Aeronautical Laboratories, Wright-Patterson AFB, Dayton, OH.

Pagano, N. J., and Pipes, R. B. (1971), "The Influence of Stacking Sequence on Laminate Strength," *J. Composite Mater.*, Vol. 5, pp. 50–57.

Perrone, N., Liebowitz, H., Mulville, D., and Pilkey, W. (1978), *Fracture Mechanics*, University Press of Virginia, Charlottesville.

Pipes, R. B., and Pagano, N. J. (1970), "Interlaminar Stresses in Composite Laminates under Uniform Axial Extension," *J. Composite Mater.*, Vol. 14, pp. 538–548.

Poe, C. C., and Sova, J. A. (1980), "Fracture Toughness of Boron Aluminum Laminates with Various Proportions of 0 degree and 45 degree Plies," NASA Technical Paper 1707, NASA Langley Research Center, Hampton, VA.

Puppo, A. H., and Evensen, H. A. (1970), "Interlaminar Shear in Laminated Composites under Generalized Plane Stress," *J. Composite Mater.*, Vol. 4, pp. 204–220.

Ramkumar, R. L. (1983), "Performance of a Quantitative Study of Instability-Related Delamination Growth," NASA Contractor Report 166046, Northrop Corporation, Aircraft Division, Hawthrone, CA, and Langley Research Center, Hampton, VA.

Ramkumar, R. L. (1985), *Delamination and Debonding of Materials*, ASTM Special Technical Publication 876, American Society for Testing and Materials, Philadelphia, pp. 315–335

Razvan, A., and Reifsnider, K. L. (1991), "Fiber Fracture Strength Degradation in Unidirectional Graphite/Epoxy Composite Materials," *Theoret. Appl. Fracture Mech.*, Vol. 16, No. 1, pp. 81–89.

Reddy, J. N., and Miravete, A. (1995), *Practical Analysis of Composite Laminates*, CRC Press, New York.

Reifsnider, K. L., Ed. (1991), *Fatigue of Composite Materials*, Elsevier Science, New York.

Reifsnider, K. L. Iyengar, N. Case, S. W., and Xu, Y. (1995), "Kinetic Methods for Durability and Damage Tolerance Design of Composite Components," Keynote Address, Conference on Composite Materials, Japan Society for Mechanical Engineers, Tokyo.

Reifsnider, K. L., and Schultz, R. W. (1984), "Fatigue of Composite Laminates under Mixed Tension-Compression Loading," in *Proceedings of the Second International Conference of Fatigue and Fatigue Thresholds*, C. J. Beevers, Ed., Engineering Materials Advisory Series, Warley Pub., West Midlands, University of Birmingham, U.K., pp. 1103–1114.

Reifsnider, K. L., Stinchcomb, W. W., Bakis, C. E., and Yih, R. Y. (1988), "The Mechanics of Micro-Damage in Notched Composite Laminates," in *Damage Mechanics in Composites*, AD Vol. 12, A. S. D. Wang and G. K. Haritos, Eds., American Society of Mechanical Engineers, New York, pp. 65–72.

Rhodes, M. D., Mikulas, M. M., and McGowan, P. E. (1982), "Effect of Orthotropic Properties and Panel Width on the Compression Strength of Graphite-Epoxy Laminates with Holes," in *Proceedings of the AIAA/ASME/ASCE/AHS 23rd Structures*, Structural Dynamics and Materials Conference: A Collection of Technical Papers, New Orleans, ASME, New York.

Rice, J. R. (1968a), "A Path Independent Integral and the Approximate Analysis of Strain Concentration by Notches and Cracks," *J. Appl. Mech.*, Vol. 35, pp. 379–386.

Rice, J. R. (1968b), "Mathematical Analysis in the Mechanics of Fracture," in *Fracture: An Advanced Treatise*, H. Leibowitz, Ed., Vol. 2, Academic, New York, p. 191.

Rice, R. C., Ed. (1988), *Fatigue Design Handbook*, SAE Publication No. AE-10, Society of Automotive Engineers, Warrendale, PA.

Rolfe, S. T., and Barsom, J. M. (1977), *Fracture and Fatigue Control in Structures: Applications of Fracture Mechanics*, Prentice-Hall, Englewood Cliffs, NJ.

Shalev, D. (1988), "Prediction Model for the Onset of Edge-Effect Delamination at Holes in Composite Laminates," Ph.D. Thesis, Department of Engineering Science and Mechanics, College of Engineering, Virginia Polytechnic Institute and State University, Blacksburg.

Shalev, D., and Reifsnider, K. L. (1989), "Analysis of the Edge Effect at Circular Hole in Composite Laminates," *Int. J. Solids Structures*, Vol. 25, No 10, pp. 1115–1132.

Shalev, D., and Reifsnider, K. L. (1990), "Study of the Onset of Delamination at Holes in Composite Laminates," *J. Composite Mater.*, Vol. 24, pp. 42–71.

Schultz, R. W. (1984), "Cumulative Damage Concepts for Composite Materials," Master's Thesis, College of Engineering, Virginia Polytechnic Institute and State University, Blacksburg, VA.

Smith, C. W., Ed. (1979), *Fracture Mechanics: Proceedings of the Eleventh National Symposium on Fracture Mechanics*, ASTM STP 677, American Society for Testing and Materials, Philadelphia.

Sneddon, I. N. (1946), "The Distribution of Stress in the Neighborhood of a Crack in an Elastic Solid," *Proc. R. Soc. Lond.*, Vol. A-187, pp. 229–260.

Srinivas, S., and Rao, A. K. (1970), "Bending, Vibration, and Buckling of Simply Supported Thick Orthotropic Rectangular Plates and Laminates," *Int. J. Solids Structures*, Vol. 6, pp. 1463–1481.

Starnes, Jr., J. H., and Williams, J. G. (1982), "Failure Characteristics of Graphite-Epoxy Structural Components Loaded in Compression," NASA Technical Memorandum 84552, Langley Research Center, Hampton, VA.

Talug, A., and Reifsnider, K. L. (1978), VPI-E-78-23, College of Engineering Report, Virgina Polytechnic Institute and State University, Blacksburg, VA.

Talug, A., and Reifsnider, K. L. (1979), "Analysis of Stress Fields in Composite Laminates with Interior Cracks," *J. Fiber Sci. Technol.*, Vol. 12, p. 201.

Wagnecz, L. (1987), "Mechanical Behavior and Damage Mechanisms of Woven Graphite-Polyimide Composite Materials," Master's Thesis, Department of Engineering Science and Mechanics, College of Engineering, Virginia Polytechnic Institute and State University, Blacksburg.

Wang, A. S. D., and Crossman, F. W. (1977), "Some New Results on Edge Effects in Symmetric Composite Laminates," *J. Composite Mater.*, Vol. 11, pp. 92–106.

Wang, A. S. D., Slomiana, N., and Bucinell, R. (1983), "A Three Dimensional Finite Element Analysis of Delamination Growth in Composite Laminates," Report No. NADC-84017-60, Naval Air Development Center, Westminster, PA.

Wang, S. S., and Choi, I. (1982a), "Boundary-Layer Effects in Composite Laminates: Part I—Free Edge Stress Singularities," *ASME J. Appl. Mech.*, Vol. 49, pp. 541–549.

Wang, S. S., and Choi, I. (1982b), "Boundary-Layer Effects in Composite Laminates: Part II—Free Edge Stress Solutions and Basic Characteristics," *ASME J. Appl. Mech.*, Vol. 49, pp. 549–560.

Westergaard, H. M. (1939), "Bearing Pressures and Cracks," *J. Appl. Mech.*, Vol. 6, pp. 49–53.

Whitcomb, J. D. (1998), "Instability-Related Delamination Growth of Embedded and Edge Delaminations," Ph.D. Dissertation, College of Engineering, Virginia Polytechnic Institute and State University, Blacksburg.

Whitney, J. M., and Sun, C. T. (1973), "A Higher Order Theory for Extensional Motion of Laminated Composites," *J. Sound Vibration*, Vol. 30, pp. 85–97.

Whitney, J. M., Browning, C. E., and Hoogsteden, W. (1982) "Double Cantilever Beam Tests for Characterizing Mode I Delamination of Composite Materials," *Journal of Reinforced Plastics and Composites*, Vol. 1, No. 4, pp. 297–313.

Whitney, J. M., and Kim, R. Y. (1977), "Effect of Stacking Sequence on the Notched Strength of Laminated Composites," in ASTM STP 617, *Composite Materials: Testing and Design* American Society for Testing and Materials, Philadelphia, pp. 229–242.

Whitney, J. M., and Knight, M., Eds. (1985), *Delamination and Debonding of Materials*, ASTM STP 876, American Society for Testing and Materials, Philadelphia.

Whitney, J. M., and Nuismer, R. F. (1974), "Stress Fracture Criteria for Laminated Composites Containing Stress Concentrations," *J. Composite Mater.*, Vol. 8, pp. 253–265.

Wilkins, D. J. (1981), "A Comparison of the Delamination and Environmental Resistance of a Graphite Epoxy and a Graphite Bismaleimide," Naval Air Systems Command Report No. NAV-GD-0037, Washington, D.C.

Wilkins, D. J. (1983), "Failure Analysis and Mechanisms of Failure of Fibrous Composite Structures," NASA CP 2278, pp. 67–93.

Wilkins, D. J., Eisenmann, J. R., Camin, R. A., Margolis, W. S., and Benson, R. A. (1982), "Characterizing Delamination Growth in Graphite Epoxy," in *Damage in Composite Materials*, K. L. Reifsnider, Ed., ASTM Special Technical Publication 775, American Society for Testing and Materials, Philadelphia, pp. 168–183.

Williams, M. L. (1957), "On the Stress Distribution at the Base of a Stationary Crack," *J. Appl. Mech.*, Vol. 24, pp. 109–114.

Wu, E. M. (1978), "Failure Analysis of Composites with Stress Gradients," UCRL-80909, Lawrence Livermore Laboratory Report, Livermore, CA.

EXERCISES

7.1 Measured values of laminate tensile strength as a function of hole diameter for graphite–epoxy materials (Bergmann and Stavrinidis, 1988) appear below:

Hole Diameter ($2R$, mm)	Strength (MPa)/Strength Ratio	
	$[0, 45, -45, 90]_S$	$[0, 0, +45, -45]_S$
0.00	502.0/1.0	802.0/1.0
0.20	491.5/0.979	—
0.40	434.0/0.865	—
0.40	463.0/0.922	—
1.60	387.9/0.733	—
1.60	380.9/0.759	—
4.80	285.6/0.570	—
4.80	308.5/0.615	—
6.40	281.3/0.562	640.3/0.800
6.40	—	412.9/0.516
6.40	270.3/0.540	598.4/0.748
12.70	260.2/0.524	480.8/0.606
12.70	244.4/0.492	461.0/0.581
12.70	—	409.6/0.516
19.10	213.6/0.436	387.2/0.495
19.10	237.0/0.484	403.7/0.516
19.10	—	353.1/0.451
25.40	199.7/0.416	379.2/0.495
25.40	218.8/0.456	359.2/0.469
25.40	—	425.2/0.555

(a) Use the Whitney–Nuismer average-stress criterion to represent the data for the two laminate types. Determine the characteristic length parameter for that criterion for this laminate. How well does the model fit the data?

(b) Use the Whitney–Nuismer point-stress criterion to represent the data for the two laminate types. Determine the characteristic length parameter for that criterion for this laminate. How well does the model fit the data?

(c) Compare the values of the characteristic length parameters for the two laminates. Are they consistently different (for the two criteria and two laminates)? What are the physical situations that are causing those differences?

7.2 Consider a graphite–epoxy material having the following lamina stiffness values: $E_1 = 129\,\text{GPa}$; $E_2 = 10.86\,\text{GPa}$; $\nu_{12} = 0.3118$; $G_{12} = 5.70\,\text{GPa}$. Consider two 50-mm-wide coupon laminates with center holes under uniaxial loading. Laminate A has the stacking sequence $[45, 0, -45, 0]_{2S}$ and laminate B has the sequence $[45, 0, -45, 90]_{2S}$. It is known that laminate A has a strength of 284 MPa when the center hole has a diameter of 7.6 mm. Use Equations (7.6)–(7.10) (the approach of Konish and Whitney) to conduct the following calculations:

(a) Predict the tensile strength of coupon laminate A having a center hole with a diameter of 15.2 mm using both the point-stress and average-stress criteria corrected for anisotropy. [Compare to measured values of 239 and 333 MPa for two specimens from Poe and Sova (1980).]

(b) Predict the tensile strength of the coupon laminate B having the following center-hole diameters and compare to the strength values given in the table below (Poe and Sova, 1980), again using both the point-stress and average-stress criteria:

Diameter of Center Hole (mm)	Average Measured Strength (MPa)
7.6	254
8.5	245
15.2	234
16.9	206
30.5	157
33.8	169

(c) Are your values systematically high or low? If so, can you construct a physical argument for that systematic difference?

(d) What strength would you expect a unidirectional coupon of this material to have under uniaxial loading if the center-hole diameter is 30 mm? Discuss all assumptions made in making this prediction, and state reasons why your estimate should be conservative or nonconservative.

7.3 From independent measurements, it is determined that the interlaminar strength of the graphite–epoxy material discussed in Exercise 7.2 is about 15 MPa. Estimate the interlaminar position at which delamination should occur using two methods for the following laminates made from that material system. For the first estimate, use the equations of O'Brien [(7.56)–(7.62)] with the assumption that delamination will initiate at the ply interface that has the largest strain energy release rate. For the second estimate, use the concept that delamination will initiate at the ply interface that has the largest interlaminar normal stress. [Use the approximate method of Pipes and Pagano (1970) to estimate those interlaminar normal stresses if necessary.] Determine the delamination initiation interface for the following laminate coupons (with no center hole): $[0, +30, -30, 0]_{2S}$; $[0_2, +45, 90, -45]_{2S}$; $[0_2, 90_2]_{2S}$; $[45, 0, -45, 90]_{2S}$.

7.4 As an engineer, suppose you want to design an engineering component using a quasi-isotropic composite laminate, that is, using an equal number of plies having orientations of 45°, −45°, 90°, and 0°.

(a) What stacking sequence should you use for the laminate to optimize the resistance of the component to edge delamination under uniaxial tensile loading? Explain the rationale and methodology used to make this specification.

(b) Would you specify the same stacking sequence if the component were to be loaded primarily in compression?

(c) Would you specify the same stacking sequence if the component were to be loaded with a complex combination of in-plane resultants, that is, with significant and similar levels of N_x, N_y, and N_{xy}?

7.5 Considering the models related to delamination discussed in the text, which is more likely to delaminate, a thick laminate or a thin one having the same stacking sequence?

7.6 It is desired to make an engineering component out of a thick-section composite laminate. To properly sustain the remote applied loads, it is decided to construct the laminate using plies having an orientation of 0° to the principal tensile load axis for 30% of the laminate, ±45° plies for 40% of the laminate, and 90° plies for the remaining 30% of the laminate. Since the laminate will be 100 plies thick, it is necessary to specify how those plies will appear in the laminate.

(a) Should all of the plies that have a given orientation be grouped together or distributed throughout the laminate? Justify your answer with a computation.

(b) What stacking sequence would you specify for this thick laminate? Justify your answer with a computation.

CHAPTER 8

EXAMPLE APPLICATIONS AND CASE STUDIES

Thesis: The residual strength approach developed in the preceding chapters is applied to the life prediction of a number of different material systems. An example application will be presented for analyzing the unnotched fatigue behavior of polymer composite laminates, including the effects of matrix cracking, viscoelasticity, and creep rupture. We will then consider five case studies: the room-temperature fatigue behavior of APC-2 laminates, the elevated-temperature fatigue behavior of graphite fiber–PPS composites, the elevated-temperature fatigue failure of Nextel 610 composites, the elevated-temperature fatigue behavior of SiC–SiC composite materials, and the fatigue failure of a structural composite shape. The experimental results obtained for those systems are compared with the available experimental data.

8.1 EXAMPLE: UNNOTCHED FAILURE OF POLYMER COMPOSITE

As an example, we will consider the residual strength analysis of an unnotched polymer composite laminate. This example will combine the many features of composite behavior that we have discussed in the previous chapters. The lamina properties are chosen to be typical of a graphite–epoxy material and are given in Table 8.1. We will consider the case in which a $[45, 0, -45, 90]_s$ laminate has been fabricated so that the stress-free temperature is 300°C. For this case, the initial

TABLE 8.1 Lamina Properties for Unnotched Example

Quantity	Value
E_1	140 GPa
E_2	7.0 GPa
G_{12}	5.5 GPa
ν_{12}	0.25
X_t	2070 MPa
X_c	1240 MPa
Y_t	55 MPa
Y_c	180 MPa
SS	180 MPa
α_1	$0.12 \times 10^{-6}/°C$
α_2	$35 \times 10^{-6}/°C$

residual thermal stresses at 120°C (in the material coordinate system) determined using classical lamination theory are given by

$$\sigma_1 = 40.9 \text{ MPa} \qquad \sigma_2 = -40.9 \text{ MPa} \qquad \sigma_{12} = 0 \tag{8.1}$$

The maximum applied cyclic fatigue stress will be 600 MPa applied at a frequency of 1 Hz. Failure of the laminate will ultimately be governed by the failure of the 0° plies (the critical element) in the fiber direction. The fatigue behavior of these plies is described by

$$\text{Fa} = \frac{\sigma_1^{0^\circ \text{ ply}}}{X_t} = 1 - 0.06 \ \log N \tag{8.2}$$

Equation (8.2) may be inverted to obtain an expression for the number of cycles to failure in terms of the failure criterion Fa:

$$N = 10^{(\text{Fa}-1)/-0.06} \tag{8.3}$$

Suppose that we wish to obtain an initial estimate of the composite lifetime at 120°C. To obtain this estimate, we first calculate the stresses in the 0° plies. These stresses are given by

$$\sigma_1 = 1527 \text{ MPa} \qquad \sigma_2 = 37.7 \text{ MPa} \qquad \sigma_{12} = 0 \tag{8.4}$$

Using the maximum fiber direction stress failure criterion and substituting into (8.3), we obtain the number of cycles to failure:

$$N = 10^{(1527/2070-1/-0.06} = 23{,}550 \tag{8.5}$$

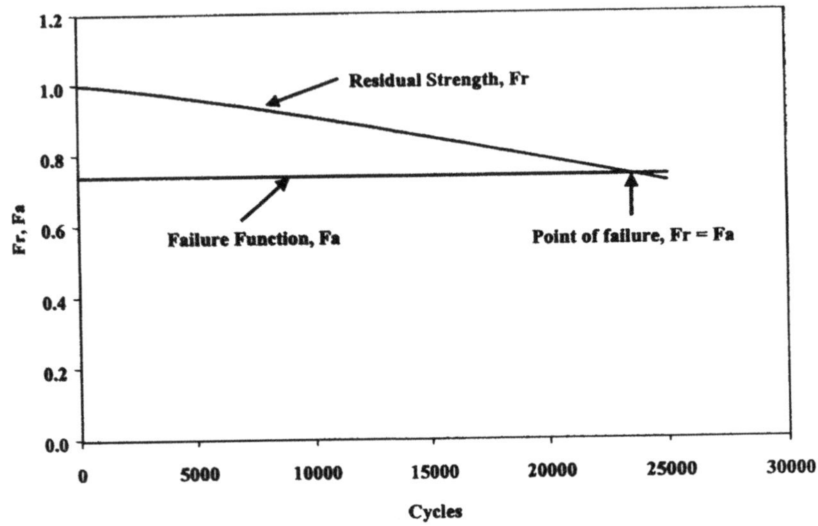

Figure 8.1 Initial residual strength and failure criterion plot for polymer composite example.

Since all of the inputs to (3.12) for this case are constant, we may readily integrate it to obtain

$$\begin{aligned} \mathrm{Fr} &= 1 - (1 - \mathrm{Fa})\left(\frac{n}{N}\right)^{j} \\ &= 1 - (1 - 0.73771)\left(\frac{n}{23550}\right)^{1.2} \end{aligned} \tag{8.6}$$

where we have assumed $j = 1.2$. A plot of (8.6) and the failure criterion Fa for this simple case is shown in Figure 8.1.

Although this does provide us with an estimate of the number of cycles to failure, our previous discussions lead us to believe that it is not the best estimate of the composite lifetime. Factors such as matrix cracking, viscoelasticity, and stress rupture can be expected to alter the behavior of the laminate. We will next consider these effects individually and collectively.

8.1.1 Matrix Cracking

For the purposes of this example, we will assume that the matrix-controlled stiffness of the $\pm 45^\circ$ and 90° plies varies linearly over the course of the estimated lifetime given by (8.5). Mathematically, this implies that

$$\frac{E_2(n)}{E_2^0} = \frac{G_{12}(n)}{G_{12}^0} = 1 - \frac{n}{23{,}550} \tag{8.7}$$

As the stiffness of the off-axis plies decreases, the load that is carried by the 0° ply increases, increasing the value of the failure criterion [i.e., Fa = Fa(n)]. Therefore, the residual strength calculation must now be performed numerically. Before we perform such a calculation, it is useful to consider two limiting cases. The first limiting case is that the stiffness of the ±45° and 90° plies has been totally reduced to a value of zero. In such a case, the value of the failure criterion is given by 0.87491. For this value of Fa, we would estimate approximately 122 cycles to failure. The other limiting case is that discussed above in which no cracking occurs (23,550 cycles to failure). With these two limits in mind, we may now calculate the residual strength and lifetime of the laminate for the case of matrix cracking. The results of such a calculation are shown in Figure 8.2. For this case, failure is predicted to occur after approximately 10,320 cycles. We note that the residual strength and failure criterion at that point each have a value of 0.77409 and the value lies between the two limiting cases.

In this example, we have used a simple method to account for stiffness changes due to cracking. For a more mechanistic treatment, we could have used the methods developed in Chapter 5.

8.1.2 Viscoelasticity

As discussed in Chapter 5, viscoelasticity may play a large role in stress redistributions in polymer composite laminates, particularly for cases in which the use temperature approaches the glass transition temperature. To incorporate this effect

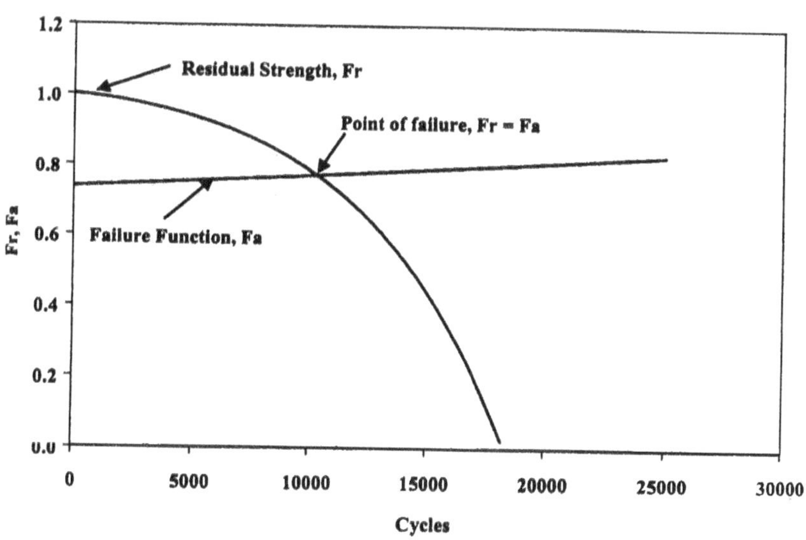

Figure 8.2 Residual strength and failure criterion plot for polymer composite example case in which only matrix cracking is active.

into the residual strength analysis, we will use the approximate approach suggested in Chapter 5, so that

$$\begin{aligned}\frac{E_2(t)}{E_2^0} = \frac{G_{12}(t)}{G_{12}^0} &= 1 - 0.0217(t^*) - 0.000797(t^*)^2 \\ &\quad + 7.24 \times 10^{-5}(t^*)^3 - 1.81 \times 10^{-6}(t^*)^4 \\ t^* &= \log t + \log t_s \\ \log t_s &= -4.8969877 + 0.2345622T - 0.0017247T^2 \\ &\quad + 1.122929 \times 10^{-5}T^3 - 2.51 \times 10^{-8}T^4\end{aligned} \tag{8.8}$$

In (8.8), t_s is a time–temperature shift factor, t is the time in minutes, and the temperature is expressed in degrees Celsius. The effect of the viscoelastic behavior is similar to that of matrix cracking: It increases the load carried by the 0° ply. Residual strength values calculated including this viscoelastic effect are shown in Figure 8.3. The effect of viscoelasticity in this case is to reduce the predicted lifetime from 23,550 cycles to 4872 cycles.

The approach described above is a first-order engineering estimate of the viscoelastic effects. If necessary, more rigorous treatments of the viscoelasticity effect may be obtained by making use of the elastic–viscoelastic correspondence principle.

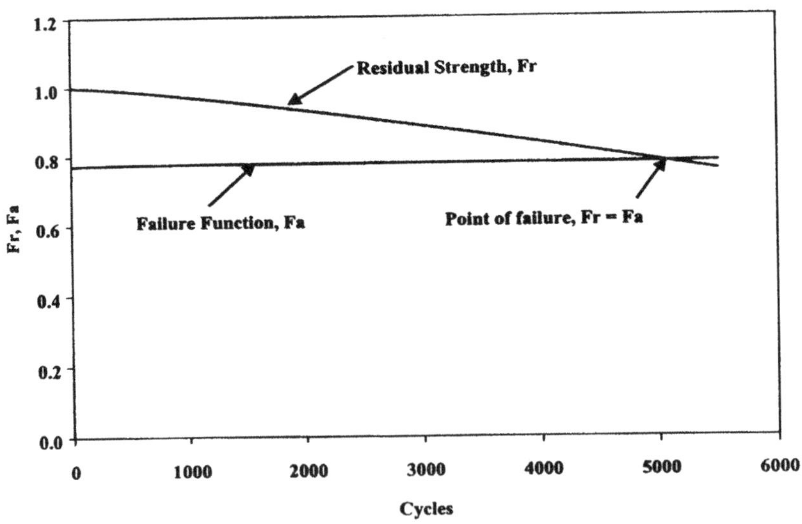

Figure 8.3 Residual strength and failure criterion for polymer composite example case in which only viscoelasticity is active.

8.1.3 Creep Rupture

The final effect to be considered in this example is creep rupture. To address the role of creep rupture, we will make use of the form suggested in Chapter 6:

$$\frac{X_t(t)}{X_t^0} = (1 - 0.0667 \log t)\left[0.4\left(\frac{T - 295}{589 - T}\right)\right] \tag{8.9}$$

where the time t is in hours and the temperature is given in Kelvin. Damage due to creep rupture enters the residual strength calculation in a different fashion than the damage due to matrix cracking and viscoelasticity. These two damage modes change the stress state in the 0° ply. Changes in stress state appear in the numerator of the failure criterion. Creep rupture changes the strength of the 0° ply and enters the denominator of the failure criterion. However, in the end, the effect is to increase the failure criterion as a function of time (or cycles) for a given applied cyclic stress.

The residual strength and failure criterion values calculated including the influence of creep rupture are shown in Figure 8.4. For this case, failure is predicted to occur after approximately 19,400 cycles.

8.1.4 Combined Matrix Cracking, Viscoelasticity, and Creep Rupture

The final aspect of this example is to examine the combination of the three effects considered previously: matrix cracking, viscoelasticity, and creep rupture. To address the matrix cracking and viscoelasticity, we assume that any stiffness changes are additive. Residual strength calculations including these interacting effects are shown in Figure 8.5. The resulting composite lifetime is 3750 cycles.

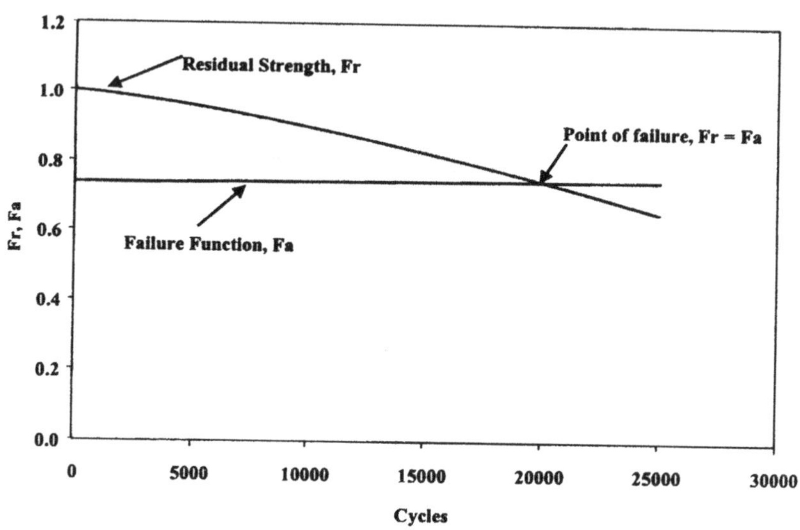

Figure 8.4 Residual strength and failure criterion plot for polymer composite example case in which only creep rupture is active.

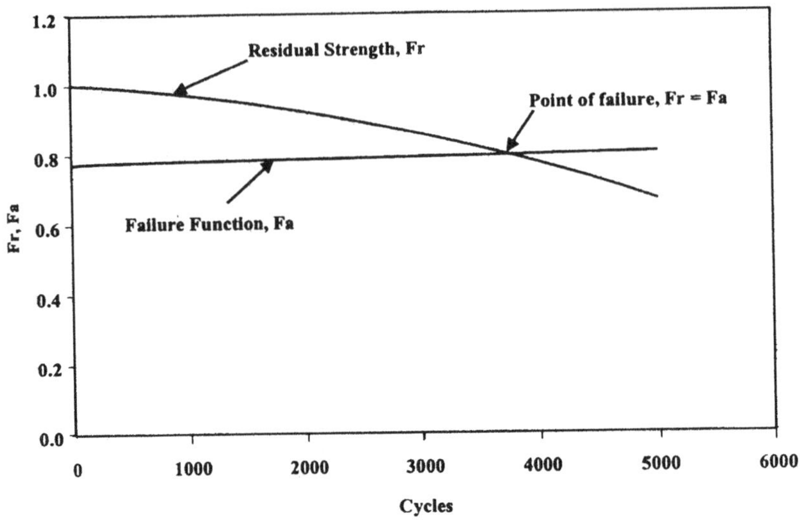

Figure 8.5 Residual strength and failure criterion plot for polymer composite example case in which matrix cracking, viscoelasticity, and creep rupture are active.

The residual strength prediction process may be repeated for a series of different loading levels to generate a traditional *S–N* curve. Such an *S–N* curve is shown in Figure 8.6. We see that for this particular example the inclusion of the matrix cracking, viscoelasticity, and creep rupture may reduce the fatigue lifetime by as much as an order of magnitude.

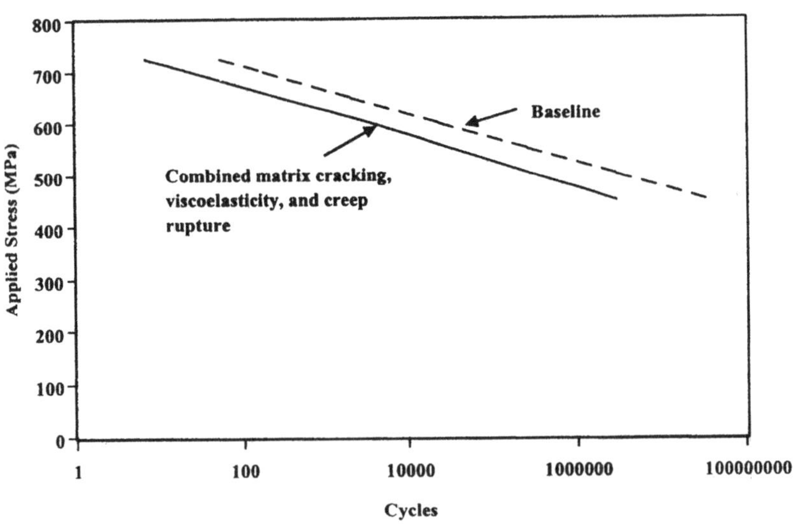

Figure 8.6 *S–N* curves for polymer composite example. Baseline curve includes only residual strength changes due to fatigue.

8.2 CASE STUDY 1: FATIGUE BEHAVIOR OF APC-2 LAMINATES

The first case study to be considered is the fatigue behavior of APC-2 laminates. APC-2 is a semicrystalline polyether ether ketone (PEEK) matrix reinforced with AS-4 carbon fibers that has typical unidirectional properties given in Table 8.2. This material system was chosen for a case study because of the large amount of available experimental characterization.

One of the most important sets of material characterization data available for this material system is the unidirectional fatigue data of Picasso and Priolo (1988). To make use of it in our analysis, we must represent it in some form so that we may determine the number of cycles to failure N from the calculated failure function Fa. A convenient representation is

$$\mathrm{Fa} = A_n + B_n \log N \tag{8.10}$$

This representation may be inverted to yield the desired form given by

$$N = 10^{(\mathrm{Fa}-A_n)/B_n} \tag{8.11}$$

To accurately represent the data of Picasso and Priolo, it is necessary to divide the data into two separate domains: $0.626 \leq \mathrm{Fa}$ and $0.626 > \mathrm{Fa}$. The curve-fitting constants for these domains are given by

$$A_n = 1.3126 \qquad B_n = -0.1818 \qquad 0.626 \leq \mathrm{Fa} \tag{8.12}$$

$$A_n = 0.7866 \qquad B_n = -0.0425 \qquad 0.626 > \mathrm{Fa} \tag{8.13}$$

TABLE 8.2 Unidirectional Properties of APC-2 Composites

Quantity	Value
E_1	134 GPa
E_2	8.9 GPa
G_{12}	5.1 GPa
ν_{12}	0.35
X_t	2090 MPa
X_c	1215 MPa
Y_t	80 MPa
Y_c	280 MPa
SS	80 MPa
α_1	$0.63 \times 10^{-6}/^\circ\mathrm{C}$
α_2	$29 \times 10^{-6}/^\circ\mathrm{C}$

Source: Vure (1993).

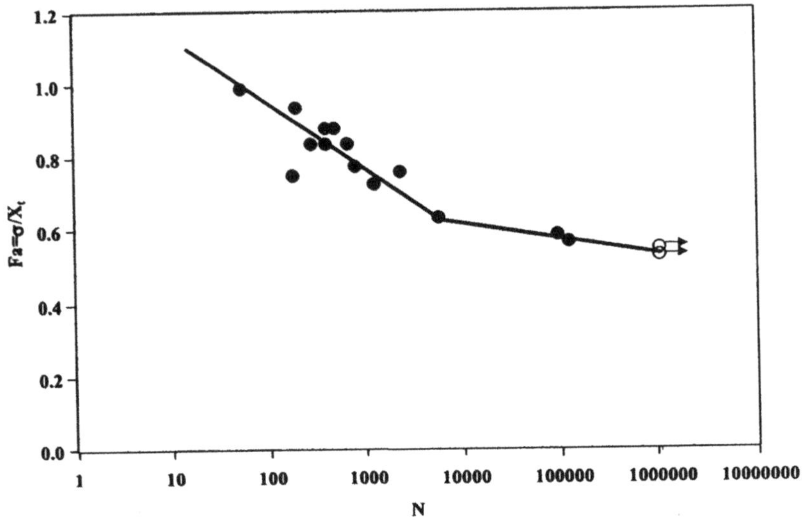

Figure 8.7 Unidirectional failure function–life curve for APC-2 (Picasso and Priolo, 1988).

The resulting fits as well as the experimental data are shown in Figure 8.7. In this case study, the effects of hygrothermal stresses will not be considered. Viscoelastic effects (with the exception of stress concentration relaxation around the hole) will not be considered.

8.2.1 Curtis Data for $[0, 90]_s$ Laminates

The first set of data to be considered is that taken from Curtis (1986), who studied the fatigue behavior of unnotched $[0, 90]_s$ laminates. Curtis presents his results in terms of initial peak strain. Using lamination theory, these results may be readily converted to applied stresses. To begin the analysis, we will consider an applied stress of 765 MPa. The stresses in the 0° ply are given by

$$\sigma_1 = 1436 \text{ MPa} \qquad \sigma_2 = 29.3 \text{ MPa} \qquad \sigma_{12} = 0 \tag{8.14}$$

Once again, we will employ a maximum fiber direction stress failure criterion. The resulting value for the failure function Fa is

$$\text{Fa} = \frac{\sigma_1}{X_t} = \frac{1436 \text{ MPa}}{2090 \text{ MPa}} = 0.687 \tag{8.15}$$

Therefore, the predicted number of cycles to failure is 2759 cycles. This process may be repeated for different applied stress levels to generate an S–N curve for the laminate, such as that shown in Figure 8.8 with the experimental data from

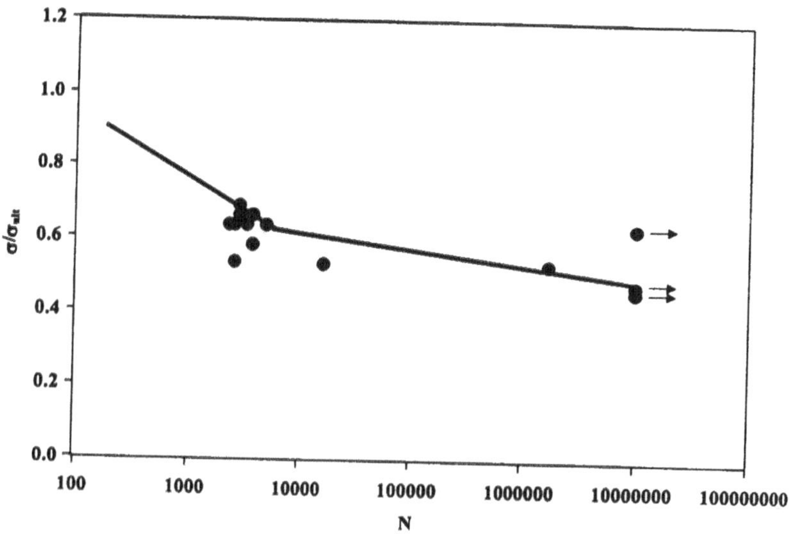

Figure 8.8 Predicted and experimental lifetime data for cross-ply laminates studied by Curtis (1986).

Curtis. It is important to remember that for predicted lives greater than 5982 the second set of *S–N* curve constants (8.13) should be employed.

8.2.2 Simonds and Stinchcomb Data for Quasi-Isotropic Laminates

The next set of APC-2 data to be considered for this case study is that of Simonds and Stinchcomb (1989), who studied the tension–compression ($R = -1$) fatigue behavior of notched quasi-isotropic laminates. The geometry of the hole was circular, with a radius of 4.7625 mm. Simonds and Stinchcomb found the compression strength of this laminate to be 289.6 MPa. To analyze this notched laminate, we will follow the procedure outlined in Chapter 7. We will use the anisotropic plate solution developed by Lekhnitskii (1968) to determine the laminate-level stresses and strains that correspond to the failure load. These stresses and strains are shown in Figures 8.9 and 8.10, respectively, for a line along the hole axis perpendicular to the direction of loading. We now wish to determine the stresses in the 0° ply—the critical element in this case. In particular, we will focus on fiber direction compression failure in this ply by making use of a maximum stress failure criterion. Failure of this ply and hence of the laminate is assumed to occur when the average value of σ_1 up to some distance (or the point-stress value of σ_1 at some distance) is equal to the compression strength. To illustrate this concept, we may consider the stress analysis results for the 0° ply presented in Figure 8.11. The upper curve corresponds to the stresses in the fiber direction at a point. By determining the point at which this curve intersects the compression strength, we may define the point-stress distance. For this laminate, the point-stress distance corresponds to a value of y/a equal to 1.349, or

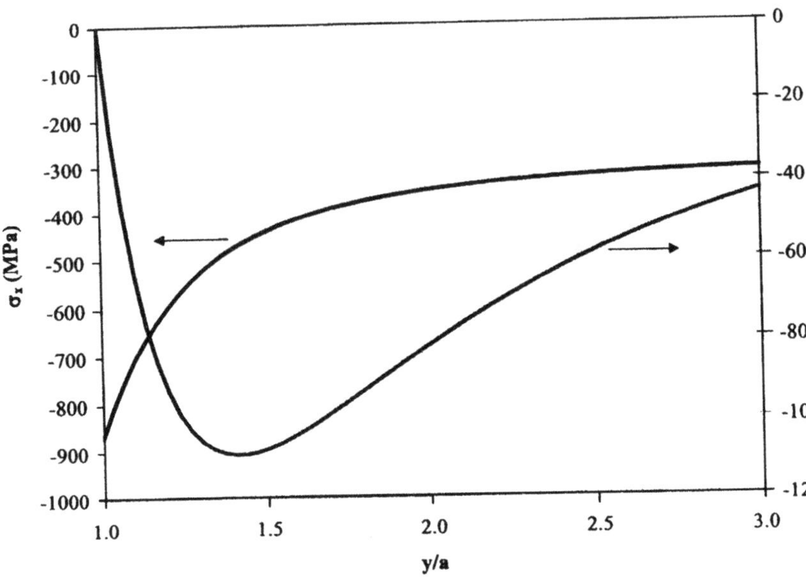

Figure 8.9 Laminate-level stresses corresponding to quasi-static failure load for Simonds and Stinchcomb (1989) data.

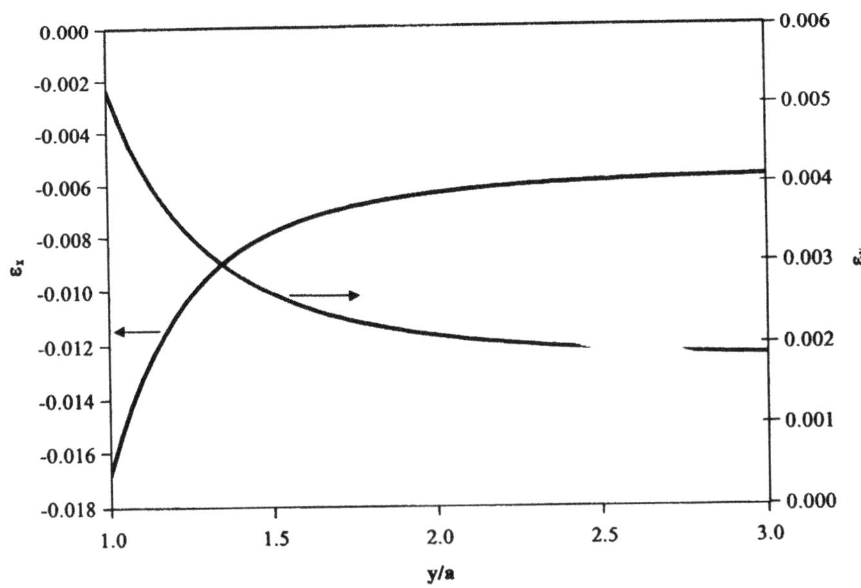

Figure 8.10 Laminate-level strains corresponding to quasi-static failure load for Simonds and Stinchcomb (1989) data.

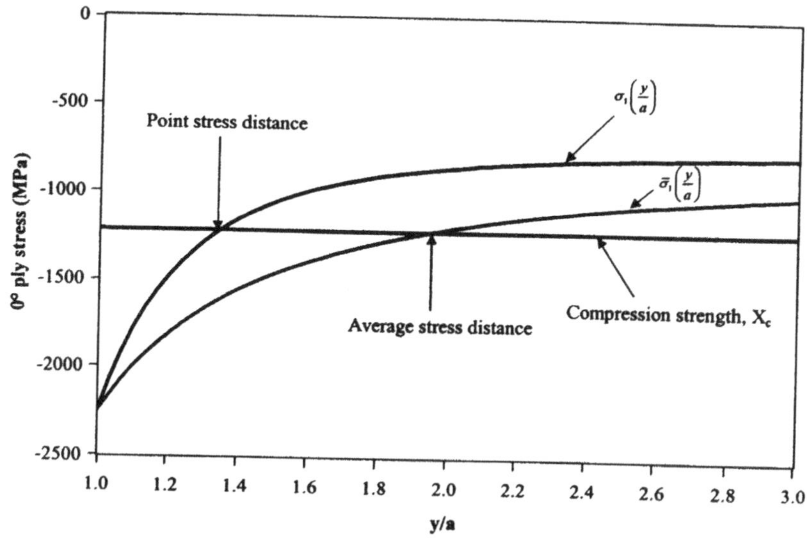

Figure 8.11 Illustration of calculation of point- and average-stress distances.

1.661 mm from the edge of the hole. The average stress distance is found in a similar fashion: We average the stresses in the fiber direction up to a point and compare this average stress to the compression strength. The point at which the compression strength and the average stress are identical defines the averaging distance. In this case, this critical averaging distance corresponds to a value of y/a equal to 1.949 (a distance of 4.520 mm from the edge of the hole), as shown in Figure 8.11.

Once the appropriate point- and average-stress distances have been determined based upon the quasi-static data, we may use them in subsequent residual strength and lifetime prediction calculations. The results are summarized in Table 8.3 and are plotted along with the experimental data in Figure 8.12.

TABLE 8.3 Predictions for Quasi-Isotropic Laminates Containing Center Hole

σ (MPa)	Estimated Fa	Estimated Lifetime
275	0.9496	99
260	0.8978	191
245	0.8460	369
230	0.7942	711
215	0.7424	1,369
200	0.6906	2,639
185	0.6388	5,085
170	0.5870	49,700
155	0.5352	822,300
140	0.4834	13,610,000

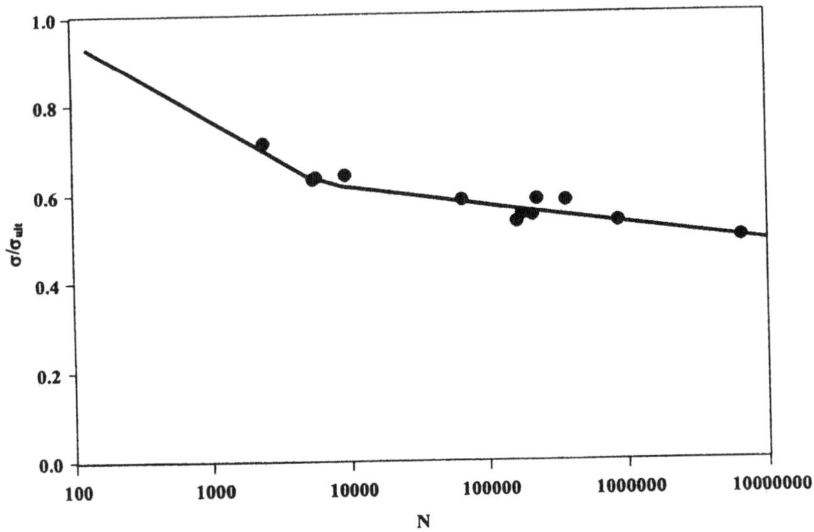

Figure 8.12 Comparison of predicted and measured lifetimes with data of Simonds and Stinchcomb (1989).

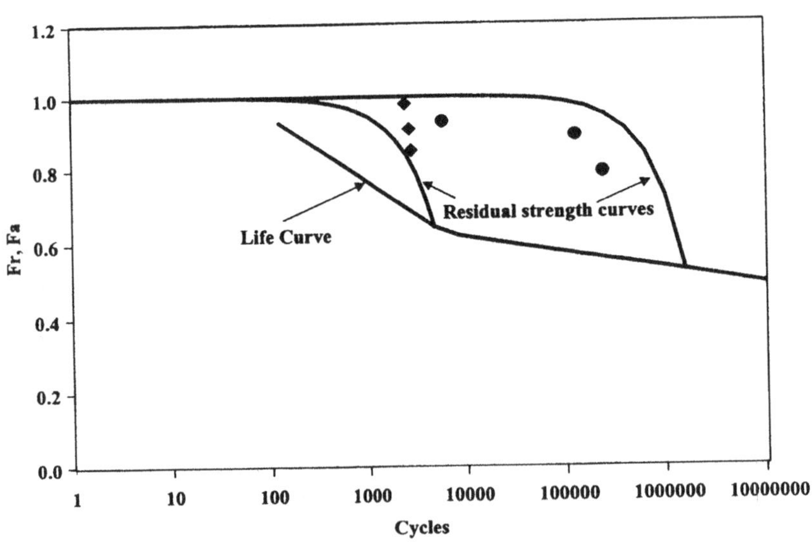

Figure 8.13 Comparison of predicted and measured residual strength for data of Simonds and Stinchcomb (1989).

In addition to the number of cycles to failure as a function of load level, residual strength data were also available for this material system. These residual strength measurements may be compared directly to the predictions. Such comparisons are shown in Figure 8.13.

8.2.3 Vure Data for Quasi-Isotropic Laminates

The final set of APC-2 data to be considered is that taken from Vure (1993), who studied the tensile fatigue ($R = 0.1$) behavior of notched quasi-isotropic APC-2 laminates fabricated using two different cooling rates. In this case, the radius of the hole was 3.175 mm. Vure's average measured laminate strength for the two different cooling rates was 451 MPa. Using the procedure outlined above for the data of Simonds and Stinchcomb (1989), we obtain an average stress distance of 2.153 mm.

To begin the residual strength and lifetime analysis, we will consider an applied stress of 361.2 MPa (this value corresponds to 80% of the laminate ultimate strength). In this case, we need to take into account the reduction in the stress concentration in the vicinity of the hole. As an engineering approach to this problem, we change the hole ellipticity in the manner described in Chapter 7. The variation in the ratio of the hole sizes for this case is shown in Figure 8.14. If we incorporate this change in the hole ellipticity into the life prediction analysis, we determine that the laminate will fail after approximately 1500 cycles at this load level. This is in favorable agreement with the experimental median value of 1800 cycles. This process may be repeated for different applied stress levels to generate a lifetime curve for the laminate, such as that shown in Figure 8.15. It is important to remember the bilinear nature of the unidirectional S–N curve when making the predictions.

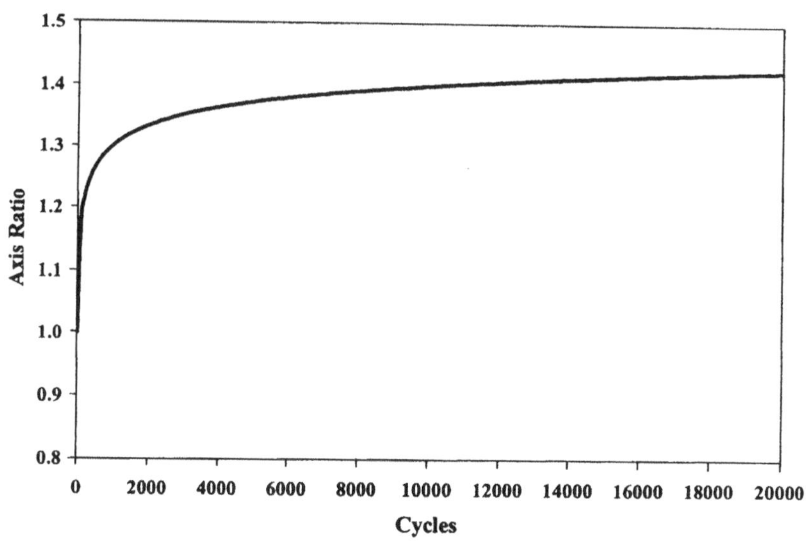

Figure 8.14 Variation of effective hole sizes with fatigue cycles.

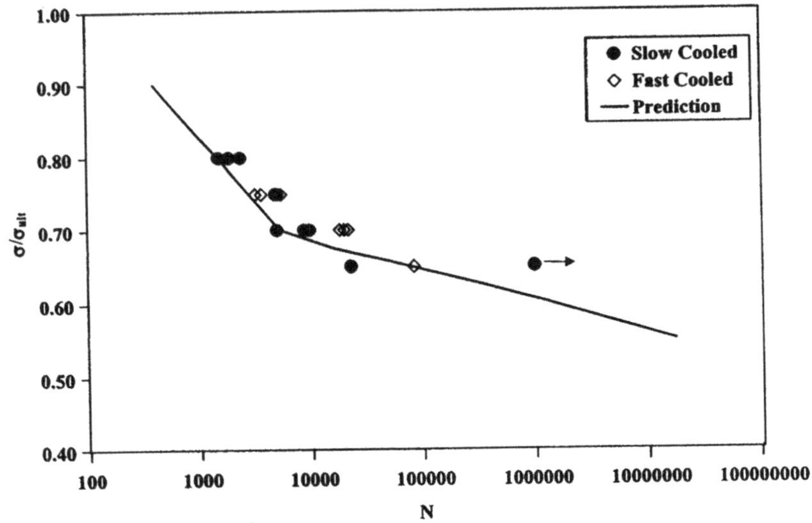

Figure 8.15 Predicted and experimental lifetime data for Vure (1993) laminates.

8.3 CASE STUDY 2: ELEVATED-TEMPERATURE FATIGUE BEHAVIOR OF GRAPHITE FIBER–PPS LAMINATES

The second case study to be considered is that of the elevated-temperature fatigue behavior of unidirectional 0° graphite fiber–PPS composite materials, as studied by Loverich et al. (2000). This material's intended application is the helical tensile strength layers, referred to as the tensile armor, in flexible risers used by the offshore oil industry. These layers support the hanging weight of the riser as well as the axial load caused by internal pressure. A typical flexible riser design is shown in Figure 8.16. To reduce the weight of flexible risers, replacement of the standard steel tensile armor with composite tensile armor is being considered. A weight reduction of 30% and increased corrosion resistance are expected for a composite armored riser designed to the specifications of a comparable conventional flexible pipe. Certification of this material for the flexible pipe application requires including the effects of fatigue and elevated temperatures up to 90°C. The life prediction approach to be developed makes use of the concept of equivalent damage implying equivalent residual strength. To begin the analysis, room-temperature fatigue data are collected at an R ratio of 0.1 and a frequency of 10 Hz. The results from these tests are represented in the form

$$\text{Fa} = \frac{\sigma_{\max}}{X_t} = 0.9207 - 0.0189 \log N \tag{8.16}$$

as shown in Figure 8.17.

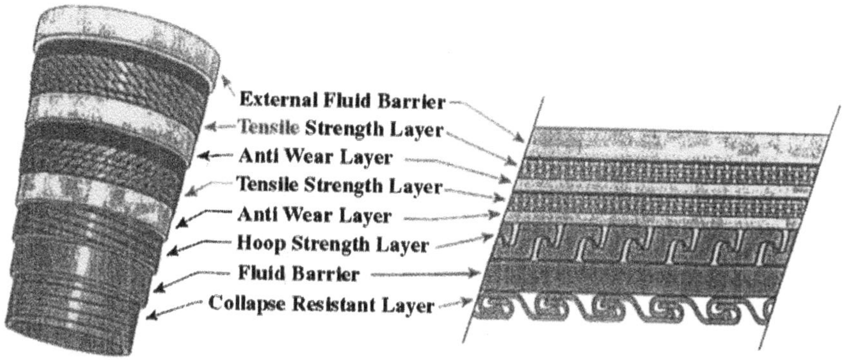

Figure 8.16 Typical flexible riser schematic.

The next step in developing data for the analysis is to conduct stress rupture tests under constant load at 90°C. The results from these tests are then represented in a form similar to that suggested by Kachanov (1986), given by

$$\text{Fa} = \frac{\sigma_a}{X_t} = \left[\frac{1}{(n+1)Bt_{\text{rupture}}}\right]^{1/n} \tag{8.17}$$

where n and B are curve-fitting constants. For the rupture time given in seconds, the constants B and n are found to be 0.0382 and 26.0, respectively, as shown in Figure 8.18.

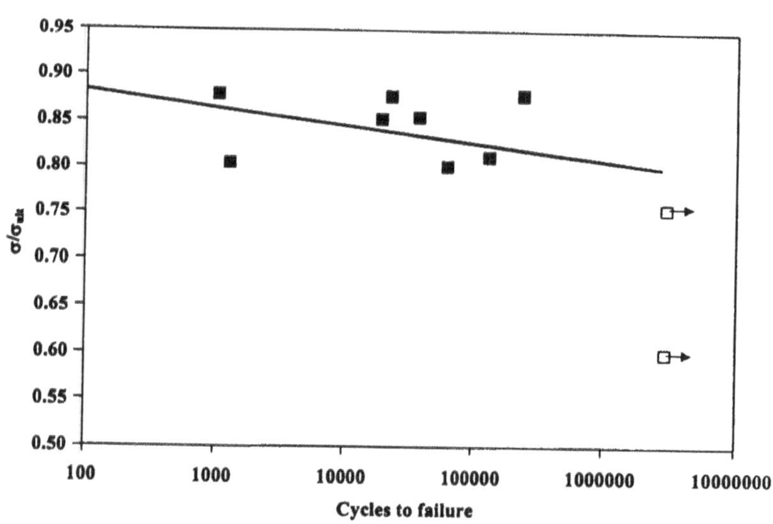

Figure 8.17 Room-temperature fatigue data and curve fit.

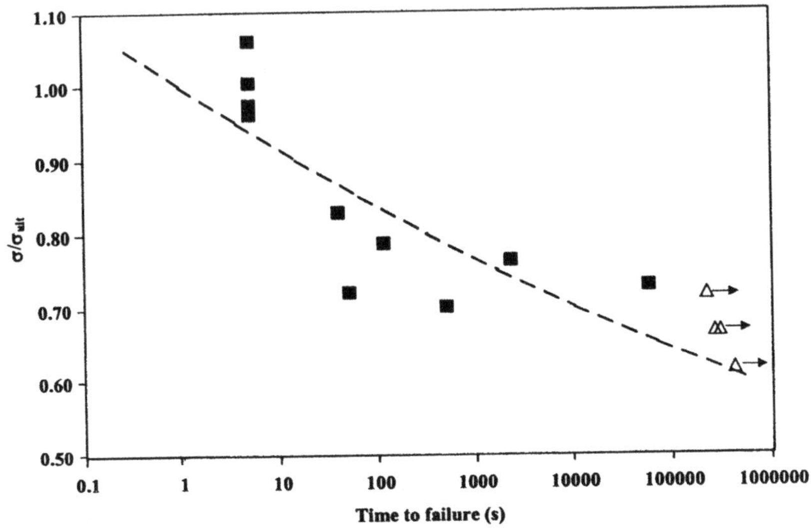

Figure 8.18 Rupture data at 90°C and curve fit.

We will assume that, because of the temperatures involved, the behavior of the material is controlled by a combination of the rupture and fatigue behaviors. We begin our analysis by calculating an average value of the failure function, $\overline{\text{Fa}}$, for a time increment as shown in Figure 8.19. A corresponding time to rupture t_{rupture} is also determined based upon (8.17). We recognize that our current value for the residual strength Fr will contain information about the past loading history. To account for this past loading history, we define a fictitious time quantity, so that

$$t_2^0 = \left(\frac{1 - \text{Fr}}{1 - \overline{\text{Fa}}}\right)^{1/j} t_{\text{rupture}} \tag{8.18}$$

We will call a time "pseudo-time" because it does not refer to actual time. The change in residual strength over the time interval $(t_1,\ t_1 + \Delta t)$ is then given by

$$\Delta\text{Fr} = -(1 - \overline{\text{Fa}})\left[\left(\frac{t_2^0 + \Delta t}{t_{\text{rupture}}}\right)^j - \left(\frac{t_2^0}{t_{\text{rupture}}}\right)^j\right] \tag{8.19}$$

By adding the result of (8.19) to our previous value of the result strength Fr, we obtain a first estimate for the residual strength at time $t_1 + \Delta t$. Next we divide the interval $(t_1,\ t_1 + \Delta t)$ into two subintervals, $(t_1,\ t_1 + \frac{1}{2}\Delta t)$ and $(t_1 + \frac{1}{2}\Delta t,\ t_1 + \Delta t)$, and calculate a remaining strength at time $t_1 + \Delta t$ by applying (8.19) twice and summing the results for those respective cases. This process of subdivision and summation is repeated until the calculated value of the remaining strength at $t_1 + \Delta t$ converges to some specified error tolerance.

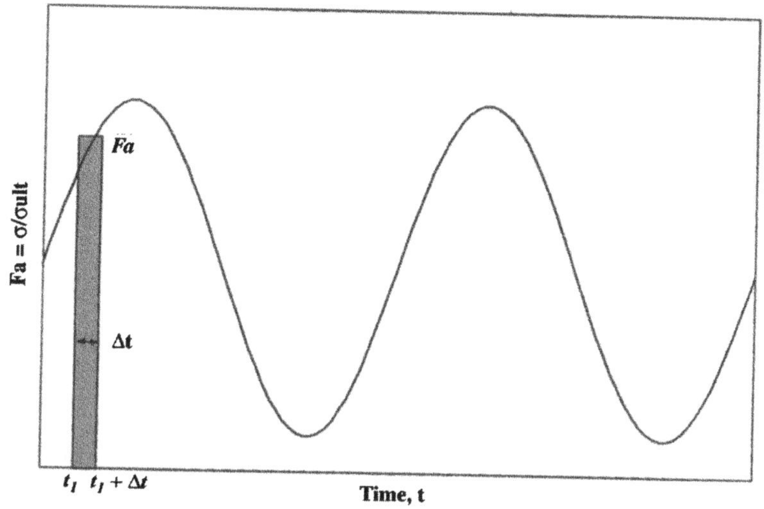

Figure 8.19 Illustration of analysis process used for Case Study 2.

In addition to the rupture process, we recognize that fatigue may also be contributing to the reduction in remaining strength and eventual failure of the component. To account for the fatigue process, we introduce a new generalized time, N, that corresponds to the number of fatigue cycles to failure at a given failure function value. At each load reversal during the loading profile, we increment the number of cycles by $1/2$ and calculate a change in remaining strength in an analogous manner to that used for the rupture process. This approach, although simplistic, does give the correct answer in the limiting cases (i.e., for a constant rupture load with no failure effect, the predicted life is equal to the rupture life, and for fatigue at a constant amplitude with no rupture effect, the predicted life is equal to the fatigue life). In addition, it also allows us to preserve the sequence of the loading profile. Case et al. (1998) and Loverich et al. (2000) give a more detailed description of this analysis process.

To illustrate the results of this analysis process, we consider the 90°C fatigue behavior of the graphite fiber–PPS composites. Unidirectional composites were tested in the fiber direction at 90°C with an R ratio of 0.1 at a frequency of 10 Hz. Lifetime results from these tests are compared to predictions made using the residual strength approach outlined above, as shown in Figure 8.20. For comparison purposes, the room-temperature fatigue curve and the elevated-temperature rupture curve are also shown. We note that, as expected, the 90°C fatigue lifetime is less than that at room temperature. Additionally, we note that the fatigue lifetimes at 90°C are greater than those from the tensile rupture tests if the comparisons are on the basis of maximum applied stress level. This makes sense physically because the samples subjected to the fatigue loading experience less time at the maximum stress level than those subjected to the rupture loading only (and in this case the rupture

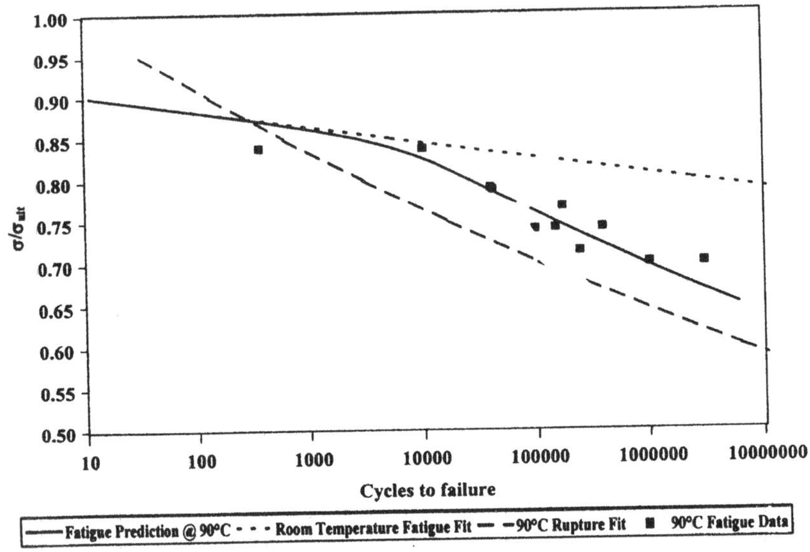

Figure 8.20 Comparison of elevated-temperature fatigue prediction and 90°C fatigue data for Case Study 2.

effect is much more severe than the fatigue effect). Finally, we note that predicted lifetimes are in close agreement with those measured experimentally.

8.4 CASE STUDY 3: ELEVATED-TEMPERATURE FATIGUE BEHAVIOR OF NEXTEL 610/ALUMINA–YTTRIA COMPOSITES

The material used for the experimental investigation is an 11-ply 8-harness satin weave composite with Nextel 610 fibers embedded in a hybrid alumina–yttria ceramic. The matrix is produced by a sol–gel process and has a residual porosity of approximately 20% and the interface is fugitive carbon. The fiber volume fraction is approximately 17% in the axial direction. Table 8.4 lists the tensile properties of the woven composite at the two temperatures of interest. Based on the measured tensile strengths, fatigue tests at ambient conditions and stress rupture tests at 1093°C were performed.

The hysteresis loop areas collected from room-temperature fatigue tests are shown in Figure 8.21. There is a definite, although slight, trend toward increasing

TABLE 8.4 Tensile Properties of Woven Nextel 610/Alumina–Yttria Composite

Temperature (°C)	Tensile Modulus (GPa)	Tensile Strength (MPa)	Failure Strain (%)
23	106	203	—
1093	84	158	0.174

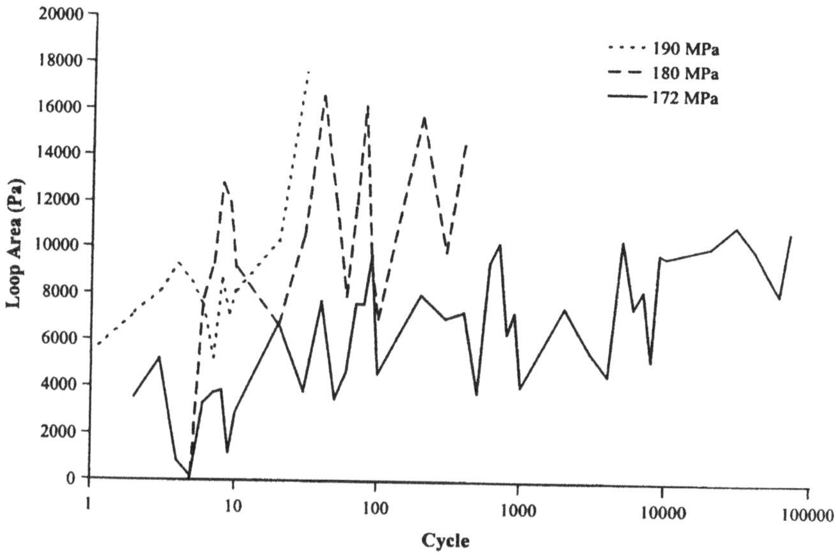

Figure 8.21 Hysteresis loop area with cycles under room-temperature fatigue for Nextel 610 composites.

loop area with cycles for the three stress levels considered. The stiffness degradation is shown in Figure 8.22.

If we consider the increase in loop area to be due to the decrease in interfacial frictional stress (Vaggaggini et al., 1995), then we can determine the degradation of τ as a function of stress and cycles. Although the woven microstructure of the compo-

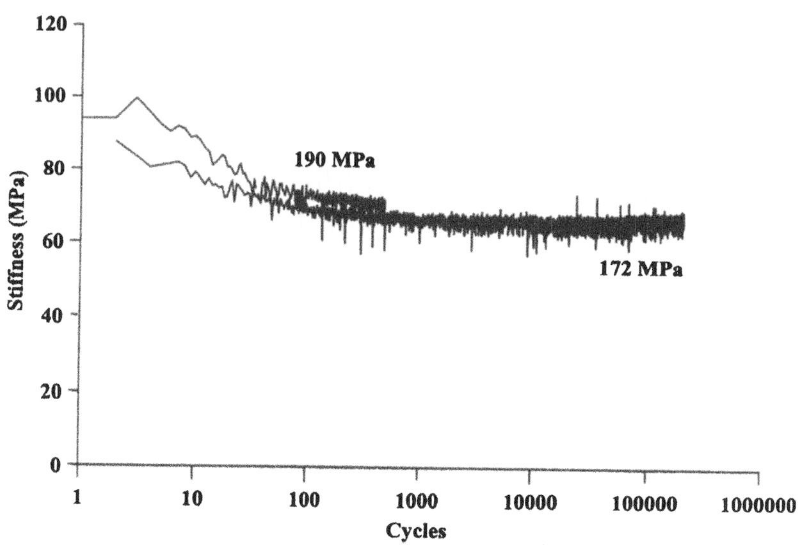

Figure 8.22 Stiffness degradation during room-temperature fatigue for Nextel 610 composites.

site makes quantitative analysis difficult, in unidirectional materials the hysteresis loop area is approximately proportional to $1/\tau$. The increase in loop area of a factor of approximately 4 at 172 MPa implies a corresponding reduction in interface frictional stress. The tensile strength of a ceramic matrix composite that exhibits statistical fiber failure and global load sharing is proportional to $\tau^{1/(m+1)}$. For $m = 5$, a typical value for ceramic fibers, this decrease in interface frictional stress results in a decrease in tensile strength of approximately 20%. This is slightly greater than that required to produce the fatigue failures measured, but the hysteresis loop data are extremely noisy and the trends difficult to quantify.

However, use of a residual strength based approach does not require detailed analysis of the effect of τ on composite lifetime and instead merely considers a general degradation of strength with time and cycles. The results of the elevated-temperature stress rupture tests are similarly fit. Additionally, two stress rupture tests with rapid unload–reload cycles were run to determine if there is any modulus degradation or interface degradation associated with stress rupture. Figure 8.23 demonstrates that there was little degradation in modulus or τ associated with the rupture process. The unload–reload cycle required 2 s and occurred once every 50 s of test time.

To apply the residual strength approach, we first require the results from the ambient fatigue tests. These results are summarized as the open diamonds in Figure 8.24 and may be well represented in the form of

$$\mathrm{Fa} = 0.966 - 0.0232\ \log N \tag{8.20}$$

Equation (8.20) is the analog to an S–N curve for the current analysis (with the failure criterion Fa replacing the stress S). Next, we represent the stress rupture data

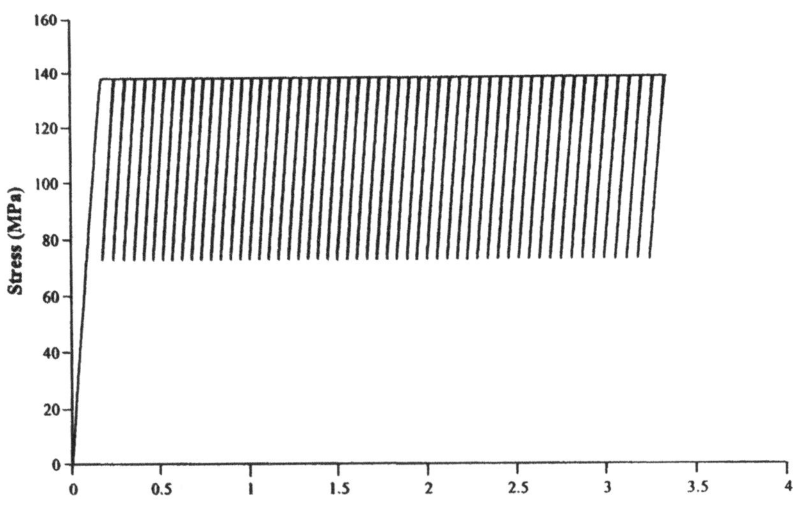

Figure 8.23 Stress rupture with unload–reload cycles for Nextel 610 composites.

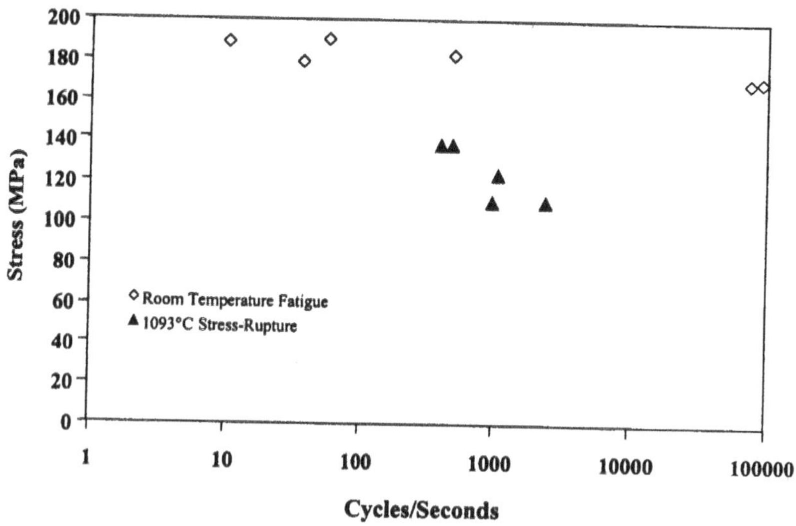

Figure 8.24 Lifetime results for room-temperature fatigue and 1093°C stress rupture of Nextel 610 composites.

(shown as the solid triangles in Fig. 8.24) in the form

$$\mathrm{Fa} = 1.013 - 0.1561 \ \log t_{\mathrm{rupture}} \tag{8.21}$$

where the time to failure is given in seconds. Using the residual strength based prediction technique detailed in the previous section, predictions for elevated-temperature fatigue tests were obtained and compared with the results of experiments. Figure 8.25 shows two residual strength curves (Fr vs. cycles) and demonstrates that the results of the lifetime prediction were virtually identical to the experiments (shown as the solid symbols). Additionally, the open symbols are results from residual strength tests conducted on samples that had previously been fatigue loaded. The agreement here between the predicted and measured residual strength values is also quite good. We note that while the fatigue lifetimes may be estimated by using an approach such as Miner's rule, this intermediate residual strength information cannot be predicted with such an approach.

8.5 CASE STUDY 4: ELEVATED-TEMPERATURE FATIGUE BEHAVIOR OF NICALON–ENHANCED SiC COMPOSITES

Our next material system that we will consider is a plain-woven Nicalon fiber–enhanced SiC matrix composite material. Mechanical properties for this material system are given in Table 8.5.

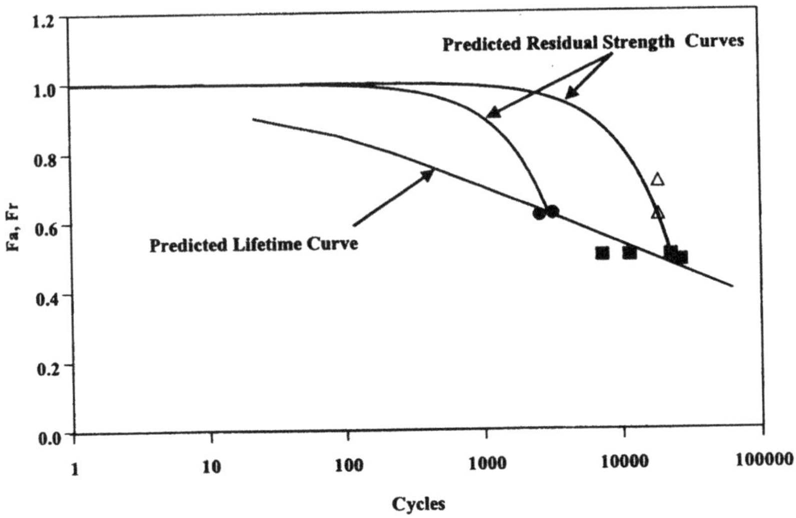

Figure 8.25 Predicted residual strength and lifetime curves for Nextel 610 composites compared with experimental lifetime results.

For this ceramic fiber–ceramic matrix composite material, experimental results have shown that the stiffness reduction due to matrix cracking is significant, unlike the case just considered for the Nextel 610/alumina–yttria composites. Prior work on ceramic matrix composites has shown that matrix crack density is a function of cycles and stress level (Vaggaggini et al., 1995). To account for these ideas and characterize this stiffness change in the material system under question, the secant

TABLE 8.5 Mechanical Properties of Nicalon–Enhanced SiC Composites

Quantity	Value
Fiber modulus, E_f	127 GPa
Fiber Poisson ratio, ν_f	0.20
Matrix modulus, E_m	146 GPa
Fiber Poisson ratio, ν_m	0.18
X_t	252 MPa
X_c	490 MPa
Y_t	252 MPa
Y_c	490 MPa
SS	192 MPa

modulus data were used. These data, as a function of cycles, were obtained from hysteresis loops produced at different load levels. These data were represented in the form of

$$\frac{E}{E_0} = (1.737 - 2.935\tilde{\sigma} + 1.601\tilde{\sigma}^2) + (0.0946 + -0.419\tilde{\sigma} + 0.347\tilde{\sigma}^2)\ \log n \quad (8.22)$$

where E is the composite stiffness, E_0 is the undamaged composite stiffness, n is the number of fatigue cycles, and $\tilde{\sigma}$ is the applied stress divided by the ultimate strength. This expression is applied to the stiffness values in the two principal material directions. We assume that no stiffness loss occurs under cyclic loading at stress levels below that of matrix cracking (approximately 30% of the ultimate strength) and that matrix crack density saturates at 70% of the ultimate strength. We also assume that all changes in stiffness are due to tensile loads only. Finally, we note that the effect of temperature on stiffness reduction has been neglected here. A comparison of the analytical representation of the stiffness loss versus the experimental data as a function of cycles for plain-woven Nicalon–enhanced SiC is shown in Figure 8.26. A key element of the analysis is that a failure criterion Fa is defined based on maximum stress *in the fibers* (i.e., Fa = fiber stress/fiber strength). To determine the average fiber stress including the effects of stiffness reduction, we next employ a simple micromechanics model. From rule of mixtures the modulus of the composite is given by

$$E_c = fE_f + (1 - f)E_m \quad (8.23)$$

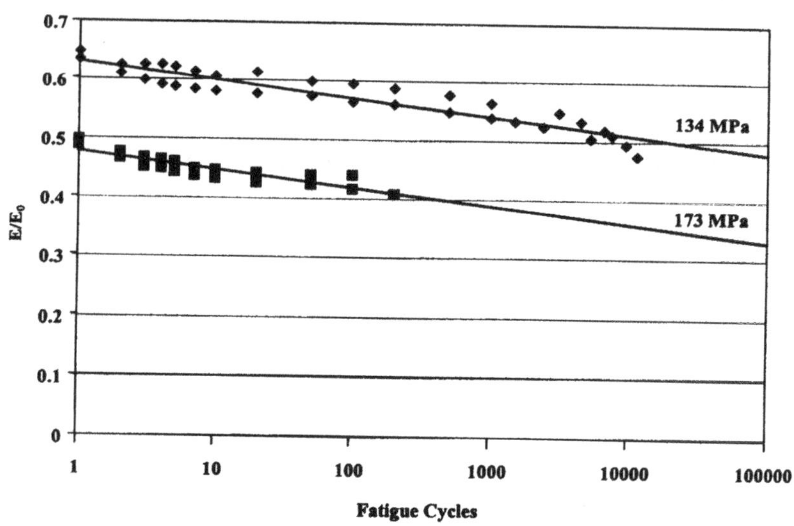

Figure 8.26 Measured and fitted stiffness reduction values for Nicalon–enhanced SiC composites.

where E_c is the modulus of the 0° lamina, f is the fiber volume fraction, and subscripts f and m refer to the fiber and matrix, respectively. The change in stiffness at the laminate level for a unidirectional material as observed in experiments is given by

$$\frac{E(n, \tilde{\sigma})}{E_c} = D(n, \tilde{\sigma} \qquad 8.24)$$

where D is a damage scalar that is a function of applied stress and cycles. Now assuming that all the damage observed at the lamina level is attributed to the matrix, (8.23) can be written as

$$E(n, \sigma) = f E_f + D_m(n, \sigma)(1 - f)E_m \qquad (8.25)$$

where D_m is the matrix damage function given by

$$D_m(n, \sigma) = \frac{D(n, \sigma)[f E_f + (1 - f)E_m] - f E_f}{(1 - f)E_m} \qquad (8.26)$$

Using laminate theory along with (8.26), the change in stress in the fibers as a function of cycles can be determined, so that the failure function Fa may be calculated.

The time-dependent effect of stress rupture at elevated temperature and as a function of applied stress level is another important process included in the model. The data used in the development of the rate equations for rupture were obtained at 982 and 1100°C. In addition, some rupture data at fixed stress levels and temperatures ranging from 500 to 1100°C were used. We assume that the time to rupture may be represented in the form of

$$\ln(\hat{\tau}) = [(17.750 - 18.890 \text{ Fa}) + (-0.611 + 0.734T_{\text{red}} + 18.945T_{\text{red}}^2)] \qquad (8.27)$$

where

$$T_{\text{red}} = \left| \frac{T - T_{\text{ref}}}{T_{\text{ref}}} \right| \qquad (8.28)$$

and T is the temperature in Celsius, T_{ref} is a reference temperature of 1550°C, and Fa is the calculated failure criterion (including the effects of matrix cracking). Since the rupture data were obtained from experiments conducted in air, the above expression is used to represent the rupture life and is a representation of the combined effects of high-temperature rupture and oxidation. The interpretation of the data from these tests is complicated by the effect of the matrix cracking. For applied stresses less than the matrix cracking, the stress in the critical element (the fibers) is proportionally less than in the case of applied stresses greater than the matrix cracking stress (as a result of the load supported by the matrix). Thus, the rupture tests measure two combined effects: the stress redistribution due to matrix cracking and the loss of strength in the fibers. For our life prediction analysis, we wish to separate the two effects. To do so, we proceed as follows. First, we assume a time to rupture–failure function curve in the critical element at 1100°C. Then, we perform the damage evolution and residual strength calculations using the stiffness reduction curves obtained from the fatigue tests expressed as a function of time. Although the use of this stiffness reduction obtained from a fatigue test is not entirely correct, it does provide an estimate of the effect of matrix cracking, as the majority of the stiffness change occurs immediately (in the first fatigue cycle). The parameters used in the time to rupture–failure function curve are adjusted based upon these results, and the analysis is repeated until satisfactory agreement between the calibrated predicted results and the experimental ones is achieved at different load levels.

Figure 8.27 displays a plot of the normalized applied stress versus rupture time for two temperatures *for the composite* (including the effects of the stiffness reduction). In this plot, the solid and dashed lines represent analytical results while the symbols represent experiment data. It should be noted here that the plot at 1100°C (solid line) lies to the right of the 982°C plot (dashed line), which indicates a longer rupture life for a given stress state at the higher temperature. This counterintuitive result is due to the fact that the material in question exhibits longer rupture times past the reference temperature T_{ref}. To illustrate this behavior, we may consider the rupture lifetimes at a constant applied stress equal to 30% of the ultimate strength, as shown in Figure 8.28.

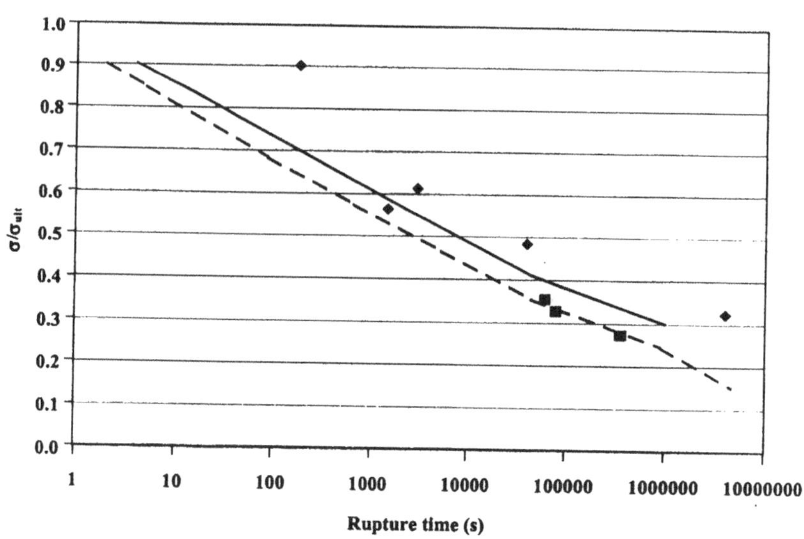

Figure 8.27 Measured and fitted rupture lifetimes for Nicalon–enhanced SiC composites tested at 982 and 1100°C.

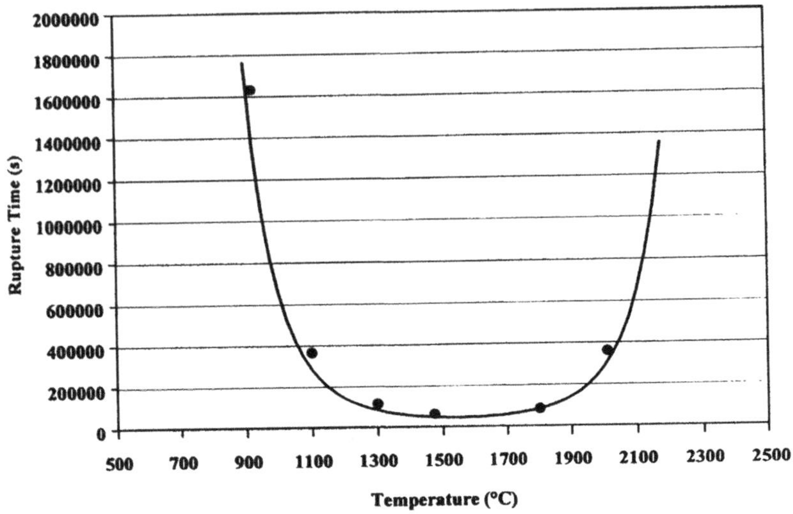

Figure 8.28 Measured and fitted rupture lifetimes as function of temperature for Nicalon–enhanced SiC composites tested at 30% of ultimate strength.

The last rate equation required for the lifetime prediction analysis is the fatigue lifetime curve (exclusive of elevated-temperature effects). This behavior is represented in the form

$$\mathrm{Fa} = 1 - 0.025 \ \log N \tag{8.29}$$

To illustrate the applicability of our life prediction approach to this material system, we will consider a number of example loading conditions. The first such loading condition is that of an $R = 0.05$, 1-Hz triangular waveform at 982°C. Our analysis procedure is very much similar to that used in Case Study 3 and Case Study 4 with the exception that the stiffness reduction due to matrix cracking is incorporated into the calculation of the failure criterion Fa. Thus, even when the global stress amplitude is constant with cycles, the value for Fa increases (in an analogous fashion to the manner in which cracking of off-axis plies increased the stress in the 0° ply of the $[45, 0, -45, 90]_s$ laminate considered previously). The predicted and experimental lifetimes for this case are summarized in Figure 8.29.

As a second illustration of the life prediction capability for this material system, we may consider the idealized mission profile loading shown schematically in Figure 8.30. In this mission profile, there is a initial loading followed by a 5400-s hold at two-thirds of this initial load value followed by a return to zero load. This waveform is repeated until failure occurs. This type of testing has been conducted at 982°C for load levels corresponding to 70, 50, 30, and 20% of the initial ultimate strength of the composite. While this loading condition is dominated by the long hold times, it also includes the effects of the overloads that introduce additional matrix cracking.

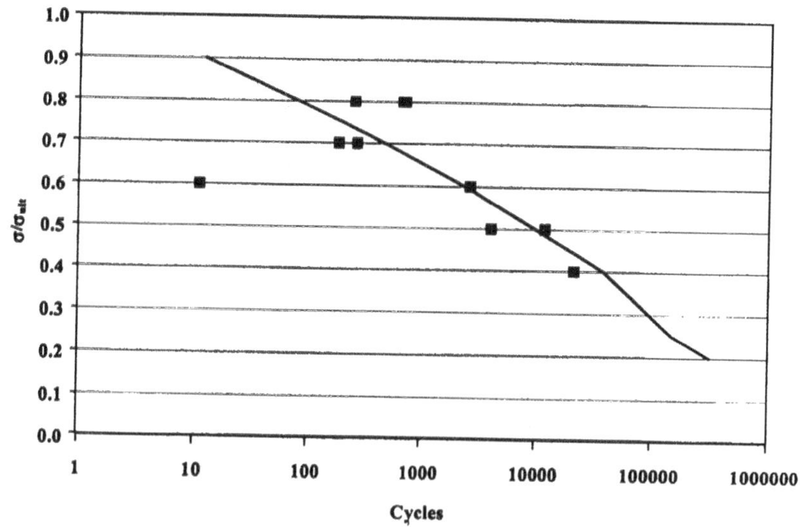

Figure 8.29 Experimental and predicted lifetimes for Nicalon–enhanced SiC composites subjected to 1-Hz triangle loading at 982°C.

The predicted and experimental lifetimes for this case are summarized in Figure 8.31.

The last loading situation that we will consider for these composites is $R = -1$ sinusoidal fatigue loading at 1 Hz as conducted by Elahi (1996). In addition to

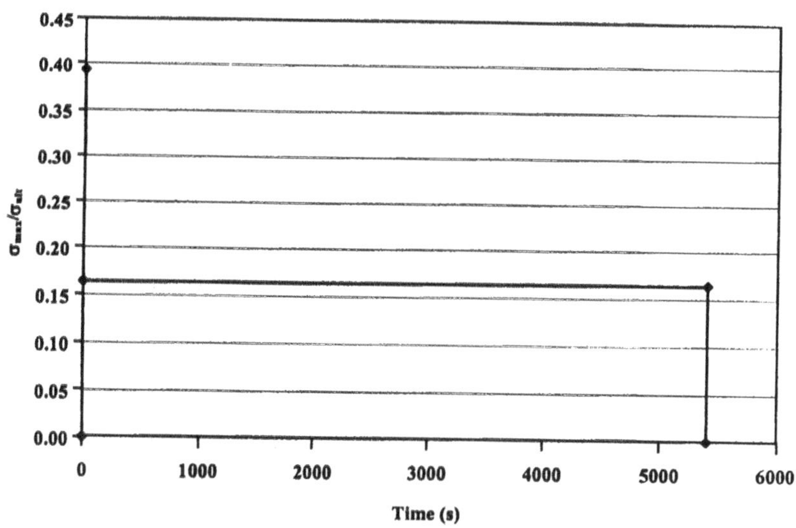

Figure 8.30 Illustration of idealized mission loading profile.

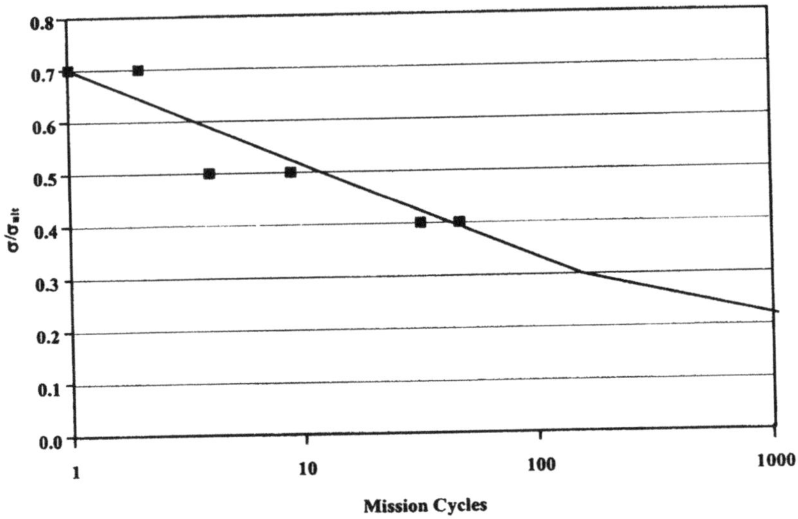

Figure 8.31 Experimental and predicted lifetimes for Nicalon–enhanced SiC composites subjected to idealized mission loading at 982°C.

measuring the fatigue lifetimes, Elahi also measured intermediate values of residual strength for fatigue load levels corresponding to 42, 36, and 28% of the initial ultimate strength values. As in the previous case study, these values provide a second means for validating the applicability of the residual strength prediction approach. These results are summarized in Figures 8.32–8.34.

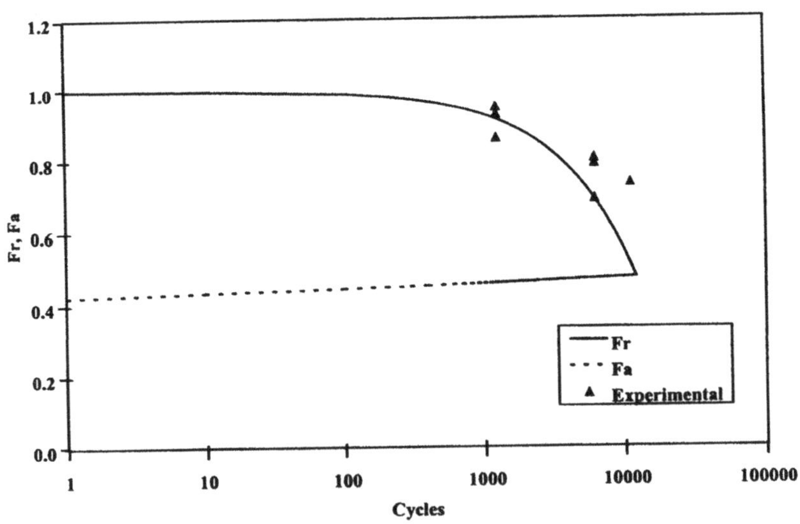

Figure 8.32 Experimental and predicted residual strength values for Nicalon–enhanced SiC composites subjected to $R = -1$ fatigue loading at 982°C and 42% of ultimate strength.

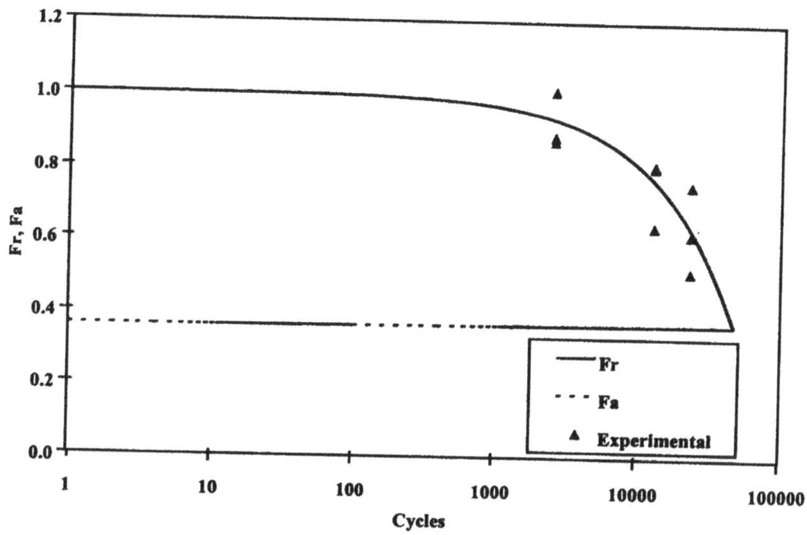

Figure 8.33 Experimental and predicted residual strength values for Nicalon–enhanced SiC composites subjected to $R = -1$ fatigue loading at 982°C and 36% of ultimate strength.

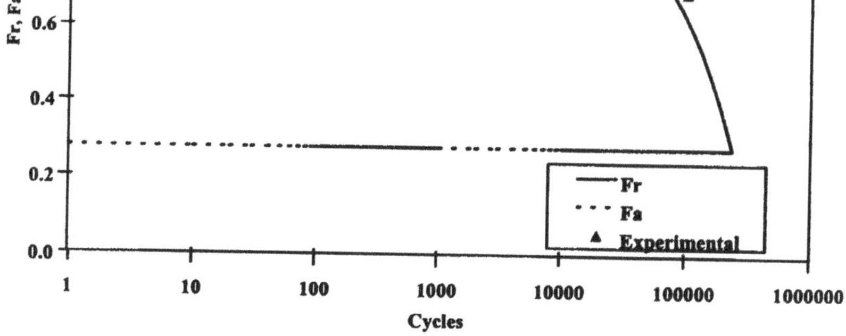

Figure 8.34 Experimental and predicted residual strength values for Nicalon–enhanced SiC composites subjected to $R = -1$ fatigue loading at 982°C and 28% of ultimate strength.

8.6 CASE STUDY 5: FATIGUE FAILURE OF A STRUCTURAL COMPOSITE SHAPE

Our previous examples and case studies have focused on in-plane, fiber-dominated damage-controlled processes. The structural members used in the civil infrastructure are often of thick section and fail due to out-of-plane or matrix-dominated damage. To illustrate the manner in which the residual strength life prediction philosophy may be applied in these situations, we will consider a specific example of a structural composite beam as studied by Senne (2000). This double-web I-beam structural shape, 20.3-cm-deep section (see Fig. 8.35) is a pultruded section composed of both E-glass and carbon fiber in a vinyl ester resin. The approximate fiber volume fraction (both glass and carbon) for the structure is 55%. The carbon fibers are located in the flanges to increase the section's bending stiffness. Glass fiber is present in the pultruded structure primarily in the form of stitched angle-ply mats, roving and continuous strand mat. Tests of individual beams at a span of 5.74 m have shown an average bending modulus of 44–48 GPa depending on the type of carbon fiber used. Failure under a four-point bending loading occured at an applied load of 129 kN. The shear contribution to deflection is approximately 5% at this span. The failure of this beam and those under investigation for fatigue performance is dominated by delamination at the hybridized region within the top flange, as shown in Figure 8.36.

We consider the case where the double-web beam is subjected to four-point bending fatigue loading. We assume that initial stiffness reduction only occurs in the tensile (bottom) flange. As the stiffness of this tensile flange is reduced, there is a redistribution of stress and an inherent shift in the neutral axis. The remaining strength approach in conjunction with an iterative stress analysis is then used to

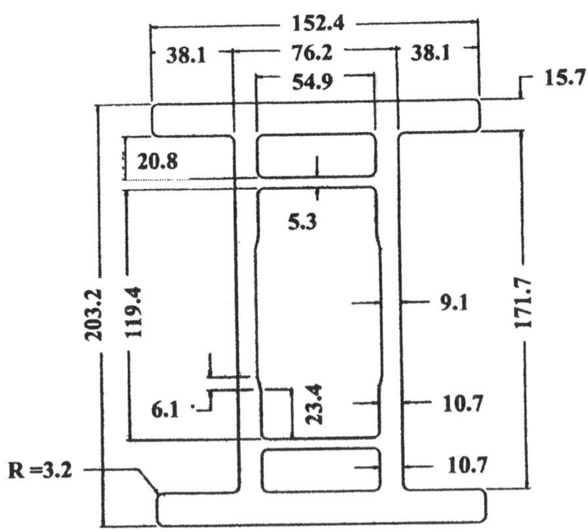

Figure 8.35 Cross section of box I-beam prototype; all dimensions in millimeters.

Figure 8.36 Failure of 20.3-cm-deep double-web I-beam under four-point bending.

determine the onset of delamination and the crack growth to failure. The assumptions employed in the residual strength model include the following:

- Reduction in tensile stiffness of the beam will be evaluated based on tensile coupon data of similar material conducted by Phifer (1999) that focus on off-axis plies.
- The unidirectional carbon plies do not experience any stiffness reduction.
- Strength reduction is uniform for both the tensile and compression flanges and is related to the in-plane strength reduction of the tensile flange.
- The carbon acts stiffer in tension than in compression; therefore, the neutral axis is initially offset toward the tensile flange but during loading shifts toward the compressive side.
- The tensile out-of-plane strength (Z_t) is calculated from the ultimate bending moment M_{ult} found from quasi-static failure testing.
- Once delamination initiates, stiffness reduction must be accounted for in the compression flange in addition to the tensile flange.
- Delamination growth is symmetric from each side of the beam, across the width of the beam (in the y direction).
- Failure occurs when the delamination propagates across the width of the beam or if the in-plane remaining strength matches the loading.

The flowchart in Figure 8.37 graphically describes the process utilized to assess lifetime. The process begins by inputting the geometry, layup, and loading. Using

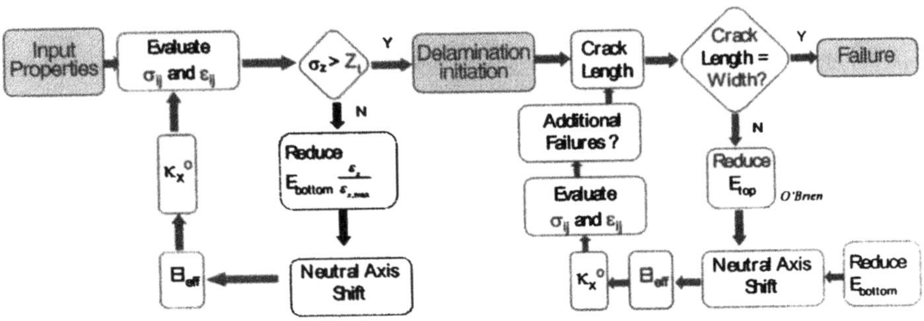

Figure 8.37 Flowchart of stress analysis and stiffness reduction, delamination determination, and stress redistribution following delamination.

this information, the stresses and strains are evaluated. The free-edge stresses are then compared to the strength of the top flange. If the stress exceeds the in-plane strength, delamination is assumed. If the stress does not exceed the strength, the stiffness in the tensile flange is reduced based on a maximum-strain criterion. The neutral axis shift corresponding to the stiffness reduction is then calculated. The new stiffness and neutral axis location are used in laminated beam theory (Swanson, 1997) to determine the new bending stiffness EI_{eff} and curvature. The curvature $\kappa_x{}^0$ becomes the new loading condition for the stress evaluation. The process is continued until delamination initiation.

Previous tensile fatigue testing of pultruded, E-glass, vinyl ester laminates are used to characterize the stiffness reduction of the bottom flange. The stiffness reduction was monitored in the tests and indicates that a linear reduction in stiffness occurs with respect to the logarithm of the fatigue cycles at a given load (Phifer, 1999). The flange is divided into sublaminates that mimic the cross-ply and quasi-isotropic coupons tested, and the stiffness is reduced on a sublaminate basis.

For modeling the beam, it seems that the initial coupon degradation is similar to what occurs in the beam itself; therefore, the data were fit to a logarithmic curve. The resulting curve fits were

$$\frac{E_x}{E_0} = -0.0124 \ln\left(\frac{n}{N}\right) + 0.8281 \quad \text{(quasi-isotropic)} \tag{8.30}$$

$$\frac{E_x}{E_0} = -0.0118 \ln\left(\frac{n}{N}\right) + 0.8933 \quad \text{(cross-ply)} \tag{8.31}$$

The reduction of the quasi-isotropic laminates is more severe initially, but both have similar attributes thereafter due to the nature of the logarithmic curve fit. In the analysis, the effective modulus of a sublaminate is found from a rule-of-mixtures approach:

$$E_{\text{eff}} = \frac{\sum E_{x,i} t_i}{\sum t_i} \tag{8.32}$$

where $E_{x,i}$ is the equivalent modulus in the axial direction for a ply of any orientation and t_i is the thickness of the respective ply. This effective stiffness was then used in the same manner to determine the stiffness of the entire flange. For the entire flange $E_{x,i}$ and t_i are the sublaminate E_{eff} and thickness, respectively.

Initially, the neutral axis shift is toward the tensile flange, as carbon acts stiffer in tension than in compression. As the stiffness of the flanges change, there is a shift in the neutral axis. This shift inherently changes the strain distribution across the section and will influence the in-plane stiffness reduction. Initially, the neutral axis moves toward the compression flange, and once delamination occurs, it begins to shift toward the bottom flange. The location of the neutral axis (NA) is simply found by considering the effective E_x for the top and bottom flanges in the standard mechanics-of-materials calculation:

$$\text{NA} = \frac{\sum E_{x,i} t_i \bar{y}_i}{\sum E_{x,i} t_i} \tag{8.33}$$

The influence of the neutral axis shift on the inertia properties is negligible and is not accounted for in the analysis. A 0.635-cm shift, representing the carbon acting 76% less stiff in compression, results in a less than 1% change in inertia values.

The quadratic delamination theory proposed by Brewer and Lagace (1988) predicts delamination initiation based on the out-of-plane stresses and strengths. Failure is assumed to occur when

$$\left(\frac{\tau_{xz}}{Z_{xz}}\right)^2 + \left(\frac{\sigma_z}{Z_t}\right)^2 \geq 1 \tag{8.34}$$

In terms of a failure criterion, (8.34) may be expressed as

$$\text{Fa}_{\text{out-of-plane}} = \sqrt{\left(\frac{\tau_{xz}}{Z_{xz}}\right)^2 + \left(\frac{\sigma_z}{Z_t}\right)^2} \tag{8.35}$$

The value of τ_{xz} is negligible in this analysis when compared to the matrix strength, allowing (8.35) to be simplified as

$$\text{Fa}_{\text{out-of-plane}} = \frac{\sigma_z}{Z_t} \tag{8.36}$$

The values for σ_z as a function of applied bending moment are calculated using a variational approach as described by Senne (2000). The out-of-plane strength in the z direction (Z_t) is assumed to be the maximum calculated σ_z at the critical glass–carbon interface at failure in the quasi-static test.

In the fatigue situation, we must predict changes in the residual through-thickness strength. To do so, we employ the failure criterion given in (8.36) with the residual strength calculation. Delamination is predicted to initiate when the residual strength calculated in this fashion is equal to the value of the failure criterion calculated using (8.36).

Following the onset of delamination, stiffness reduction of the compressive flange must also be considered with the tensile in-plane effects. These effects are also coupled with the crack growth and propagation to predict the ultimate failure of the beam. The reduction scheme is shown schematically in Figure 8.37. This stiffness reduction is used with the continued modulus reduction in the bottom (tensile) flange to determine the neutral axis shift. The new stiffness values and neutral axis location are then used to determine EI_{eff} and $\kappa_x{}^0$ that allow for calculation of the stress state. The drop in stiffness and increase in curvature will inherently raise the stresses and may cause additional failures. The initial crack and any newly formed cracks are then monitored and continue through this evaluation cycle until failure. The process used to calculate the stiffness reduction in the compression flange as well as the rate of delamination growth is outlined below.

The new modulus of the compression flange is determined through a rule-of-mixtures approach developed by O'Brien (1987):

$$E_x = (E^* - E_{\text{lam}})\frac{a}{b} + E_{\text{lam}} \tag{8.37}$$

where a is the crack length of the largest crack in the laminate, b is the half width of the laminate, E^* represents the effective modulus of the laminate if the layers are completely delaminated from each other, and E_{lam} is the initial effective modulus value of the laminate. The rule of mixtures is also used to determine E^*:

$$E^* = \frac{\sum E_{x,i} t_i}{t} \tag{8.38}$$

where $E_{x,i}$ and t_i represent the effective modulus and thickness of the sublaminates formed by the cracks.

Once delamination initiates, its growth is assumed to be symmetric from each free edge of the beam. O'Brien has shown a good estimation of crack growth is based on the relation (O'Brien, 1993)

$$\frac{da}{dn} = \left(\frac{b}{E^* - E_{\text{lam}}}\right)\left(\frac{dE}{dn}\right) \tag{8.39}$$

where dE/dn is the change in modulus over the step size and all other terms are consistent with their definitions above. The crack growth rate (da/dn) is not constant, since it is dependent on the number of layers that have delaminated at a given time; thus, as more layers delaminate, the rate of crack growth increases.

The model predicts failure due to in-plane effects and also delamination. Failure is assumed when either of the following criteria are met:

1. Delamination initiates and the crack completely propagates across the width of the beam.
2. The in-plane remaining strength of the beam matches the loading (Fa=Fr).

These criteria are met when the remaining strength curve intersects the respective loading curve intersection of the remaining strength curve and the applied load curve.

An *S–N* curve was created for the beam considering no neutral axis shift and using the overall average ultimate moment data to determine the strength of the beam. The experimental points and the predicted *S–N* curve are plotted in Figure 8.38, normalized to the average ultimate moment of all of the hybrid beams tested from both batches. The beam failure at 53% is about six orders of magnitude below the prediction, but the beam that survived 8 million cycles at that stress level (before the test was stopped at "runout") did not fail. A runout at 10 million cycles at a lower applied stress is also shown. The beam that failed after 370,000 cycles at 71% of the average ultimate moment agrees well with the prediction of 300,000 cycles at

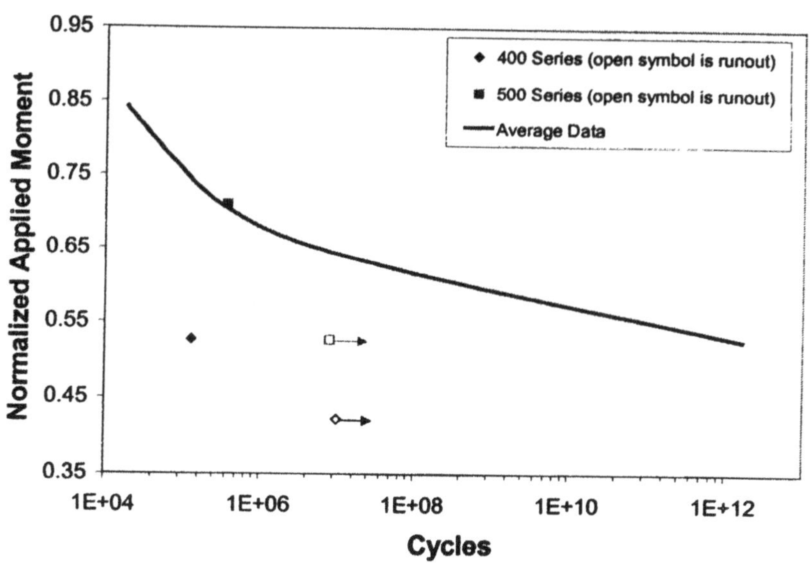

Figure 8.38 Comparison of predicted *S–N* curve for beam fatigue loading to experimental data.

the same load. Further data are needed to fully validate the model, but the predictions provide valuable estimates of behavior in this complex situation.

8.7 SUMMARY

Further comparisons with experimental observations for a wide range of applications can be found in Reifsnider (1992, 1995, 1996), Reifsnider et al. (1997), Reifsnider and Stinchcomb (1986), and Reifsnider et al. (2000). Various embodiments of the method have appeared in the MRLife code that are used by numerous investigators and application development groups. Several versions of MRLife have been fully integrated into commercial analysis codes, including ANSYS (ANSYS, Canonsburg, PA) and ABAQUS (Hibbitt, Karlsson & Sorensen, Pawtucket, RI). Established application areas include fuel cells, turbines (especially combustor liners and ceramic turbine components), automotive structures, infrastructure (especially bridge structures), aircraft and ship structures, composite pipe, sports equipment, energy storage and conversion devices, biomedical systems (especially prosthetic devices), tires (especially heavy-duty tires), manufacturing equipment (such as paper mill components), power systems (including motors and fans), radiant burners, and off-shore equipment (including flexible pipe). The concept of remaining strength as a damage metric has also been extended to other performance metrics, including power performance metrics for energy conversion equipment and devices.

Other advances to be covered in subsequent publications include recent work in simulation, especially our efforts to provide robust micro- and nanoscience and engineering foundations for materials behavior to continue to reduce the need for case-specific laboratory characterization, a major cost driver. Finally, we have been forced to defer comprehensive discussion of the manner in which statistical and stochastic behavior enters the method outlined in this text, but that subject is rather well in hand.

The foundation of understanding and methodology for predicting remaining strength and life of material systems is, however, fully established by the substance of the present text. The approach uses mechanistic representations of material behavior, based on explanations rather than descriptions, when they are available, and provides a means of integrating understandings of the interactions of physics, chemistry, and mechanics when they act together, and interact, to control the life of material systems. Finally, the approach is valuable as a guide to what is needed when experimental characterization is needed, so that consistent and complete data sets can be generated at minimum cost to ensure safe and reliable service of material systems.

REFERENCES

Brewer, J. C., and Lagace, P. A. (1988), "Quadratic Stress Criterion for Initiation of Delamination," *J. Composite Mater.*, Vol. 22, pp. 1141–1155.

Case, S., Iyengar, N., and Reifsnider, K. (1998), "Life Prediction Tool for Ceramic Matrix Composites at Elevated Temperatures," in *Composite Materials: Fatigue and Fracture, Seventh Volume*, ASTM STP 1330, R. B. Bucinell, Ed., American Society for Testing and Materials, pp. 165–178.

Curtin, W. A. (1991), "Theory of Mechanical Properties of Ceramic-Matrix Composites," *J. Am. Ceram. Soc.*, Vol. 74, pp. 2837–2845.

Curtis, P. T. (1986), "An Investigation of the Mechanical Properties of Improved Carbon Fibre Composite Materials," RAE Technical Report 86021, Royal Aircraft Establishment.

Elahi, M. (1996), "Fatigue Behavior of Ceramic Matrix Composites at Elevated Temperatures Under Cyclic Loading," Ph.D. Dissertation, Virginia Polytechnic Institute and State University, Blacksburg, VA.

Kachanov, L. M. (1986), *Introduction to Continuum Damage Mechanics*. Martinus Nijhoff, Boston.

Lekhnitskii, S. G. (1968), *Anisotropic Plates*, S. W. Tsai and T. Cheron, Transl., Gordon and Breach.

Loverich, J. S., Russell, B. E., Case, S. W., and Reifsnider, K. L. (2000), "Life Prediction of PPS Composites Subjected to Cyclic Loading at Elevated Temperatures," in *Time Dependent and Nonlinear Effects in Polymers and Composites*, ASTM STP 1357, R. A. Schapery and C. T. Sun, Eds., American Society for Testing and Materials, Philadelphia, pp. 310–317.

O'Brien, T. K. (1987), "Generic Aspects of Delamination in Fatigue of Composite Materials," *J. Am. Helicopter Soc.*, Vol. 32, pp. 13–18.

O'Brien, T. K. (1993), "Stacking Sequence Effect on Local Delamination Onset in Fatigue," in *International Conference on Advanced Composite Materials*, Minerals, Metals and Materials Society, pp. 399–406.

Phifer, S. P. (1999), "Quasi-Static and Fatigue Evaluation of Pultruded Vinyl Ester/E-Glass Composites," M.S. Thesis, Virginia Tech.

Picasso, B., and Priolo, P. (1988), "Damage Assessment and Life Prediction for Graphite-PEEK Quasi-Isotropic Composites," in *American Society of Mechanical Engineers, Pressure Vessels and Piping Division (Publication) PVP v 146*, American Society of Mechanical Engineers, New York, pp. 183–188.

Reifsnider, K. L. (1992), "Use of Mechanistic Life Prediction Methods for the Design of Damage Tolerant Composite Material Systems," in ASTM STP 1157, M. R. Mitchell and O. Buck, Eds., American Society for Testing and Materials, Philadelphia, pp. 205–223.

Reifsnider, K. L. (1995), "Evolution Concepts for Microstructure-Property Interactions in Composite Systems," in *Proceedings of the IUTAM Symposium on Microstructure-Property Interactions in Composite Materials*, R. Pyrz, Ed., Kluwer, New York, pp. 327–348.

Reifsnider, K. L. (1996), "A Micro-Kinetic Approach to Durability Analysis: The Critical Element Method," in *Progress in Durability of Composite Systems*, A. H. Cardon, K. L. Reifsnider, and H. Fukuda, Eds., Balkema, Rotterdam, pp. 3–11.

Reifsnider, K. L., Case, S. W. and Elahi, M. (1997), "Predicitve Methodologies for Elevated Temperature High-Cycle Fatigue of Structural Composite Materials," *Structural Materials Symposium Proceedings in Honor of Professor Paul C. Paris*, W. O. Soboyejo and T. S. Srivatsan, Eds., The Minerals, Metals and Materials Society, pp. 461–478.

Reifsnider, K., and Case, S. (1998), "Strength-Based Predictive Methods for Large-Scale Combined Degradative Effects," in *Proceedings of the 1998 Thirty-Ninth AIAA/ASME/ASCE/AHS/ASC Structures, Structural Dynamics, and Materials Conference and Exhibit and AIAA/ASME/AHS Adaptive Structures*, AIAA-98-1739, American Institute of Aeronautics and Astronautics, pp. 352–362.

Reifsnider, K. L., Case, S. W., and Duthoit, J. (2000), "The Mechanics of Composite Strength Evolution," *Composites Sci. Technol.*, Vol. 60, pp. 2539–2546.

Reifsnider, K. L., and Stinchcomb, W. W. (1986), "A Critical Element Model of the Residual Strength and Life of Fatigue-Loaded Composite Coupons," in *Composite Materials: Fatigue and Fracture*, ASTM STP 907, H. T. Hahn, Ed., American Society for Testing and Materials, Philadelphia, pp. 298–313.

Senne, J. L. (2000), "Fatigue Life of Hybrid FRP Composite Beams," M.S. Thesis, Virginia Polytechnic Institute and State University, Blacksburg, VA.

Simonds, R. A., and Stinchcomb, W. W. (1989), "Response of Notched AS4/PEEK Laminates to Tension/Compression Loading," in *Advances in Thermoplastic Matrix Composite Materials*, ASTM STP 1044, G. M. Newaz, Ed., American Society for Testing and Materials, Philadelphia, pp. 133–145.

Swanson, S. R. (1997), *Introduction to Design and Analysis with Advanced Composite Materials*. Prentice-Hall, Englewood Cliffs, NJ, pp. 191–224.

Vaggaggini, E., Domergue, J.-M., and Evans, A. G., (1995), "Relationships between Hysteresis Measurements and the Constituent Properties of Ceramic Matrix Composites: I, Theory," *J. Am. Ceram. Soc.*, Vol. 78, No. 10, pp. 2709–2720.

Vure, N. R. S. (1993), "Effect of Cooling Rate and Stacking Sequence on the Fatigue Behavior of Notched Quasi-Isotropic APC-2 Laminates," M.S. Thesis, Virginia Polytechnic Institute and State University.

APPENDIX TO CHAPTER 1

TABLE 1-1. Mechanical Properties of Common Fiber Reinforcement Materials

Material	Specific Gravity	Tensile Modulus (GPa)	Tensile Strength (GPa)	Elongation (%)	Poisson's Ratio	Specific Modulus (GPa)	Specific Strength (MPa)
Al_2O_3							
Fiber FP (DuPont)	3.95	379	1.90	0.4		95.9	481.0
$Al_2O_3SiC_2$							
Fiberfrax (Carborundum)	2.73	103	1.03–1.72			37.7	476.2
Aramid							
Kevlar 49 (Dupont)	1.45	131	3.62	2.8	0.35	90.3	2496.6
Kevlar 149	1.47	179	3.45	1.9		121.8	2346.9
Technora (Teijin)	1.39	70	3.0	4.4		50.4	2158.3
Boron	2.7	393	3.1	0.79	0.2	145.6	1148.1
Carbon							
AS-1 (Hercules)	1.80	228	3.10	1.32		126.7	1722.2
AS-4 (Hercules)	1.80	248	4.07	1.65		137.8	2261.1
GY-70 (BASF)	1.96	483	1.52	0.38		246.4	775.5
HMS-4 (Hercules)	1.80	345	2.48	0.70		191.7	1377.8
IM-7 (Hercules)	1.78	301	5.31	1.81		169.1	2983.1
P-55 (Amoco)	2.0	380	1.90	0.5		190.0	950.0
P-100 (Amoco)	2.15	758	2.41	0.32		352.5	1120.9
T-40 (Amoco)	1.81	290	5.65	1.8		160.2	3121.5
T-300 (Amoco)	1.76	231	3.65	1.4	0.02	131.3	2073.9
Glass							
E-glass	2.54	72.4	3.45	4.8	0.2	28.5	1358.3
S-glass	2.49	86.9	4.30	5.0	0.22	34.9	1726.9
Polyethylene							
Spectra 900 (Allied-Signal)	0.97	117	2.59	3.5		120.6	2670.1
Spectra 1000	0.97	172	3.0	2.7		177.3	3092.8
SiC							
Monofilament	3.08	400	3.44	0.86		129.8	1116.8
Nicalon (Nippon Carbon)	2.55	196	2.75	1.4		76.9	1078.4
Steel (1%, hot rolled)	7.9	206	1.0	8	0.3	26.1	126.6

Source: Mallik (1993). *Fiber Reinforced Composites*, Marcel Dekker, Inc., New York.

TABLE 1-2. Mechanical Properites of Common Matrix Materials

Material	Specific Gravity	Tensile Modulus (GPa)	Tensile Strength (GPa)	Elongation (%)	Poisson's Ratio	Specific Modulus (GPa)	Specific Strength (GPa)
Metals							
Aluminum Alloys							
2024-T6	2.78	70	468.9	11		25.2	168.9
6061-T6	2.70	70	275.9	17		25.9	102.2
7075-T6	2.80	70	503.5	11	0.33	25.0	179.8
8009	2.92	88	407	17		30.1	139.4
380	2.71	70	165	4		25.8	60.9
Titanium Alloys							
Ti-6Al-4V	4.43	110	1068	8	0.33	24.8	241.1
Magnesium Alloys							
AZ91A	1.81	45	158.6	3		24.9	87.6
Zinc-Aluminum Alloys							
ZA-27 (Allied Metal Company)	5	78	370	3		15.6	74.0
Ceramics							
α-Al_2O_3	3.95	380	200–310			96.2	64.6
SiC	3.17	414				130.6	
Si_3N_4	3.19	304	350–580			95.3	145.8
ZrO_2	5.80	138				23.8	
Boro-silicate glass	2.3	60	100			26.1	43.5
Lithium-alumino-silicate glass-ceramic	2.0	100	100–150			50.0	62.5
Polymers							
Thermosetting							
Cast Epoxy	1.2–1.3	2.75–4.10	55–130		0.2–0.33	5.5	74.0

Epoxy 1	1.2–1.3	3.854	140.7			3.1	112.56
Epoxy 2 (Epon HPT 1072, Shell Chemical)	1.2–1.3	3.378	111.7			2.7	89.4
Epoxy 3 (Tactix 742, Dow Chemical)		2.965	124.1			2.4	99.3
Polyster	1.1–1.4	2.1–3.45	34.5–103.5	1–5		2.22	55.2
Cast Vinyl Ester	1.12–1.32	3–3.5	73–81	3.5–5.5		2.64	60.8
PMR-15	1.32	3.9	38.6	1.5		2.95	29.2
ACTP (Thermid 600)	1.34	4.1	82.7	1.5		3.1	61.7
BSL 914	1.2	3.5	60–80			2.9	58.3
Thermoplastics							
PEEK (Victrex, ICI)	1.30–1.32	3.24	100	50	0.4	2.5	76.3
PPS (Ryton, Phillips Petroleum)	1.36	3.3	82.7	4		2.4	60.8
PSUL (Udel, Amoco)	1.24	2.48	70.3	75	0.37	2.0	56.7
PEI (Ultem, GE)	1.27	3	105	60		2.4	82.7
PAI (Torlon, Amoco)	1.40	3.03	185.5	12		2.2	132.5
K-III (Avimid, DuPont)	1.31	3.76	102	14	0.365	2.9	77.9
LARC-TPI (Durimid, Rogers)	1.37	3.45	138	5	0.36	2.5	100.7

TABLE 1-3. Mechanical Properties of Common Fiber Reinforced Material Systems

Material System	Specific Gravity	Tensile Modulus (GPa) E_1	Tensile Modulus (GPa) E_2	Shear Modulus in plain (GPa) G_{12}	Shear Strength in plain (MPa) S	Tensile Strength (MPa) X_t	Tensile Strength (MPa) Y_t	Compresive Strength (MPa) X_e	Poisson's Ratio V_{12}	Specific Modulus (GPa)	Specific Strength (GPa)
Epoxy Matrix Composites											
Boron-Epoxy	1.99	207	19	6.4	131	1585	62.7	2481	0.21		
AS Carbon-Epoxy	1.54	127.5	9.0	5.7	60	1447	62.0	1172	0.25	82.3	939.6
T-300-Epoxy	1.55	138	10	6.5	62	1447	44.8	1447	0.21		
HMS Carbon-Epoxy	1.63	207	13.8	5.9	72	827	86.2	620	0.20		
GY-70-Epoxy	1.69	276	8.3	4.1	96.5	586	41.3	517	0.25	163.3	246.7
Kevlar 49-Epoxy	1.38	76	5.5	2.1	60	1379	28.3	276	0.34		
E-Glass-Epoxy	1.80	39	4.8	4.8	83	1130	96.5	620	0.30	21.6	627.7
S-Glass-Epoxy	1.82	43			83	1214	758				
Carbon-BSL 914	1.5	130	9	5–6	90	1600	60	1200			
Metal Matrix Composites											
Boron-Aluminum	2.7	235	138	66	69	1100	110	1215	0.23		
Ceramic Matrix Composites											
SiC/1723		128	101	66		734	27		0.24		
SiC/LAS-11		121	45			586	25				
SiC/CAS-11		124	117	48		450	28				

Source: Mallik (1993). Corning Glass.

TABLE 1-4. Carbonfiber-BSL 914 Laminates

Property	Units	Resin	0° Laminate	90° Laminate	0/90° Laminate	±45° Laminate
Fiber Content	%	0	60	60	60	60
Specific Gravity		1.2–1.3	1.5	1.5	1.5	1.5
Tensile Strength	MPa	60–80	1600	60	700	180
E-Modulus (tension)	GPa	3.5	130	9	70	23
Bending Strength	MPa	80–120	1700	100	600	
E-Modulus (bending)	GPa	3.5	120	9.5	60	
Interlaminar Shear Strength	MPa		90	10	40	
Compression Strength	MPa	90	1200	220	700	180
E-Modulus (compression)	GPa	3.5	130	9	70	23
Shear Strength	MPa	35	90	90		
Shear Modulus	GPa	1.3	5–6	5–6	5–6	30–35
Maximum Strain	%		1.1	0.6	1.0	5–6
Coefficient of Thermal Expansion	10^{-6}/K	40–60	0.5–1.0	30–50	0.8–2.0	3–5

TABLE 1-5. Temperature Effects on Material Properties Boron Aluminum (Unidirectional, $V_f = 50\%$)

Properties	Temperature				
	−320°F	75°F	400°F	600°F	800°F
Longitudinal Tensile Strength (ksi)	218	215	212	193	171
Longitudinal Tensile Modulus (msi)	31.1	30.5	29.9	28.9	27.7
Transverse Tensile Strength (ksi)	26	19	17	9	4.5
Transverse Tensile Modulus (msi)	22	19	13	11	9
Longitudinal Compressive Strength (ksi)	271	283	231	160	87
Longitudinal Compressive Modulus (msi)	38	37	32	25	21
Transverse Compressive Strength (ksi)	61	41	30	15	8
Transverse Compressive Modulus (msi)	22	33.5	24	14.4	12.5
Shear Strength (ksi)	23.5	22.6	19.1	8.5	5.3
Shear Modulus (msi)		9.5	8.3	7.1	4.5

Source: "Boron Composites—Development and Application Status," W. D. Dittmer, Editor, International Society of Allied Weight Engineers, Inc., Los Angeles, CA, 1973.

INDEX

Printed and bound by CPI Group (UK) Ltd, Croydon, CR0 4YY
16/09/2026
14963441-0006